Current Natural Sciences

Max CERF

Optimization Techniques II

Discrete and Functional Optimization

Science Press | edp sciences

EDP Sciences – ISBN(print): 978-2-7598-3163-0 – ISBN(ebook): 978-2-7598-3165-4
DOI: 10.1051/978-2-7598-3163-0

I would like to express my sincere thanks to the people who helped me to produce this book, especially:

France Citrini *for having accepted the publication of the book by EDP Sciences;*

Sophie Hosotte *and* ***Magalie Jolivet*** *for their assistance in the editorial process;*

Thomas Haberkorn *for his verification of scientific content, and for the countless numerical tricks he shared with me;*

Emmanuel Trélat *for his preface, proofreading and presentation advices, and also for our long-standing collaboration and the expertise I have acquired owing to him;*

Claude Blouvac*†* *my high school teacher who set me on the path of mathematics in 1982 and who would have been very happy to see this book.*

Preface

Algorithms are omnipresent in our modern world. They show us the optimal routes, prevent and assess risks, provide forecasts, anticipate or assist our decisions. They have become an essential part of our daily lives. These algorithms are mostly based on optimization processes and consist in minimizing or maximizing a criterion under certain constraints, thus indicating feasible and intelligent solutions, allowing us to plan a process to be carried out in the best possible way.

There are many different optimization methods, based on various heuristics, sometimes simple and intuitive, sometimes more elaborate and requiring fine mathematical developments.

It is through this jungle of algorithms, resulting from decades of research and development, that Max Cerf guides us with all his expertise, his intuition and his pragmatism.

Max Cerf has been a Senior engineer at ArianeGroup for 30 years. As a recognized specialist in trajectography, he designs and optimizes spacecraft trajectories under multiple constraints. He has thus acquired and developed a comprehensive knowledge of the best algorithms in continuous optimization, discrete optimization (on graphs) and optimal control. He is also an exceptional teacher, with a real talent for explaining complicated concepts in a clear and intuitive way.

With this double book, he offers an invaluable guide to the non-specialist reader who wishes to understand or solve optimization problems in the most efficient way possible.

Emmanuel Trélat, Sorbonne University, Paris

Table of contents

Introduction

Mathematical optimization has undergone a continuous development since the Second World War and the advent of the first computers. Facing the variety of available algorithms and software, it is sometimes difficult for a non-specialist to find his way around. The aim of this two-volume book is to provide an overview of the field. It is intended for students, teachers, engineers and researchers who wish to acquire a general knowledge of mathematical optimization techniques.

Optimization aims to control the inputs (variables) of a system, process or model to obtain the desired outputs (constraints) at the best cost. Depending on the nature of the inputs to be controlled, a distinction is made between continuous, discrete or functional optimization.

Continuous optimization (chapters 1 to 5 of Volume I) deals with real variable problems. Chapter 1 presents the optimality conditions for differentiable functions and the numerical calculation of derivatives. Chapters 2, 3 and 4 give an overview of gradient-free, unconstrained and constrained optimization methods. Chapter 5 is devoted to continuous linear programming, with the simplex and interior point methods.

Discrete optimization (chapters 1 and 2 of Volume II) deals with integer variable problems. Chapter 1 deals with mixed-variable linear programming by cutting or tree methods. Chapter 2 presents an overview of combinatorial problems, their modelling by graphs and specific algorithms for path, flow or assignment problems.

Functional optimization (chapters 3 to 5 of Volume II) deals with infinite dimension problems. The input to be controlled is a function and no longer a finite number of variables. Chapter 3 introduces the notions of functional and calculus of variations. Chapter 4 presents optimal control problems and optimality conditions. Chapter 5 is devoted to numerical methods (integration of differential equations, direct and indirect methods).

In each chapter, the theoretical developments and the demonstrations are limited to the essentials. The algorithms are accompanied by detailed examples to facilitate their understanding.

1. Mixed linear programming

Mixed linear programming concerns linear problems where some variables are real and others are integer or binary. Many practical problems can be formulated as a mixed linear problem. The solution methods are mainly based on the simplex algorithm.

Section 1 deals with the problem formulation. Linearization techniques transform a nonlinear real-variable problem into an equivalent linear mixed-variable problem that is easier to solve. Reduction techniques reduce the size of the problem by fixing variables and eliminating constraints. These pre-processing stages speed up the numerical solution.

Section 2 presents the cutting methods, which consist in solving the relaxed problem in real variables and then adding constraints to force an integer solution. These cutting constraints progressively reduce the feasible domain. They can be generated by different approaches. For large problems, the number of cuts can become prohibitive and these methods are generally used in addition to tree methods.

Section 3 presents tree search methods that explore the tree of possible combinations. The exploration consists in progressively fixing the binary or integer variables and estimating the best possible solution resulting from this choice. The estimation can be obtained quickly thanks to the properties of the dual simplex algorithm. With an exploration strategy adapted to the problem, it is possible to obtain the optimal integer solution by exploring only a small fraction of the set of possible combinations.

Section 4 presents applications of mixed linear programming to classical problems: travelling salesman, assignment, coloring, flow, knapsack. Although these problems can be solved efficiently by dedicated algorithms, the linear formulation remains interesting because it allows to easily add constraints specific to the problem to be solved.

Section 5 presents mixed quadratic programming. Tree search methods are applied in the same way as in linear programming, provided that the relaxed problem is made convex so that it can be solved safely and quickly.

1.1 Formulation

This section introduces the standard formulation of a mixed-variable linear problem, the linearization techniques for nonlinear problems and the reduction techniques for eliminating variables and/or constraints before solving.

1.1.1 Mixed-variable linear problem

The standard form of a mixed-variable linear problem (LPM) is as follows.

$$\min_{x} z = c^T x \quad \text{s.t.} \begin{cases} Ax = b \\ x_j \in \mathbb{N},\ j \in J_E \\ x_j \geq 0,\ j \in J_C \end{cases} ,\ A \in \mathbb{R}^{m \times n},\ b \in \mathbb{R}^m,\ c \in \mathbb{R}^n \tag{1.1}$$

The vector $x \in \mathbb{R}^n$ represents the n variables of the problem:
- the indexes J_E correspond to integer variables;
- the indexes J_C correspond to positive real (also called continuous) variables.

The **relaxed problem** associated with problem (1.1) is obtained by considering all variables as real (we say that the integrity constraints are relaxed).

$$\min_{x} z = c^T x \quad \text{s.t.} \begin{cases} Ax = b \\ x_j \geq 0,\ j \in J_E \cup J_C \end{cases} \tag{1.2}$$

Since the relaxed problem is less constrained than the integer problem, its solution is better and it provides a lower bound of the minimum cost of problem (1.1).

It is tempting to solve the relaxed problem and then round up the variables that must be integers. But, this rounded solution may be non-feasible or very far from the true mixed-variable solution as the example 1-1 shows.

Example 1-1: Solution of the relaxed and the mixed problem (from [R5])

Consider the problem in integer variables:

$$\min_{x_1, x_2 \in \mathbb{N}} z = -10x_1 - 11x_2 \quad \text{s.t.} \quad 10x_1 + 12x_2 \leq 59$$

The relaxed solution is: $\begin{cases} x_1 = 5.9 \\ x_2 = 0 \end{cases} \rightarrow z = -59$.

The integer solution is: $\begin{cases} x_1 = 1 \\ x_2 = 4 \end{cases} \rightarrow z = -54$ (found by enumeration).

Let us try to round up the relaxed solution:

- the closest integer point would be: $\begin{cases} x_1 = 6 \\ x_2 = 0 \end{cases} \rightarrow z = -60$.

  This point is not feasible (the inequality constraint is not satisfied);

- the closest <u>feasible</u> integer point is: $\begin{cases} x_1 = 5 \\ x_2 = 0 \end{cases} \rightarrow z = -50$.

This rounded point is far from the exact integer solution, although close in cost.

Figure 1-1 shows the constraint (solid line), the relaxed solution C, the integer solution E and the level line of the cost function through the integer solution E.

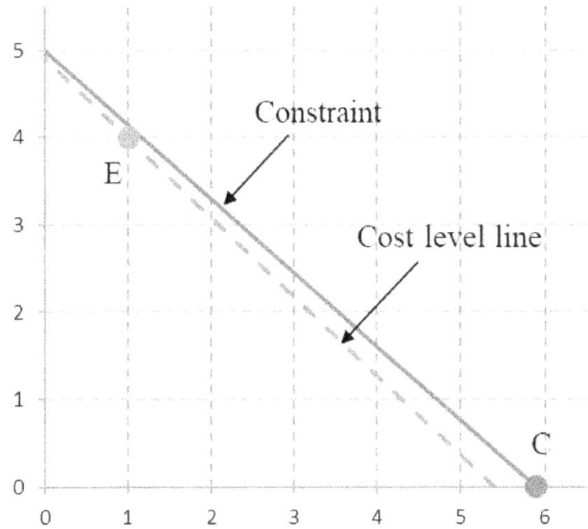

Figure 1-1: Relaxed and integer solution.

In some cases, it is useful to reformulate the problem in **binary variables.**
Any integer variable x bounded between 0 and m can be replaced by p + 1 binary variables $(y_0, y_1, \cdots, y_p)$ corresponding to the binary writing of x.

$$x = \underset{\text{binary}}{\overline{y_p y_{p-1} \cdots y_1 y_0}} = y_p 2^p + y_{p-1} 2^{p-1} + \ldots + y_1 2^1 + y_0 2^0 \quad \text{with} \quad m < 2^{p+1} \tag{1.3}$$

For binary variables, we will note in the following: $U = \{0 ; 1\}$.

The solution of a continuous linear problem is a vertex of the constraint polytope, associated with one of the possible C_n^m bases (Volume I, section 5.1.2).
This simple characterization of the solution allows the construction of efficient solution algorithms such as the simplex or interior point methods.

With mixed variables, the solution lies inside the constraint polytope and it does not admit a simple characterization. For these so-called NP-complete problems, there is no polynomial time solution algorithm. It is therefore important to simplify the problem by the techniques presented in section 1.1.2.

1.1.2 Linearization techniques

Some nonlinear expressions can be linearized at the cost of additional variables and constraints. These reformulations allow mixed linear programming algorithms to be applied to initially nonlinear problems. This section based on reference [R1] presents the basic techniques.

Product of two binary variables
The product of two binary variables s and t is linearized by introducing an additional real variable f and four inequality constraints.

$$f = s.t \quad \text{with} \quad \begin{cases} s \in U \\ t \in U \end{cases} \quad \text{is equivalent to} \quad \begin{cases} f \le s \\ f \le t \\ s + t - f \le 1 \\ f \ge 0 \end{cases} \tag{1.4}$$

Verification: all possible combinations of s and t are examined:

- case $s = 0 \Rightarrow \begin{cases} f \le 0 \\ f \le t \\ t - f \le 1 \\ f \ge 0 \end{cases} \Rightarrow \begin{cases} 0 \le f \le 0 \\ 0 \le t \le 1 \end{cases} \Rightarrow \begin{cases} f = 0 \\ t \in U \text{, t free} \end{cases}$;
- case $t = 0 \Rightarrow$ calculations similar to the case $s = 0$ above;
- case $\begin{cases} s = 1 \\ t = 1 \end{cases} \Rightarrow \begin{cases} f \le 1 \\ f \le 1 \\ 2 - f \le 1 \\ f \ge 0 \end{cases} \Rightarrow 1 \le f \le 1 \Rightarrow f = 1$.

The value of f is equal to the product of s and t in all cases.

Product of a binary variable and a bounded real variable

The product of a binary variable s and a real variable x comprised between x_L and x_U is linearized by introducing a real variable f and four inequality constraints.

$$f = s.x \ \text{ with } \begin{cases} s \in U \\ x \in \mathbb{R} \\ x_L \le x \le x_U \end{cases} \quad \text{is equivalent to} \quad \begin{cases} f \le s.x_U \\ f \le x-(1-s).x_L \\ f \ge x-(1-s).x_U \\ f \ge s.x_L \end{cases} \tag{1.5}$$

Verification: all possible values of s are examined:

- case $s=0 \Rightarrow \begin{cases} f \le 0 \\ f \le x-x_L \\ f \ge x-x_U \\ f \ge 0 \end{cases} \Rightarrow \begin{cases} 0 \le f \le 0 \\ x-x_U \le f \le x-x_L \end{cases} \Rightarrow \begin{cases} f=0 \\ x_L \le x \le x_U \text{ free} \end{cases}$;
- case $s=1 \Rightarrow \begin{cases} f \le x_U \\ f \le x \\ f \ge x \\ f \ge x_L \end{cases} \Rightarrow \begin{cases} x \le f \le x \\ x_L \le f \le x_U \end{cases} \Rightarrow \begin{cases} f=x \\ x_L \le x \le x_U \text{ free} \end{cases}$.

The value of f is either 0 or x depending on the value of s.

Linear approximation of a bounded real variable

A real variable x comprised between 0 and x_U can be approximated within ε by a linear combination of m binary variables $(s_i)_{i=1 \text{ to } m}$ of the form:

$$x \approx \varepsilon \sum_{i=1}^{m} 2^{i-1} s_i \quad \text{with } (s_i)_{i=1 \text{ to } m} \in U \ , \ m = E\left(\log_2 \frac{x_U}{\varepsilon}\right)+1 \tag{1.6}$$

Verification

The variable $E\left(\frac{x}{\varepsilon}\right)$ is integer and less than 2^m because $x \le x_U$ and $\log_2 \frac{x_U}{\varepsilon} < m$.

It is written in binary as: $E\left(\frac{x}{\varepsilon}\right) = \sum_{i=1}^{m} 2^{i-1} s_i$.

Let us note: $y = \varepsilon E\left(\frac{x}{\varepsilon}\right) = \varepsilon \sum_{i=1}^{m} 2^{i-1} s_i$.

This variable y is a linear combination of the binary variables $(s_i)_{i=1 \text{ to } m}$ and it satisfies: $\frac{x-y}{\varepsilon} = \frac{x}{\varepsilon} - E\left(\frac{x}{\varepsilon}\right) \Rightarrow 0 \le \frac{x-y}{\varepsilon} < 1 \Rightarrow 0 \le x-y < \varepsilon$.

The variable y is therefore an approximation of x within ε.

Discrete-valued variable or function

A variable x taking a set of m discrete values $\{x_1;\cdots;x_m\}$ can be replaced by m binary variables $(s_i)_{i=1 \text{ to } m}$ and an equality constraint.

$$x = \sum_{i=1}^{m} x_i s_i \quad \text{with} \quad \sum_{i=1}^{m} s_i = 1 \,,\; s_i \in U \tag{1.7}$$

The constraint imposes that only one of the variables s_i is 1, giving x the value x_i.
Similarly, a function $f : x \in \mathbb{R} \mapsto f(x)$ taking a set of m discrete values $\{f_1;\cdots;f_m\}$ can be replaced by m binary variables $(s_i)_{i=1 \text{ to } m}$ and an equality constraint.

$$f(x) = \sum_{i=1}^{m} f_i s_i \quad \text{with} \quad \sum_{i=1}^{m} s_i = 1 \,,\; s_i \in U \tag{1.8}$$

Variable either worth zero or above a threshold

Suppose that the real variable x takes either the value 0 or any value comprised between x_L and x_U. These constraints are linearized by introducing a binary variable s and two inequality constraints.

$$\begin{cases} x = 0 \\ \text{or} \\ 0 < x_L \le x \le x_U \end{cases} \quad \text{is equivalent to} \quad \begin{cases} x \ge s.x_L \\ x \le s.x_U \\ s \in U \end{cases} \tag{1.9}$$

Verification: all possible values of s are examined:

- case $s = 0 \Rightarrow \begin{cases} x \ge 0 \\ x \le 0 \end{cases} \Rightarrow x = 0$;
- case $s = 1 \Rightarrow \begin{cases} x \ge x_L \\ x \le x_U \end{cases} \Rightarrow x_L \le x \le x_U$.

The value of x is either 0 or between x_L and x_U depending on the value of s.

Integer variables taking distinct values

Suppose that two integer variables x and y are bounded and must take different values. These constraints are linearized by introducing a binary variable s and four inequality constraints.

$$\begin{cases} x \neq y \\ x, y \in \mathbb{N} \\ x \leq x_U \;,\; y \leq y_U \end{cases} \quad \text{is equivalent to} \quad \begin{cases} x \geq y + 1 - (y_U + 1).s \\ y \geq x + 1 - (x_U + 1).(1 - s) \\ x \leq x_U \;,\; y \leq y_U \;,\; s \in U \end{cases} \qquad (1.10)$$

Verification: all possible values of s are examined:

- case $s = 0 \Rightarrow \begin{cases} x \geq y + 1 & \Rightarrow x > y \;\; \text{(because } x, y \text{ are integers)} \\ y \geq x - x_U & \rightarrow \text{always satisfied, because } y \geq 0 \text{ and } x \leq x_U; \\ x \leq x_U, y \leq y_U \end{cases}$
- case $s = 1 \Rightarrow \begin{cases} x \geq y - y_U & \rightarrow \text{always satisfied, because } x \geq 0 \text{ and } y \leq y_U \\ y \geq x + 1 & \Rightarrow y > x \;\; \text{(because } x, y \text{ are integers)} \\ x \leq x_U, y \leq y_U \end{cases}$.

The values of x and y are different, with $x > y$ if $s = 0$, or $x < y$ if $s = 1$.

Variable belonging to the union of intervals

Suppose that the real variable x must belong to the union of m intervals.

$$\begin{array}{l} x \in [a_1; b_1] \cup [a_2; b_2] \cup \ldots \cup [a_m; b_m] \\ \quad \text{with } \; a_1 \leq b_1 \leq a_2 \leq b_2 \leq \ldots \leq a_m \leq b_m \end{array} \qquad (1.11)$$

These constraints are linearized by introducing m binary variables $(s_i)_{i=1 \text{ to } m}$ and 2m + 1 inequality constraints.

$$x \in \bigcup_{i=1}^{m} [a_i, b_i] \quad \text{is equivalent to} \quad \begin{cases} x \geq a_i s_i + a_1 (1 - s_i) \\ x \leq b_i s_i + b_n (1 - s_i) \\ \sum_{i=1}^{m} s_i = 1 \;,\; (s_i)_{i=1 \text{ to } m} \in U \end{cases} \qquad (1.12)$$

Verification

The sum constraint imposes that only one of the variables s_i is 1, all the others being 0. Let j be the number of the non-zero variable.

Let us look at the consequences for the constraints (1.12):

- $s_j = 1 \quad \Rightarrow \begin{cases} x \geq a_j \\ x \leq b_j \end{cases} \Rightarrow a_j \leq x \leq b_j$;
- $s_i = 0 \, , i \neq j \Rightarrow \begin{cases} x \geq a_1 \\ x \leq b_n \end{cases} \Rightarrow a_1 \leq x \leq b_n \rightarrow$ always satisfied .

We have thus $x \in [a_j ; b_j]$ for the interval selected by s_j.

Satisfying p constraints among m

Suppose that p constraints: $g_i(x) \leq 0$ must be satisfied among m and that the functions g_i are bounded: $\forall x \, , g_i(x) \leq g_{Ui}$. These constraints are linearized by introducing m binary variables $(s_i)_{i=1 \text{ to } m}$ and m + 1 inequality constraints.

$$\left(g_i(x) \leq 0\right)_{i=1 \text{ to } m} \quad \text{is equivalent to} \quad \begin{cases} g_i(x) \leq s_i . g_{Ui} \\ \sum_{i=1}^{m} s_i \leq m - p \; , \; (s_i)_{i=1 \text{ to } m} \in U \end{cases} \tag{1.13}$$

Verification

The constraint on the sum of s_i requires that at least p variables s_i are zero:

- case $s_j = 0 \quad \Rightarrow g_i(x) \leq 0$;
- case $s_j = 1 \quad \Rightarrow g_i(x) \leq g_{Ui}$.

We obtain p constraints satisfied among the m inequality constraints.

Implication between constraints

Consider two inequality constraints on functions of real variables.

$$\begin{array}{ll} f(x) \leq a & \text{with } \forall x \, , f_L \leq f(x) \\ g(y) \leq b & \text{with } \forall y \, , g(y) \leq g_U \end{array} \tag{1.14}$$

The function f is lower bounded, the function g is upper bounded.
The constraints (1.14) are linked by the implication: "If $f(x) \leq a$, then $g(y) \leq b$ ", which linearizes by introducing a binary variable s and two inequality constraints.

$$\left(f(x) \leq a \Rightarrow g(y) \leq b\right) \quad \text{is equivalent to} \quad \begin{cases} f(x) \geq a + (f_L - a).s \\ g(y) \leq b + (g_U - b).(1 - s) \\ s \in U \end{cases} \tag{1.15}$$

Verification

The implication means that: either $f(x) \ge a$ and $g(y)$ has a free value,
or $g(y) \le b$ and $f(x)$ has a free value.

- case $s=0 \;\Rightarrow\; f(x) \ge a$ and $g(y) \le g_U$;
- case $s=1 \;\Rightarrow\; g(y) \le b$ and $f_L \le f(x)$.

This gives us one or the other of the two possibilities associated to the implication.

Maximum of m bounded functions

Suppose that the function f is defined as the maximum of m bounded functions. The common upper bound of these functions is denoted g_U.

$$f(x) = \max_{i=1,\dots,m} g_i(x) \quad \text{with } \forall x\,,\; g_{Li} \le g_i(x) \le g_U \tag{1.16}$$

This expression for f is linearized by introducing m binary variables $(s_i)_{i=1 \text{ to } m}$, one real variable f and 2m + 1 inequality constraints.

$$\begin{cases} f(x) = \max\limits_{i=1,\dots,m} g_i(x) \\ \forall x\,,\; g_{Li} \le g_i(x) \le g_U \end{cases} \text{ is equivalent to } \begin{cases} f \le g_i(x) + s_i(g_U - g_{Li}) \\ f \ge g_i(x) \\ \sum\limits_{i=1}^{m} s_i = m-1 \;,\; (s_i)_{i=1 \text{ to } m} \in U \end{cases} \tag{1.17}$$

Verification

The sum constraint imposes that only one of the variables s_i is 0, all the others being 1. Let j be the number of the zero variable.
Let us look at the consequences for the constraints (1.17):

- $s_j = 0 \;\Rightarrow\; \begin{cases} f \le g_j(x) \\ f \ge g_j(x) \end{cases} \;\Rightarrow\; f = g_j(x)$;
- $s_i = 1\,,\, i \ne j \;\Rightarrow\; \begin{cases} f \le g_i(x) + (g_U - g_{Li}) \\ f \ge g_i(x) \end{cases} \;\Rightarrow\; g_i(x) \le f$.

We have $\left(f \ge g_i(x)\right)_{i=1 \text{ to } m}$ with equality $f = g_j(x)$ for the function number j.

Minimum of m bounded functions

Suppose that the function f is defined as the minimum of m bounded functions. The common lower bound of these functions is denoted g_L.

$$f(x) = \min_{i=1,\dots,m} g_i(x) \quad \text{with } \forall x, \; g_L \le g_i(x) \le g_{Ui} \tag{1.18}$$

This expression for f is linearized by introducing m binary variables $(s_i)_{i=1 \text{ to } m}$, one real variable f and 2m + 1 inequality constraints.

$$\begin{cases} f(x) = \min\limits_{i=1,\dots,m} g_i(x) \\ \forall x, \; g_L \le g_i(x) \le g_{Ui} \end{cases} \text{ is equivalent to } \begin{cases} f \ge g_i(x) - s_i(g_{Ui} - g_L) \\ f \le g_i(x) \\ \sum\limits_{i=1}^{m} s_i = m-1 \;,\; (s_i)_{i=1 \text{ to } m} \in U \end{cases} \tag{1.19}$$

Verification

The sum constraint imposes that only one of the variables s_i is 0, all the others being 1. Let j be the number of the zero variable:

- $s_j = 0 \;\Rightarrow\; \begin{cases} f \ge g_j(x) \\ f \le g_j(x) \end{cases} \;\Rightarrow\; f = g_j(x)$;
- $s_i = 1, i \ne j \;\Rightarrow\; \begin{cases} f \ge g_i(x) - (g_{Ui} - g_L) \\ f \le g_i(x) \end{cases} \;\Rightarrow\; f \le g_i(x)$.

We have $\left(f \le g_i(x)\right)_{i=1 \text{ to } m}$ with equality $f = g_j(x)$ for the function number j.

1.1.3 Reduction techniques

A mixed linear problem usually has a large number of variables and constraints. A preliminary analysis can allow fixing some variables and/or eliminating some constraints. This section, inspired by reference [R1], presents the usual techniques for reducing the problem size.

Bounds on the variables

Let us assume that the variable $x \in \mathbb{R}^n$ is bounded: $x_L \le x \le x_U$ and consider an inequality constraint of the form

$$g(x) = \sum_{j \ne i} a_j x_j + a_i x_i \le b \tag{1.20}$$

By calculating the minimum value of g on variables other than x_i

$$g_L = \min_{\substack{x_{Lj} \le x_j \le x_{Uj} \\ j \ne i}} g(x) = \sum_{\substack{a_j<0 \\ j\ne i}} a_j x_{Uj} + \sum_{\substack{a_j>0 \\ j\ne i}} a_j x_{Lj} \tag{1.21}$$

we can derive bounds on the variable x_i due to the inequality constraint (1.20).

$$a_i x_i \le b - \sum_{j\ne i} a_j x_j \le b - g_L \quad \Rightarrow \quad \begin{cases} x_i \le \dfrac{b-g_L}{a_i} & \text{if } a_i > 0 \\ x_i \ge \dfrac{b-g_L}{a_i} & \text{if } a_i < 0 \end{cases} \tag{1.22}$$

By proceeding in this way for all pairs (variable–inequality constraint), we can tighten the bounds x_L, x_U on all variables. This treatment should preferably be carried out before the search for bounds on the constraints presented below.

<u>Bounds on the constraints</u>

Let us consider a linear function: $g(x) = a^T x$. If the variable $x \in \mathbb{R}^n$ is bounded: $x_L \le x \le x_U$, we can calculate the extreme values that the function g can take.

$$g(x) = \sum_j a_j x_j \quad \Rightarrow \quad \begin{cases} g_L = \min\limits_{x_L \le x \le x_U} g(x) = \sum\limits_{a_j<0} a_j x_{Uj} + \sum\limits_{a_j>0} a_j x_{Lj} \\ g_U = \max\limits_{x_L \le x \le x_U} g(x) = \sum\limits_{a_j<0} a_j x_{Lj} + \sum\limits_{a_j>0} a_j x_{Uj} \end{cases} \tag{1.23}$$

These extreme values are used to check before solving whether a constraint on g is feasible, infeasible or redundant:

- an equality constraint: $g(x) = a^T x = b$ is feasible if $g_L \le b \le g_U$;
- an inequality constraint: $g(x) = a^T x \le b$ is feasible if $g_L \le b$;
- an inequality constraint: $g(x) = a^T x \le b$ is redundant if $g_U \le b$.

If a constraint is detected as infeasible, the problem has no solution.
A constraint detected as redundant can be removed.

<u>Fixing binary variables</u>

By alternately setting each binary variable to 0 and then to 1, and calculating the bound on the constraints as above, we can detect certain infeasibilities that will allow us to fix the binary variable to 0 or 1. This preliminary variable fixing often allow to significantly reduce the size of the problem before starting to solve it.

1.2 Cutting methods

Cutting methods consist in solving a series of relaxed problems by progressively adding inequality constraints until an integer solution is obtained.

1.2.1 Cut on a variable

Consider a linear problem in integer variables, denoted (LPI).

$$\min_x z = c^T x \quad \text{s.t.} \begin{cases} Ax = b \\ x \in \mathbb{N}^n \end{cases} \tag{1.24}$$

The constraints $x \in \mathbb{N}^n$ are called **integrity constraints.**
The relaxed problem (LPR) is obtained from (LPI) by replacing the integrity constraints by positivity constraints.

$$\min_x z = c^T x \quad \text{s.t.} \begin{cases} Ax = b \\ x \geq 0 \end{cases} \tag{1.25}$$

This standard linear problem is solved by the simplex method (Volume I, section 5.1). Its solution corresponds to an optimal basis B. The simplex array denoted T represents the canonical form of the problem (LPR) in this basis.

$$T = \begin{array}{c} \quad x_B \qquad x_N \qquad \\ \begin{array}{|cc|c|} \hline I & B^{-1}N & \bar{b} \\ \hline 0 & \bar{c}_N^T & -\bar{z} \\ \hline \end{array} \end{array} \begin{array}{l} \rightarrow x_B + B^{-1}N x_N = \bar{b} \\ \rightarrow \bar{c}_N^T x_N = z - \bar{z} \end{array} \quad \rightarrow \text{solution} \begin{cases} x_B = \bar{b} \\ x_N = 0 \end{cases} \tag{1.26}$$

If by chance the basic variables x_B are all integer, the problem (LPI) is solved.
Assume that the basic variable x_i is not an integer.
The canonical form (1.26) gives the expression of the basic variables x_B in terms of the non-basic variables x_N.

$$x_i + \sum_{j \in N} \bar{a}_{ij} x_j = \bar{b}_i \quad , \quad i \in B \tag{1.27}$$

<u>*Remark on the notation*</u>

The notation N above refers to the set of non-basic indexes, not to be confused with the set $\mathbb{N}$ *of natural numbers.*

We note respectively $E(y)$ and $F(y)$, the integer and fractional part of a real y.

$$\begin{cases} E(y) \le y < E(y)+1 \ , \ E(y) \in \mathbb{N} \\ F(y) = y - E(y) \end{cases} \tag{1.28}$$

Since the variables x must be positive, we deduce from (1.27) the inequality

$$x_i + \sum_{j \in N} E(\bar{a}_{ij}) x_j \le \bar{b}_i \tag{1.29}$$

Since the variables x must be integers, the first member of (1.29) must be an integer. The second member of the inequality can then be replaced by the lower whole number without changing the solution set.

$$x_i + \sum_{j \in N} E(\bar{a}_{ij}) x_j \le E(\bar{b}_i) \tag{1.30}$$

Let us replace in this inequality the basic variable x_i by its canonical expression (1.27) in terms of the non-basic variables x_N : $x_i = \bar{b}_i - \sum_{j \in N} \bar{a}_{ij} x_j$.

By grouping the terms with the fractional part defined by (1.28), we obtain

$$\sum_{j \in N} F(\bar{a}_{ij}) x_j \ge F(\bar{b}_i) \tag{1.31}$$

This inequality constraint is called a **cut** or **truncation** on the variable x_i .

The integer solution of the problem (LPI) must satisfy this new inequality. If the relaxed problem solution $x_i = \bar{b}_i$ is not integer, inequality (1.31) is not satisfied, because the non-basic variables are zero $(x_N = 0)$ whereas the second member is strictly positive. We can then resume the solution of the relaxed problem by adding the constraint (1.31) in order to exclude the non-integer solution $x_i = \bar{b}_i$.

This constraint can be reformulated by introducing a positive slack variable y.

$$-\alpha_N^T x_N + y = -\beta \quad \text{with} \quad \begin{cases} \alpha_j = F(\bar{a}_{ij}) \ \ge 0 \\ \beta \ = F(\bar{b}_i) \ > 0 \end{cases} \tag{1.32}$$

Returning to (1.30), the new constraint takes the form

$$x_i + \sum_{j \in N} E(\bar{a}_{ij}) x_j + y = E(\bar{b}_i) \tag{1.33}$$

which shows that the new variable y is necessarily integer (because the other terms are all integers).

Section 1.2.3 explains how to add this constraint into the simplex algorithm starting from the basis solution already found.

1.2.2 Cut on the cost

A cut can also be added to the value of the cost if it is to be an integer. Assume that the cost $\bar{z}$ of the relaxed solution (1.26) is not integer.

$$E(\bar{z}) < \bar{z} \tag{1.34}$$

The canonical form (1.26) gives the expression of the cost z as a function of the non-basic variables x_N.

$$z - \sum_{j \in N} \bar{c}_j x_j = \bar{z} \tag{1.35}$$

Since the variables x must be positive, we deduce from (1.35) the inequality

$$z - \sum_{j \in N} E(\bar{c}_j) x_j \geq \bar{z} \tag{1.36}$$

Since the variables x and the cost z must be integers, the first member of (1.36) must be an integer greater than $\bar{z}$. If $\bar{z}$ is not an integer, the second member can be replaced by the higher whole number without changing the solution set.

$$z - \sum_{j \in N} E(\bar{c}_j) x_j \geq E(\bar{z}) + 1 \tag{1.37}$$

Let us replace in this inequality the cost z by its canonical expression (1.35) as a function of the non-basic variables x_N: $z = \bar{z} + \sum_{j \in N} \bar{c}_j x_j$.

By grouping the terms with the fractional part defined by (1.28), we obtain

$$\sum_{j \in N} F(\bar{c}_j) x_j \geq 1 - F(\bar{z}) \tag{1.38}$$

This inequality constraint is a **cut** or **truncation** on the cost. If the relaxed problem cost $z = \bar{z}$ is not integer, the inequality (1.38) is not satisfied, because the non-basic variables are zero whereas the second member is strictly positive. We can then resume the solution the relaxed problem by adding the constraint (1.38) to exclude the previous solution.
This constraint can be reformulated by introducing a positive slack variable y.

$$-\alpha_N^T x_N + y = -\beta \quad \text{with} \quad \begin{cases} \alpha_j = F(\bar{c}_j) & \geq 0 \\ \beta = 1 - F(\bar{z}) > 0 \end{cases} \tag{1.39}$$

Returning to (1.37), the new constraint takes the form

$$z - \sum_{j \in N} E(\bar{c}_j) x_j - y = E(\bar{z}) + 1 \tag{1.40}$$

which shows that the new variable y is necessarily integer (because the other terms are all integers).

1.2.3 Gomory's method

A cut on a variable or on the cost adds to the relaxed problem a new integer variable y and a new equality constraint (1.32) or (1.39) of the form

$$-\alpha_N^T x_N + y = -\beta \tag{1.41}$$

The previous relaxed problem has been solved by the simplex method (Volume I, section 5.1.4). Its canonical form in the optimal basis B gives the simplex array T.

$$T = \begin{array}{c} \begin{matrix} x_B & \quad x_N & \end{matrix} \\ \begin{array}{|cc|c|} \hline I & B^{-1}N & \bar{b} \\ \hline 0 & \bar{c}_N^T & -\bar{z} \\ \hline \end{array} \end{array} \quad \rightarrow \text{ solution } \begin{cases} x_B = \bar{b} \geq 0 \,,\; x_N = 0 \\ \bar{c}_N \geq 0 \end{cases} \tag{1.42}$$

Let us add the new variable y and the new constraint (1.41) to the simplex array.

$$T' = \begin{array}{c} \begin{matrix} x_B & y & \quad x_N & \end{matrix} \\ \begin{array}{|ccc|c|} \hline I & 0 & B^{-1}N & \bar{b} \\ 0 & 1 & -\alpha_N^T & -\beta \\ \hline 0 & 0 & \bar{c}_N^T & -\bar{z} \\ \hline \end{array} \end{array} \tag{1.43}$$

We observe that the new problem is directly in canonical form in the basis formed by the variables (x_B, y). The corresponding basis solution is: $\begin{cases} x_B = \bar{b} \geq 0 \\ y = -\beta < 0 \end{cases}$.

The variable y is negative, because the previous relaxed solution does not satisfy the added cutting constraint. The base (x_B, y) is therefore not primal-feasible. On the other hand, it is **dual-feasible**, as the reduced costs are positive: $\bar{c}_N \geq 0$. We can therefore apply the dual simplex algorithm (Volume I, section 5.1.8) directly from the base (x_B, y) and the array T'.

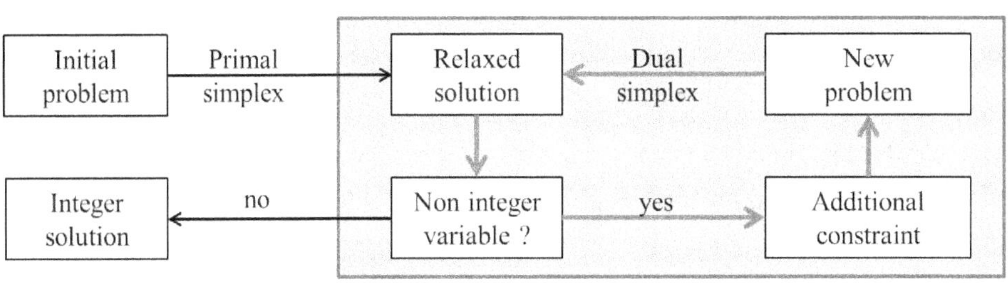

Figure 1-2: Gomory's method.

Gomory's method (1958) shown in figure 1-2 consists in solving the initial relaxed problem by the primal simplex algorithm. If the relaxed solution is not integer, a cutting constraint is added and the new problem is solved by the dual simplex algorithm. The addition of a cut is repeated until an integer solution is obtained.

Example 1-2 illustrates the application of Gomory's method to the solution of a linear problem in integer variables.

Example 1-2: Gomory's method

Consider the linear problem: $\min_{x_1,x_2} z = -x_2 \quad \text{s.t.} \begin{cases} 3x_1 + 2x_2 \le 6 \\ -3x_1 + 2x_2 \le 0 \\ x_1 \in \mathbb{N}\,,\ x_2 \in \mathbb{N} \end{cases}$

This problem is put into standard form by adding 2 slack variables x_3, x_4.
Since the inequality constraints have only integer terms, the variables x_3, x_4 are also integer.

$$\min_{\substack{x_1,x_2\\x_3,x_4}} z = -x_2 \quad \text{s.t.} \begin{cases} 3x_1 + 2x_2 + x_3 = 6 \\ -3x_1 + 2x_2 + x_4 = 0 \\ x_1, x_2, x_3, x_4 \in \mathbb{N} \end{cases} \quad \textbf{(problem } \mathbf{P_0}\textbf{)}$$

The first stage is to solve the relaxed problem P_{0r} obtained by replacing integrity constraints with positivity constraints.

Stage 0: solving the relaxed problem P_{0r} by the primal simplex

Initial array of P_{0r}

Basis: (x_3, x_4)

Basis solution: $\begin{cases} x_3 = 6 \\ x_4 = 0 \end{cases}$

x_1	x_2	x_3	x_4			
3	2	1	0	6	x_3	$\rightarrow s_{23}=3$
-3	2	0	1	0	x_4	$\rightarrow s_{24}=0$
0	-1	0	0	0	z	

The problem is directly in canonical form in the primal-feasible basis (x_3, x_4).
We solve the relaxed problem P_{0r} by the primal simplex algorithm.

Pivoting 1: $x_4 \leftrightarrow x_2$

New basis: (x_3, x_2)

Basis solution: $\begin{cases} x_3 = 6 \\ x_2 = 0 \end{cases}$

x_1	x_2	x_3	x_4			
6	0	1	-1	6	x_3	$\rightarrow s_{13}=1$
-3/2	1	0	1/2	0	x_2	
-3/2	0	0	1/2	0	z	

Pivoting 2: $x_3 \leftrightarrow x_1$

New basis: (x_1, x_2)

Basis solution: $\begin{cases} x_1 = 1 \\ x_2 = \dfrac{3}{2} \end{cases}$

x_1	x_2	x_3	x_4		
1	0	1/6	-1/6	1	x_1
0	1	1/4	1/4	3/2	x_2
0	0	1/4	1/4	3/2	z

The solution of the relaxed problem P_{0r} is obtained. This solution is not integer because of x_2. The next stage is to generate a cut excluding the solution $x_2 = \frac{3}{2}$, and then solve the new relaxed problem with the cutting constraint.

Stage 1: solving the relaxed problem P_{1r} with a cut on x_2

To generate a cut on x_2, we start from the solution array of the problem P_{0r}.

The 2nd row gives x_2 as a function of (x_3, x_4): $x_2 + \frac{1}{4}x_3 + \frac{1}{4}x_4 = \frac{3}{2}$ **(equality E₁)**

An inequality is generated from (E_1).

$$(E_1) \Rightarrow x_2 + E\left(\frac{1}{4}\right)x_3 + E\left(\frac{1}{4}\right)x_4 \le E\left(\frac{3}{2}\right) \Rightarrow x_2 \le 1$$

Substituting x_2 given by (E_1), we obtain: $-\frac{1}{4}x_3 - \frac{1}{4}x_4 \le 1 - \frac{3}{2}$

which is rewritten with a slack variable x_5 : $-\frac{1}{4}x_3 - \frac{1}{4}x_4 + x_5 = -\frac{1}{2}$ **(cut C_1)**

The new relaxed problem P_{1r} is formed from P_{0r} to which we add the variable x_5 and the constraint C_1. The simplex array of P_{1r} is formed from the solution array of P_{0r} by adding a column for x_5 and a row for C_1.

Initial array of P_{1r}

Basis: (x_1, x_2, x_5)

Basis solution:

$$\begin{cases} x_1 = 1 \\ x_2 = 3/2 \\ x_5 = -1/2 \end{cases}$$

x_1	x_2	x_3	x_4	x_5			
1	0	1/6	-1/6	0	1	x_1	
0	1	1/4	1/4	0	3/2	x_2	
0	0	-1/4	-1/4	1	-1/2	x_5	$\rightarrow$ s=-1/2
0	0	1/4	1/4	0	3/2	z	

The problem is directly in canonical form in the basis (x_1, x_2, x_5).
This basis is not primal-feasible $(x_5 < 0)$, but it is dual-feasible (positive reduced costs). We solve the relaxed problem P_{1r} by the **dual simplex** algorithm. Recall that the dual pivoting consists in choosing:

- as entering variable the first negative basic variable: $\bar{b}_e < 0 \,,\, e \in B$;

- as leaving variable the first non-basic variable to cancel: $s \rightarrow \max\limits_{\substack{j \in N \\ \bar{a}_{ej} < 0}} \dfrac{\bar{c}_{Nj}}{\bar{a}_{ej}}$.

Pivoting 1: $x_5 \leftrightarrow x_3$

New basis: (x_1, x_2, x_3)

Basis solution:

$$\begin{cases} x_1 = 2/3 \\ x_2 = 1 \\ x_3 = 2 \end{cases}$$

x_1	x_2	x_3	x_4	x_5		
1	0	0	-1/3	2/3	2/3	x_1
0	1	0	0	1	1	x_2
0	0	1	1	-4	2	x_3
0	0	0	0	1	1	z

The solution of the relaxed problem P_{1r} is obtained. This solution is not integer because of x_1. The next stage is to generate a cut excluding the solution $x_1 = \dfrac{2}{3}$, and then solve the new relaxed problem with the cutting constraint.

Stage 2: solving the relaxed problem P_{2r} with a cut on x_1

To generate a cut on x_1, we start from the solution array of the problem P_{1r}.

The 1st row gives x_1 as a function of (x_4, x_5): $x_1 - \dfrac{1}{3}x_4 + \dfrac{2}{3}x_5 = \dfrac{2}{3}$ **(equality E_2)**

An inequality is generated from (E_2).

$$(E_2) \Rightarrow x_1 + E\left(-\frac{1}{3}\right)x_4 + E\left(\frac{2}{3}\right)x_5 \le E\left(\frac{2}{3}\right) \Rightarrow x_1 - x_4 \le 0$$

Substituting x_1 given by (E_2), we obtain: $-\frac{2}{3}x_4 - \frac{2}{3}x_5 \le 0 - \frac{2}{3}$

which is rewritten with a slack variable x_6: $-\frac{2}{3}x_4 - \frac{2}{3}x_5 + x_6 = -\frac{2}{3}$ **(cut C_2)**

The new relaxed problem P_{2r} is formed from P_{1r} to which we add the variable x_6 and the constraint C_2. The simplex array of P_{2r} is formed from the solution array of P_{1r} by adding a column for x_6 and a row for C_2.

<u>Initial array of P_{2r}</u>
Basis: (x_1, x_2, x_3, x_6)
Basis solution:
$$\begin{cases} x_1 = 2/3 \\ x_2 = 1 \\ x_3 = 2 \\ x_6 = -2/3 \end{cases}$$

x_1	x_2	x_3	x_4	x_5	x_6			
1	0	0	-1/3	2/3	0	2/3	x_1	
0	1	0	0	1	0	1	x_2	
0	0	1	1	-4	0	2	x_3	
0	0	0	-2/3	-2/3	1	-2/3	x_6	→ s=0
0	0	0	0	1	0	1	z	

The problem is directly in canonical form in the basis (x_1, x_2, x_3, x_6).
This basis is not primal-feasible $(x_6 < 0)$, but it is dual- feasible (positive reduced costs). We solve the relaxed problem P_{2r} by the dual simplex algorithm.

<u>Pivoting 1</u>: $x_6 \leftrightarrow x_4$
New basis: (x_1, x_2, x_3, x_4)

Basis solution: $\begin{cases} x_1 = 1 \\ x_2 = 1 \\ x_3 = 1 \\ x_4 = 1 \end{cases}$

x_1	x_2	x_3	x_4	x_5	x_6		
1	0	0	0	1	-1/2	1	x_1
0	1	0	0	1	0	1	x_2
0	0	1	0	-5	3/2	1	x_3
0	0	0	1	1	-3/2	1	x_4
0	0	0	0	1	0	1	z

We obtain the solution of the relaxed problem P_{2r}. This solution is integer and therefore gives the solution of the problem P_0. Its cost is $z = -1$.

Figure 1-3 shows the geometric interpretation of the cuts in the space of the original variables (x_1, x_2). These cuts successively eliminate the non-integer solutions S_0 and S_1 (table 1-1) to end at the integer solution S_2.

Initial problem: $\min\limits_{x_1,x_2\in N} z=-x_2$ s.t. $\begin{cases} 3x_1+2x_2\le 6 \\ -3x_1+2x_2\le 0 \end{cases}$

Cutting equations:

- cut C_1 : $x_2 \le 1$;

- cut C_2 : $x_1 - x_4 \le 0$
$\Rightarrow x_1-(3x_1-2x_2)\le 0$
$\Rightarrow -x_1+x_2\le 0$.

Point	x_1	x_2	z
S_0	1	3/2	−3/2
S_1	2/3	1	−1
S_2	1	1	−1

Table 1-1: Successive solutions.

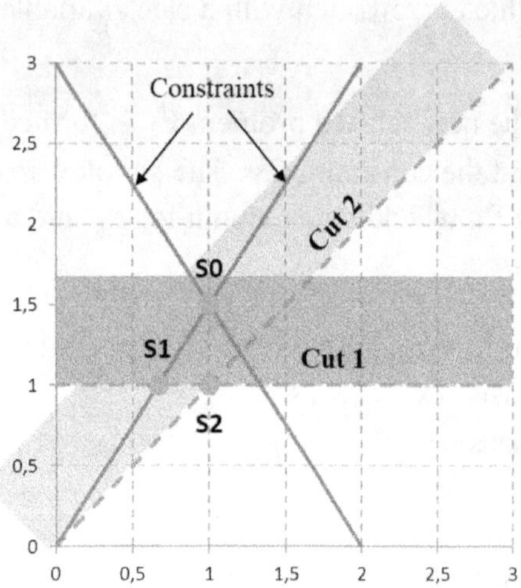

Figure 1-3: Cutting constraint effect.

The efficiency of Gomory's method depends on the choice and order of the cuts. If several cuts are possible, we can choose the one that produces the maximum increase in cost at the first pivoting. Let us retrieve the array T' (1.43) obtained by adding a cut of the form: $-\alpha_N^T x_N + y = -\beta$.

$$T' = \begin{array}{ccc|c} x_B & y & x_N & \\ \hline I & 0 & B^{-1}N & \bar{b} \\ 0 & 1 & -\alpha_N^T & -\beta \\ \hline 0 & 0 & \bar{c}_N^T & -\bar{z} \end{array} \tag{1.44}$$

The variable y is the only negative basic variable. It will leave the basis at the first dual pivoting and be replaced by the first non-basic variable to cancel.

The entering variable x_j is determined by: $\max\limits_{\substack{j\in N \\ -\alpha_j<0}} \dfrac{\bar{c}_j}{-\alpha_j}$.

This pivoting leads to the following cost change.

$$-\bar{z} \rightarrow -\bar{z} - \frac{\bar{c}_j}{-\alpha_j}(-\beta) \quad \Rightarrow \Delta z = \frac{\bar{c}_j}{\alpha_j}\beta \ge 0 \tag{1.45}$$

This formula gives the cost increase at the first pivoting associated with the chosen cut (values of α, β). It can be used to select the most efficient initial cut, although this does not guarantee the final efficiency.

If the cut is on the cost, the coefficients α are given by (1.39): $\alpha_j = F(\bar{c}_j)$.

Since the entering variable satisfies: $\alpha_j > 0$, its reduced cost is non-zero: $\bar{c}_j > 0$.

The increase in cost at the first pivoting is given by

$$\Delta z = \frac{\bar{c}_j}{\alpha_j}\beta \quad \text{with} \quad \begin{cases} \alpha_j = F(\bar{c}_j) & > 0 \\ \beta = 1 - F(\bar{z}) > 0 \end{cases} \tag{1.46}$$

This increase is strictly positive.

It can be demonstrated that Gomory's method achieves an integer solution in a finite number of cuts, provided that these cuts induce a strict increase in the cost of the relaxed problem. The cut on the cost fulfils this condition.
Despite this theoretical property of convergence, the number of cuts required can be exponential. Furthermore, the method is sensitive to rounding errors. For example, it will be necessary to decide whether a solution such as $x = 1.0001$ should be considered as integer or requires the addition of a cut. These numerical difficulties make the method ineffective on large problems. On the other hand, the principle of adding cuts is useful as a complement to the tree methods presented in section 1.3. These cuts constrain the relaxed problem and allow us to find better minorants of the optimal cost.

Cuts on variables or on the cost are called **simple cuts**. Depending on the problem, stronger cuts can be generated as shown below.

1.2.4 Integral cut

Let us retrieve the integer linear problem (LPI) and assume that the coefficients of the matrices/vectors A, b and c are integers.

$$\min_x z = c^T x \quad \text{s.t.} \quad \begin{cases} Ax = b \\ x \in \mathbb{N}^n \end{cases} \quad \text{with } A \in \mathbb{Z}^{m \times n}, b \in \mathbb{Z}^m, c \in \mathbb{Z}^n \tag{1.47}$$

The solution of the problem in canonical form is

$$x_B = B^{-1}b - B^{-1}Nx_N \tag{1.48}$$

where B is the sub-matrix of A associated with the optimal basis and N is the sub-matrix of A associated with the non-basic variables x_N.

Since the matrix B has integer coefficients, its determinant $d = \det(B)$ is integer and its inverse B^{-1} is of the form: $B^{-1} = \frac{1}{d}B'$, where B' is a matrix with integer elements. The relaxed problem solution (1.48) is therefore rational of the form

$$x_i = \bar{b}_i - \sum_{j\in N} \bar{a}_{ij} x_j = \frac{\beta_i}{d} - \sum_{j\in N} \frac{\alpha_{ij}}{d} x_j \quad , \quad i \in B \tag{1.49}$$

where the coefficients α_{ij}, β_i, d are all integers.

Let $[y]_d$ be the integer congruent to y modulo d between 0 and $d-1$.

$$[y]_d \equiv y\,[d] \iff y = [y]_d + kd \quad \text{with} \begin{cases} 0 \le [y]_d \le d-1 \\ k \in \mathbb{N} \end{cases} \tag{1.50}$$

For any integer p that is prime to d, the integer solution of problem (1.47) must satisfy the following inequalities called **integral cuts**.

$$\sum_{j\in N} [p\alpha_{ij}]_d x_j \ge [p\beta_i]_d \quad , \quad i \in B \tag{1.51}$$

Demonstration (see [R15])

The basic variable x_i given by (1.49) must take an integer value noted k.

$$x_i = \frac{\beta_i}{d} - \sum_{j\in N} \frac{\alpha_{ij}}{d} x_j = k \in \mathbb{N} \;\Rightarrow\; \beta_i - \sum_{j\in N} \alpha_{ij} x_j = kd$$

Let us multiply each member by an integer p prime to d: $p\beta_i - \sum_{j\in N} p\alpha_{ij} x_j = pkd$.

By expressing $p\beta_i$ and $p\alpha_{ij}$ modulo d: $\begin{cases} p\beta_i = [p\beta_i]_d + k_1 d \\ p\alpha_{ij} = [p\alpha_{ij}]_d + k_{2j} d \end{cases}$,

the previous equality yields: $[p\beta_i]_d - \sum_{j\in N} [p\alpha_{ij}]_d x_j = Kd$

where $K = pk - k_1 + \sum_{j\in N} k_{2j} x_j$ is an integer (because the variables x_j are integers).

In addition, we have

$$\begin{cases} [p\beta_i]_d \le d-1 \\ [p\alpha_{ij}]_d \ge 0 \end{cases} \Rightarrow [p\beta_i]_d - \sum_{j\in N} [p\alpha_{ij}]_d x_j \le d-1 \Rightarrow Kd \le d-1 \Rightarrow K \le 0$$

from which the inequality (1.51) can be deduced: $[p\beta_i]_d - \sum_{j\in N} [p\alpha_{ij}]_d x_j = Kd \le 0$.

The basis solution defined by (1.49) is given by

$$x_N = 0 \ , \ x_i = \frac{\beta_i}{d} \ , \ i \in B \tag{1.52}$$

If this solution is not integer, then β_i is not divisible by d and $[p\beta_i]_d > 0$ (because p is prime to d). The inequality (1.51) is not satisfied and allows to exclude this non-integer solution.

Integral cuts on the cost can be defined in a similar way if the cost is to be integer. The canonical form of the cost is written in rational form

$$z = \bar{z} - \sum_{j \in N} \bar{c}_j x_j = \frac{\delta}{d} - \sum_{j \in N} \frac{\gamma_j}{d} x_j \tag{1.53}$$

and the resulting cut is similar to (1.51) with (δ, γ_j) replacing (β_i, α_{ij}).

The integral cuts (1.51) allow to generate a set of inequalities by varying the value of the integer p, that must be prime to $d = \det(B)$. The *a priori* most efficient cut is the one for which the second member $[p\beta_i]_d$ is the largest.

1.2.5 Mixed cut

Let us now consider a mixed-variable linear problem (LPM).

$$\min_x z = c^T x \ \text{ s.t. } \begin{cases} Ax = b \\ x_j \in \mathbb{N}, \ j \in J_E \\ x_j \geq 0 \ , \ j \in J_C \end{cases} , \ A \in \mathbb{R}^{m \times n}, b \in \mathbb{R}^m, c \in \mathbb{R}^n \tag{1.54}$$

The indices J_E correspond to integer variables.
The indices J_C correspond to positive real (or continuous) variables.
The solution of the problem is expressed in the canonical form as

$$x_i + \sum_{j \in N} \bar{a}_{ij} x_j = \bar{b}_i \ , \ i \in B \tag{1.55}$$

The non-basic variables $(j \in N)$ are split into 4 subsets:
- the integer variables are split into J_E^+ and J_E^-;
- the real variables are split into J_C^+ and J_C^-.

These subsets are defined according to the signs of the coefficients as follows.

$$\begin{aligned} J_E^+ &= \left\{ j \in N \cap J_E \ , \ F(\bar{a}_{ij}) > F(\bar{b}_i) \right\} \ , \ J_C^+ = \left\{ j \in N \cap J_C \ , \ \bar{a}_{ij} \geq 0 \right\} \\ J_E^- &= \left\{ j \in N \cap J_E \ , \ F(\bar{a}_{ij}) \leq F(\bar{b}_i) \right\} \ , \ J_C^- = \left\{ j \in N \cap J_C \ , \ \bar{a}_{ij} < 0 \right\} \end{aligned} \tag{1.56}$$

The solution of problem (1.54) must satisfy the following **mixed cuts**.

$$\sum_{j\in J_E^-} F(\bar a_{ij})x_j + \sum_{j\in J_E^+} F(\bar b_i)\frac{1-F(\bar a_{ij})}{1-F(\bar b_i)}x_j + \sum_{j\in J_C^+} \bar a_{ij}x_j - \sum_{j\in J_C^-}\frac{F(\bar b_i)}{1-F(\bar b_i)}\bar a_{ij}x_j \ge F(\bar b_i)\ ,\ i\in B\cap J_E \tag{1.57}$$

Demonstration (see [R15])

Let us first assume that each subset $J_E^+, J_E^-, J_C^+, J_C^-$ contains only one non-basic variable and consider the canonical form of a basic variable x_0, which has to be integer. This canonical form is written as the following equality E_1.

$$\underset{J_E^+}{x_0 + a_1x_1} + \underset{J_E^-}{a_2x_2} + \underset{J_C^+}{y_1} - \underset{J_C^-}{y_2} = b_0 \quad \text{with} \begin{cases} x_0, x_1, x_2 \in \mathbb{N} \\ y_1, y_2 \ge 0 \end{cases} \text{and } \underset{J_E^-}{F(a_2) \le} F(b_0) \underset{J_E^+}{<} F(a_1)$$

(equality E_1)

$$\Rightarrow x_0 + a_1x_1 + a_2x_2 - y_2 \le b_0 \quad \text{because } y_1 \ge 0$$

$$\Rightarrow x_0 + [E(a_1)+F(a_1)]x_1 + [E(a_2)+1-(1-F(a_2))]x_2 \le b_0 + y_2$$

$$\Rightarrow x_0 + E(a_1)x_1 + [E(a_2)+1]x_2 \le b_0 + y_2 + [1-F(a_2)]x_2 \quad \text{because } F(a_1)x_1 \ge 0$$

The first member of the above inequality is an integer denoted z.

$$z = x_0 + E(a_1)x_1 + [E(a_2)+1]x_2 \in \mathbb{N}$$

$$\Rightarrow z - b_0 \le y_2 + [1-F(a_2)]x_2 \quad \textbf{(inequality } I_1\textbf{)}$$

If $z \le E(b_0)$, the inequality I_1 is satisfied. Indeed, the first member is negative and the second member is positive because $x_2 \ge 0$, $y_2 \ge 0$.

If $z \ge E(b_0)+1$, we transform the inequality I_1 as follows.

$$z - [E(b_0)+1] + [1-F(b_0)] \le y_2 + [1-F(a_2)]x_2$$

$$\Rightarrow \frac{z-[E(b_0)+1]}{1-F(b_0)} + 1 \le \frac{y_2+[1-F(a_2)]x_2}{1-F(b_0)} \quad \text{(dividing by } 1-F(b_0) \text{ that is positive)}$$

$$\Rightarrow z - [E(b_0)+1] + 1 \le \frac{y_2+[1-F(a_2)]x_2}{1-F(b_0)}$$

$$\text{because } \begin{cases} z \ge E(b_0)+1 \\ 1-F(b_0) \le 1 \end{cases} \Rightarrow \frac{z-[E(b_0)+1]}{1-F(b_0)} \ge z - [E(b_0)+1]$$

$$\Rightarrow z \le E(b_0) + \frac{y_2+[1-F(a_2)]x_2}{1-F(b_0)} \quad \textbf{(inequality } I_2\textbf{)}$$

Note that I_2 established for $z \geq E(b_0)+1$ is also satisfied for $z \leq E(b_0)$.

Let us return to the initial variables: $z = x_0 + E(a_1)x_1 + [E(a_2)+1]x_2$

$$\Rightarrow \; x_0 + E(a_1)x_1 + [E(a_2)+1]x_2 \leq E(b_0) + \frac{y_2 + [1-F(a_2)]x_2}{1-F(b_0)}$$

$$\Rightarrow \; x_0 + E(a_1)x_1 + \left[E(a_2)+1-\frac{1-F(a_2)}{1-F(b_0)}\right]x_2 - \frac{1}{1-F(b_0)}y_2 \leq E(b_0)$$

Let us replace $x_0 = b_0 - a_1x_1 - a_2x_2 - y_1 + y_2$ given by the equality E_1.

$$\Rightarrow \; b_0 + [E(a_1)-a_1]x_1 + \left[E(a_2)-a_2+1-\frac{1-F(a_2)}{1-F(b_0)}\right]x_2 - y_1 + \left[1-\frac{1}{1-F(b_0)}\right]y_2 \leq E(b_0)$$

$$\Rightarrow \; F(a_1)x_1 + F(b_0)\frac{1-F(a_2)}{1-F(b_0)}x_2 + y_1 + \frac{F(b_0)}{1-F(b_0)}y_2 \geq F(b_0) \quad \textbf{(inequality I}_3\textbf{)}$$

In the equality E_1 considered at the beginning of the demonstration:

- $y_1 \geq 0$ is a real non-basic variable $+\bar{a}_{ij}x_j$ such that $\bar{a}_{ij} \geq 0$ (set J_C^+);
- $y_2 \geq 0$ is a real non-basic variable $-\bar{a}_{ij}x_j$ such that $\bar{a}_{ij} \leq 0$ (set J_C^-).

The inequality I_3 thus corresponds well to the inequality (1.57) for the basic variable x_0 when each subset $J_E^+, J_E^-, J_C^+, J_C^-$ contains only one variable.

In the case where the subsets $J_E^+, J_E^-, J_C^+, J_C^-$ contain several variables, the previous demonstration applies identically by replacing the terms in x_1, x_2, y_1, y_2 by the corresponding sums on $J_E^+, J_E^-, J_C^+, J_C^-$. We then obtain the general form (1.57) which applies to each basic variable that must be integer.

$$\Rightarrow \; \sum_{j\in J_E^-} F(\bar{a}_{ij})x_j + \sum_{j\in J_E^+} F(\bar{b}_i)\frac{1-F(\bar{a}_{ij})}{1-F(\bar{b}_i)}x_j + \sum_{j\in J_C^+} \bar{a}_{ij}x_j - \sum_{j\in J_C^-} \frac{F(\bar{b}_i)}{1-F(\bar{b}_i)}\bar{a}_{ij}x_j \geq F(\bar{b}_i) \;,\; i \in B \cap J_E$$

The basis solution defined by (1.49) is given by

$$x_N = 0 \;,\; x_i = \bar{b}_i \;,\; i \in B \tag{1.58}$$

If this solution is not integer, then $F(\bar{b}_i) > 0$. The inequality (1.51) is not satisfied and allows to exclude this non-integer solution.

The mixed cuts (1.57) are defined directly from the simplex array and apply to any mixed-variable linear problem.

Example 1-3 compares the efficiency of the simple, integral and mixed cuts.

Example 1-3: Simple, integral and mixed cuts (from [R15])

Consider the problem: $\min_{x_1,x_2} z = -x_1 - x_2$ s.t. $\begin{cases} -6x_1 + 4x_2 \le 1 \\ 6x_1 - 8x_2 \le 9 \\ 2x_1 + 4x_2 \le 13 \\ x_1 \in \mathbb{N}\ ,\ x_2 \in \mathbb{N} \end{cases}$

This problem is put into standard form by introducing 3 slack variables x_3, x_4, x_5
Since the constraints have only integer terms, the slack variables are also integer.

$$\min_{\substack{x_1,x_2\\x_3,x_4,x_5}} z = -x_1 - x_2 \text{ s.t. } \begin{cases} -6x_1 + 4x_2 + x_3 = 1 \\ 6x_1 - 8x_2 + x_4 = 9 \\ 2x_1 + 4x_2 + x_5 = 13 \\ x_1, x_2, x_3, x_4, x_5 \in \mathbb{N} \end{cases} \quad \text{(problem } P_0\text{)}$$

The relaxed problem P_{0r} associated with the problem P_0 is solved by the primal simplex.

Initial array of P_{0r}
Basis: (x_3, x_4, x_5)

Basis solution: $\begin{cases} x_3 = 1 \\ x_4 = 9 \\ x_5 = 13 \end{cases}$

x_1	x_2	x_3	x_4	x_5		
-6	4	1	0	0	1	x_3
6	-8	0	1	0	9	x_4
2	4	0	0	1	13	x_5
-1	-1	0	0	0	0	-z

Pivoting 1: $x_3 \leftrightarrow x_2$
New basis: (x_2, x_4, x_5)
Basis solution:

$\begin{cases} x_2 = 1/4 \\ x_4 = 11 \\ x_5 = 12 \end{cases}$

x_1	x_2	x_3	x_4	x_5		
-3/2	1	1/4	0	0	1/4	x_2
-6	0	2	1	0	11	x_4
8	0	-1	0	1	12	x_5
-5/2	0	1/4	0	0	1/4	-z

Pivoting 2: $x_5 \leftrightarrow x_1$
New basis: (x_2, x_4, x_1)
Basis solution:

$\begin{cases} x_2 = 5/2 \\ x_4 = 20 \\ x_1 = 3/2 \end{cases}$

x_1	x_2	x_3	x_4	x_5		
0	1	1/16	0	3/16	5/2	x_2
0	0	5/4	1	3/4	20	x_4
1	0	-1/8	0	1/8	3/2	x_1
0	0	-1/16	0	5/16	4	-z

Pivoting 3: $x_4 \leftrightarrow x_3$

New basis: (x_2, x_3, x_1)

Basis solution:

$$\begin{cases} x_2 = 3/2 \\ x_3 = 16 \\ x_1 = 7/2 \end{cases}$$

x_1	x_2	x_3	x_4	x_5		
0	1	0	-1/20	3/20	3/2	x_2
0	0	1	4/5	3/5	16	x_3
1	0	0	1/10	1/5	7/2	x_1
0	0	0	1/20	7/20	5	-z

The solution of the relaxed problem P_0 is not integer because of x_1 and x_2.
The last simplex array gives the canonical form of the solution in terms of the non-base variables (x_4, x_5).

$$\begin{cases} x_1 + \dfrac{1}{10}x_4 + \dfrac{1}{5}x_5 = \dfrac{7}{2} & \rightarrow \text{equality } E_1 \\ x_2 - \dfrac{1}{20}x_4 + \dfrac{3}{20}x_5 = \dfrac{3}{2} & \rightarrow \text{equality } E_2 \end{cases}$$

Figure 1-4 shows in the plane (x_1, x_2) the feasible domain limited by the three constraints and the two axes. The relaxed solution corresponds to the vertex of the polytope located at R. There are 5 integer solutions in the feasible domain (marked by points) and the optimal integer solution is at E.

Relaxed solution

$$\text{R: } \begin{cases} x_1 = 3.5 \\ x_2 = 1.5 \end{cases} \rightarrow z = -5$$

Integer solution

$$\text{E: } \begin{cases} x_1 = 2 \\ x_2 = 2 \end{cases} \rightarrow z = -4$$

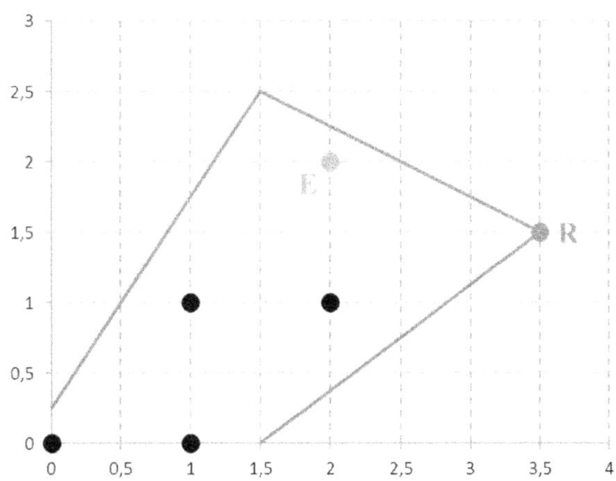

Figure 1-4: Relaxed and integer solution.

The equalities E_1 and E_2 are now used to generate successively simple, integral or mixed cuts in order to compare their efficiency.

Simple cuts

Simple cuts are of the form: $\sum_{j\in N} F(\bar{a}_{ij})x_j \geq F(\bar{b}_i)$.

- a cut on x_1 yields: $\frac{1}{10}x_4 + \frac{1}{5}x_5 \geq \frac{1}{2} \Rightarrow x_1 \leq 3$

  We add a slack variable and the cutting constraint to the solution array of P0, and then solve the new problem by the dual simplex.

  The new relaxed solution is: $\begin{cases} x_1 = 3 \\ x_2 = 1.75 \end{cases} \rightarrow z = -4.75$

- a cut on x_2 yields: $\frac{19}{20}x_4 + \frac{3}{20}x_5 \geq \frac{1}{2} \Rightarrow 6x_1 - 7x_2 \leq 10$

  We add a slack variable and the cutting constraint to the solution array of P0, and then solve the new problem by the dual simplex.

  The new relaxed solution is: $\begin{cases} x_1 = 3.45 \\ x_2 = 1.53 \end{cases} \rightarrow z = -4.98$

These simple cuts are shown in dotted lines in figure 1-5. They reduce the feasible domain very little and improve the relaxed solution only slightly.

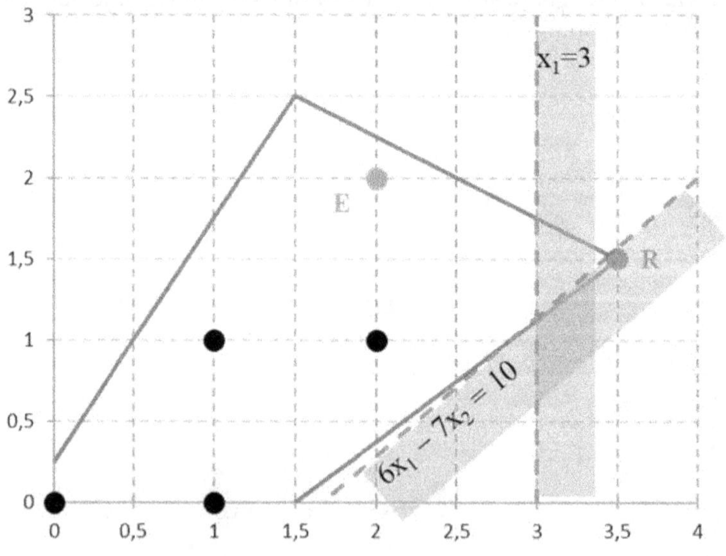

Figure 1-5: Simple cuts.

Integral cuts

Integral cuts are of the form: $\sum_{j\in N}[p\alpha_{ij}]_d x_j \geq [p\beta_i]_d$ with an integer p prime to d.

d is the determinant of the basis matrix B and corresponds to the denominator of the equations E_1 or E_2 .

- for a cut on x_1 : $x_1+\frac{1}{10}x_4+\frac{1}{5}x_5=\frac{7}{2} \;\rightarrow\; x_4+2x_5 \equiv 5 \;[d=10]$.

  If we take $p=1$: $p=1 \rightarrow x_4+2x_5 \geq 5 \;\Rightarrow\; x_1 \leq 3$.

  We retrieve the previous simple cut that was obtained for x_1 ;

- for a cut on x_2 : $x_2-\frac{1}{20}x_4+\frac{3}{20}x_5=\frac{3}{2} \;\rightarrow\; x_4+17x_5 \equiv 10 \;[d=20]$.

  If we take $p=1$: $p=1 \rightarrow x_4+17x_5 \geq 10 \;\Rightarrow\; 2x_1+3x_2 \leq 11$.

  The new relaxed solution is: $\begin{cases} x_1=3,38 \\ x_2=1,41 \end{cases} \rightarrow z=-4,79$.

  If we take $p=3$: $p=3 \rightarrow 3x_4+11x_5 \geq 10 \;\Rightarrow\; 2x_1+x_2 \leq 8$.

  The new relaxed solution is: $\begin{cases} x_1=3.17 \\ x_2=1.67 \end{cases} \rightarrow z=-4.83$.

These integral cuts on x_2 are shown in dotted lines in figure 1-6. They reduce the feasible domain very little and improve the solution only slightly.

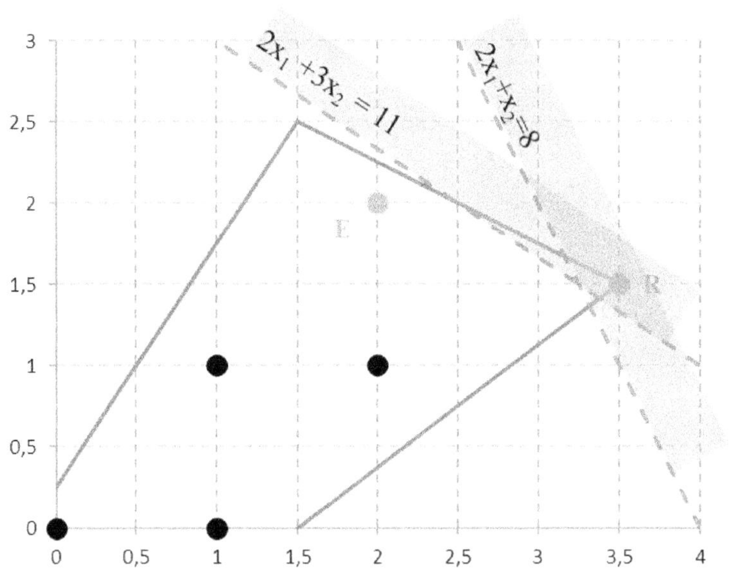

Figure 1-6: Integral cuts.

<u>Mixed cuts</u>

Mixed cuts are of the form: $\sum_{j\in J_E^-} F(\bar{a}_{ij})x_j + \sum_{j\in J_E^+} \frac{F(\bar{b}_i)\left(1-F(\bar{a}_{ij})\right)}{1-F(\bar{b}_i)} x_j \ge F(\bar{b}_i)$.

All non-basic variables are integer: $J_E = \{4\,,5\}\,,\,J_C = \varnothing$.

- cut on x_1 : $x_1 + \frac{1}{10}x_4 + \frac{1}{5}x_5 = \frac{7}{2} \;\rightarrow\; F(\bar{a}_{14}), F(\bar{a}_{15}) \le F(\bar{b}_1) = \frac{1}{2}$.

  $J_E^- = \{4\,,5\}\,,\,J_E^+ = \varnothing \;\rightarrow\; \frac{1}{10}x_4 + \frac{1}{5}x_5 \ge \frac{1}{2} \quad \Rightarrow\; x_1 \le 3$

  We retrieve the previous simple cut that was obtained for x_1 ;

- cut on x_2 : $x_2 - \frac{1}{20}x_4 + \frac{3}{20}x_5 = \frac{3}{2} \;\rightarrow\; F(\bar{a}_{25}) \le F(\bar{b}_2) = \frac{1}{2} \le F(\bar{a}_{24})$.

  $J_E^- = \{5\}\;,\;J_E^+ = \{4\} \;\rightarrow\; \frac{3}{20}x_5 + \frac{1}{20}x_4 \ge \frac{1}{2} \;\Rightarrow\; 6x_1 + 2x_2 \le 19$

  The new relaxed solution is: $\begin{cases} x_1 = 2.5 \\ x_2 = 2 \end{cases} \;\rightarrow\; z = -4.5$.

This mixed cut on x_2 is shown in dotted lines in figure 1-7. It restricts the feasible domain much more than the simple or integral cuts and produces therefore a stronger improvement in the relaxed solution.

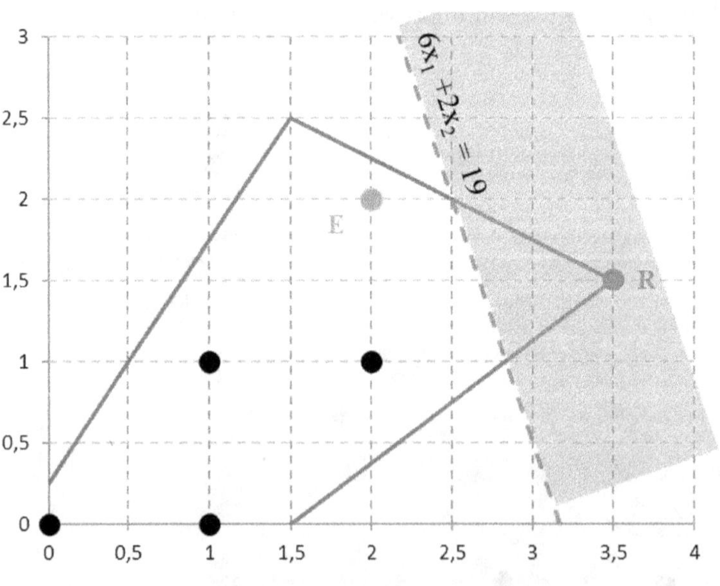

Figure 1-7: Mixed cut.

1.3 Tree methods

Tree methods explore the set of possible combinations of integer values. The evaluation of relaxed solutions allows for the elimination of subsets and avoids examining the complete tree of possibilities.

1.3.1 Implicit enumeration

Consider a linear problem in binary variables, denoted (LPB).

$$\min_{x} z = c^T x \quad \text{s.t.} \quad \begin{cases} Ax = b \\ x \in U^n \quad , \quad U = \{0 \,;\, 1\} \end{cases} \tag{1.59}$$

The set of possible combinations $X = U^n$ is finite with cardinal $\text{Card}(X) = 2^n$.

Explicit enumeration consists of evaluating each of the 2^n elements of X. This method is guaranteed to find the optimum, but quickly has limitations as the problem size increases. For $n = 60$, and evaluating one billion solutions per second, it would take 30 years of computation to obtain the solution.
The binary formulation of an integer problem with the decomposition (1.3) leads to thousands of binary variables. Explicit enumeration is then no longer possible.

Implicit enumeration consists in partitioning the set X by forming a tree of possibilities (separation stage), and then moving down the tree by eliminating non-optimal branches (evaluation stage). Elimination reduces the number of possibilities to be examined, while ensuring that the optimal solution is found. These methods are called tree methods or **branch and bound** methods.

1.3.2 Separation

The set X of possibilities is defined by

$$X = \{x \in U^n\} \tag{1.60}$$

Each component of x can take the value 0 or 1.
The set X is partitioned into 2 subsets by setting x_1 to 0 or 1.
This first level partition has 2^1 subsets noted X_0 and X_1.

$$X_0 = \{x \in X \,,\, x_1 = 0\} \quad , \quad X_1 = \{x \in X \,,\, x_1 = 1\} \tag{1.61}$$

Both subsets X_0, X_1 are in turn partitioned into 2 subsets by setting x_2 to 0 or 1. This second level partition has 2^2 subsets noted X_{00}, X_{01}, X_{10} and X_{11}.

$$\begin{aligned} X_{00} &= \{x \in X , x_1 = 0 , x_2 = 0\} \quad , \quad X_{10} = \{x \in X , x_1 = 1 , x_2 = 0\} \\ X_{01} &= \{x \in X , x_1 = 0 , x_2 = 1\} \quad , \quad X_{11} = \{x \in X , x_1 = 1 , x_2 = 1\} \end{aligned} \tag{1.62}$$

The partitioning continues until the last variable x_n. The final partition at level n has 2^n single element subsets. Each element is one of the possible combinations of the set X.

Figure 1-8 schematizes the first level partition. The set X is **separated** into two subsets X_0 and X_1:

- X is the predecessor or parent of X_0 and X_1;
- X_0 and X_1 are the successors or children of X.

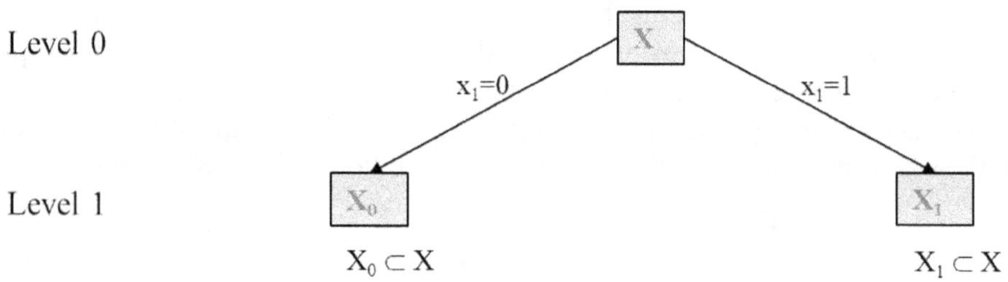

Figure 1-8: First level partition.

The successive partitions form a tree structure.

Each subset $X, X_0, X_1, X_{10}, X_{11}, \cdots$ is a **vertex** (or **node**) of the tree. The vertices of successive levels are connected by arrows indicating inclusion.

The initial set X (corresponding to level 0) is the **root** of the tree structure.
The vertices of level n (containing only 1 element) are the **terminal** vertices.
The tree is composed of $n+1$ levels. The level k has 2^k vertices. The total number of vertices is $2^{n+1}-1$.

Figure 1-9 shows a tree structure with 3 binary variables x_1, x_2, x_3.
The eight terminal vertices $X_{000}, X_{001}, X_{010}, X_{011}, X_{100}, X_{101}, X_{110}, X_{111}$ represent the eight possible combinations of values of x_1, x_2, x_3.

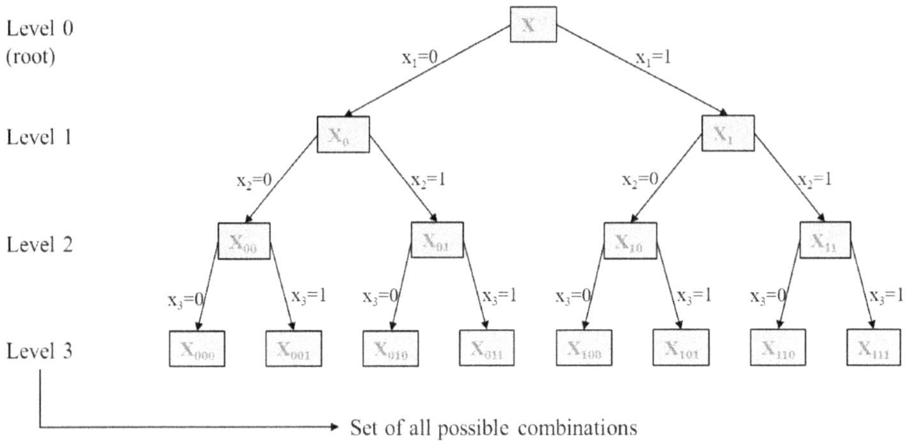

Figure 1-9: Tree of possibilities.

The separation procedure presented above concerns binary variables, which are set to 0 or 1. For an integer variable x_j, the procedure consists in choosing an integer threshold δ_j and separating the vertex X into a vertex X^- associated with the values $x_j \le \delta_j$ and a vertex X^+ associated with the values $x_j \ge \delta_j + 1$. Each vertex is thus associated with a variation interval of x_j which will be reduced as the separations are made. This procedure is more complex to implement than reformulating the problem in binary variables, but it can be much more effective in eliminating possibilities.

1.3.3 Evaluation

Let us note z^* the optimal cost of the problem (LPB) we are trying to solve.

$$\min_x z = c^T x \quad \text{s.t.} \begin{cases} Ax = b \\ x \in U^n \quad , \quad U = \{0\,;1\} \end{cases} \tag{1.63}$$

The optimal cost z^* is not known (we are looking for it).

Suppose we know a feasible solution of (LPB) of cost z_a. This binary solution satisfies the constraints. Its cost z_a is a majorant of the optimal cost z^* of (LPB).

A vertex S of level k in the tree corresponds to the problem (1.63) with the first k variables set to 0 or 1. The problem (LPB)s associated with vertex S has $n-k$ binary variables and its (unknown) optimal cost is denoted z_S.

$$\min_x z = c^T x \quad \text{s.t.} \begin{cases} Ax = b \\ x \in U^n \\ x_1, x_1, \cdots, x_k \text{ fixed} \end{cases} \qquad (\text{PLB})_S \qquad (1.64)$$

The (LPB)s problem cannot be solved directly in binary variables. However, it is possible to solve the relaxed problem (LPR)s obtained by changing the integrity constraints $x_j \in U$ into positivity constraints $x_j \geq 0$.

$$\min_x z = c^T x \quad \text{s.t.} \begin{cases} Ax = b \\ x \geq 0 \\ x_1, x_1, \cdots, x_k \text{ fixed} \end{cases} \qquad (\text{PLR})_S \qquad (1.65)$$

This problem is solved by the simplex method. Its cost z_S^r is lower than z_S because the relaxed problem (LPR)s is less constrained than the binary problem (LPB)s. The relaxed cost z_S^r is a minorant of the cost of (LPB)s, but also of all problems derived from (LPB)s by fixing additional variables $x_{k+1}, x_{k+2}, \cdots$. A solution with a cost lower than z_S^r cannot be found by exploring the vertices from S.

Suppose that this relaxed cost z_S^r is greater than the known value z_a, associated with a feasible solution of (LPB). For any vertex S' derived from S, we have:

$$z^* \leq z_a < z_S^r \leq z_{S'} \qquad (1.66)$$

It is then useless to explore the vertices below S, because they cannot give any binary solution with a cost lower than z_a. We can thus eliminate completely this part of the tree.

Any minorant of the problem cost (LPB)s is called the **valuation** of the vertex S. The efficiency of the evaluation depends on the quality of the minorant. A high minorant favors the elimination of branches from the tree. The evaluation function is usually the relaxed problem cost (RPC)s as above (1.65). This minorant is simple to obtain by the simplex method, and its computation can be accelerated by using the solution of the parent vertex (**warm start** procedure).

Suppose indeed that the relaxed problem of the parent vertex P has been solved by the simplex method. Its canonical form in the optimal basis B is known and it corresponds to the following array.

$$T = \begin{array}{c} \quad x_B \quad\quad x_N \\ \begin{array}{|cc|c|} \hline I & B^{-1}N & \bar{b} \\ \hline 0 & \bar{c}_N^T & -\bar{z} \\ \hline \end{array} \end{array} \quad \begin{array}{l} \to x_B + B^{-1}Nx_N = \bar{b} \\ \to \bar{c}_N^T x_N = z - \bar{z} \end{array} \quad \to \text{solution} \quad \begin{cases} x_B = \bar{b} \\ x_N = 0 \end{cases} \tag{1.67}$$

The separation of the vertex P consists either in fixing a binary variable or in adding an inequality constraint on an integer variable. Let us examine these two possibilities.

<u>Separation on a binary variable</u>

Assume that the vertex is separated by setting the binary variable x_e to 0 or 1.

$$x_e = \delta \quad \text{with} \quad \delta = 0 \text{ or } 1 \tag{1.68}$$

If this variable is in the basis B, we start by forcing a dual pivoting with a non-basic variable x_s. By making the variables x_e and x_s appear explicitly in the array T (1.67), we have the form

$$T = \begin{array}{c} \quad x_B \;\; x_e \;\;\; x_N \quad\; x_s \\ \begin{array}{|cc:cc|c|} \hline I & 0 & B^{-1}N & \bar{a}_{is} & \bar{b} \\ 0 & 1 & \bar{a}_{ej} & \bar{a}_{es} & \bar{b}_e \\ \hline 0 & 0 & \bar{c}_N^T & \bar{c}_s & -\bar{z} \\ \hline \end{array} \end{array} \tag{1.69}$$

Since the basis B is optimal for the parent problem, the reduced costs and the basic variables are all positive. The variable x_s for the dual pivoting is chosen by

$$s \to \max \frac{\bar{c}_{Nj}}{\bar{a}_{ej}} \quad \text{for} \quad j \in N \text{ such that } \bar{a}_{ej} < 0 \tag{1.70}$$

The reduced costs and basic variables become

$$\bar{c}_j \to \bar{c}_j - \bar{c}_s \frac{\bar{a}_{ej}}{\bar{a}_{es}} \quad , \quad \bar{b}_i \to \bar{b}_i - \bar{b}_e \frac{\bar{a}_{is}}{\bar{a}_{es}} \tag{1.71}$$

By construction, this dual pivoting keeps the reduced costs positive. On the other hand, the basic variables may become negative, as the purpose of this forced pivot is to move the variable x_e out of the basis. After this pivoting, we find ourselves with a dual-feasible basis (positive reduced costs) and the non-basic variable x_e which we wish to fix at the value δ (0 or 1).

The resulting array is of the form

$$T' = \begin{array}{c} \begin{array}{ccc|c} x_B & x_N & x_e & \\ \hline I & B^{-1}N & \bar{a}_{ie} & \bar{b} \\ \hline 0 & \bar{c}_N^T & \bar{c}_e & -\bar{z} \end{array} \end{array} \tag{1.72}$$

Next, let us perform the change of variable: $x'_e = x_e - \delta$. The objective is that the new non-basic variable x'_e is set to 0, which corresponds to its value in the basis solution of the array T'.

With this change of variable, the constraints and cost become

$$\begin{array}{ll} x_i + \sum\limits_{\substack{j\in N \\ j\neq e}} \bar{a}_{ij}x_j + \bar{a}_{ie}x_e = \bar{b}_i & \Rightarrow\ x_i + \sum\limits_{\substack{j\in N \\ j\neq e}} \bar{a}_{ij}x_j + \bar{a}_{ie}x'_e = \bar{b}_i - \bar{a}_{ie}\delta \\ z - \sum\limits_{\substack{j\in N \\ j\neq e}} \bar{c}_j x_j - \bar{c}_e x_e = \bar{z} & \Rightarrow\ z - \sum\limits_{\substack{j\in N \\ j\neq e}} \bar{c}_j x_j - \bar{c}_e x'_e = \bar{z} + \bar{c}_e\delta \end{array} \tag{1.73}$$

We can then form the simplex array with this new canonical form

$$T'' = \begin{array}{ccc|c} x_B & x_N & x'_e & \\ \hline I & B^{-1}N & \bar{a}_{ie} & \bar{b} - \bar{a}_{ie}\delta \\ \hline 0 & \bar{c}_N^T & \bar{c}_e & -\bar{z} + \bar{c}_e\delta \end{array} \tag{1.74}$$

then remove the non-basic variable x'_e which must remain set to 0.

This new problem associated to the vertex S (variable x_e fixed) is in canonical form in the basis B which is dual-feasible (positive reduced costs). It is therefore ready to be solved by the dual simplex algorithm.

Example 1-4 illustrates the separation-evaluation procedure on a binary variable problem.

Example 1-4: Separation-evaluation with binary variables (after [R15])

Consider the linear problem with 6 binary variables

$$\min_{\substack{x_1,x_2,x_3 \\ x_4,x_5,x_6}} z = -20x_1 - 16x_2 - 11x_3 - 9x_4 - 7x_5 - x_6$$

$$\text{s.t.} \begin{cases} 9x_1 + 8x_2 + 6x_3 + 5x_4 + 4x_5 + x_6 \le 12 \\ x_j = 0 \text{ or } 1\,,\ j = 1 \text{ to } 6 \end{cases}$$

This problem is of the knapsack type (section 1.4.5).

$$\min_{x_j \in U} z = \sum_j c_j x_j \quad \text{s.t.} \quad \sum_j a_j x_j \le b$$

The variables are numbered in order of increasing utility (ratios c_j/a_j).
There are $2^6 = 64$ possible combinations. The tree has 7 levels and a total of $2^7 - 1 = 127$ vertices.

A tree-search method is applied with a depth exploration strategy:
- the vertices are treated in increasing valuation (best vertex first);
- the variables are treated in increasing utility (= in their numbered order).

Initialization: initial feasible solution
The initialization stage aims at building a feasible solution. This is not always possible in a simple way when there are many constraints.

For this knapsack problem, we apply a greedy method by taking the variables in order of increasing utility and setting them to 1 if possible (as long as the constraint remains satisfied).
This yields the following initial solution: $x_1 = 1$, $x_2 = x_3 = x_4 = x_5 = 0$, $x_6 = 1$
whose cost is: $z_a = -21$.
This value is a majorant of the cost of the binary problem: $z^* \le z_a$.

Iteration 0: vertex X (root of the tree)

The relaxed problem associated with the vertex X is

$$\min_{\substack{x_1,x_2,x_3\\x_4,x_5,x_6}} z = -20x_1 - 16x_2 - 11x_3 - 9x_4 - 7x_5 - x_6$$

$$\text{s.t.} \begin{cases} 9x_1 + 8x_2 + 6x_3 + 5x_4 + 4x_5 + x_6 \le 12 \\ x_j \ge 0 \text{ , } j = 1 \text{ to } 6 \end{cases}$$

The solution is: $\begin{cases} x_1 = 1 \text{ , } x_2 = \dfrac{3}{8} \\ x_3 = x_4 = x_5 = x_6 = 0 \end{cases} \quad \rightarrow \quad z = -26$

This solution is not feasible, as the variable x_2 is not binary.
The valuation of X is –26. The vertex X is separated by setting the variable x_1 to 0 (vertex X_0) or 1 (vertex X_1).

Iteration 1: vertex X_0

The relaxed problem associated with the vertex X_0 ($x_1 = 0$) is

$$\min_{\substack{x_2,x_3,x_4 \\ x_5,x_6}} z = -16x_2 - 11x_3 - 9x_4 - 7x_5 - x_6 \quad \text{s.t.} \begin{cases} 8x_2 + 6x_3 + 5x_4 + 4x_5 + x_6 \le 12 \\ x_j \ge 0 \ , \ j = 2 \text{ to } 6 \end{cases}$$

The solution of this problem is: $\begin{cases} x_2 = 1 \ , \ x_3 = \dfrac{2}{3} \\ x_4 = x_5 = x_6 = 0 \end{cases} \rightarrow z = -\dfrac{70}{3} \approx -23.3$

This solution is not feasible, as the variable x_3 is not binary.
The valuation of X_0 is -23.3 and it is lower than the reference $z_a = -21$.
The vertex is retained.

Iteration 1: vertex X_1

The relaxed problem associated with the vertex X_1 ($x_1 = 1$) is

$$\min_{\substack{x_2,x_3,x_4 \\ x_5,x_6}} z = -16x_2 - 11x_3 - 9x_4 - 7x_5 - x_6 \quad \text{s.t.} \begin{cases} 8x_2 + 6x_3 + 5x_4 + 4x_5 + x_6 \le 12 \\ x_j \ge 0 \ , \ j = 2 \text{ to } 6 \end{cases}$$

This problem has the same solution as the problem of the root vertex X. This is logical, as the relaxed solution of X already gave: $x_1 = 1$.
The valuation of X_1 is -26 and it is lower than the reference $z_a = -21$.
The vertex is retained.

The exploration of the tree at iteration 1 is shown in figure 1-10.

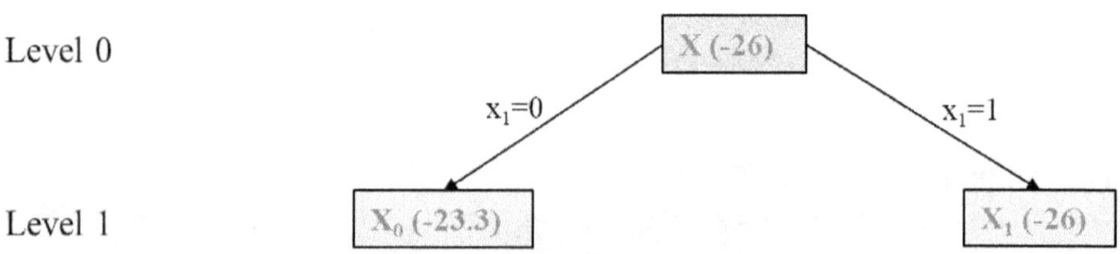

Figure 1-10: Exploration at iteration 1.

The best vertex is X_1 (lowest valuation). This vertex is separated by setting the variable x_2 to 0 (vertex X_{10}) or to 1 (vertex X_{11}).

Iteration 2: vertex X_{10}

The relaxed problem associated with the vertex X_{10} $(x_1 = 1, x_2 = 0)$ is

$$\min_{x_3,x_4,x_5,x_6} z = -11x_3 - 9x_4 - 7x_5 - x_6 - 20 \quad \text{s.t.} \begin{cases} 6x_3 + 5x_4 + 4x_5 + x_6 \le 3 \\ x_j \ge 0 \ , j = 3 \text{ to } 6 \end{cases}$$

The solution of this problem is: $\begin{cases} x_3 = \dfrac{1}{2} \\ x_4 = x_5 = x_6 = 0 \end{cases} \rightarrow z = -\dfrac{51}{2} \approx -25.5$

This solution is not feasible, as the variable x_3 is not binary.
The valuation of X_{10} is –25.5 and it is lower than the reference $z_a = -21$.
The vertex is retained.

Iteration 2: vertex X_{11}

The relaxed problem associated with the vertex X_{11} $(x_1 = 1, x_2 = 1)$ is

$$\min_{x_3,x_4,x_5,x_6} z = -11x_3 - 9x_4 - 7x_5 - x_6 - 36 \quad \text{s.t.} \begin{cases} 6x_3 + 5x_4 + 4x_5 + x_6 \le -5 \\ x_j \ge 0 \ , j = 3 \text{ to } 6 \end{cases}$$

The constraint cannot be satisfied because the variables are positive.
This problem has no solution. The valuation $+\infty$ is assigned to X_{11}.
The exploration stops at this vertex which contains no feasible solution.

The exploration of the tree at iteration 2 is shown in figure 1-11.

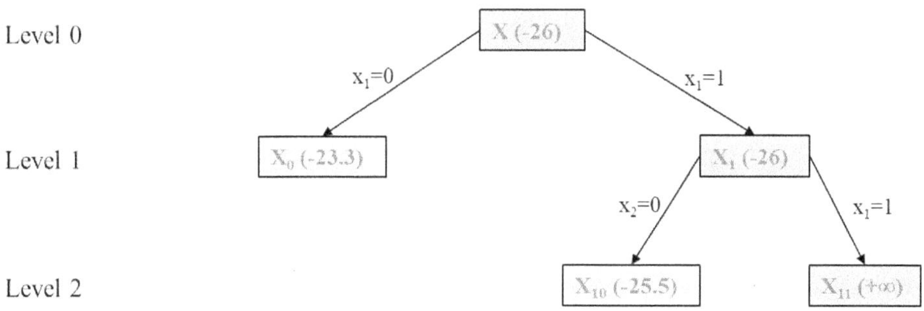

Figure 1-11: Exploration at iteration 2.

The best vertex is X_{10} (lowest valuation). This vertex is separated by setting the variable x_3 to 0 (vertex X_{100}) or 1 (vertex X_{101}).

Iteration 3: vertex X_{100}

The relaxed problem associated with the vertex X_{100} $(x_1 = 1, x_2 = 0, x_3 = 0)$ is

$$\min_{x_4,x_5,x_6} z = -9x_4 - 7x_5 - x_6 - 20 \quad \text{s.t.} \begin{cases} 5x_4 + 4x_5 + x_6 \le 3 \\ x_j \ge 0 \;,\; j = 4 \text{ to } 6 \end{cases}$$

The solution of this problem is: $x_4 = \frac{3}{5}$, $x_5 = x_6 = 0 \rightarrow z = -\frac{127}{5} \approx -25.4$

This solution is not feasible, as the variable x_4 is not binary.
The valuation of X_{100} is -25.4 and it is lower than the reference $z_a = -21$.
The vertex is retained.

Iteration 3: vertex X_{101}

The relaxed problem associated with the vertex X_{101} $(x_1 = 1, x_2 = 0, x_1 = 0)$ is

$$\min_{x_4,x_5,x_6} z = -9x_4 - 7x_5 - x_6 - 31 \quad \text{s.t.} \begin{cases} 5x_4 + 4x_5 + x_6 \le -3 \\ x_j \ge 0 \;,\; j = 4 \text{ to } 6 \end{cases}$$

The constraint cannot be satisfied because the variables are positive.
This problem has no solution. The valuation $+\infty$ is assigned to X_{101}.
The exploration stops at this vertex which contains no feasible solution.

The following iterations are summarized in table 1-2.
Figure 1-12 shows the result of the complete exploration of the tree structure.

The optimal solution was obtained by evaluating the vertex X_{0100}.

$$\begin{cases} x_2 = x_5 = 1 \\ x_1 = x_3 = x_4 = x_6 = 0 \end{cases} \rightarrow z = -23$$

This solution is confirmed at the end of the exploration, which requires the evaluation of only 19 vertices out of the 127 in the tree structure.

Separated vertex	Evaluated vertex	Solution	Evaluation	Best solution	Branch cut	Remaining vertices
	X	non-integer	−26	−21		
	X_0	non-integer	−23.3	−21		
X	X_1	non-integer	−26	−21		X_0, X_1
	X_{10}	non-integer	−25.5	−21		
X_1	X_{11}	non-feasible	$+\infty$	−21	X_{11}	X_0, X_{10}
	X_{100}	non-integer	−25.4	−21		
X_{10}	X_{101}	non-feasible	$+\infty$	−21	X_{101}	X_0, X_{100}
	X_{1000}	non-integer	−25.2	−21		
X_{100}	X_{1001}	non-feasible	$+\infty$	−21	X_{1001}	X_0, X_{1000}
	X_{10000}	non-integer	−23	−21		
X_{1000}	X_{10001}	non-feasible	$+\infty$	−21	X_{10001}	X_0, X_{10000}
	X_{00}	non-integer	−21.7	−21		
X_0	X_{01}	non-integer	−23.3	−21		$X_{00}, X_{01}, X_{10000}$
	X_{010}	non-integer	−23.2	−21		
X_{01}	X_{011}	non-feasible	$+\infty$	−21	X_{011}	$X_{00}, X_{010}, X_{10000}$
	X_{0100}	integer	−23	−23	X_{0100}, X_{00}	
X_{010}	X_{0101}	non-feasible	$+\infty$	−23	X_{0101}	X_{10000}
	X_{100000}	integer	−20	−23	X_{100000}	Solution
X_{10000}	X_{100001}	integer	−21	−23	X_{100001}	X_{0100}

Table 1-2: Summary of the iterations.

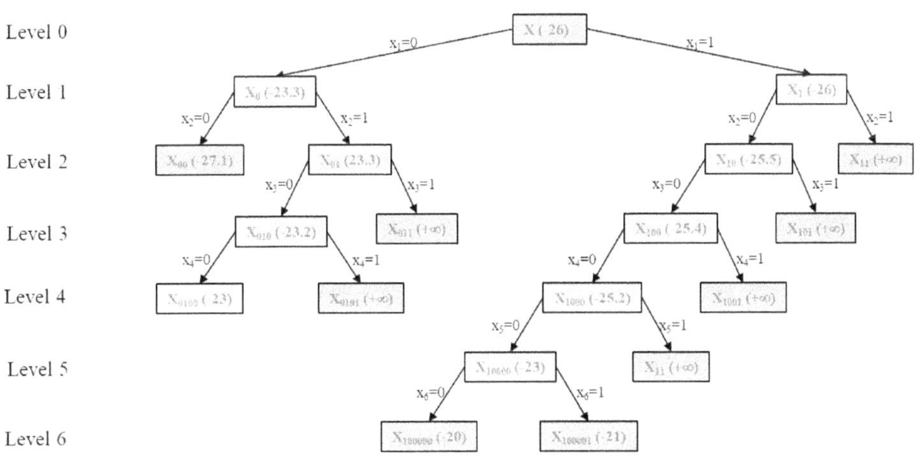

Figure 1-12: Exploration of the tree structure.

Separation on an integer variable

Now consider the separation on an integer variable x_e. The solution of the parent problem P gave the canonical expression for the variable x_e.

$$x_e + \sum_{j\in N} \bar{a}_{ej} x_j = \bar{b}_e \tag{1.75}$$

Assume that the basis solution takes a non-integer value: $\bar{b}_e \notin N$.

The vertex P is then separated by adding on the one hand the inequality constraint: $x_e \le E(\delta)$ and on the other hand the inequality constraint: $x_e \ge E(\delta)+1$.

The inequality $x_e \le E(\delta)$ is changed to equality with a slack variable $y^- \ge 0$.

$$\begin{cases} x_e + \sum_{j\in N} \bar{a}_{ej} x_j = \bar{b}_e \\ x_e + y^- = E(\bar{b}_e) \end{cases} \Rightarrow y^- - \sum_{j\in N} \bar{a}_{ej} x_j = -F(\bar{b}_e) \tag{1.76}$$

The inequality $x_e \ge E(\delta)+1$ is changed to equality with a slack variable $y^+ \ge 0$.

$$\begin{cases} x_e + \sum_{j\in N} \bar{a}_{ej} x_j = \bar{b}_e \\ x_e - y^+ = E(\bar{b}_e)+1 \end{cases} \Rightarrow y^+ + \sum_{j\in N} \bar{a}_{ej} x_j = F(\bar{b}_e)-1 \tag{1.77}$$

In both cases, the new constraint is of the form: $y + \alpha_N^T x_N = \beta \ , \beta < 0$.

The simplex array T of the parent vertex P (1.67) is completed with the new variable and the new constraint to give the array T' of vertex S.

$$T = \begin{array}{cc|c} x_B & x_N & \\ \hline I & B^{-1}N & \bar{b} \\ \hline 0 & \bar{c}_N^T & -\bar{z} \end{array} \longrightarrow T' = \begin{array}{ccc|c} x_B & y & x_N & \\ \hline I & 0 & B^{-1}N & \bar{b} \\ 0 & 1 & \alpha_N^T & \beta \\ \hline 0 & 0 & \bar{c}_N^T & -\bar{z} \end{array} \tag{1.78}$$

The new problem is in canonical form in the basis (x_B, y). This basis is not primal-feasible, because the variable y takes the negative value β. On the other hand, it is dual-feasible, because the reduced costs resulting from the solution of the vertex P are all positive. The new problem associated to the vertex S (with the constraint inequality on the variable x_e) is thus ready to be solved by the dual simplex algorithm.

Example 1-5 illustrates the separation-evaluation procedure on a problem in integer variables, detailing the dual simplex solution procedure.

Example 1-5: Separation-evaluation in integer variables

Consider the problem: $\min\limits_{x_1,x_2 \in \mathbb{N}} z = -4x_1 + x_2 \quad$ s.t. $\begin{cases} 7x_1 - 2x_2 \le 14 \\ x_2 \le 3 \\ 2x_1 - 2x_2 \le 3 \end{cases}$

The problem is put into standard form with 3 positive slack variables x_3, x_4, x_5. Note that these variables and the cost will be integers.
The problem in standard form is

$$\min\limits_{\substack{x_1,x_2 \in \mathbb{N} \\ x_3,x_4,x_5 \ge 0}} z = -4x_1 + x_2 \quad \text{s.t.} \begin{cases} 7x_1 - 2x_2 + x_3 = 14 \\ x_2 + x_4 = 3 \\ 2x_1 - 2x_2 + x_5 = 3 \end{cases}$$

A tree-search method is applied with a depth exploration strategy:
- the vertices are treated in increasing valuation (best vertex first);
- the variables are treated in numbered order.

Iteration 0: vertex X (root of the tree)
The relaxed problem associated with the vertex X is

$$\min\limits_{x_1,x_2,x_3,x_4,x_5 \ge 0} z = -4x_1 + x_2 \quad \text{s.t.} \begin{cases} 7x_1 - 2x_2 + x_3 = 14 \\ x_2 + x_4 = 3 \\ 2x_1 - 2x_2 + x_5 = 3 \end{cases}$$

We form the simplex array of this relaxed problem.

Initial array

Basis: (x_3, x_4, x_5)

Basis solution: $\begin{cases} x_3 = 14 \\ x_4 = 3 \\ x_5 = 3 \end{cases}$

x_1	x_2	x_3	x_4	x_5		
7	-2	1	0	0	14	x_3
0	1	0	1	0	3	x_4
2	-2	0	0	1	3	x_5
-4	1	0	0	0	0	z

The problem is directly in canonical form in the basis (x_3, x_4, x_5) which is primal-feasible. We solve the relaxed problem of X by the primal simplex algorithm.

Pivoting 1: $x_5 \leftrightarrow x_1$
New basis: (x_3, x_4, x_1)
Basis solution:
$$\begin{cases} x_3 = 7/2 \\ x_4 = 3 \\ x_1 = 3/2 \end{cases}$$

x_1	x_2	x_3	x_4	x_5		
0	5	1	0	-7/2	7/2	x_3
0	1	0	1	0	3	x_4
1	-1	0	0	1/2	3/2	x_1
0	-3	0	0	2	6	z

Pivoting 2: $x_3 \leftrightarrow x_2$
New basis: (x_2, x_4, x_1)
Basis solution:
$$\begin{cases} x_2 = 7/10 \\ x_4 = 23/10 \\ x_1 = 11/5 \end{cases}$$

x_1	x_2	x_3	x_4	x_5		
0	1	1/5	0	-7/10	7/10	x_2
0	0	-1/5	1	7/10	23/10	x_4
1	0	1/5	0	-1/5	11/5	x_1
0	0	3/5	0	-1/10	81/10	z

Pivoting 3: $x_4 \leftrightarrow x_5$
New basis: (x_2, x_5, x_1)
Basis solution:
$$\begin{cases} x_2 = 3 \\ x_5 = 23/7 \\ x_1 = 20/7 \end{cases}$$

x_1	x_2	x_3	x_4	x_5		
0	1	0	1	0	3	x_2
0	0	-2/7	10/7	1	23/7	x_5
1	0	1/7	2/7	0	20/7	x_1
0	0	4/7	1/7	0	59/7	z

The solution is: $x_1 = \frac{20}{7}$, $x_2 = 3 \rightarrow z = -\frac{59}{7}$.

This solution is not feasible, as the variable x_1 is not integer.

The valuation of X (−59/7) gives a minorant of the cost: $z^* \geq E\left(-\frac{59}{7}\right) + 1 = -8$.

The vertex X is separated by imposing a constraint on the variable x_1 :

- $x_1 \leq E\left(\frac{20}{7}\right) = 2 \quad \rightarrow$ vertex X_0;
- $x_1 \geq E\left(\frac{20}{7}\right) + 1 = 3 \quad \rightarrow$ vertex X_1.

These constraints are expressed from the canonical form of the solution of X $x_1 + \frac{1}{7}x_3 + \frac{2}{7}x_4 = \frac{20}{7}$ and introducing a positive slack variable x_6.

Iteration 1: vertex X_0

The constraint $x_1 \le 2$ yields: $x_1 + x_6 = 2 \quad \Rightarrow x_6 - \frac{1}{7}x_3 - \frac{2}{7}x_4 = -\frac{6}{7}$

The new variable and the associated constraint are added to the simplex array.

<u>Solution array of X</u>

Optimal basis: (x_2, x_5, x_1)

x_1	x_2	x_3	x_4	x_5		
0	1	0	1	0	3	x_2
0	0	-2/7	10/7	1	23/7	x_5
1	0	1/7	2/7	0	20/7	x_1
0	0	4/7	1/7	0	59/7	z

<u>Initial array of X_0</u>

Initial basis:
(x_2, x_5, x_1, x_6)

x_1	x_2	x_3	x_4	x_5	x_6		
0	1	0	1	0	0	3	x_2
0	0	-2/7	10/7	1	0	23/7	x_5
1	0	1/7	2/7	0	0	20/7	x_1
0	0	-1/7	-2/7	0	1	-6/7	x_6
0	0	4/7	1/7	0	0	59/7	z

The problem is directly in canonical form in the basis (x_2, x_5, x_1, x_6).
This basis is not primal-feasible $(x_6 < 0)$, but it is dual- feasible (positive reduced costs).
The relaxed problem of X_0 is solved by the dual simplex algorithm.

<u>Initial array</u>
Basis: (x_2, x_5, x_1, x_6)
Basis solution:

$$\begin{cases} x_2 = 3 \\ x_5 = 23/7 \\ x_1 = 20/7 \\ x_6 = -6/7 \end{cases}$$

x_1	x_2	x_3	x_4	x_5	x_6		
0	1	0	1	0	0	3	x_2
0	0	-2/7	10/7	1	0	23/7	x_5
1	0	1/7	2/7	0	0	20/7	x_1
0	0	-1/7	-2/7	0	1	-6/7	x_6
0	0	4/7	1/7	0	0	59/7	z

Pivoting 1: $x_6 \leftrightarrow x_4$

New basis: (x_2, x_5, x_1, x_4)

Basis solution: $\begin{cases} x_2 = 0 \\ x_5 = -1 \\ x_1 = 2 \\ x_4 = 3 \end{cases}$

x_1	x_2	x_3	x_4	x_5	x_6		
0	1	-1/2	0	0	7/2	0	x_2
0	0	-1	0	1	5	-1	x_5
1	0	0	0	0	1	2	x_1
0	0	1/2	1	0	-7/2	3	x_4
0	0	1/2	0	0	1/2	8	z

Pivoting 2: $x_5 \leftrightarrow x_3$

New basis: (x_2, x_3, x_1, x_4)

Basis solution: $\begin{cases} x_2 = 1/2 \\ x_3 = 1 \\ x_1 = 2 \\ x_4 = 5/2 \end{cases}$

x_1	x_2	x_3	x_4	x_5	x_6		
0	1	0	0	-1/2	1	1/2	x_2
0	0	1	0	-1	-5	1	x_3
1	0	0	0	0	1	2	x_1
0	0	0	1	1/2	-1	5/2	x_4
0	0	0	0	1/2	3	15/2	z

The solution is: $x_1 = 2$, $x_2 = \frac{1}{2} \rightarrow z = -\frac{15}{2}$.

This solution is not feasible, as the variable x_2 is not integer.

The solutions contained in X_0 have a cost bounded by: $z^* \geq E\left(-\frac{15}{2}\right) + 1 = -7$.

Iteration 1: vertex X_1

The constraint $x_1 \geq 3$ yields: $x_1 - x_6 = 3 \quad \Rightarrow x_6 + \frac{1}{7}x_3 + \frac{2}{7}x_4 = -\frac{1}{7}$.

The new variable and the associated constraint are added to the simplex array.

Solution array of X

Optimal basis: (x_2, x_5, x_1)

x_1	x_2	x_3	x_4	x_5		
0	1	0	1	0	3	x_2
0	0	-2/7	10/7	1	23/7	x_5
1	0	1/7	2/7	0	20/7	x_1
0	0	4/7	1/7	0	59/7	z

<u>Initial array of X_1</u>

Initial basis: (x_2, x_5, x_1, x_6)

x_1	x_2	x_3	x_4	x_5	x_6		
0	1	0	1	0	0	3	x_2
0	0	-2/7	10/7	1	0	23/7	x_5
1	0	1/7	2/7	0	0	20/7	x_1
0	0	1/7	2/7	0	1	-1/7	x_6
0	0	4/7	1/7	0	0	59/7	z

The problem is directly in canonical form in the basis (x_2, x_5, x_1, x_6), which is dual-feasible. The leaving variable is x_6 (only negative basic variable). No dual pivoting is possible, because there is no negative coefficient on the row of x_6. This means that the problem X_1 has no solution (incompatible constraints).

Indeed, the constraint: $x_6 + \frac{1}{7}x_3 + \frac{2}{7}x_4 = -\frac{1}{7}$ cannot be satisfied ($x \geq 0$).

The tree under X_1 is no longer explored, as it does not contain any solutions. The exploration of the tree at iteration 1 is shown in figure 1-13.

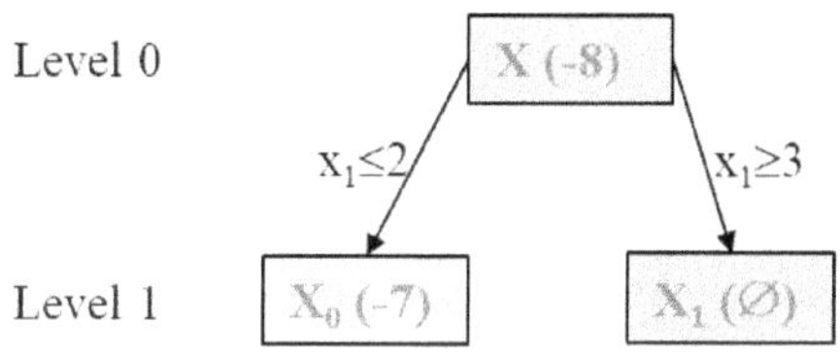

Figure 1-13: Exploration at iteration 1.

The only vertex to separate is X_0. It is therefore known that the solutions have a cost lower bounded by –7. Moreover, no feasible solution is yet known.

The vertex X_0 is separated by a constraint on the non-integer variable $x_2 = \frac{1}{2}$:

- $x_2 \leq E\left(\frac{1}{2}\right) = 0$ $\rightarrow$ vertex X_{00};

- $x_2 \geq E\left(\frac{1}{2}\right) + 1 = 1$ $\rightarrow$ vertex X_{01}.

These constraints are expressed from the canonical form of the solution of X_0

$x_2 - \frac{1}{2}x_5 + x_6 = \frac{1}{2}$ and introducing a positive slack variable x_7.

Iteration 2: vertex X_{00}

The constraint $x_2 \le 0$ yields: $x_2 + x_7 = 0 \Rightarrow x_7 + \frac{1}{2}x_5 - x_6 = -\frac{1}{2}$.

The new variable and the associated constraint are added to the simplex solution array. It would be possible to remove the fixed variable $x_2 = 0$ from the array, but for this illustrating example, the complete array with all variables is kept.

Solution array of X_0

Optimal basis:
(x_2, x_3, x_1, x_4)

x_1	x_2	x_3	x_4	x_5	x_6		
0	1	0	0	-1/2	1	1/2	x_2
0	0	1	0	-1	-5	1	x_3
1	0	0	0	0	1	2	x_1
0	0	0	1	1/2	-1	5/2	x_4
0	0	0	0	1/2	3	15/2	z

Initial array of X_{00}

Initial basis:
$(x_2, x_3, x_1, x_4, x_7)$

x_1	x_2	x_3	x_4	x_5	x_6	x_7		
0	1	0	0	-1/2	1	0	1/2	x_2
0	0	1	0	-1	-5	0	1	x_3
1	0	0	0	0	1	0	2	x_1
0	0	0	1	1/2	-1	0	5/2	x_4
0	0	0	0	1/2	-1	1	-1/2	x_7
0	0	0	0	1/2	3	0	15/2	z

The problem is directly in canonical form in the basis $(x_2, x_3, x_1, x_4, x_7)$.
This basis is not primal-feasible $(x_7 < 0)$, but it is dual-feasible (positive reduced costs).
The relaxed problem of X_{00} is solved by the dual simplex algorithm.

Initial array

Basis:

$(x_2, x_3, x_1, x_4, x_7)$

Basis solution:

$$\begin{cases} x_2 = 1/2 \\ x_3 = 1 \\ x_1 = 2 \\ x_4 = 5/2 \\ x_7 = -1/2 \end{cases}$$

x_1	x_2	x_3	x_4	x_5	x_6	x_7		
0	1	0	0	-1/2	1	0	1/2	x_2
0	0	1	0	-1	-5	0	1	x_3
1	0	0	0	0	1	0	2	x_1
0	0	0	1	1/2	-1	0	5/2	x_4
0	0	0	0	1/2	-1	1	-1/2	x_7
0	0	0	0	1/2	3	0	15/2	z

Pivoting 1: $x_7 \leftrightarrow x_6$

New basis:

$(x_2, x_3, x_1, x_4, x_6)$

Basis solution:

$$\begin{cases} x_2 = 0 \\ x_3 = 7/2 \\ x_1 = 3/2 \\ x_4 = 3 \\ x_6 = 1/2 \end{cases}$$

x_1	x_2	x_3	x_4	x_5	x_6	x_7		
0	1	0	0	0	0	1	0	x_2
0	0	1	0	-7/2	0	-5	7/2	x_3
1	0	0	0	1/2	0	1	3/2	x_1
0	0	0	1	0	0	-1	3	x_4
0	0	0	0	-1/2	1	-1	1/2	x_6
0	0	0	0	2	0	3	6	z

The solution is: $x_1 = \dfrac{3}{2}$, $x_2 = 0 \rightarrow z = -6$.

This solution is not feasible, as the variable x_1 is not integer.

The solutions contained in X_{00} have a cost lower bounded by –6.

Iteration 2: vertex X_{01}

The constraint $x_2 \geq 1$ yields: $x_2 - x_7 = 1 \quad \Rightarrow -\dfrac{1}{2}x_5 + x_6 + x_7 = -\dfrac{1}{2}$

The new variable and the associated constraint are added to the simplex array.

Solution array of X_0

Optimal basis:

(x_2, x_3, x_1, x_4)

x_1	x_2	x_3	x_4	x_5	x_6		
0	1	0	0	-1/2	1	1/2	x_2
0	0	1	0	-1	-5	1	x_3
1	0	0	0	0	1	2	x_1
0	0	0	1	1/2	-1	5/2	x_4
0	0	0	0	1/2	3	15/2	z

Initial array of X_{01}

Initial basis:
$(x_2, x_3, x_1, x_4, x_7)$

x_1	x_2	x_3	x_4	x_5	x_6	x_7		
0	1	0	0	-1/2	1	0	1/2	x_2
0	0	1	0	-1	-5	0	1	x_3
1	0	0	0	0	1	0	2	x_1
0	0	0	1	1/2	-1	0	5/2	x_4
0	0	0	0	-1/2	1	1	-1/2	x_7
0	0	0	0	1/2	3	0	15/2	z

The problem is directly in canonical form in the basis $(x_2, x_3, x_1, x_4, x_7)$.
This basis is not primal-feasible $(x_7 < 0)$, but it is dual- feasible (positive costs).
The relaxed problem of X_{01} is solved by the dual simplex algorithm.

Initial array
Basis:
$(x_2, x_3, x_1, x_4, x_7)$
Basis solution:

$$\begin{cases} x_2 = 1/2 \\ x_3 = 1 \\ x_1 = 2 \\ x_4 = 5/2 \\ x_7 = -1/2 \end{cases}$$

x_1	x_2	x_3	x_4	x_5	x_6	x_7		
0	1	0	0	-1/2	1	0	1/2	x_2
0	0	1	0	-1	-5	0	1	x_3
1	0	0	0	0	1	0	2	x_1
0	0	0	1	1/2	-1	0	5/2	x_4
0	0	0	0	-1/2	1	1	-1/2	x_7
0	0	0	0	1/2	3	0	15/2	z

Pivoting 1: $x_7 \leftrightarrow x_5$
New basis:
$(x_2, x_3, x_1, x_4, x_5)$
Basis solution:

$$\begin{cases} x_2 = 1 \\ x_3 = 2 \\ x_1 = 2 \\ x_4 = 2 \\ x_5 = 1 \end{cases}$$

x_1	x_2	x_3	x_4	x_5	x_6	x_7		
0	1	0	0	0	0	-1	1	x_2
0	0	1	0	0	-7	-2	2	x_3
1	0	0	0	0	1	0	2	x_1
0	0	0	1	0	0	1	2	x_4
0	0	0	0	1	-2	-2	1	x_5
0	0	0	0	0	4	1	7	z

The solution is: $x_1 = 2$, $x_2 = 1 \rightarrow z = -7$.
This solution is feasible (integer variables) so that there is no need to explore the tree beyond X_{01}. The feasible solution gives a majorant of the cost: $z^* \le z_a = -7$.

Table 1-3 recaps the vertices explored at iteration 2.
The tree structure is shown in figure 1-14.

Vertex	Solution	Cost
X	non-integer	$-8 \leq z^*$
X_0	non-integer	$-7 \leq z^*$
X_1	non-feasible	
X_{00}	non-integer	$-6 \leq z^*$
X_{01}	integer $z_a = -7$	$z^* \leq -7$

Table 1-3: Vertices explored at iteration 2.

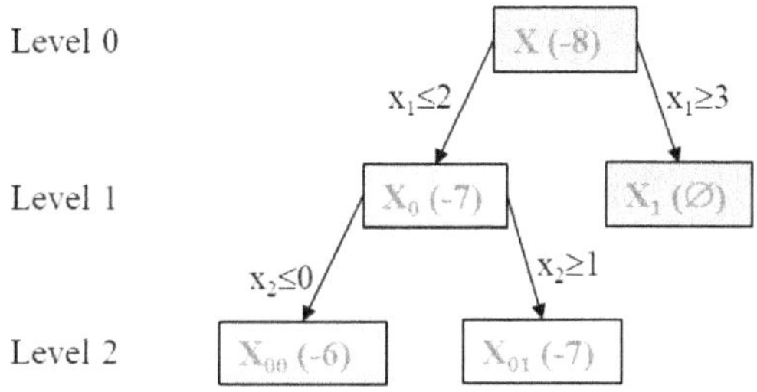

Figure 1-14: Exploration at iteration 2.

The feasible solution found at X_{01} gives the reference cost $z_a = -7$.
All vertices with a valuation greater than z_a are eliminated, as their valuation is a minorant of the cost of the solutions they may contain.

The exploration of the vertex X_{00} can thus be stopped. There are no more vertices left to separate and the exploration of the tree is finished. The solution was obtained by evaluating the vertex X_{01}: $x_1 = 2$, $x_2 = 1 \rightarrow z = -7$.
The exploration involved the assessment of only 4 vertices.

1.3.4 Exploration strategy

The exploration of the tree structure requires to store the list $L = \{X_i\}$ of vertices to explore, called active vertices. Each active vertex is stored with its relaxed solution (canonical form, variables and cost). The cost of the relaxed solution is the evaluation of the vertex. In addition, the reference cost z_a of the best known feasible solution is also stored.

The exploration is initialized with the root vertex $L = \{X\}$ and the reference value $z_a = +\infty$ (if no feasible solution is known). An iteration consists in processing a vertex and updating the list L.

Each iteration involves the following stages.

- choice of the vertex $X_i \in L$ to be separated;
- choice of the non-integer (or non-binary) variable x_j to constrain (or to fix);
- separation of the vertex X_i on the variable x_j.
  The 2 child vertices X_{i0} and X_{i1} are added to the list L.
  The parent vertex X_i is removed from the list L;
- evaluation of the vertex X_{i0} giving a relaxed solution x_0 cost z_0.
  The assessment leads to the following three possibilities:
  - if the relaxed solution x_0 is feasible (integer or binary), the reference cost is updated: $z_a \rightarrow \min(z_a, z_0)$. The vertex X_{i0} does not need further separation and it is removed from the list L;
  - if $z_0 > z_a$, the vertex X_{i0} contains no solution better than z_a. Its exploration is abandoned and it is removed from the list L;
  - if $z_0 \le z_a$, the vertex X_{i0} must be explored further. It is kept in the list L;
- evaluation of the vertex X_{i1} in a similar way to X_{i0}.

Figure 1-15 depicts the stages of an iteration.

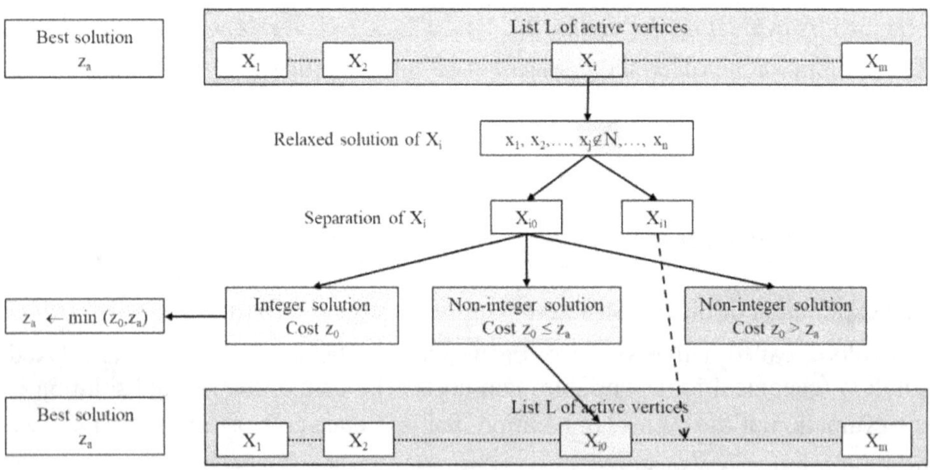

Figure 1-15: Separation and evaluation process.

The efficiency of a separation-evaluation method depends on the choice of the vertex to be separated, the choice of the separation variable, the necessary memory storage and the solution method of the relaxed problems.

Let us examine these aspects.

Choice of the vertex to be separated

The two strategies are depth exploration and breadth exploration.

The **depth search** systematically selects the lowest level vertex. The objective is to quickly arrive at a feasible terminal vertex in order to determine a first reference cost z_a. This value is indeed essential to eliminate branches during the search. Figure 1-16 illustrates the depth exploration strategy on a 5-level tree. The tree structure is shown in figure 1-16a, with arrows indicating the order in which the vertices are processed. The evolution of the list of active vertices is shown in figure 1-16b. The descent in the tree structure is carried out by starting with the leftmost branch (successive variables fixed to 0 for example) until meeting, either an infeasibility, or a feasible solution. Then the exploration resumes at the previous vertex to go down again as far as possible in the tree.

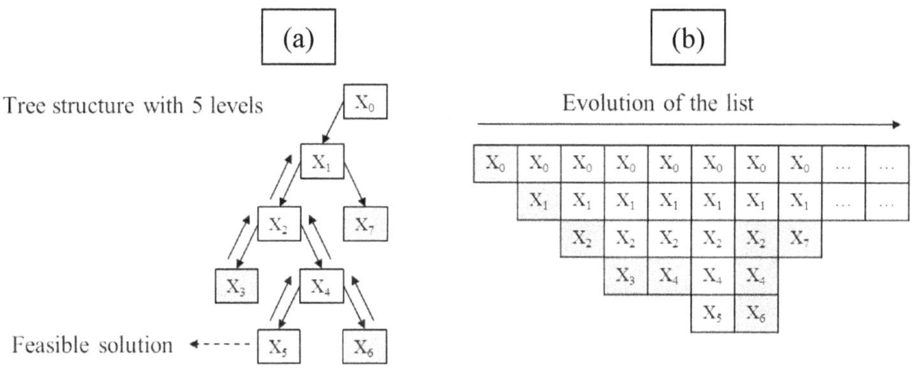

Figure 1-16: Depth search.

Depth search is economical in terms of memory space, as the list contains only one branch of the tree, *i.e.* at most n vertices for a problem with n binary variables. However, it may require the evaluation of many vertices, as it is necessary to wait until a feasible solution is found before branches can be eliminated from the tree. Furthermore, the first feasible solutions may be of poor quality (high cost z_a), which does not allow for efficient branch elimination.

The **breadth search** systematically selects the vertex with the best (lowest) valuation. The objective is to determine a first reference cost z_a that is as low as possible, allowing the elimination of branches at a high level.

Figure 1-17 illustrates the updating of the list for a breadth exploration strategy. The list at the beginning of the iteration has k vertices sorted in decreasing cost value. These vertices all have a valuation lower than the reference cost z_a (best known feasible cost). The last vertex X_k (best valuation vertex) is separated into two vertices X_a and X_b, which are evaluated.

Suppose that X_a yields a feasible (integer or binary) solution of cost z'_a with $z_i > z'_a \geq z_{i+1}$. The vertices $X_1, X_2, \cdots, X_i$ can be removed from the list. The vertex X_a is also removed, as its exploration is complete.

Suppose besides that X_b yields a relaxed solution of cost z_b with $z_j \geq z_b \geq z_{j+1}$. This vertex is inserted in the list between the vertices X_j and X_{j+1}.

At the end of this iteration, the vertex X_k that has been separated is removed from the list.

The new list contains the previous $k-i-1$ vertices from X_{i+1} to X_{k-i} plus the new vertex X_b.

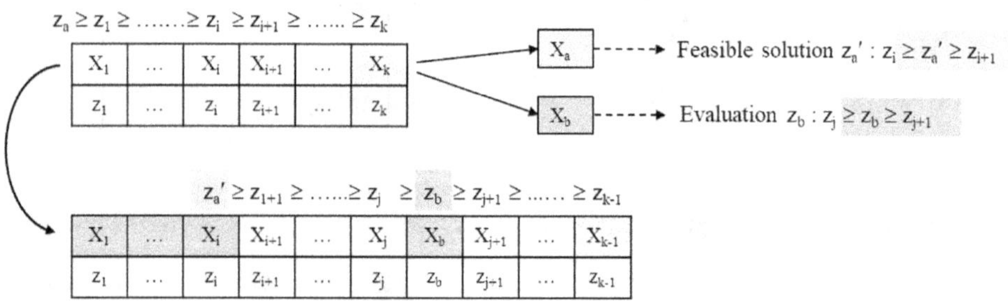

Figure 1-17: Breadth search.

Breadth search is generally more efficient than depth search (fewer vertices to evaluate), as it allows for early elimination of branches from the tree. However, it requires many more vertices to be stored in the list, which can lead to memory space problems.

Choice of the separation variable

The choice of the non-integer variable to separate the vertex is arbitrary. It is usually based on one of the following empirical criteria:

- choice based on the cost function:
  The variable with the most negative coefficient c_j in the cost function is chosen. It is hoped to find a new feasible solution with a large decrease in the reference cost;
- choice based on the constraints:
  The variable present in the largest number of constraints is chosen.
  It is hoped that an infeasibility will appear in order to eliminate the vertex;
- choice based on the dual function:
  The variable giving the largest initial value to the dual function (before solving the dual problem) is chosen. It is hoped to obtain a high relaxed cost to eliminate the vertex.

These empirical criteria can have very different efficiency. They should be tried and adapted on a case-by-case basis.

Memory storage

The implementation of the algorithm requires precautions to save memory space. Indeed, the amount of information to be stored (list of active vertices and their relaxed solutions) can become very large depending on the size of the problem to be solved.

The active vertices are stored in a chained list. This list is updated at each iteration by dynamic allocation/deallocation of the vertices. The links between successive vertices are defined by pointers. For a depth exploration with n binary variables, the list contains at most n vertices.
The best known feasible solution is stored separately.

Each vertex is a data structure comprising at least:
- a Boolean per variable indicating whether that variable is fixed;
- a Boolean per variable indicating that the variable is in the optimal basis;
- the value of each variable (for fixed variables);
- the value of the evaluation function;
- a pointer to the next vertex in the list.

For a problem with n binary variables, the values of the variables are Boolean.
Storing a vertex requires a total of 3n Booleans, one real and one pointer.
In practice, it is possible to solve linear problems with a few dozen integer variables or a few hundred binary variables. The number of real variables can be as high as 100 000, as these variables do not lead to the generation of additional branches.

Relaxed problem solving

A tree-based method requires the evaluation of a large number of vertices. The relaxed problem must be solved as quickly as possible.
The following elements contribute to the speed of resolution:

- preliminary reductions of the relaxed problem (section 1.1.3) allow for infeasibility detection and/or variable fixing;
- the initialization from the solution of the parent vertex (section 1.3.3) allows to directly construct a dual-feasible basis. This warm start procedure saves a lot of computation time;
- the preliminary knowledge of an initial feasible solution speeds up the elimination of branches from the tree. This solution can in some cases be provided directly by the user or determined by a greedy method adapted to the linear problem to be solved;
- the evaluation of the relaxed solution can be stopped early when the dual function becomes greater than the feasible reference cost z_a. This is because the dual function gives a minorant of the optimal cost and it grows with each iteration of the dual simplex. It is therefore not necessary to complete the solution to eliminate the vertex when it exceeds z_a.

Most separation-evaluation methods are based on the dual simplex algorithm with a warm start and the early termination of an evaluation. The primal simplex algorithm or interior point methods do not have these advantages and are much less efficient in this context. However, they can be implemented as "fallback" methods in cases where the dual simplex method fails, for example due to numerical accuracy problems.

The relaxed cost is a minorant of the integer solutions contained in a vertex. Ideally, it should be as large as possible to favor the elimination of vertices. In some cases, it can be strengthened by adding variables and/or constraints not present in the initial problem formulation, as discussed below.

Adding variables

Adding an integer variable to a problem can make the separation of vertices much more selective. This is particularly the case for knapsack problems as shown in example 1-6.

Example 1-6: Adding an integer variable

Consider the problem: $\max_{x_i \in U} z = \sum_{i=1}^{2n} 3x_i$ s.t. $\sum_{i=1}^{2n} 2x_i \le 2n+1$

The constraint imposes that at most n variables out of the 2n are equal to 1. The maximum cost is obtained by setting n variables to 1 and the other n to 0. The choice of variables is indifferent. There are C_{2n}^n equivalent binary solutions.

The solution of the relaxed problem (real variables in [0 ; 1]) has n variables worth 1, n–1 variables worth 0 and one variable worth 0.5. The cost of the relaxed solution remains z = 3n + 0.5 as long as fewer than n variables are set to 0. It will thus be necessary to go down to level n of the tree to obtain a feasible solution, then it will be necessary to find the C_{2n}^n equivalent solutions to guarantee optimality. The separation-evaluation method is inefficient for this problem.

Let us introduce an additional integer variable y defined by: $y = \sum_{i=1}^{2n} x_i$.

The new problem is: $\max_{x_i \in U, y \in \mathbb{N}} z = \sum_{i=1}^{2n} 3x_i$ s.t. $\sum_{i=1}^{2n} 2x_i \le 2n+1$ and $\sum_{i=1}^{2n} x_i = y$

This problem with 2n binary variables and one integer variable admits the same solutions as the original problem, because the constraint $y \in \mathbb{N}$ is satisfied by all original solutions. The solution of the new relaxed problem has, as before, n variables worth 1, n–1 variables worth 0 and one variable worth 0.5. The cost is: z = 3n + 0.5 and the variable y takes the value: y = n + 0.5.

Let us separate the root vertex X on the variable y, which must be integer:

- the vertex X_0 receives the constraint: $y \le n$. Its relaxed solution has n variables worth 1 and n variables worth 0. This solution is feasible of cost: $z = 3n$ and the exploration stops at vertex X_0;
- the vertex X_1 receives the constraint: $y \ge n+1$. This constraint imposes at least n +1 variables to 1. It is incompatible with the other constraint. The vertex X_1 has no solution and the exploration stops at the vertex X_1.

The separation on the additional integer variable y thus makes it possible to obtain the binary solution with certainty by evaluating only 3 vertices instead of the C_{2n}^n possible combinations.

Adding constraints

The addition of constraints can enhance the evaluation of the vertices. The purpose of the additional constraints is to reduce the feasible domain of the relaxed problem. The relaxed solution is thus closer to the integer solution, which favors the elimination of branches. Additional constraints can be generated by integral or mixed cuts (**branch and cut** methods), linear combinations of constraints from the original problem, or congruence relations of the form
$x+y=z \Rightarrow [x]_p+[y]_p \geq [z]_p$.

Example 1-7: Adding a constraint

Consider the constraint on 5 integer variables: $5x_1+2x_2+6x_3+x_4+4x_5=7$
By taking each member modulo p, valid inequality constraints are generated.

- $p=2 \Rightarrow x_1+x_4 \geq 1$

  This constraint implies that x_1 and x_4 cannot be simultaneously zero. Otherwise, the 1st member would be even, while the 2nd member is odd.

- $p=4 \Rightarrow x_1+2x_2+2x_3+x_4 \geq 3$

These inequality constraints help to fix variables or to make infeasibilities appear.

1.4 Applications

Mixed linear programming is applicable to many problems. This section presents the formulation of classical applications. Many examples are detailed in references [R1] and [R8].

1.4.1 Travelling salesman problem

Problem statement

The Travelling Salesman Problem (TSP) consists in visiting n cities, passing through each city once and returning to the starting point. The objective is to find the shortest path.

The n cities are denoted $S_1, \dots, S_n$. The distance from S_i to S_j is denoted d_{ij}.
The matrix of city-to-city distances is called the **distancer**: $D=(d_{ij})_{i,j=1 \text{ to } n}$.

In the wording of combinatorial optimization, the cities and distances form a graph $G = (S, D)$, where the cities are the nodes and the distances are the valuations of the edges. If the matrix D is symmetric, the direction of travel is indifferent and the graph is undirected. The problem is then a symmetric TSP. Otherwise, the problem is an asymmetric TSP.

Before formulating the TSP as a mixed linear problem, let us show some possible paths.

Example 1-8: Travelling salesman problem

Let us consider a case with 5 cities as shown in figure 1-18. The matrix D is of size 5×5.

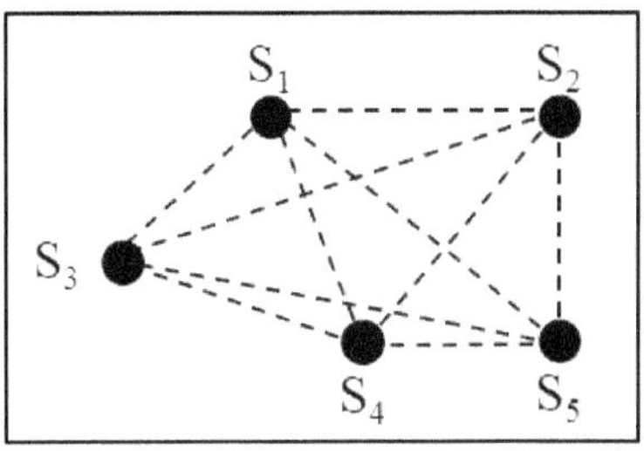

$$D = \begin{pmatrix} 0 & d_{12} & d_{13} & d_{14} & d_{15} \\ d_{21} & 0 & d_{23} & d_{24} & d_{25} \\ d_{31} & d_{32} & 0 & d_{34} & d_{35} \\ d_{41} & d_{42} & d_{43} & 0 & d_{45} \\ d_{51} & d_{52} & d_{53} & d_{54} & 0 \end{pmatrix}$$

Figure 1-18: TSP problem with 5 cities.

Figure 1-19a shows a path passing through the 5 cities, but not feasible. This is because this circuit has 2 sub-circuits. Figure 1-19b shows a feasible, but non-optimal path. The optimal circuit is shown in figure 1-19c.

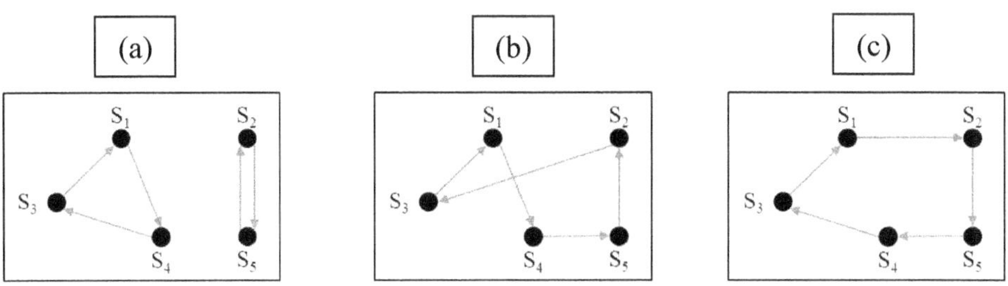

Non-feasible path (sub-circuits)

Non-optimal path $d = d_{31} + d_{14} + d_{45} + d_{52} + d_{23}$

Optimal path $d = d_{31} + d_{12} + d_{25} + d_{54} + d_{43}$

Figure 1-19: Non-feasible or non-optimal TSP path.

Formulation

The TSP is formulated as a mixed linear problem.

- Variables

The binary variable s_{ij} is 1 if the path from S_i to S_j is selected, and 0 otherwise.

With the variables s_{ii} set to 0, there are a total of $n \times (n-1)$ binary variables.

- Cost

The cost of the path is the total distance: $z = \sum_{i=1}^{n} \sum_{j=1}^{n} s_{ij} d_{ij}$.

- Constraints

There is one departure from each city S_i: $\sum_{j=1}^{n} s_{ij} = 1 \ , \ i = 1 \text{ to } n$ (n constraints).

There is one arrival at each city S_j: $\sum_{i=1}^{n} s_{ij} = 1 \ , \ j = 1 \text{ to } n$ (n constraints).

These constraints are not sufficient, as they do not prohibit sub-circuits, as in figure 1-19a.

A first idea is to prohibit sub-circuits by explicit constraints.

To prohibit the sub-circuit passing through the p cities $S_{k1}, S_{k2}, \ldots, S_{kp}$, we can introduce the inequality constraint: $s_{k_1k_2} + s_{k_2k_3} + \cdots + s_{k_{p-1}k_p} + s_{k_pk_1} \le p-1$.

The drawback of this approach is that it will require an exponential number of constraints (of the order of 2^n) to prohibit all possible sub-circuits.

One way to get around this difficulty is to iteratively introduce the prohibition constraints. The initial problem is solved without any sub-circuit constraints. If the solution contains a sub-circuit, the corresponding constraint is added and the new problem is solved. By continuing in this way, one eventually arrives at a solution without sub-circuit. In the worst case, this method may lead to adding an exponential number of constraints and solving the new problem. There is therefore no guarantee of a result in a reasonable time.

A more efficient method is to introduce constraints on the order of visit. Each city S_k is associated with a real variable t_k between 1 and n. This variable represents the "time of passage" at the city S_k. The initial city S_1 is chosen arbitrarily and its time is fixed: $t_1 = 1$.

Consider then the following $n \times (n-1)$ constraints.

$$t_i - t_j + n s_{ij} \le n - 1 \ , \ i = 1 \text{ to } n \ , \ j = 2 \text{ to } n \tag{1.79}$$

If the path from S_i to S_j is selected $(s_{ij} = 1)$, the constraint gives: $t_i \le t_j - 1$. The time at S_i is earlier than the time at S_j.

If the path from S_i to S_j is not selected $(s_{ij} = 0)$, the constraint gives: $t_i - t_j \le n - 1$. This constraint is always satisfied, as the times are between 1 and n.

The constraints (1.79) therefore impose increasing times along the path, except for the return in S_1 (no constraint for j = 1). The <u>real</u> variables t_k increase by at least 1 at each city visited and they are comprised between 1 and n. Their solution value is necessarily equal to the number of the city on the path.

These constraints prohibit sub-circuits. Indeed, a path through the n cities with sub-circuits has at least two sub-circuits, as in figure 1-19a. One of the sub-circuits does not pass through S_1, and it will not be able to satisfy the increasing time constraints (1.79).

Constraints can also be added to allow more efficient separations in a tree method. The additional constraint: $\sum_{i=1}^{n}\sum_{j=1}^{n} s_{ij} = n$ imposes, for example, that the number of selected edges is exactly equal to n.

Mixed linear programming can be used to solve TSP for up to 10 000 cities.

Figure 1-20 illustrates the problem called Att532, which consists in visiting the 532 largest cities in the US.

Figure 1-20: Att532 problem.

Extensions

The TSP formulation lends itself to various extensions, especially for optimizing vehicle routes.

The cost function can represent the consumption or the travel time:

$z=\sum_{i=1}^{m}\sum_{j=1}^{p}s_{ij}c_{ij}$ where c_{ij} is the cost (fuel, time) of the travel from S_i to S_j

Constraints can be added to the route, such as:

- impose an order of visit between the city S_i and the city S_j : $t_i < t_j$;
- limit the total time: $\sum_{i=1}^{n}\sum_{j=1}^{n}s_{ij}T_{ij} \le T_{max}$ where T_{ij} is the travel time from S_i to S_j.

1.4.2 Assignment problem

Problem statement

The problem consists in assigning m resources to p tasks, while minimizing the total cost. The cost of assigning resource i to task j is c_{ij}. The number of tasks to be performed is less than or equal to the number of available resources: $p \le m$. This assumption is not restrictive: if p > m, the inverse problem of assigning tasks to resources can be handled.

Formulation

The assignment problem is formulated as a mixed linear problem.

- Variables

The binary variable s_{ij} is 1 if resource i is assigned to task j, and 0 otherwise. There is a total of $m \times p$ binary variables.

- Cost

The total cost of the assignments is: $z=\sum_{i=1}^{m}\sum_{j=1}^{p}s_{ij}c_{ij}$.

- Constraints

Task j is performed only once: $\sum_{i=1}^{m}s_{ij}=1$, $j=1$ to p (p constraints).

Resource i is assigned once at most: $\sum_{j=1}^{p}s_{ij}\le 1$, $i=1$ to m (m constraints).

By introducing slack variables $(y_i)_{i=1 \text{ to } m}$, the problem is formulated as

$$\min_{s_{ij},y_i \in U} z = \sum_{i=1}^{m} \sum_{j=1}^{p} s_{ij} c_{ij} \quad \text{s.t.} \quad \begin{cases} \sum_{i=1}^{m} s_{ij} = 1 \;, & j = 1 \text{ to } p \\ \sum_{j=1}^{p} s_{ij} + y_i = 1 \;, & i = 1 \text{ to } m \end{cases} \tag{1.80}$$

As the variables s_{ij} are binary, so are the variables y_i.

The constraints (1.80) are of the form: $Ax = b$, where the matrix A has only elements being 0 or 1 and the vector b has all its components equal to 1.

This particular form of matrix facilitates the solution of problem (1.80).

Resolution

A rectangular matrix A is **unimodular** if the determinant of any sub-square matrix extracted from A is -1, 0 or 1.

The following properties can be demonstrated:

- a matrix whose elements are -1, 0 or 1, and whose each column contains at most two non-zero elements is unimodular;
- if A is unimodular, then any submatrix B of A is also unimodular;
- if B is a unimodular square matrix and b is a vector of integers, then the solution of the system $Bx = b$ is integer.

It can be shown that the matrix A of the assignment problem (1.80) is unimodular. Consider then the relaxed problem where the variables s_{ij}, y_i are real positive, noting the $m \times p + m$ variables: $x = (s_{ij}, y_i)$ and the $m + p$ constraints: $Ax = b$.

The solution of this continuous linear problem has $m + p$ basic variables x_B associated with a sub-matrix B and $(m-1) \times p$ zero non-basic variables associated with a sub-matrix N. The basic variables are given by

$$Ax = Bx_B + Nx_N = b \;\Rightarrow\; Bx_B = b \;\text{ because }\; x_N = 0 \tag{1.81}$$

The matrix B is unimodular (sub-matrix of A) and the vector b is integer (components equal to 1). The solution x_B of the relaxed problem is therefore integer. It is obtained directly by a standard linear programming method.

Extensions

The assignment problem (1.80) assumes independent tasks and associates one resource per task. The formulation can be adapted to more complex problems, including the maximum cost task, set covering or scheduling problems presented below.

Extension 1: Maximum cost task

Suppose that we want to minimize the cost of the most expensive task, instead of the total cost. The cost of task j is: $z_j = \sum_{i=1}^{m} s_{ij}c_{ij}$. We seek to minimize

$$z = \max_{j=1 \text{ to } p} z_j = \max_{j=1 \text{ to } p} \sum_{i=1}^{m} s_{ij}c_{ij} \tag{1.82}$$

This cost function is not linear. To formulate a linear problem, we introduce a real variable z and p constraints inequality $z_j \le z$.

$$\min_{z,s_{ij}} z \quad \text{s.t.} \begin{cases} \sum_{i=1}^{m} s_{ij} = 1 \ , & j=1 \text{ to } p \\ \sum_{j=1}^{p} s_{ij} \le 1 \ , & i=1 \text{ to } m \\ \sum_{i=1}^{m} s_{ij}c_{ij} \le z \ , & j=1 \text{ to } p \end{cases} \tag{1.83}$$

The first p equalities control the completion of each task. The next m inequalities control the allocation of each resource. The last p inequalities ensure that the cost function z represents at least the most expensive task. The problem has $m \times p$ binary variables s_{ij} and one real variable z.

Extension 2: Set covering

Assume that tasks can be shared between resources according to the capabilities of each. This is a problem of set covering of p tasks with m resources.

The applications concern, for example, the implementation of distribution sites that have to supply several customers.

The variable s_{ij} is the capacity of resource i assigned to task j. This variable is real or integer, depending on the nature of the problem.

The total cost of the assignments is: $z = \sum_{i=1}^{m}\sum_{j=1}^{p} s_{ij}c_{ij}$.

The task j requires a capacity t_j : $\sum_{i=1}^{m} s_{ij} = t_j$, $j = 1$ to p .

The resource i has a capacity r_i : $\sum_{j=1}^{p} s_{ij} \le r_i$, $i = 1$ to m .

The set covering problem is formulated as

$$\min_{s_{ij}} z = \sum_{i=1}^{m}\sum_{j=1}^{p} s_{ij}c_{ij} \quad \text{s.t.} \begin{cases} \sum_{i=1}^{m} s_{ij} = t_j \ , \ j = 1 \text{ to } p \\ \sum_{j=1}^{p} s_{ij} \le r_i \ , \ i = 1 \text{ to } m \end{cases} \tag{1.84}$$

The linear problem (1.84) may be continuous or mixed depending on whether the capacities s_{ij} are real or integer.

Extension 3: Scheduling problem
Let us assume that tasks require a certain duration and that some tasks should not overlap (disjointed tasks). This is a task scheduling problem. Applications concern time schedule, aircraft landing, connections, telecommunications...
The real variable t_j represents the start time of the task j of fixed duration d_j .

The binary variable u_{jk} is 1 if task j is performed before task k, 0 otherwise.

The problem has p real variables t_j and $p \times p$ binary variables u_{jk} .

If tasks j and k are to be disjoint, an order constraint is imposed on the variables t_j, t_k and u_{jk} . This constraint is formulated as

$$\begin{cases} u_{jk} + u_{kj} = 1 \\ u_{jk}(t_j + d_j) \le t_k \end{cases} , \ j = 1 \text{ to } p \ , \ k = 1 \text{ to } p \tag{1.85}$$

The cost function is usually the total time to be minimized: $z = \max_j (t_j + d_j)$.

The constraints give rise to products of variables that are linearized using the techniques of section 1.1.2. The cost function is linearized by a real variable z subject to the constraints: $t_j + d_j \le z$, $j = 1$ to p.

We can apply temporal constraints similar to (1.85) on resources or constraints on a minimum temporal spacing between tasks.

1.4.3 Coloring problem

Problem statement

A graph is defined by a set of nodes and edges connecting the nodes.
The m-coloring problem consists in coloring the graph with m colors, knowing that adjacent nodes (connected by an edge) must be of different colors.

Formulation

The coloring problem is formulated as a mixed linear problem. We note e_{jk} the binary variable worth 1 if node j is adjacent to node k, and 0 otherwise.

- Variables

The binary variable s_{ij} is 1 if color i is assigned to node j, and 0 otherwise.
There is a total of $m \times n$ binary variables.

- Constraints

The node j is assigned a single color: $\sum_{i=1}^{m} s_{ij} = 1$ (n constraints).

Adjacent nodes j and k have different colors: $e_{jk}(s_{ij} + s_{ik}) \le 1$.

The m-coloring problem consists in finding a feasible solution with m colors. We consider a constant cost, as there is no criterion to minimize.

$$\min_{s_{ij}} z = 1 \quad \text{s.t.} \begin{cases} \sum_{i=1}^{m} s_{ij} = 1 \ , & j = 1 \text{ to } n \\ e_{jk}(s_{ij} + s_{ik}) \le 1 \ , & i = 1 \text{ to } m \ , \ j = 1 \text{ to } n \ , \ k = 1 \text{ to } n \end{cases} \tag{1.86}$$

We are generally interested in the minimum number of colors needed to color the graph. This minimum number m_c is the **chromatic number** of the graph. A minorant of m_c can be found by solving the maximum cardinal clique problem shown below.

Clique problem

A **clique** of the graph G is a complete subgraph, *i.e.* a subgraph whose nodes are all connected by edges. If G has a clique with p nodes, then at least p colors are needed to color G, because each node is connected to all the others.

To find a minorant of m_c, we look for the largest clique of G, called the clique of maximum cardinal. This problem is formulated as follows.

- Variables

The binary variable s_i is 1 if node i belongs to the clique C, and 0 otherwise. There is a total of n binary variables.

- Constraints

Non-adjacent nodes j and k cannot simultaneously belong to the clique. These constraints take the form: $(1-e_{jk})(s_j+s_k)\le 1$.

The maximum cardinal clique problem is formulated as

$$\max_{s_i} z=\sum_{i=1}^{n} s_i \quad \text{s.t.} \quad (1-e_{jk})(s_j+s_k)\le 1 \ , \ j=1 \text{ to } n \ , k=1 \text{ to } n \qquad (1.87)$$

<u>Stable set</u>

Another problem of practical interest is the problem of the stable set of maximum cardinal. A stable set S of the graph G is a subgraph with no connected nodes. This is the "inverse" clique problem.

The maximum cardinal stable set problem is formulated like the maximum cardinal clique problem, simply replacing e_{jk} by $1-e_{jk}$.

$$\max_{s_i} z=\sum_{i=1}^{n} s_i \quad \text{s.t.} \quad e_{jk}(s_j+s_k)\le 1 \ , \ j=1 \text{ to } n \ , k=1 \text{ to } n \qquad (1.88)$$

Figure 1-21 illustrates the problems of coloring and maximum cardinal clique.

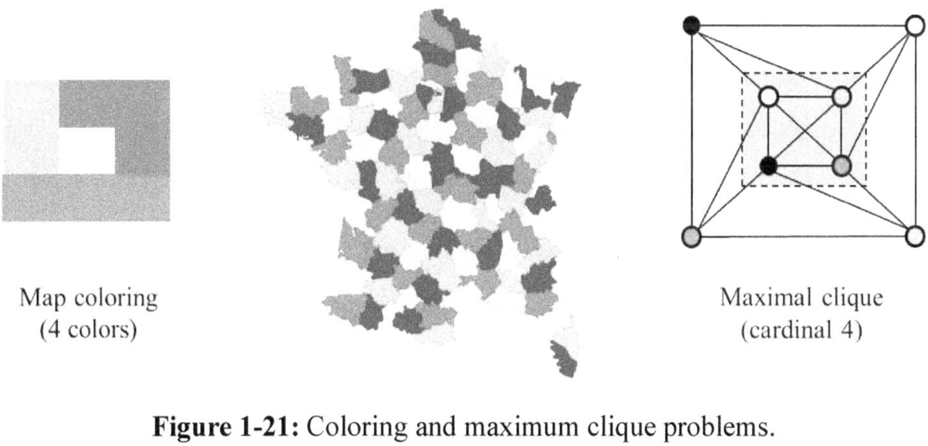

Figure 1-21: Coloring and maximum clique problems.

1.4.4 Flow problem

Problem statement

A network is a graph defined by a set of nodes and (directed) arcs connecting the nodes. The capacity and cost of the arc from node i to node j are denoted by f_{ij} and c_{ij} respectively. The nodes s and t are the entry (source) and exit (sink) points of the network, respectively. The objective is to get a given flow φ from the source to the sink with the lowest total cost.

The set of arcs is represented by the nodes-arcs **incidence matrix**. This matrix A has n rows (n = number of nodes) and m columns (m = number of arcs). If arc k goes from node i to node j, the values in column k are zero except $a_{ik} = +1$ (departure node) and $a_{jk} = -1$ (arrival node).

The incidence matrix is unimodular (section 1.4.2), as its elements are equal to −1, 0 or 1, and each column has at most two non-zero elements. The cost vector c has m elements representing the costs of the arcs.

Example 1-9: Node-arcs incidence matrix of a network

Consider the 6-node, 9-arc network in figure 1-22 where the costs of the arcs are indicated on the arrows. The incidence matrix A is of size 6×9 and the cost vector c has 9 elements.

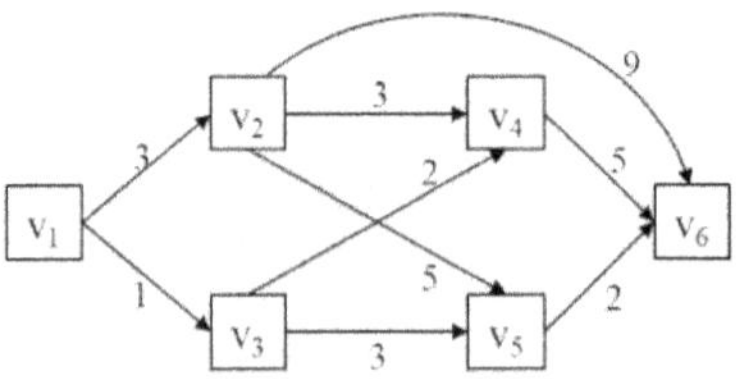

$$
\begin{array}{c} \text{arcs} \\ \begin{array}{ccccccccc} 1\text{-}2 & 1\text{-}3 & 2\text{-}4 & 2\text{-}5 & 2\text{-}6 & 3\text{-}4 & 3\text{-}5 & 4\text{-}6 & 5\text{-}6 \end{array} \end{array} \quad \text{nodes}
$$

$$
A = \begin{pmatrix} -1 & -1 & 0 & 0 & 0 & 0 & 0 & 0 & 0 \\ 1 & 0 & -1 & -1 & -1 & 0 & 0 & 0 & 0 \\ 0 & 1 & 0 & 0 & 0 & -1 & -1 & 0 & 0 \\ 0 & 0 & 1 & 0 & 0 & 1 & 0 & -1 & 0 \\ 0 & 0 & 0 & 1 & 0 & 0 & 1 & 0 & -1 \\ 0 & 0 & 0 & 0 & 1 & 0 & 0 & 1 & 1 \end{pmatrix} \begin{array}{c} v_1 \\ v_2 \\ v_3 \\ v_4 \\ v_5 \\ v_6 \end{array}
$$

$$
\begin{array}{ccccccccc} 1\text{-}2 & 1\text{-}3 & 2\text{-}4 & 2\text{-}5 & 2\text{-}6 & 3\text{-}4 & 3\text{-}5 & 4\text{-}6 & 5\text{-}6 \end{array}
$$

$$
c = \begin{pmatrix} 3 & 1 & 3 & 5 & 9 & 2 & 3 & 5 & 2 \end{pmatrix}
$$

Figure 1-22: Network with 6 nodes and 9 arcs.

Formulation

The flow problem can be formulated as a mixed linear problem. We denote e_{ij} the binary variable worth 1 if there is an arc from node i to node j, and 0 otherwise.

- Variables

The variable x_{ij} is the flow on the arc from node i to node j. This variable is real or integer depending on the nature of the problem. There is a total of m "useful" variables x_{ij} associated with the arcs of the network.

- Cost

The total cost of the flow through the network is: $z = \sum_{i,j=1}^{n} e_{ij} c_{ij} x_{ij}$.

- Constraints

The flow on the arc i-j is limited by its capacity: $x_{ij} \le f_{ij}$.

The flow is conserved at each node (Kirchhoff's law).

The flow arriving at node j is equal to the flow leaving: $\sum_{i=1}^{n} e_{ij} x_{ij} = \sum_{k=1}^{n} e_{jk} x_{jk}$.

The flow from the source s is φ : $\sum_{k=1}^{n} e_{sk} x_{sk} = \varphi$.

If the graph is sparse $(m \ll n^2)$, memory space is saved by keeping only the variables x_{ij} associated with the existing arcs. The linear problem is formulated in a compact way using the node-arc incidence matrix of the network.

$$\min_{x} z = c^T x \quad \text{s.t.} \begin{cases} Ax = b \\ x \le f \end{cases} \tag{1.89}$$

$x = (x_{ij}) \in \mathbb{R}^m$ is the vector of variables associated with the arcs.

$f = (f_{ij}) \in \mathbb{R}^m$ and $c = (c_{ij}) \in \mathbb{R}^m$ are the vectors of the arc capacities and costs.

The constraints $Ax = b$ express the conservation of the flow at each node as well as the flow $-\varphi$ leaving the source and the flow $+\varphi$ arriving at the sink.

The vector $b \in \mathbb{R}^m$ is defined by

$$b^T = \begin{pmatrix} \overset{\text{source}}{-\varphi} & \overset{\text{node 1}}{0} & \cdots & \overset{\text{node m}}{0} & \overset{\text{sink}}{+\varphi} \end{pmatrix} \tag{1.90}$$

The linear problem (1.89) is in integer variables if the flows x_{ij} on each arc are integer quantities. Since the constraint matrix is unimodular, the solution of the relaxed problem will be directly integer.

Extensions
The formulation of the flow problem can be adapted to various problems.

Extension 1: Minimum cost path
By fixing the flow $\varphi = 1$ and removing the constraints on capacities, problem (1.89) gives the minimum cost path between the source and the sink. The costs of the arcs can represent distances or travel times between nodes. Since the constraint matrix A is unimodular, the solution is directly integer and the variables x_{ij} take the value 0 or 1.

Extension 2: Maximum flow
To search for the maximum flow that can go from the source to the sink, we add the variable φ, which becomes the cost function. The costs c_{ij} of the arcs are not used and the linear problem is formulated as

$$\max_{x,\varphi} z = \varphi \quad \text{s.t.} \begin{cases} Ax = b \\ x \le f \end{cases} \tag{1.91}$$

Extension 3: Multiple sources and sinks
Assume that the network has several sources and several sinks. The source s_i emits a flow φ_i and the sink t_j receives a flow φ_j. The sum of the flows from the sources is equal to the sum of the flows to the sinks: $\Phi = \sum_{s_i} \varphi_i = \sum_{t_j} \varphi_j$.

We define a super-source σ linked to each source s_i by an arc of capacity φ_i.
The flow from the super-source is: $\Phi = \sum_{s_i} \varphi_i$.

We define a super-sink τ linked to each sink t_j by an arc of capacity φ_j.
The flow to the super-sink is: $\Phi = \sum_{t_j} \varphi_j$.

This brings us back to a simple flow problem of value Φ between the super-source and the super-sink.

1.4.5 Knapsack problem

Problem statement

We have a set of objects with their volumes and prices. The objects to be carried in a knapsack of a given volume must be chosen in order to maximize the price. The object i has the volume a_i and the price c_i. The knapsack has a volume b.

Formulation

The knapsack problem can be formulated as a mixed linear problem.

- Variables

The binary variable x_i is 1 if object i is selected, and 0 otherwise.

- Cost

The total price of the selected objects is: $z = \sum_{i=1}^{n} c_i x_i$.

- Constraints

The chosen objects must fit in the knapsack: $\sum_{i=1}^{n} a_i x_i \le b$.

The problem of the so-called one-dimensional knapsack is formulated as

$$\max_x z = c^T x \quad \text{s.t.} \begin{cases} a^T x \le b \\ x \in U^n \end{cases} \tag{1.92}$$

A good solution can be constructed by a greedy method based on the utility u_i of each object, defined as the price/volume ratio: $u_i = \frac{c_i}{a_i}$.

This solution allows branches to be eliminated in a tree-based method.

Example 1-10: Greedy solution to the knapsack problem

Consider the following knapsack problem with 8 objects.

$$\max_x z = 14x_1 + 13x_2 + 15x_3 + 10x_4 + 4x_5 + 13x_6 + 10x_7 + 10x_8$$
$$\text{s.t.} \quad 7x_1 + 7x_2 + 10x_3 + 7x_4 + 3x_5 + 10x_6 + 8x_7 + 9x_8 \le 30$$

These 8 objects are already numbered in order of decreasing utility.

The greedy solution $x = (1\ \ 1\ \ 1\ \ 0\ \ 1\ \ 0\ \ 0\ \ 0)$ is priced at $z = 46$ with a volume of 27.
The optimal solution $x = (1\ \ 1\ \ 0\ \ 1\ \ 0\ \ 0\ \ 1\ \ 0)$ is priced at $z = 47$ with a volume of 29.

<u>Extensions</u>
The formulation adapts to the case where several objects of the same type can be chosen (the variable x_i is then an integer) and to the problem of several knapsacks to fill. This so-called "multidimensional" knapsacks problem is formulated as follows.

- Variables

The binary variable x_{ij} is 1 if object i is placed in knapsack j, and 0 otherwise.

There is a total of $n \times m$ binary variables (n = number of objects, m = number of knapsacks).

- Cost

The total price of the selected objects is: $z = \sum_{i=1}^{n} c_i \left(\sum_{j=1}^{m} x_{ij} \right)$.

- Constraints

The volume of the knapsack j is b_j : $\sum_{i=1}^{n} a_i x_{ij} \le b_j$.

The object i is in one knapsack at most: $\sum_{j=1}^{m} x_{ij} \le 1$.

To improve the separation efficiency (tree method), we introduce the n variables: $s_i = \sum_{j=1}^{m} x_{ij}$.

The binary variable s_i is 1 if object i is selected, and 0 otherwise.
The multidimensional knapsack problem is formulated as

$$\max_{x,s} z = \sum_{i=1}^{n} c_i s_i \quad \text{s.t.} \begin{cases} \sum_{i=1}^{n} a_i x_{ij} \le b_j \ , \ j = 1 \text{ to } m \\ \sum_{j=1}^{m} x_{ij} \ \ = s_i \ , \ i = 1 \text{ to } n \end{cases} \quad \text{with } x \in U^{n \times m}, s \in U^{n} \quad (1.93)$$

This formulation applies to placement problems (saving files on disks), loading problems (filling delivery vehicles) or cutting problems (minimizing waste).

1.5 Quadratic problem

Mixed linear programming techniques extend to problems where the cost function is quadratic convex.

1.5.1 Tree method

The standard form of a quadratic problem in binary variables is as follows.

$$\min_{x} z = \frac{1}{2}x^T Q x + c^T x \quad \text{s.t.} \quad \begin{cases} Ax = b \\ x \in U^n \ , \ U = \{0\,;\,1\} \end{cases} \tag{1.94}$$

We can always return to the case where the matrix Q is symmetric.

By posing $Q' = \frac{1}{2}(Q + Q^T)$, we have indeed: $x^T Q' x = \frac{1}{2}x^T Q x + \frac{1}{2}x^T Q^T x = x^T Q x$.

A first approach is to apply linearization techniques (section 1.1.2) to come back to a linear problem. This approach has the disadvantage of increasing the number of variables. It is not efficient on high dimensional problems.

A second approach is to directly apply a tree method similar to those used for linear problems (section 1.3):
- we explore the tree structure by progressively setting the variables to 0 or 1;
- the relaxed solution at each node gives a minorant of the branch cost;
- branches with a higher valuation than the best known solution are eliminated.

This approach is developed in this section. It requires the ability to efficiently solve the relaxed problem obtained by replacing integrity constraints with positivity constraints.

$$\min_{x} z = \frac{1}{2}x^T Q x + c^T x \quad \text{s.t.} \quad \begin{cases} Ax = b \\ x \ge 0 \end{cases} \tag{1.95}$$

Since the relaxed problem is less constrained than the original problem, its solution is better and gives a minorant of the optimal cost of problem (1.94).

Before tackling the solution of the relaxed problem, let us illustrate on an example the comparison between the integer solution and the relaxed solution of a quadratic problem.

Example 1-11: Solution of the relaxed quadratic problem

Consider the problem with 2 integer variables

$$\min_{x_1,x_2} z = 2x_1^2 + 4x_2^2$$

$$\text{s.t.} \begin{cases} x_1 + x_2 \geq 5 \\ x_1 \geq 2 \\ x_2 \geq 1 \\ 2x_1 + x_2 \leq 10,5 \\ x_1, x_2 \in \mathbb{N} \end{cases}$$

The relaxed solution is

$$x_1 = \frac{10}{3} \ , \ x_2 = \frac{5}{3} \ \rightarrow \ z_r = \frac{100}{3}$$

The integer solution is

$$x_1 = 3 \ , \ x_2 = 2 \ \rightarrow \ z^* = 34$$

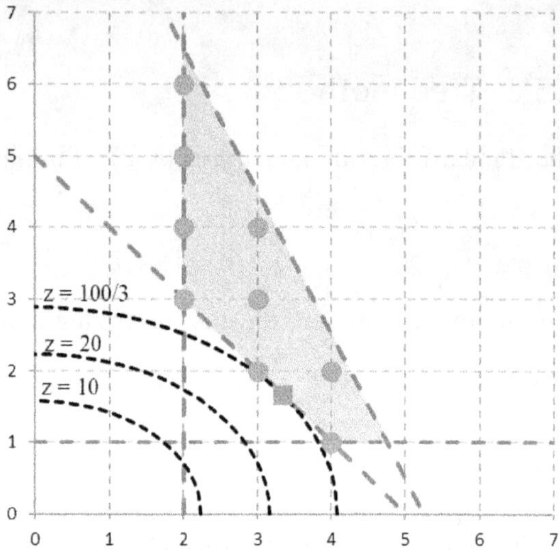

Figure 1-23: Relaxed solution and integer solution.

Figure 1-23 shows the cost level lines (dotted), the feasible domain (shaded), the integer solutions (circles) and the relaxed solution (square).

Let us now turn our attention to solving the relaxed problem (1.95).
This relaxed problem is put into standard form

$$\min_x z = \frac{1}{2} x^T Q x + c^T x \quad \text{s.t.} \begin{cases} b - Ax = 0 \\ -x \leq 0 \end{cases} \tag{1.96}$$

The Lagrangian is defined with multipliers λ for the equality constraints and multipliers μ for the inequality constraints.

$$L(x, \lambda, \mu) = \frac{1}{2} x^T Q x + c^T x + \lambda^T (b - Ax) - \mu^T x \tag{1.97}$$

The solution of the relaxed problem satisfies the KKT conditions of order 1.

$$\begin{cases} \nabla_x L = 0 \\ \nabla_\lambda L = 0 \\ \nabla_\mu L \geq 0 \end{cases} \Rightarrow \begin{cases} Qx + c - A^T \lambda - \mu = 0 \\ Ax = b \\ x \geq 0 \end{cases} \tag{1.98}$$

The Lagrangian Hessian is constant and the KKT condition of order 2 yields

$$\nabla_{xx} L = Q \geq 0 \tag{1.99}$$

A sufficient condition for optimality is that the matrix Q is positive semidefinite. In this case, the cost function is convex and the relaxed problem admits a unique global minimum given by the KKT conditions (1.98). This minimum can be obtained in polynomial time by an interior point algorithm.

Otherwise, the KKT conditions are not sufficient to guarantee the global optimality of a solution and there is no algorithm that ensures that this minimum is found in a reasonable time. In view of a tree method based on relaxed problem solving, it is necessary to make the problem convex in order to be able to solve it safely and quickly.

1.5.2 Convexification

Convexification consists in transforming the quadratic problem (1.96) into a convex problem having the same binary solutions.
The transformations are based on the property

$$x \in \{0\,;1\} \;\Rightarrow\; x^2 - x = 0 \tag{1.100}$$

Terms of the form $\alpha(x^2 - x)$ can be added to the quadratic cost function, without changing the values it takes for binary variables. The two main convexification techniques are product transformation and the addition of a diagonal matrix.

Product transformation

The quadratic part of the cost function is of the form: $q = x^T Q x = \sum_{i,j=1}^{n} q_{ij} x_i x_j$. The terms of the form: $t(x,y) = \alpha xy = \frac{1}{2}\begin{pmatrix} x \\ y \end{pmatrix}\begin{pmatrix} 0 & \alpha \\ \alpha & 0 \end{pmatrix}\begin{pmatrix} x \\ y \end{pmatrix}$ are not convex, because the eigenvalues of the matrix are $\pm\alpha$.

Each term $t(x,y)$ is replaced by $t_c(x,y)$ defined according to the sign of α.

$$\begin{cases} t_c(x,y) = +\frac{1}{2}\alpha(x+y)^2 - \frac{1}{2}\alpha(x+y) & \text{if } \alpha > 0 \\ t_c(x,y) = -\frac{1}{2}\alpha(x-y)^2 + \frac{1}{2}\alpha(x+y) & \text{if } \alpha < 0 \end{cases} \tag{1.101}$$

The term $t_c(x,y)$ is indeed convex (quadratic part with a positive coefficient). It takes the same values as $t(x,y)$ when x and y are binary. Indeed, we have

$$(x \pm y)^2 = x^2 + y^2 \pm 2xy = x + y \pm 2xy \quad \text{because} \begin{cases} x \in U \Rightarrow x^2 = x \\ y \in U \Rightarrow y^2 = y \end{cases}.$$

and the resulting terms of degree 1 are eliminated in the expression of $t_c(x,y)$. The transformation (1.101) is applied to each term of the cost function according to the sign of the coefficient. Developing these terms, we get

$$\begin{aligned}
\sum_{i,j=1}^{n} q_{ij}x_ix_j &= \sum_{q_{ij}>0} q_{ij}x_ix_j + \sum_{q_{ij}<0} q_{ij}x_ix_j \\
&= \frac{1}{2}\sum_{q_{ij}>0} q_{ij}\left[+(x_i+x_j)^2-(x_i+x_j)\right] + \frac{1}{2}\sum_{q_{ij}<0} q_{ij}\left[-(x_i-x_j)^2+(x_i+x_j)\right] \\
&= \frac{1}{2}\sum_{q_{ij}>0} q_{ij}\left[+(x_i^2+x_j^2)-(x_i+x_j)\right] + \frac{1}{2}\sum_{q_{ij}<0} q_{ij}\left[-(x_i^2+x_j^2)+(x_i+x_j)\right] \\
&\quad + \frac{1}{2}\sum_{q_{ij}>0} q_{ij}\left[2x_ix_j\right] + \frac{1}{2}\sum_{q_{ij}<0} q_{ij}\left[2x_ix_j\right]
\end{aligned}$$

hence the expression of the cost

$$\begin{aligned}
q &= \frac{1}{2}\sum_{q_{ij}>0} q_{ij}\left[+(x_i^2+x_j^2)-(x_i+x_j)\right] \\
&\quad + \frac{1}{2}\sum_{q_{ij}<0} q_{ij}\left[-(x_i^2+x_j^2)+(x_i+x_j)\right] + \sum_{i,j=1}^{n} q_{ij}x_ix_j
\end{aligned} \tag{1.102}$$

The quadratic problem with the cost function (1.102) is convex and has the same binary solutions as the original problem.

Example 1-12: Convexification by product transformation

The quadratic function: $z=3x_1-2x_2+x_3+2x_4-4x_1x_3+2x_1x_4-4x_2x_3+6x_3x_4$ is not convex.

Applying the transformation (1.101), we obtain

$z'=-4x_2-6x_3-2x_4+3x_1^2+2x_2^2+7x_3^2+4x_4^2-4x_1x_3+2x_1x_4-4x_2x_3+6x_3x_4$

The function z' is convex and takes the same values as z for binary variables.

Adding a diagonal matrix

The symmetric matrix Q has real eigenvalues and it admits an orthogonal basis of eigenvectors. Let D be the diagonal matrix of the eigenvalues of Q and P the orthogonal matrix $(P^{-1} = P^T)$ of transformation in this basis.

$$Q = PDP^T \tag{1.103}$$

Assume that Q is not positive definite and consider the matrix $Q' = Q + \tau I$ where $\tau > |\lambda_m|$ and λ_m is the most negative eigenvalue of Q. Then we have

$$Q' = Q + \tau I = PDP^T + \tau PP^T = P(D + \tau I)P^T = PD'P^T \tag{1.104}$$

The matrix D' is diagonal and has all its terms positive, because $\tau > |\lambda_m|$. The matrix Q' is thus positive definite and has the same basis of eigenvectors as Q. The cost function is made convex by adding terms of the form $\tau(x^2 - x)$.

$$q' = q + \tau\sum_{i=1}^{n}(x_i^2 - x_i) = \sum_{i,j=1}^{n} q_{ij}x_ix_j + \tau\sum_{i=1}^{n}(x_i^2 - x_i) \tag{1.105}$$

The new quadratic cost function q' is convex and takes the same values as q when the variables are binary.

Example 1-13: Convexification by adding a diagonal matrix

The function: $z = \frac{1}{2}(x_1^2 - x_2^2)$ of matrix $Q = \begin{pmatrix} 1 & 0 \\ 0 & -1 \end{pmatrix}$ is not convex.

By choosing $\tau > \lambda_m = -1$, the function: $z' = \frac{1}{2}(x_1^2 - x_2^2) + \tau(x_1^2 - x_1) + \tau(x_2^2 - x_2)$ is convex and takes the same values as z for binary variables.

The previous convexification techniques allow the relaxed problem to be solved safely and quickly, for example by an interior point method.

The efficiency of a separation-evaluation method depends on the quality of the relaxed solution. A large relaxed cost (and thus close to the cost of the binary solution) favors the elimination of branches by comparison to the best known feasible solution. We must therefore try to formulate the quadratic problem in a way that maximizes the cost of the relaxed solution. This process of reconvexification of the cost function is useful, even if it is already convex.

For this purpose, we can use the linear constraints of problem (1.94)

$$\begin{cases} \sum_{j=1}^{n} a_{ij} x_j = b_i \ , i = 1 \text{ to } m \\ x_i \in \{0\,;1\} \end{cases} \tag{1.106}$$

to add convex terms to the cost function:

$$z' = z + \sum_{i=1}^{n}\sum_{k=1}^{m} \alpha_{ki} x_i \left(\sum_{j=1}^{n} a_{kj} x_j - b_k \right) + \sum_{i=1}^{n} u_i \left(x_i^2 - x_i \right) \tag{1.107}$$

with coefficients α_{ki} and u_i chosen to maximize the relaxed cost.

No convexification method is systematically better than the others. The choice must therefore be adapted to the quadratic problem to be solved.

Example 1-14 compares the quality of the relaxed cost obtained with the two previous convexification techniques.

Example 1-14: Comparison of the two convexification methods

Consider the problem: $\min\limits_{x_1,x_3} z = x_1^2 - x_2^2$ s.t. $\begin{cases} x_1 + x_2 \le b \\ x_1, x_2 \in \{0;1\} \end{cases}$ with $b > 1$.

The quadratic function: $z = x_1^2 - x_2^2$ of matrix $Q = \begin{pmatrix} 1 & 0 \\ 0 & -1 \end{pmatrix}$ is not convex.

The binary solution is: $\begin{cases} x_1 = 0 \\ x_2 = 1 \end{cases} \rightarrow z_b = -1$ (by examining the 4 combinations).

The relaxed solution is obtained from the KKT conditions: $\begin{cases} x_1 = 0 \\ x_2 = b \end{cases} \rightarrow z_r = -b^2$

Let us compare the relaxed solutions of the two convexification methods.

Method 1: Product transformation
Each term is replaced by an equivalent convex expression in binary variables.

$$\left. \begin{aligned} x_1^2 &\rightarrow \frac{1}{2}\left((2x_1)^2 - 2x_1\right) \\ -x_2^2 &\rightarrow -\frac{1}{2}(2x_2) \end{aligned} \right\} \rightarrow z_{c1} = 2x_1^2 - x_1 - x_2$$

The relaxed convexified problem is: $\min_{x_1,x_2} z_{c1}=2x_1^2-x_1-x_2$ s.t. $\begin{cases} x_1+x_2\le b \\ x_1,x_2\ge 0\end{cases}$

The Lagrangian of the problem is: $L(x_1,x_2,\mu)=2x_1^2-x_1-x_2+\mu(x_1+x_2-b)$ (treating the constraints $x_1,x_2\ge 0$ apart).

The KKT conditions yield: $\begin{cases} 4x_1-1+\mu=0 \\ -1+\mu=0 \\ \mu(x_1+x_2-b)=0\end{cases} \Rightarrow \begin{cases} x_1=0 \\ \mu=1 \\ x_2=b\end{cases} \rightarrow z_{c1}=-b$

Method 2: Adding a diagonal matrix

The most negative eigenvalue of the quadratic form is -1.

The terms $a(x_1^2-x_1)+a(x_2^2-x_2)$ with $a>1$ are added to the cost function.

The relaxed convexified problem is

$$\min_x z_{c2}=(a+1)x_1^2+(a-1)x_2^2-ax_1-ax_2 \quad \text{s.t.} \quad \begin{cases} x_1+x_2\le b \\ x_1,x_2\ge 0\end{cases}$$

The Lagrangian of the problem is (treating the constraints $x_1,x_2\ge 0$ apart)

$$L(x_1,x_2,\mu)=(a+1)x_1^2+(a-1)x_2^2-ax_1-ax_2+\mu(x_1+x_2-b)$$

The KKT conditions yield: $\begin{cases} 2(a+1)x_1-a+\mu=0 \\ 2(a-1)x_2-a+\mu=0 \\ \mu(x_1+x_2-b)=0\end{cases} \Rightarrow \begin{cases} x_1=\dfrac{a-\mu}{2(a+1)} \\ x_2=\dfrac{a-\mu}{2(a-1)} \\ \mu\left(\dfrac{a(a-\mu)}{a^2-1}-b\right)=0\end{cases}$

The complementarity condition (3rd equation) gives two possible cases:

- first case: $\mu=0 \rightarrow x_1=\dfrac{a}{2(a+1)}$, $x_2=\dfrac{a}{2(a-1)} \rightarrow z_{c2}=-\dfrac{a^3}{2(a^2-1)}$

The variables x_1 and x_2 are positive, because $a>1$. The solution exists in all cases;

- second case: $x_1+x_2-b=0$

The solution is

$$\mu=\frac{a^2(1-b)+b}{a} \ , \ x_1=\frac{b(a-1)}{2a} \ , \ x_2=\frac{b(a+1)}{2a} \rightarrow z_{c2}=\frac{1}{2}b^2\left(a-\frac{1}{a}\right)-ab$$

The variables x_1 and x_2 are positive, because $a>1$.

The multiplier μ must be positive: $\mu = \dfrac{a^2(1-b)+b}{a} \geq 0 \Rightarrow a^2 \leq \dfrac{b}{b-1}$.
The solution exists only for values of the real a giving a positive multiplier.

The relaxed cost is the better of the two solutions above.

$$z_{c2} = \min\left(-\frac{a^3}{2(a^2-1)} \;,\; \frac{1}{2}b^2\left(a-\frac{1}{a}\right)-ab \right)$$

We can now compare the relaxed costs given by the 2 methods as a function of the value of a. For example, let us assume that b = 1.5 and a = 2 or 4.
Recall that we want the cost of the relaxed solution to be as high as possible.
The relaxed cost with method 1 is: $z_{c1} = -b = -1.5$.

The relaxed cost with method 2 is: $z_{c2} = -\dfrac{a^3}{2(a^2-1)}$ because $a^2 > \dfrac{b}{b-1}$.

For $a = 2$, we get $z_{c2} = -1.33 > z_{c1}$ and method 2 is more efficient.
For $a = 4$, we get $z_{c2} = -2.13 < z_{c1}$ and method 1 is more efficient.

1.5.3 Quadratic assignment problem

Problem statement
The quadratic assignment problem illustrated in figure 1-24 consists in assigning n objects to n sites knowing that:
- the objects i and j exchange a fixed flow f_{ij};
- the sites u and v are at a fixed distance d_{uv};
- the cost is the product of the flow by the distance $f_{ij}d_{uv}$.

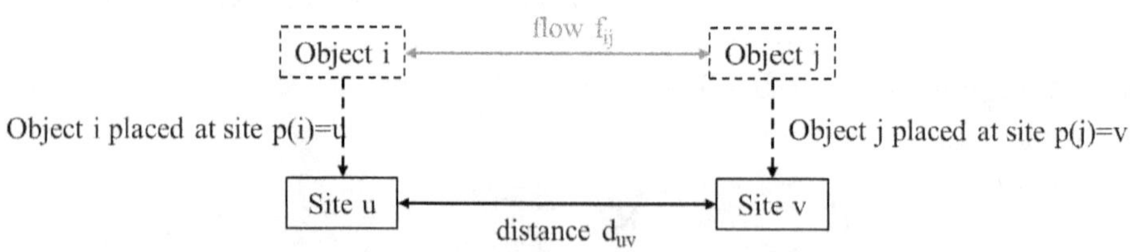

Figure 1-24: Quadratic assignment problem.

<u>Formulation</u>

This problem is formulated as a mixed linear problem.

- Variables

The binary variable s_{iu} is 1 if object i is assigned to site u, and 0 otherwise.
There is a total of $n \times n$ binary variables.

- Cost

The total cost of the assignments is: $z = \sum_{i,j,u,v=1}^{n} s_{iu} s_{jv} f_{ij} d_{uv}$.

- Constraints

The object i is assigned to a single site: $\sum_{u=1}^{n} s_{iu} = 1 \ , \ i = 1 \text{ to } n$.

The site u receives a single object: $\sum_{i=1}^{n} s_{iu} = 1 \ , \ u = 1 \text{ to } n$.

The cost is a quadratic function of the binary variables. To solve this problem by a tree-based method, one can either apply a linearization technique to reduce to a linear problem, or directly treat the quadratic problem after convexification.

1.6 Conclusion

1.6.1 The key points

- Various linearization techniques can be used to transform a continuous nonlinear problem into a mixed-variable linear problem;
- it is possible to reduce the dimension of the mixed problem significantly (fixing variables, removing constraints) before solving it;
- cutting methods consist in iteratively adding constraints to the relaxed problem in order to arrive at an integer solution. These methods are used in conjunction with tree methods;
- tree-based methods explore the tree of possible combinations by eliminating as soon as possible branches that cannot contain the optimal solution. The exploration strategies differ in the choice of the node to be separated and the choice of the variable to be fixed;
- mixed linear programming is applicable to many practical problems (travelling salesman, assignment, flow...).

1.6.2 To go further

- *Programmation mathématique (M. Minoux, Lavoisier 2008, 2e édition)*

  Chapter 2 presents the theoretical results of linear programming, the simplex method and the dual simplex method. Chapter 7 is devoted to integer linear programming. The tree and cutting methods are detailed in a very clear way with their theoretical justifications.

- *Optimization discrète (A. Billionnet, Dunod 2007)*

  The book is devoted to mixed-variable optimization problems. Chapters 1 and 2 introduce the principles of integer linear and quadratic programming. Chapters 3, 4 and 6 detail the formulation and linearization techniques to reduce the problem to a mixed-variable linear problem, as well as the preprocessing to reduce the problem dimension. Chapter 5 presents several practical applications with their formulation and solution by commercial software. Numerous examples are given to illustrate the formulation of the problems and their numerical solution. This book is highly educational and well-supplied and is recommended for users of linear programming.

- *Programmation linéaire (C. Guéret, C. Prins, M. Sevaux, Eyrolles 2003)*

  The book aims to show how to use linear programming in practical applications. Each chapter is devoted to a particular type of application (scheduling, planning, loading, transport, telecommunication, timetabling, etc.). The emphasis is on the formulation and use of commercial software.

2. Discrete optimization

Discrete optimization concerns problems with integer variables. The difficulty lies in the exponential number of combinations and the absence of optimality conditions as in continuous optimization. Although combinatorial problems can be solved by linear programming, it is more efficient to apply specific algorithms, which are the subject of this chapter.

Section 1 introduces the notions associated with graphs and complexity. Many combinatorial problems can be formulated as the search for a path in a graph. The graph formalism unifies the approaches and allows the development of problem-specific algorithms.

Section 2 presents path problems. The objective is to find the minimum cost path in a graph between a starting point and an ending point. The main path algorithms are Ford's algorithm, Bellman's algorithm, and Dijkstra's and A* algorithms, which are the most efficient.

Section 3 presents the scheduling problems. These problems can be considered as particular path problems. The standard methods are the PERT method and the MPM method.

Section 4 presents the flow problems. The two standard problems are the search for the maximum flow that can traverse the network and the search for the maximum flow of minimum cost (when several maximum flow solutions exist). These problems can be solved by Ford-Fulkerson and Roy-Busacker-Gowen algorithms, respectively.

Section 5 presents assignment problems. These problems can be considered as flow problems in a bipartite resource-task graph or they can be solved by the Hungarian method.

Section 6 presents heuristics adapted to the different problems. These so-called greedy empirical methods produce an approximate solution of non-guaranteed quality. They offer the advantage of speed when the problem is too complex to be solved exactly and they can be used in some cases to initialize the tree methods presented in chapter 1.

2.1 Combinatorial problem

Discrete or combinatorial optimization problems can be mostly modeled as graph routing problems. This section based on reference [R12] introduces the main notions associated with graphs and problem complexity.

2.1.1 Graph

A graph is a set of **nodes** or **vertices**, some of which are connected by **edges** or **arcs**. We use the term **edge** if the graph is undirected and the term **arc** if the graph is directed.
An edge is a pair of unordered nodes, denoted with square brackets: $e = [v, w]$.
An arc is an ordered pair of nodes, denoted with parentheses: $e = (v, w)$.

Two nodes connected to each other are neighbors or adjacent. A loop connects a node to itself. For an arc $e = (v, w)$, the following wording is used:
- v is the initial end or origin of the arc e, and v is the predecessor of w;
- w is the terminal end or end of the arc e, and w is the successor of v;
- the arc e starting from v is outgoing at v;
- the arc e arriving at w is incoming at w.

There can be several links between two nodes. Figure 2-1 shows from left to right an arc, an edge, a loop and a triple link between two nodes.

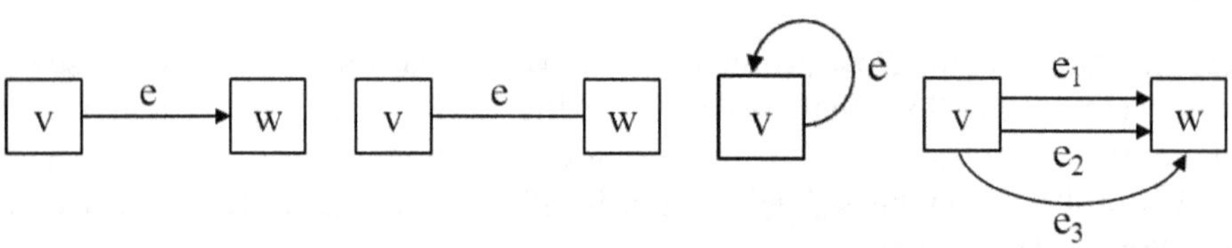

Figure 2-1: Arc, edge, loop.

The set of n nodes or vertices is noted: $V = (v_i)_{i=1 \text{ to } n}$.

The set of m edges or arcs is noted: $E = (e_k)_{k=1 \text{ to } m}$.

The graph G is defined by the two sets V and E: $G = (V, E)$.

<u>*Note on the wording*</u>

The term "arc" is often used, even for an undirected graph.

Number of arcs between nodes

In a multigraph or p-graph, there can be up to p arcs between two nodes.
In a simple graph or 1-graph, there is at most one arc between two nodes.
The density of a simple graph is the ratio $\frac{m}{n^2}$, between the actual number of arcs and the theoretical maximum number (if all nodes were connected two by two).

The outer half degree of node v is the number of arcs starting from v.
The inner half degree of node v is the number of arcs arriving at v.
The degree of the node v is the number of arcs starting and arriving at v.
The average degree of a graph is the average of the degrees of the nodes (often around 3 in practice). An isolated node is a node of zero degree.

Network

A network is an oriented valued graph: each arc is assigned a cost (price of passing through the arc) and a capacity (quantity that can pass through the arc).
The matrix $F = (f_{ij})_{i,j=1 \text{ to } n}$ represents the capacities of the arcs.
The matrix $C = (c_{ij})_{i,j=1 \text{ to } n}$ represents the costs of the arcs.
The node s is the source (entry point of the network). No arcs arrive at s.
The node t is the sink (exit point of the network). No arc starts from t.

The network R is defined by the sets V and E, the matrices F and C, the source s and the sink t: $R = (V, E, F, C, s, t)$.

Representation

A graph can be represented either by the successors of each node, or by the adjacency matrix, or by the incidence matrix.
The adjacency matrix A is a Boolean matrix with n rows and n columns:
- the rows correspond to the starting nodes;
- the columns correspond to the arrival nodes.

If there is an arc from node v_i to node v_j, the element a_{ij} is 1, otherwise a_{ij} is 0.

The incidence matrix I is a Boolean matrix with n rows and m columns:
- the rows correspond to the nodes;
- the columns correspond to the arcs.

If there is an arc e_k from node v_i to node v_j, the element a_{ik} is +1 and the element a_{jk} is −1 (otherwise $a_{ik} = a_{jk} = 0$). For a loop on node v_i, the element a_{ii} is +1.

Figure 2-2 shows the filling of adjacency and incidence matrices for an arc e_k from node v_i to node v_j. Example 2-1 illustrates the three possible representations of a graph.

$$\begin{array}{c} \\ v_i \\ \\ \end{array}\begin{pmatrix} & & v_j & \\ & & \vdots & \\ \cdots & \cdots & 1 & \cdots \\ & & \vdots & \\ & & \vdots & \end{pmatrix} \qquad \begin{array}{c} \\ v_i \\ v_j \\ \end{array}\begin{pmatrix} & & e_k & \\ & & \vdots & \\ \cdots & \cdots & +1 & \cdots \\ \cdots & \cdots & -1 & \cdots \\ & & \vdots & \end{pmatrix}$$

Figure 2-2: Adjacency matrix and incidence matrix.

Example 2-1: Representation of a graph

Consider the graph with 5 nodes and 10 arcs shown in figure 2-3.
A first representation of this graph is the list giving the successors of each node.

$v_1 \rightarrow v_2, v_3, v_5$
$v_2 \rightarrow v_2, v_3$
$v_3 \rightarrow v_4$
$v_4 \rightarrow v_1, v_5$
$v_5 \rightarrow v_1, v_3$

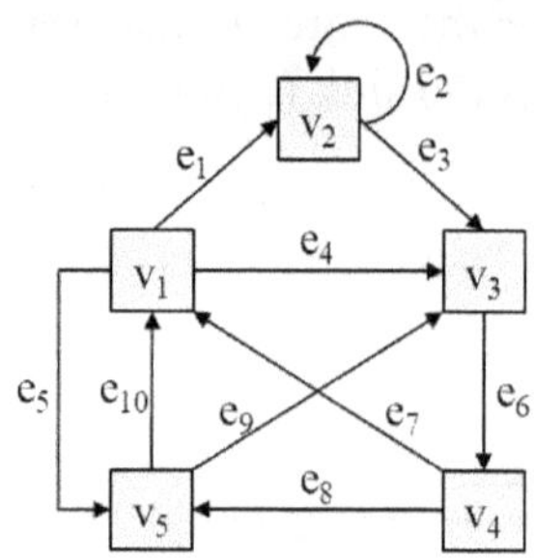

Figure 2-3: 5-node, 10-arc graph.

A second representation of this graph is its adjacency matrix (left).
A third representation of this graph is its incidence matrix (right).

	v_1	v_2	v_3	v_4	v_5
v_1	0	1	1	0	1
v_2	0	1	1	0	0
v_3	0	0	0	1	0
v_4	1	0	0	0	1
v_5	1	0	1	0	0

	e_1	e_2	e_3	e_4	e_5	e_6	e_7	e_8	e_9	e_{10}
v_1	+1	0	0	+1	+1	0	−1	0	0	−1
v_2	−1	+1	+1	0	0	0	0	0	0	0
v_3	0	0	−1	−1	0	+1	0	0	−1	0
v_4	0	0	0	0	0	−1	+1	+1	0	0
v_5	0	0	0	0	−1	0	0	−1	+1	+1

2.1.2 Route in a graph

Many problems involve finding an optimal route in a graph.
A route in a graph is called:
- a path if the route is oriented;
- a circuit if the route is oriented and closed;
- a string if the path is undirected;
- a cycle if the path is non-oriented and closed.

A route is represented by the sequence of its arcs or by the sequence of its nodes (if the graph is simple). The length of the route is the number of consecutive arcs.

Specific routes

An elementary route passes through its nodes only once.
An Eulerian route passes exactly once through each arc of the graph.
A Hamiltonian route passes exactly once through each node of the graph.
An Eulerian route can pass through a node several times, whereas a Hamiltonian route passes through each arc at most once. Eulerian or Hamiltonian routes are used in vehicle touring problems.

The house game shown here consists in drawing a house without lifting the pencil and without passing over the same line twice.

The result is an Eulerian, but not Hamiltonian, route.

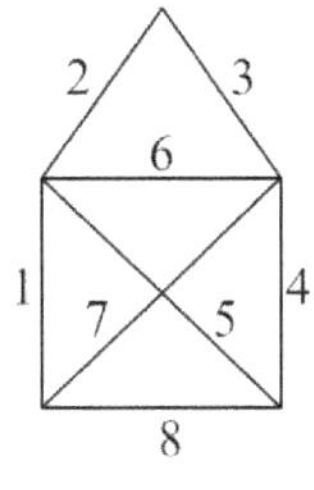

Special graphs

A graph is complete if there is an arc between any pair of nodes.
A graph is connected if there is a chain between any pair of nodes.
A connected graph has at least $n-1$ edges.
A tree is an undirected graph with no cycles. The number of edges in a tree is equal to the number of nodes $-$ 1: $m = n-1$.
A rooted tree is a directed tree (connected directed graph without cycle).
All nodes in a tree are descendants of a single node called the root of the tree.

A graph is planar if it can be represented in the plane without crossing arcs.
The number of edges in a planar graph is: $m \leq 3n-6$. Planar graphs are used in electronic component placement and map coloring problems.

Connectivity

The study of connectivity is a prerequisite for any graph problem. It relies on the adjacency matrix. This Boolean matrix A defined in section 2.1.1 indicates whether two nodes are connected (value 1) or not (value 0).

A first question of connectivity is the existence of a path between any two nodes of the graph. This question is solved by computing the Boolean powers of the matrix $I+A$ as explained below:
- Boolean multiplication represents the logical product "and";
- Boolean addition represents the logical sum "or".

These operations are shown in figure 2-4.

×	0	1
0	0	0
1	0	1

+	0	1
0	0	1
1	1	1

Figure 2-4: Boolean multiplication ("and") and addition ("or").

The Boolean matrix A indicates whether 2 nodes are connected by a 1-arc path.
The product $A.A \underset{\text{noted}}{=} A^2$ indicates whether 2 nodes are connected by a 2-arc path.

Continuing the reasoning, the Boolean matrix A^k indicates whether 2 nodes are connected by a path with k arcs.

The sum $A^0 + A^1 + A^2 + ... + A^k$ indicates whether there is a path with at most k arcs between 2 nodes (the matrix $A^0 = I$ indicates that each node is connected to itself). Using the Boolean identity $A+A=A$, we factor the sum into the form

$$I + A + A^2 + ... + A^k = (I+A)^k \tag{2.1}$$

The Boolean matrix $(I+A)^k$ indicates for each node (row) which nodes can be reached (columns) by a path of length k at most.
In a graph with n nodes, an elementary path has length $n-1$ at most.
We therefore stop the calculation at the matrix $(I+A)^{n-1}$ or at the first power k such that $(I+A)^k = (I+A)^{k-1}$.

The matrix $(I+A)^{n-1}$ indicates for each node (row) which nodes are accessible (columns). A node w accessible from node v is a descendant of v and v is an ascendant of w. The set of descendants of v is noted $\Gamma(v)$. The graph is connected if and only if all elements of $(I+A)^{n-1}$ are 1.

Example 2-2: Existence of paths

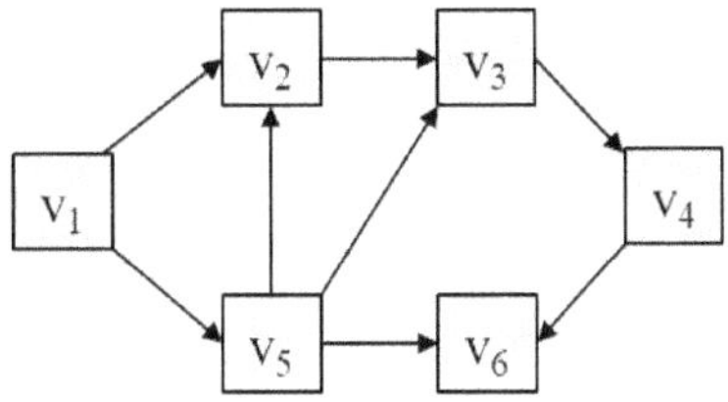

Consider the graph with 6 nodes shown opposite.

Let us raise $(I+A)$ to successive powers.

$$A=\begin{pmatrix}0&1&0&0&1&0\\0&0&1&0&0&0\\0&0&0&1&0&0\\0&0&0&0&0&1\\0&1&1&0&0&1\\0&0&0&0&0&0\end{pmatrix},\quad I+A=\begin{pmatrix}1&1&0&0&1&0\\0&1&1&0&0&0\\0&0&1&1&0&0\\0&0&0&1&0&1\\0&1&1&0&1&1\\0&0&0&0&0&1\end{pmatrix},\quad (I+A)^2=\begin{pmatrix}1&1&1&0&1&1\\0&1&1&1&0&0\\0&0&1&1&0&1\\0&0&0&1&0&1\\0&1&1&1&1&1\\0&0&0&0&0&1\end{pmatrix}$$

$$(I+A)^3=\begin{pmatrix}1&1&1&1&1&1\\0&1&1&1&0&1\\0&0&1&1&0&1\\0&0&0&1&0&1\\0&1&1&1&1&1\\0&0&0&0&0&1\end{pmatrix},\quad (I+A)^4=\begin{pmatrix}1&1&1&1&1&1\\0&1&1&1&0&1\\0&0&1&1&0&1\\0&0&0&1&0&1\\0&1&1&1&1&1\\0&0&0&0&0&1\end{pmatrix},\quad \hat{A}=(I+A)^5=\begin{pmatrix}1&1&1&1&1&1\\0&1&1&1&0&1\\0&0&1&1&0&1\\0&0&0&1&0&1\\0&1&1&1&1&1\\0&0&0&0&0&1\end{pmatrix}$$

We observe that: $(I+A)^4=(I+A)^3$. We therefore stop the calculation at the power $k=4$.
In the matrix $\hat{A}=(I+A)^5$, the rows indicate the successors of each node (when the value is 1) and the columns indicate the predecessors of each node (when the value is 1).
The node v_1 admits all other nodes as successors (first row formed by 1). The node v_1 admits no predecessors (first column formed by 0 except on v_1). Such a node with no predecessors is a **source** and the graph is a **tree** of **root** v_1 .
The node v_6 admits all other nodes as predecessors (sixth column formed by 1). The node v_6 admits no successors (sixth row formed by 0 except on v_6). Such a node with no successors is called a **sink**.

A second question of connectivity is the number of paths between any two nodes in the graph. This question is solved by calculating the algebraic powers of the adjacency matrix A.

The matrix A^k (A power k) indicates the number of paths of length k between each pair of nodes, as shown by the following recurrence reasoning.

- Initialization

For $k=1$, the element a_{ij} of the matrix A^1 gives the number of paths of length 1 between node v_i and node v_j. This number is either 0 or 1.

- Recurrence

Note $B=A^k$ and assume that the element b_{ij} of the matrix B gives the number of paths of length k between node v_i and node v_j.

The matrix product $C=BA$ is calculated by: $c_{ij}=\sum_{l=1}^{n} b_{il}a_{lj}$, $i,j=1$ to n.

For two given nodes v_i and v_j, the term c_{ij} sums up all possible paths of length k starting from v_i (term b_{il}) and which can be extended to v_j by an arc between v_l and v_j (term a_{lj}). This term gives the number of paths of length k+1 between the node v_i and the node v_j.

In a graph with n nodes, an elementary path is of length $n-1$ at most.
We therefore stop the calculation at the matrix A^{n-1}.

Example 2-3: Number of paths

Let us retrieve the graph with 5 nodes and 10 arcs from example 2-2 shown opposite.

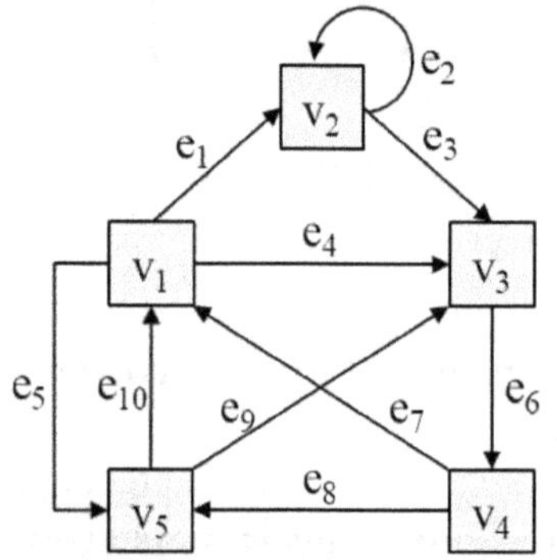

Let us raise its adjacency matrix A to successive powers.

$$A=\begin{pmatrix}0&1&1&0&1\\0&1&1&0&0\\0&0&0&1&0\\1&0&0&0&1\\1&0&1&0&0\end{pmatrix},\ A^2=\begin{pmatrix}1&1&2&1&0\\0&1&1&1&0\\1&0&0&0&1\\1&1&2&0&1\\0&1&1&1&1\end{pmatrix},\ A^3=\begin{pmatrix}1&2&2&2&2\\1&1&1&1&1\\1&1&2&0&1\\1&2&3&2&1\\2&1&2&1&1\end{pmatrix},\ A^4=\begin{pmatrix}4&3&5&2&3\\2&2&3&1&2\\1&2&3&2&1\\3&3&4&3&3\\2&3&4&2&3\end{pmatrix}$$

The element in row 3 column 1 of the matrix A^4 indicates that there is only one path of length 4 between node v_3 and node v_1. Indeed, we observe that the only path of length 4 between these two nodes is: $v_3 \to v_4 \to v_1 \to v_5 \to v_1$.

Computations based on the adjacency matrix characterize the connectivity of a graph in a simple way. However, this approach is not the most efficient for low density graphs, because the adjacency matrix is then very sparse.

2.1.3 Complexity

Combinatorial problems fall into different categories:

- path problems:
  - What is the minimum cost path between two nodes?
  - What is the minimum cost Hamiltonian circuit? (travelling salesman)
- flow problems:
  - What is the maximum flow that can pass from the source to the sink?
  - What is the minimum cost of the maximum flow?
- problems of assignment, coloring, stacking...

The difficulty comes from the number of combinations, which is an exponential function of the problem size. The size of a problem is the memory space required to store the data. It is measured by characteristic parameters:
- for a graph with n nodes and m arcs, the size is measured by the pair (n,m);
- for a square matrix n×n, the size is measured by n.

An algorithm is a sequence of elementary operations applied to data.
The complexity of an algorithm is measured by the number of operations as a function of the size of the data.
For example, consider the multiplication of two matrices of size n×n.

$$C = AB \quad \Leftrightarrow \quad c_{ij} = \sum_{k=1}^{n} a_{ik} b_{kj} \quad , \quad i = 1 \text{ to } n \, , \; j = 1 \text{ to } n \tag{2.2}$$

The calculation of each element of the matrix C requires n multiplications and n additions. For the n×n matrix elements, the total number of operations is $2n^3$.

The complexity of an algorithm is the order of magnitude of the number of operations as a function of the characteristic size n of the data:
- a polynomial algorithm has a complexity in powers of n:
  For example: $O(n)$, $O(n \log n)$, $O(n^2)$
- a non-polynomial algorithm has a complexity in exponential of n:
  For example: $O(2^n)$, $O(n^n)$, $O(n!)$ with $n! \approx \left(\frac{n}{e}\right)^n \sqrt{2\pi n}$ (Stirling formula)

A polynomial algorithm is considered efficient because it can handle large data sizes. A composition of polynomial algorithms remains polynomial. Table 2-1 shows the computation times as a function of the complexity of the algorithm and the size of the problem (from 10 to 10 000). The times are estimated for a computer performing one billion operations per second.

	n	n.log n	n^2	n^3	n^5	2^n	3^n	n!
10	10^{-8} s	10^{-8} s	10^{-7} s	10^{-6} s	10^{-4} s	10^{-6} s	6.10^{-5} s	3.10^{-3} s
50	5.10^{-8} s	8.10^{-8} s	2.10^{-6} s	10^{-4} s	0,3 s	13 days	10^{7} years	10^{48} years
100	10^{-7} s	2.10^{-7} s	10^{-5} s	10^{-3} s	10 s	10^{13} years	10^{31} years	
1000	10^{-6} s	3.10^{-6} s	10^{-3} s	1 s	11 days			
10000	10^{-5} s	4.10^{-5} s	0,1 s	1000 s	3100 years			

Table 2-1: Computation time as a function of algorithm complexity.

Combinatorial problems are classified into optimization problems (OP) and existence problems (EP):

- an optimization problem consists in minimizing a function on a finite set:
  (PO): Find $x^* \in X$ such that $f(x^*) = \min_{x \in X} f(x)$.
  The answer to an optimization problem is the solution x^*;
- an existence problem consists in answering a question of the type:
  (PE): There exists $x \in X$ such that $f(x) \le c$.
  The answer to a problem of existence is 'yes' or 'no'.

The existence problem (EP) and the optimization problem (OP) are of comparable difficulty. Indeed, an algorithm solving (PO) can answer (PE). Conversely, an algorithm solving (PE) can solve (PO) by dichotomy on the value of c (the dichotomy being a polynomial algorithm).
A combinatorial problem is said to be "easy" or "hard" depending on whether a polynomial algorithm for solving the existence problem is known. The problems are classified as NP, P and NPC.

The class NP (Polynomial Non-Deterministic) contains problems for which a polynomial algorithm is used to verify a solution. The solution x_0 is given (we do not care how it is obtained) and we have only to check that it solves problem (PE): $f(x_0) \le c$.
The class NP contains almost all combinatorial problems encountered in practice.

The class P (Polynomial Deterministic) contains problems for which a polynomial algorithm can find a solution. These problems are the easy NP problems.

The NPC (NP-Complete or NP-Hard) class contains NP problems that do not belong to P and of "equivalent" difficulty. Two problems are said of equivalent difficulty if we can pass from one to the other by a polynomial transformation. NPC problems have no known polynomial solution algorithm. But, if a polynomial algorithm can be developed for one of them, then all NPC problems will fall into class P.

NP problems that do not belong to either P or NPC are said to be indeterminate. Example 2-4 illustrates the different classes.

Example 2-4: NP, P and NPC problems (from [R12])

NP problems
NP problems are those for which a polynomial algorithm for verifying a solution is known. NP problems are easy (P), hard (NPC) or indeterminate:

- example of NP problem:

Stacking problem: Are there p integers among n whose sum is b?
Verification of a solution: We check that the p integers are among the n and that their sum is b. The verification algorithm is polynomial;

- example of non-NP problem:

Chess game: Is there a winning strategy in a given position?
Verification of a solution: All possible games should be examined from the position. The verification algorithm is not polynomial;

- example of NP indeterminate problem:

Graph isomorphism: Can we pass from a graph G_1 to a graph G_2 by renumbering the nodes?
The verification algorithm is polynomial, but it is not known if there is a polynomial solution algorithm. The problem is indeterminate.

Problems of P
P problems are those for which a polynomial solution algorithm is known.
These problems are the easy NP problems:

- connectivity of a graph:
  Is there a path between any two nodes in a graph?;
- minimum cost path in a valued graph:
  What is the minimum cost path between two nodes in a graph?;
- maximum flow in a network:
  What is the maximum flow between the source and the sink?;
- Eulerian path in a graph:
  Is there a path that passes through each edge once and only once?;
- assignment (coupling):
  What is the minimum cost assignment between n resources and n tasks?;
- sorting n numbers.

<u>NPC problems</u>
NPC problems are those for which no polynomial solution algorithm is known. These NP-complete problems are the NP-hard problems:

- coloring:
  Find the minimum number of colors to color a graph.
  The 4-color theorem for a planar graph conjectured in 1880 was proved in 1977 (computer-assisted demonstration);
- travelling salesman:
  Find the minimum cost Hamiltonian circuit in a valued graph.
  This problem serves as a test for combinatorial optimization algorithms;
- stacking (knapsack):
  Select items to fill a knapsack maximizing the total value;
- bin packing:
  Arrange objects using as few boxes as possible.

A combinatorial problem can be solved by mixed linear programming (chapter 1) or by a problem-specific algorithm (path, flow, assignment). These specific algorithms are presented from section 2.2 onwards.
For large problems (easy or difficult), it is not possible to obtain the exact solution in a reasonable time. Approximate methods are either general metaheuristics (Volume I, chapter 2) or heuristics specific to a type of problem (stacking, bin packing, set covering, coloring). These heuristics, which generally use greedy approaches, are presented in section 2.6.

2.2 Path problem

We place ourselves in a valued graph $G = (V, E, C)$ defined by its nodes V, its arcs E and the costs of the arcs C. The classical problems consist in:
- finding the minimum cost path between two given nodes;
- finding the minimum cost path between a node and all other nodes;
- finding the minimum cost path between all pairs of nodes.

These problems can be solved by mixed linear programming (section 1.4.4) or by the following specific algorithms:
- the Ford algorithm can be applied in any graph;
- the Bellman algorithm applies in graphs without circuits;
- the Dijkstra and A* algorithms apply if the valuations are positive.

Table 2-2 summarizes the characteristics of these algorithms, which are detailed in the following sections. Their complexity is expressed as a function of the number of nodes n and the number of arcs m.

Algorithm	Valuations (sign)	Possible circuit	Processing (nodes, arcs)	Choice of the node to process	Complexity
Ford	any sign	non absorbing	several times	free	2^n
Bellman	any sign	no	once per arc	topological order	m
Dijkstra	positive	yes	once per node	minimal cost	m.log(n)
A*	positive	yes	once per node if h consistent	minimale value	≤ Dijkstra

Table 2-2: Path algorithms.

2.2.1 Ford's algorithm

Ford's algorithm determines the minimum cost path from the source to each node of the graph. The valuations of the arcs are of any sign.

The minimum costs from the source to each node are improved iteratively.
At each iteration, the node v_i is marked with the value z_i of the best cost found.

The stages of the algorithm are as follows.

Initialization
A value $z_s = 0$ is assigned to the source and values $z_i = \infty$ to the other nodes.

Iteration
An iteration consists in traversing all the arcs of the graph in any order.
Each node v_i has a value z_i. The arc from v_i (value z_i) to v_j (value z_j) has a cost c_{ij}.

If $z_i + c_{ij} < z_j$,
passing through v_i improves the solution arriving at v_j.
The value of v_j is updated: $z'_j = z_i + c_{ij}$.
Each node is updated once per iteration.

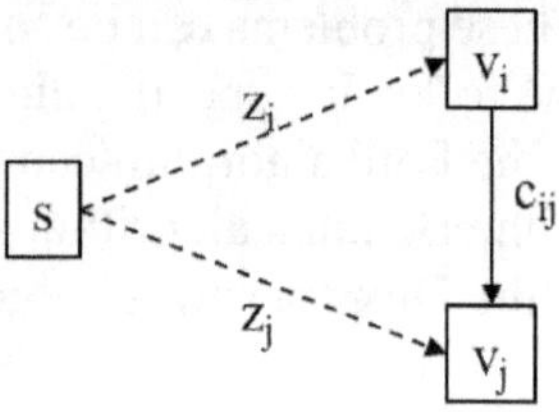

End of the algorithm
We stop when no costs can be improved.
Each node is then marked with the value of the best path from the source.

In practice, it is not useful to go through all the arcs at each iteration. It is sufficient to examine the arcs starting from the nodes updated in the previous iteration.
Example 2-5 illustrates how Ford's algorithm works.

Example 2-5: Ford's algorithm (from [R6])

Consider the 6-node valued graph shown opposite.
The source is at v_1. Some valuations are positive, others negative.

Let us apply Ford's algorithm to find the minimum cost path from the source v_1 to each of the nodes v_2, v_3, v_4, v_5, v_6.

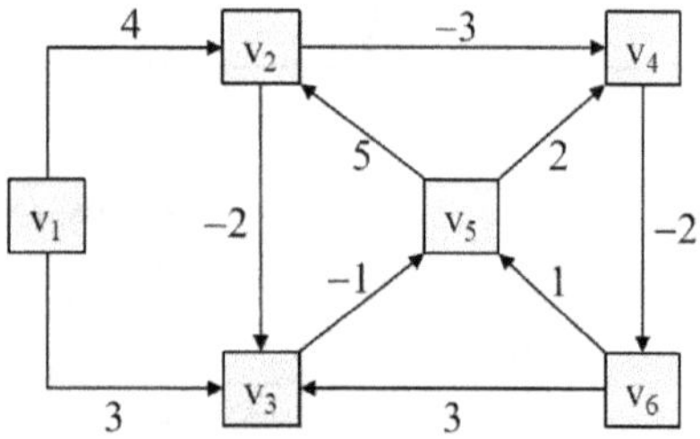

Initialization

The source v_1 is given the value $z_i = 0$.
The other nodes receive the value $z_i = \infty$.

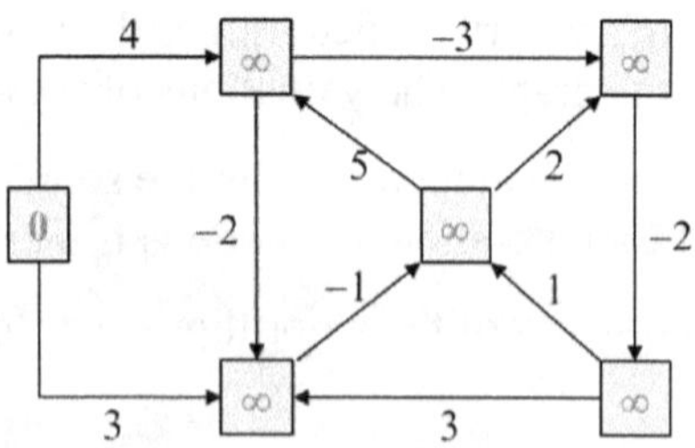

Iteration 1

We examine all arcs in an arbitrary order, starting from the source v_1.
Each node is updated once at most.
The arcs used for this iteration are in thick lines.

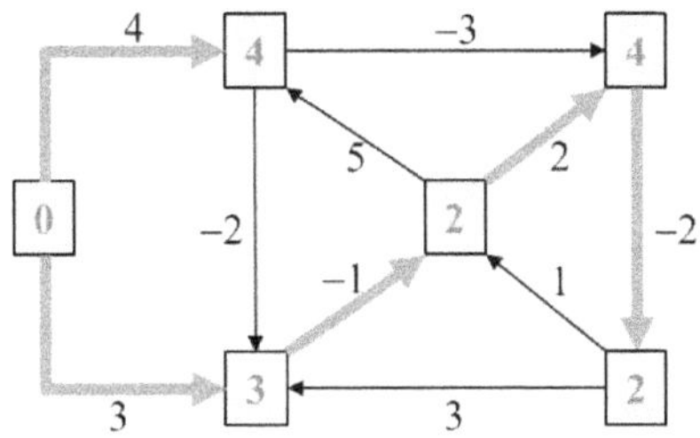

Iteration 2

Unused arcs from iteration 1 are examined in an arbitrary order.
Each node is updated once at most.
The arcs used for this iteration are in thick lines.
This iteration improves 2 nodes (v_3,v_4).

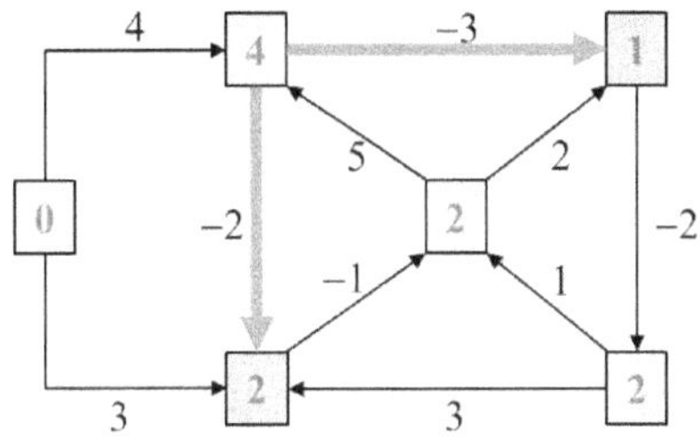

Iteration 3

We examine the successors of the nodes updated at iteration 2 (v_3,v_4).
Each successor is updated once at most.
The arcs used for this iteration are in thick lines.
This iteration improves 2 nodes (v_5,v_6).

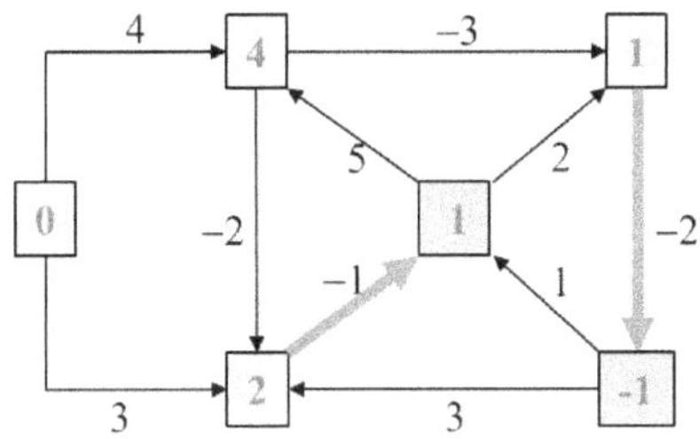

Iteration 4

We examine the successors of the nodes updated at iteration 3 (v_5,v_6).
Each successor is updated once at most.
The arcs used for this iteration are in thick lines.
This iteration improves 1 node (v_5).

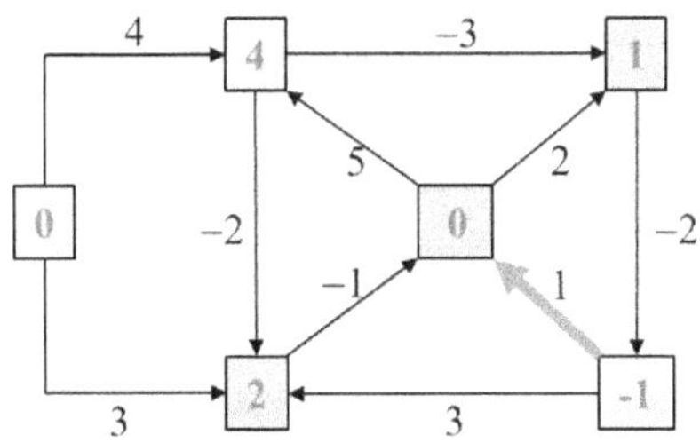

End of the algorithm

There is no further improvement possible from the last updated node (v_5).

The value (or mark) of each node is the cost of the best path from the source to that node. The optimal arcs in thick lines are those whose cost equals the difference in marks between the start and end nodes.

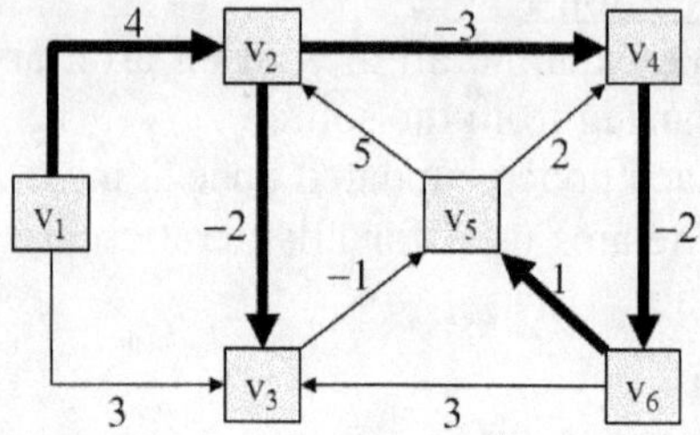

The minimum costs obtained in iteration 4 are summarized in the table opposite.

V_1-V_2	V_1-V_3	V_1-V_4	V_1-V_5	V_1-V_6
4	2	1	0	−1

Ford's algorithm is applicable in a graph with valuations of any sign. The graph may include circuits provided that these are not absorbing.

The cost of an **absorbing circuit** is negative.

The one shown here has a cost of −1.

An absorbing circuit allows an unbounded decrease in the cost of a path.

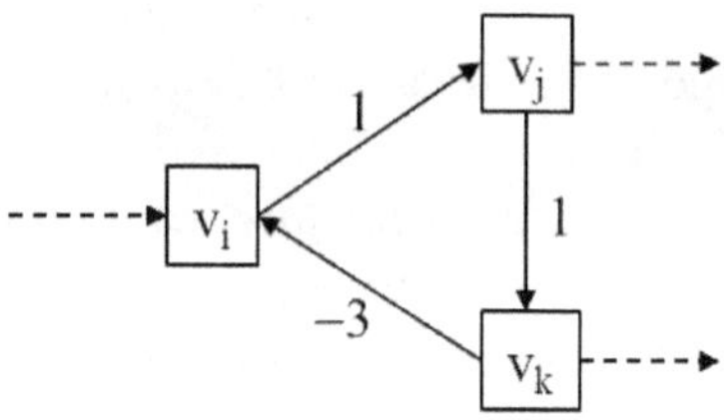

In the absence of an absorbing circuit, Ford's algorithm terminates in a finite number of iterations. The number of iterations depends on the order in which the arcs are processed.

In a graph with n nodes, the complexity is in $O(2^n)$ and Ford's algorithm is non-polynomial. However, it is possible to derive polynomial algorithms by optimizing the order of processing of the arcs. This is the case of the Bellman and Dijkstra algorithms presented in the following sections.

2.2.2 Bellman's algorithm

Bellman's algorithm applies in a graph without circuit (tree) with valuations of any sign. It proceeds in the same way as Ford's algorithm, but imposes an order of treatment of the nodes according to their position.

In a graph without circuits, the nodes can be numbered in such an order that any arc connects nodes of increasing number. This order is called **topological**.

The Bellman algorithm processes the nodes in topological order. Each node is processed only once. The mark z_i of node v_i is the value of the best cost found. The stages of the algorithm are as follows.

Initialization

A value $z_s = 0$ is assigned to the source and values $z_i = \infty$ to the other nodes.

The n nodes $(v_1, v_2, \cdots, v_n)$ are arranged in topological order.

Treatment of nodes in topological order

For each node v_i, the values of the successors v_j are updated.

If $z_i + c_{ij} < z_j$,

passing through v_i improves the solution arriving at v_j.

The value of v_j is updated: $z'_j = z_i + c_{ij}$.

The node v_j is processed.

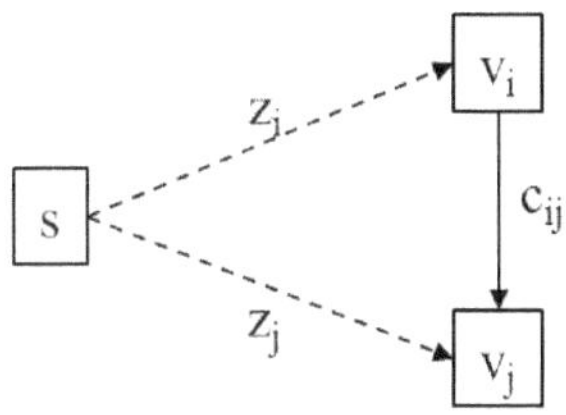

End of the algorithm

We stop when the n nodes have been processed.

Each node is then marked with the value of the best path from the source.

Example 2-6 illustrates how Bellman's algorithm works.

Example 2-6: Bellman's algorithm (from [R6])

Consider the 5-node valued graph shown opposite.

The source is at v_1.

The graph is circuit-free (tree).

The valuations are of any sign.

The nodes are numbered in topological order.

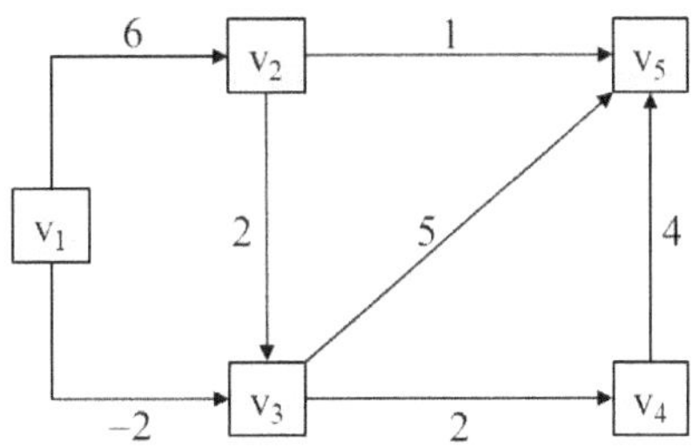

Let us apply Bellman's algorithm to find the minimum cost paths from the source. The arcs used in each iteration are in thick lines.

Initialization

The source v_1 is given the value $z_i = 0$.

The other nodes receive the value $z_i = \infty$.

The nodes are unprocessed.

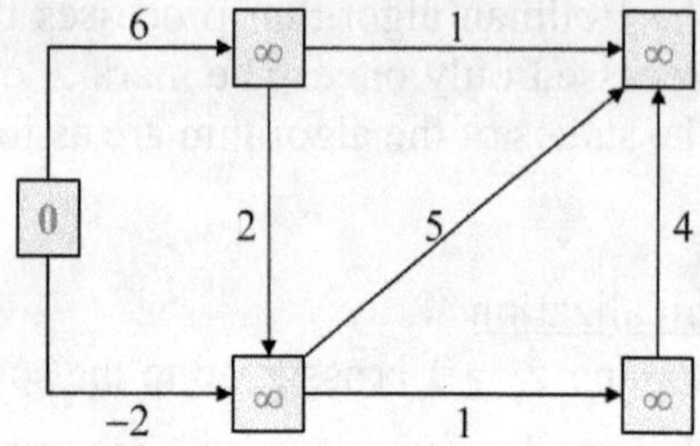

Treatment of the first node

The first node in topological order is v_1.
The successors of v_1 are updated.
The node v_1 is processed.

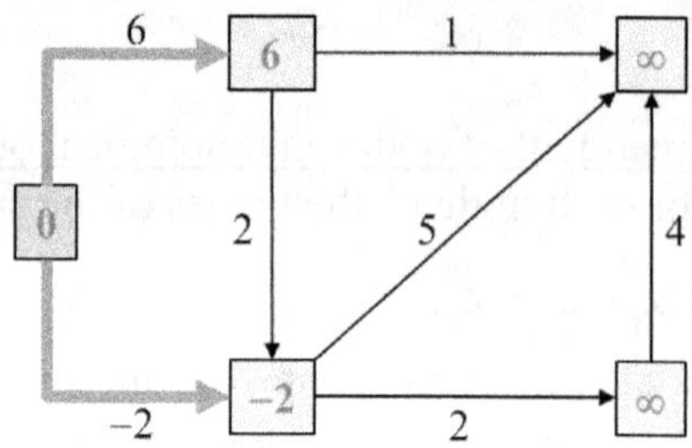

Treatment of the second node

The second node in topological order is v_2.
The successors of v_2 are updated.
The node v_2 is processed.

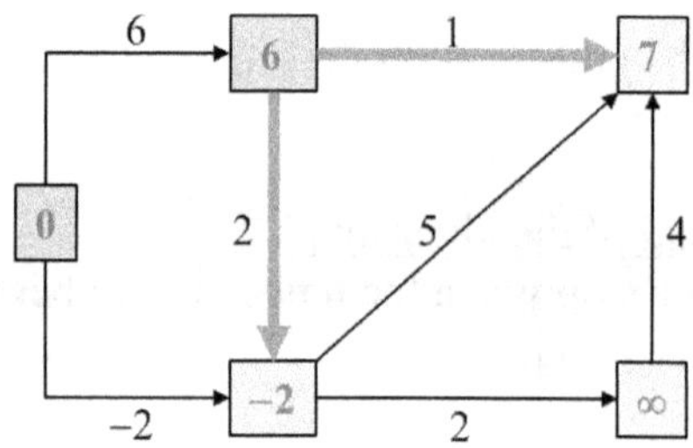

Treatment of the third node

The third node in topological order is v_3.
The successors of v_3 are updated.
The node v_3 is processed.

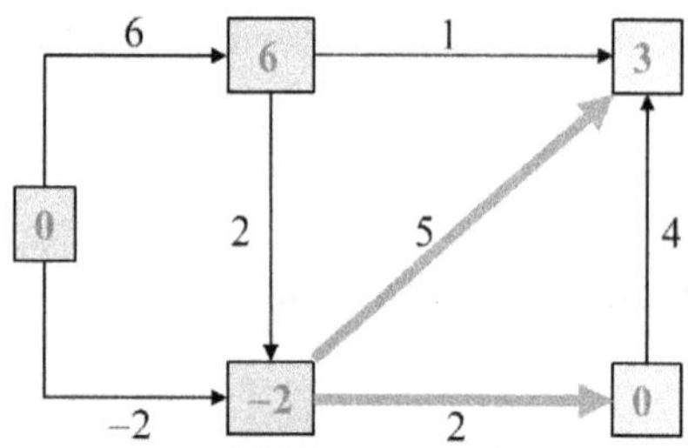

Treatment of the fourth node

The fourth node in topological order is v_4.
The successors of v_4 are updated.
The node v_4 is processed.

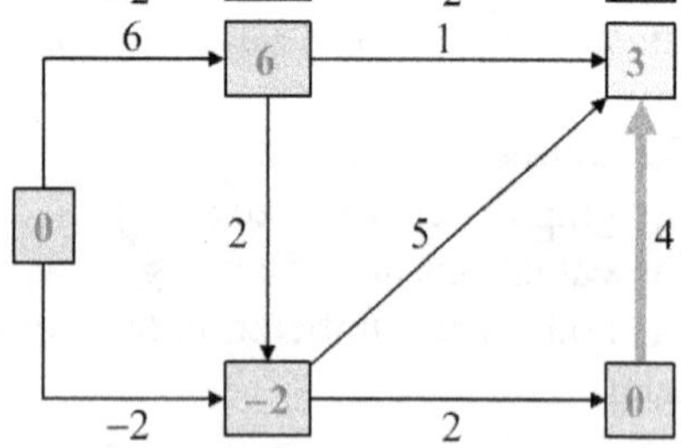

Treatment of the fifth node

The fifth node in topological order is v_5.
This node has no successor.
The node v_5 is processed.

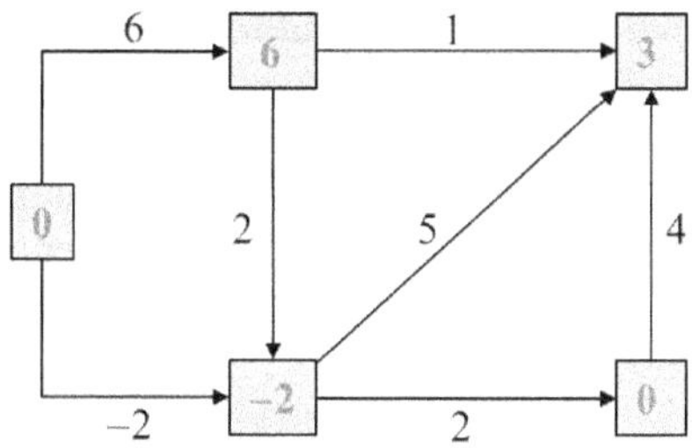

End of the algorithm

All nodes have been processed.
The value (or mark) of each node is the cost of the best path from the source to that node.
The optimal arcs in thick lines are those whose cost equals the difference in marks between the start and end nodes.

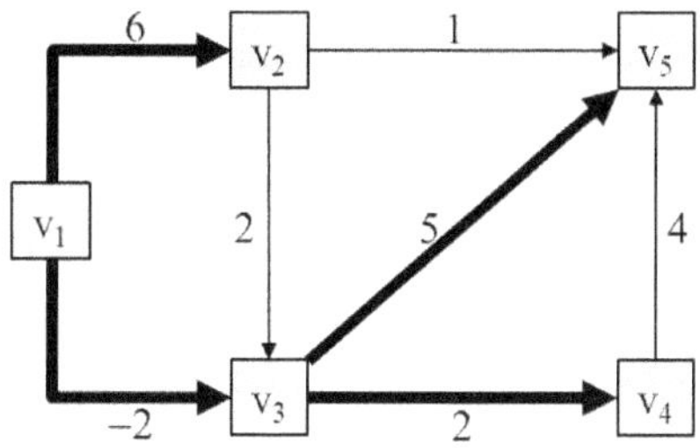

The minimum costs are summarized in the table opposite.

v_1-v_2	v_1-v_3	v_1-v_4	v_1-v_5
6	–2	0	3

The topological order of a tree is not unique. A systematic way of constructing a topological order is to group the nodes by level:
- level 0 is the set V_0 of nodes without predecessors (sources);
- level 1 is the set V_1 of successors of V_0;
- level k is the set V_k of successors of V_{k-1}.

A node can belong to several levels, as in figure 2-5b.

Bellman's algorithm then consists in processing the nodes in successive levels. This level-based updating procedure is called **dynamic programming**.
It relies on Bellman's principle of optimality.

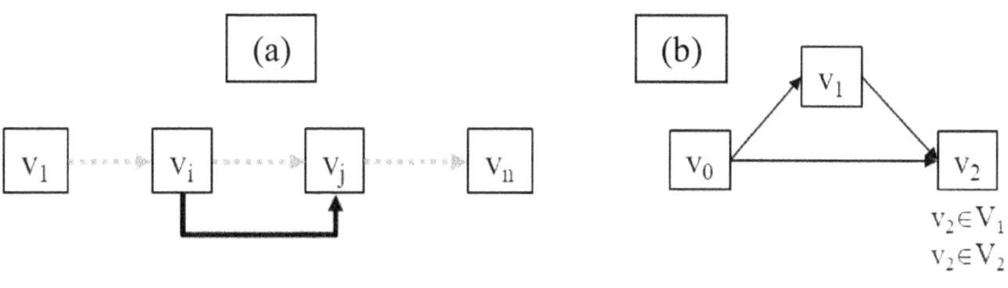

Figure 2-5: Bellman's optimality principle.

Bellman's optimality principle

Any sub-path of an optimal path is itself optimal.

Suppose that the optimal path from v_1 to v_n passes through v_i and v_j, then the sub-path from v_i to v_j is optimal. Indeed, if there were a better path from v_i to v_j (solid line in figure 2-5), the global path from v_1 to v_n would be improved by passing through this path.

To apply the dynamic programming algorithm, the nodes in level 0 are initialized to $z = 0$ and the others to $z = \infty$. At each stage, the value $z(v_i)$ is the cost of the best known path from the source to node v_i.
The nodes of level k are updated from their predecessors by

$$z(v_j) = \min_{v_i \in V_{k-1}} \left(z(v_i) + c_{ij} \right) \tag{2.3}$$

The value $z(v_j)$ will thus be the cost of the best path from the source to node v_j.
The relationship (2.3) called the functional equation of dynamic programming is illustrated in figure 2-6.

Passing through v_{i1} : $z_1 = z(v_{i1}) + c_{i1-j}$

Passing through v_{i2} : $z_2 = z(v_{i2}) + c_{i2-j}$

Passing through v_{i3} : $z_3 = z(v_{i3}) + c_{i3-j}$

Best path: $z(v_j) = \min(z_1, z_2, z_3)$

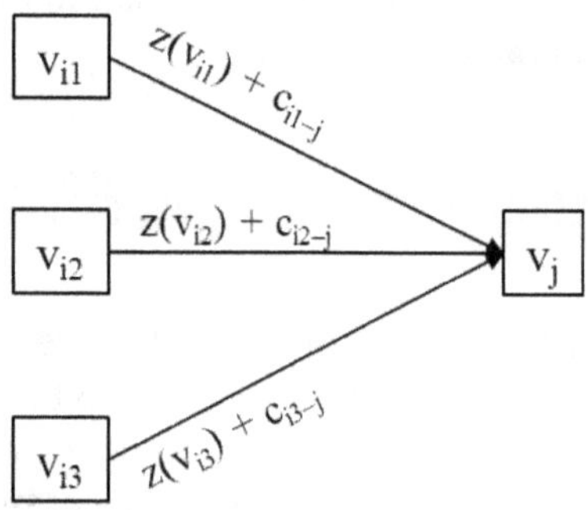

Figure 2-6: Functional equation for dynamic programming.

Since each of the m arcs of the graph is processed only once, the complexity of the dynamic programming algorithm is $O(m)$.

Dynamic programming can be applied forwards or backwards:

- the objective of the forward procedure is to determine the optimal paths from the source to each node. We start from the first level and try to <u>minimize</u> the cost by <u>adding up</u> the costs of the examined arcs;
- the objective of the backward procedure is to determine the optimal paths from each node to the sink. It starts from the last level and seeks to <u>maximize</u> the cost by <u>subtracting</u> the costs of the examined arcs.

These two procedures are illustrated in example 2-7.

Example 2-7: Dynamic programming algorithm (from [R6])

Consider the 5-level valued graph in figure 2-7.

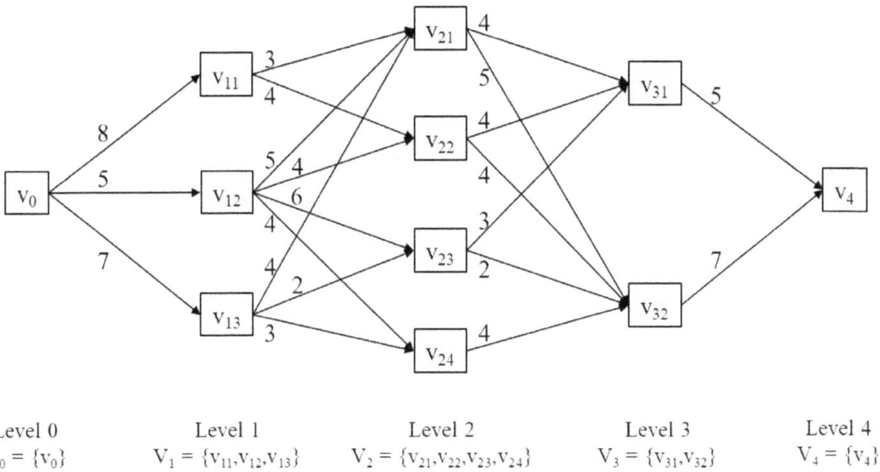

Figure 2-7: 5-level graph.

Let us apply the forward dynamic programming algorithm to find the minimum cost paths from the source to each node. The levels are processed in ascending order. The optimal arcs selected at each level are in thick lines.

Treatment of level 1

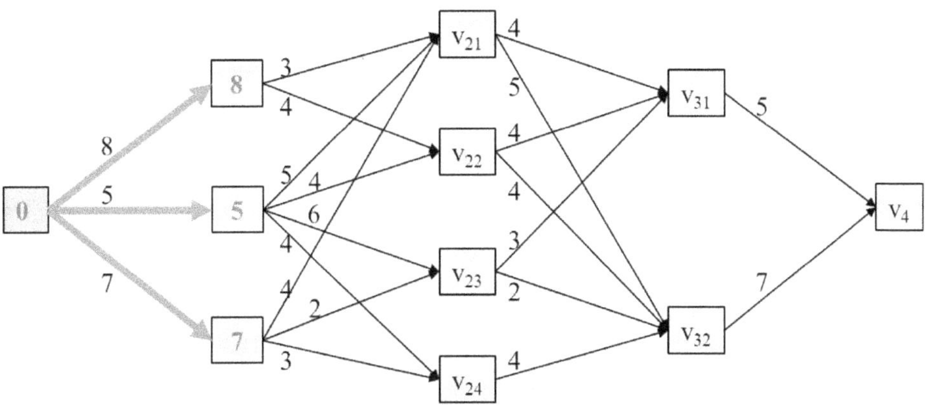

Treatment of level 2

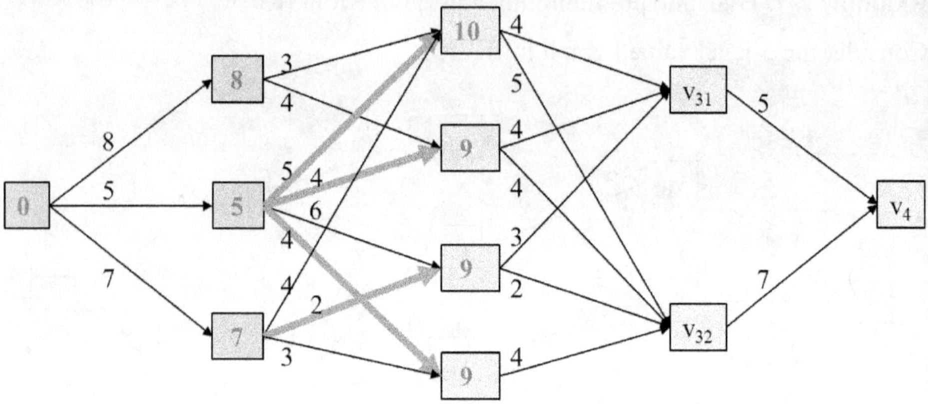

Treatment of level 3

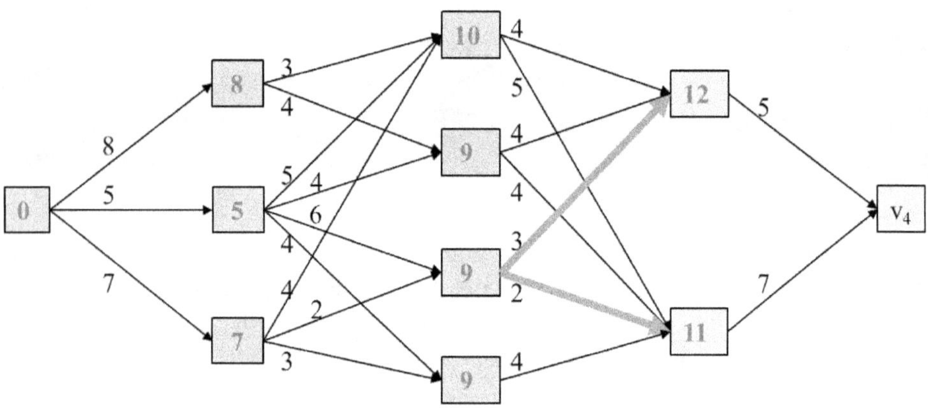

Treatment of level 4

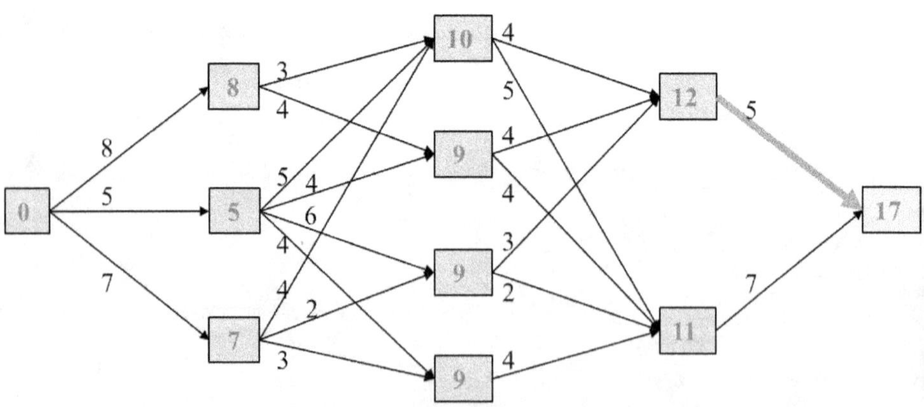

Figure 2-8 shows the optimal arcs in thick lines. The optimal paths from the source to each node are listed below.

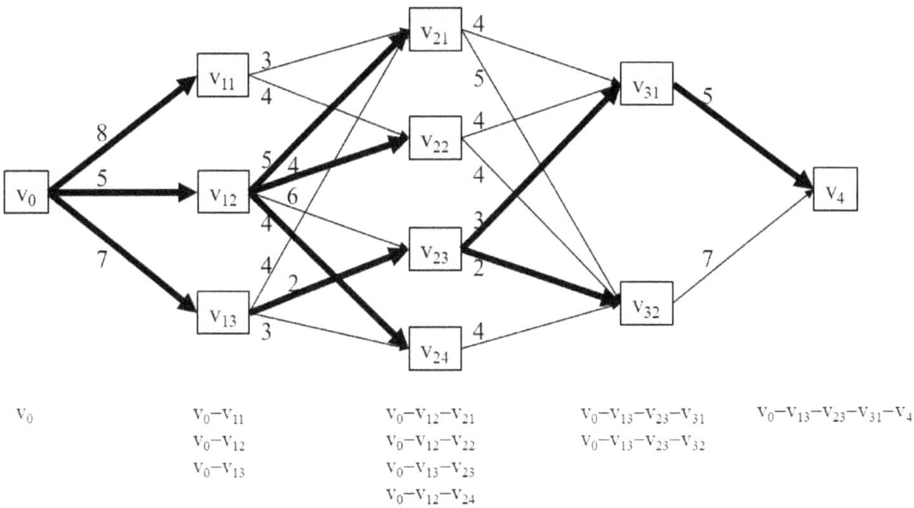

Figure 2-8: Optimal forward paths.

Let us apply the backward dynamic programming algorithm to the same graph to find the minimum cost paths from each node to the sink. The levels are processed in descending order. The optimal arcs selected at each level are in thick lines.

<u>Treatment of level 3</u>

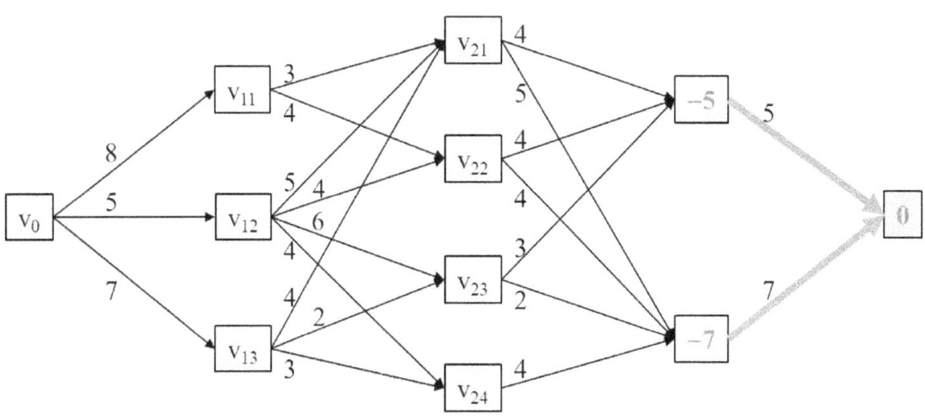

Treatment of level 2

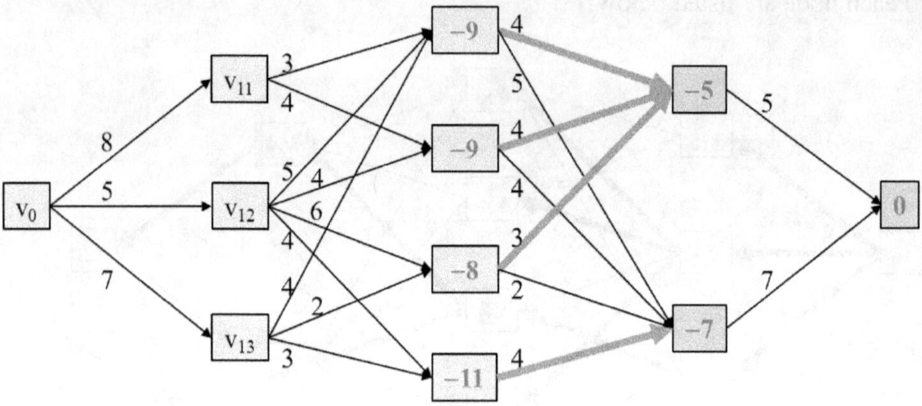

Treatment of level 1

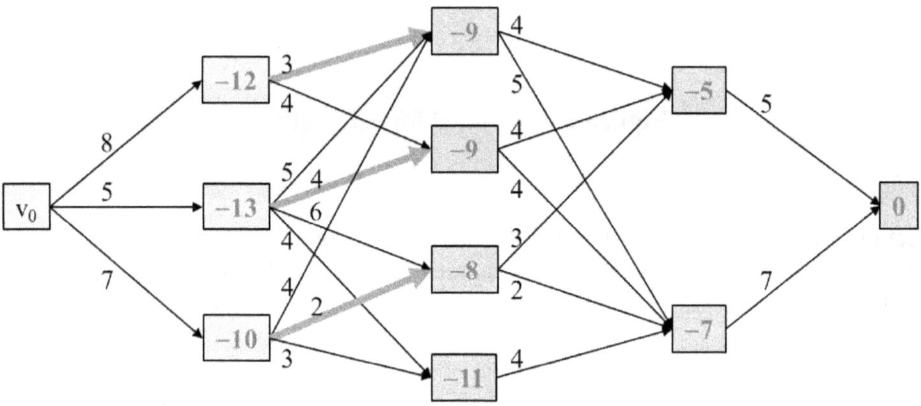

Treatment of level 0

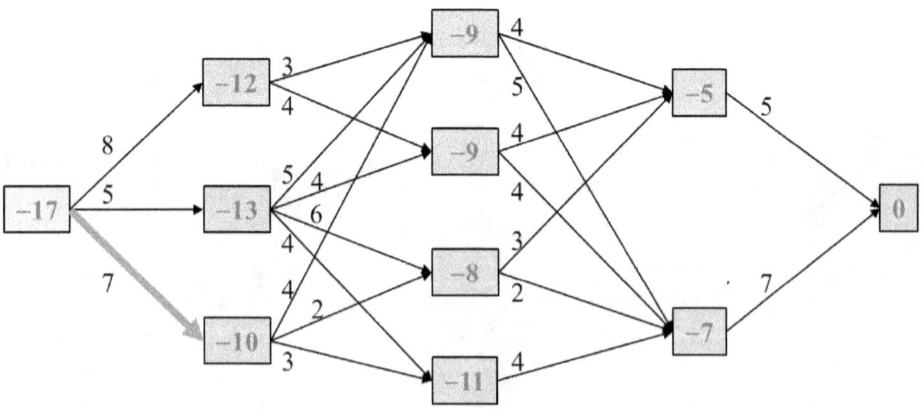

Figure 2-9 shows the optimal arcs in thick lines. The optimal paths from each node to the sink are listed below.

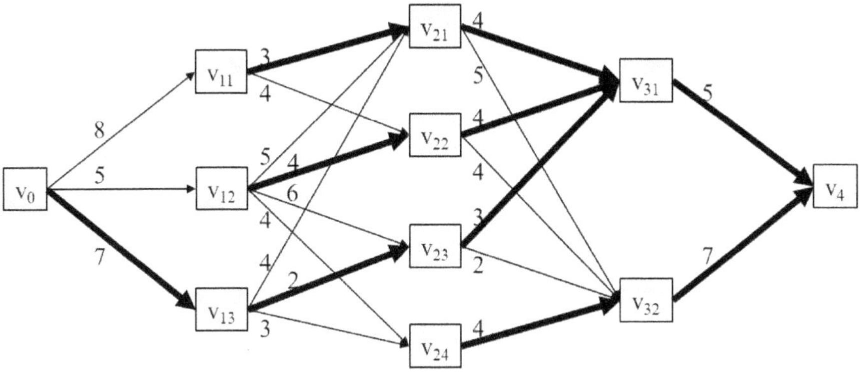

$v_0-v_{13}-v_{23}-v_{31}-v_4$	$v_{11}-v_{21}-v_{31}-v_4$	$v_{21}-v_{31}-v_4$	$v_{31}-v_4$	v_4
	$v_{12}-v_{22}-v_{31}-v_4$	$v_{22}-v_{31}-v_4$	$v_{32}-v_4$	
	$v_{13}-v_{23}-v_{31}-v_4$	$v_{23}-v_{31}-v_4$		
		$v_{24}-v_{32}-v_4$		

Figure 2-9: Optimal backward paths.

2.2.3 Dijkstra's algorithm

Dijkstra's algorithm applies in a graph with positive valuations. It is identical to Ford's algorithm in that it treats the nodes by increasing value. Therefore, each node has to be processed only once.

Initialization
A value $z_s = 0$ is assigned to the source and values $z_i = \infty$ to the other nodes. The nodes are initially unprocessed.

Treatment of a node
From the unprocessed nodes, we choose the node v_i with the minimum value z_i. The values of the nodes v_j successors of node v_i are updated.

If $z_i + c_{ij} < z_j$,
passing through v_i improves the solution arriving at v_j.
The value of v_j is updated: $z'_j = z_i + c_{ij}$.
The node v_j is processed.

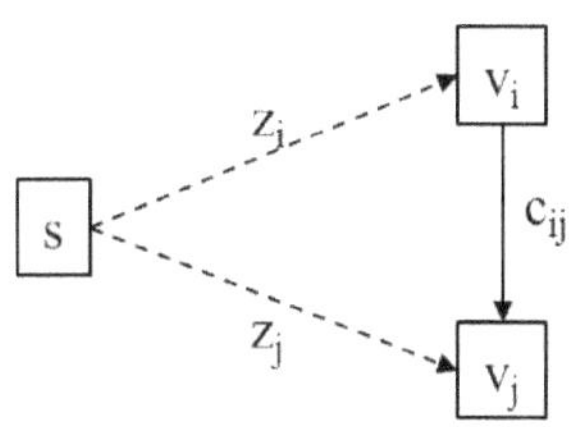

<u>End of the algorithm</u>

We stop when the n nodes have been processed.
Each node is then marked with the value of the best path from the source.

Example 2-8 illustrates how Dijkstra's algorithm works.

Example 2-8: Dijkstra's algorithm

Consider the 5-node valued graph shown opposite.

The source is at v_1.
The valuations are positive.

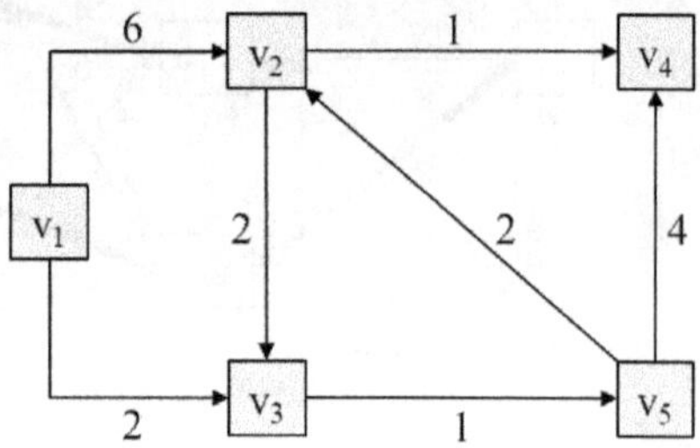

Let us apply Dijkstra's algorithm to find the minimum cost paths from the source. The arcs used in each iteration are in thick lines.

<u>Initialization</u>

The source v_1 has the value: $z_i = 0$.

The other nodes have the value: $z_i = \infty$.

The nodes are unprocessed.

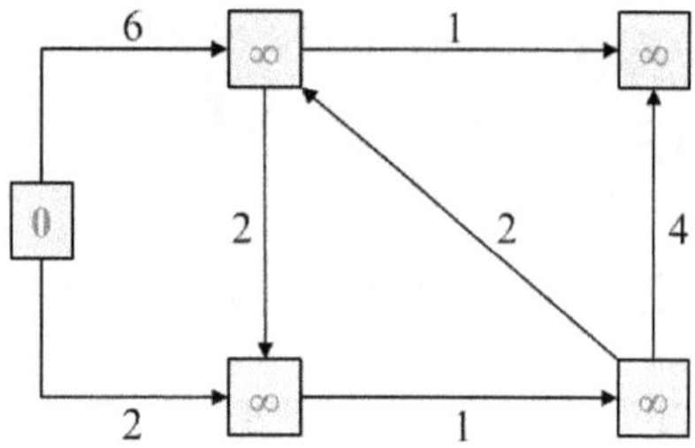

<u>Treatment of the first node</u>

The unprocessed nodes are: v_1, v_2, v_3, v_4, v_5.
The minimum value node is v_1.

The successors of v_1 are updated.
The node v_1 is processed.

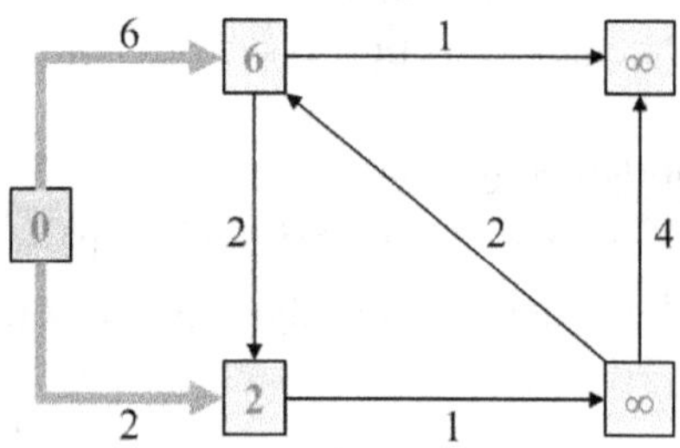

<u>Treatment of the second node</u>

The unprocessed nodes are: v_2, v_3, v_4, v_5.
The minimum value node is v_3.

The successors of v_3 are updated.
The node v_3 is processed.

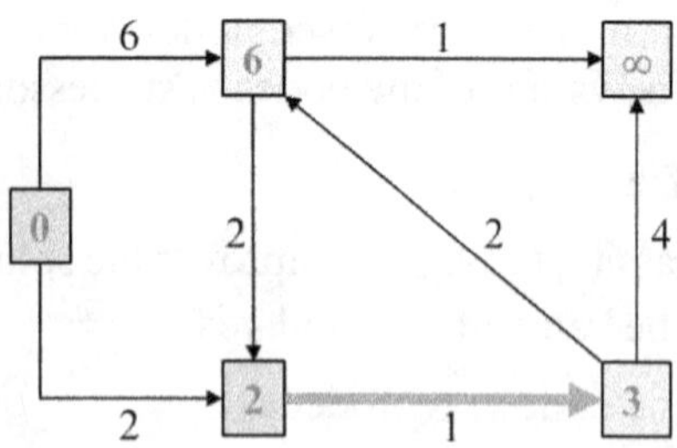

<u>Treatment of the third node</u>

The unprocessed nodes are: v_2,v_4,v_5.
The minimum value node is v_5.

The successors of v_5 are updated.
The node v_5 is processed.

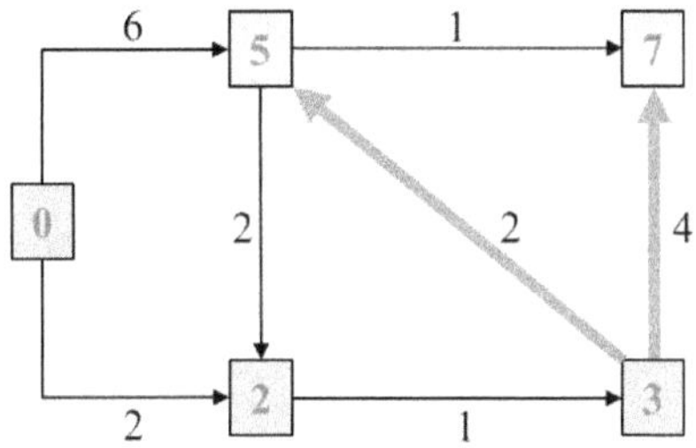

<u>Treatment of the fourth node</u>

The unprocessed nodes are: v_2,v_4.
The minimum value node is v_2.

The successors of v_2 are updated.
The node v_2 is processed.

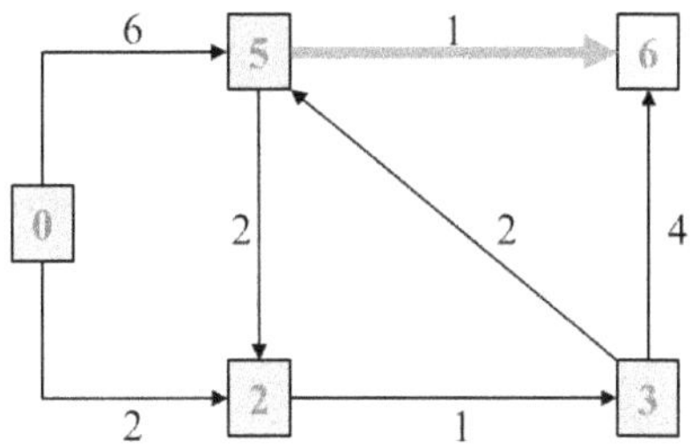

<u>Treatment of the fifth node</u>

The unprocessed nodes are: v_4.
The minimum value node is v_4.

This node has no successor.
The node v_4 is processed.

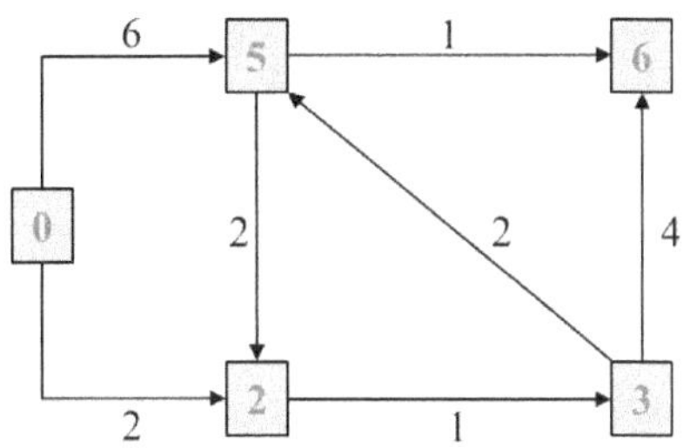

<u>End of the algorithm</u>

All nodes have been processed.
The value (or mark) of each node is the cost of the best path from the source to that node.
The optimal arcs in thick lines are those whose cost equals the difference in marks between the start and end nodes.

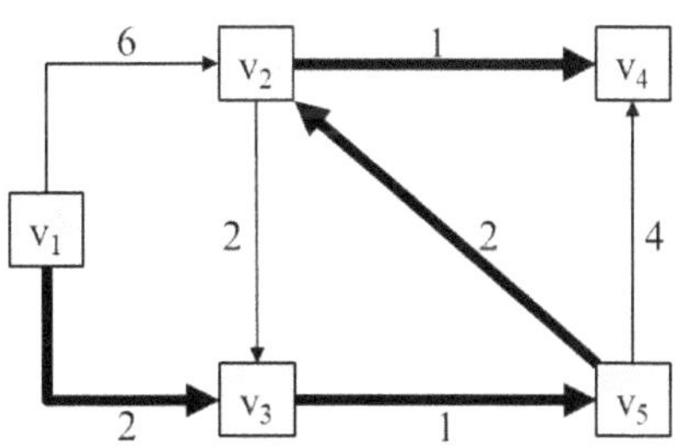

The minimum costs are summarized in the table opposite.

v_1-v_2	v_1-v_3	v_1-v_4	v_1-v_5
5	2	6	3

Dijkstra's algorithm is faster than Ford's algorithm, because each node is processed only once. Its complexity is in $O(m.\log(n))$ where n and m are the number of nodes and arcs, respectively. This polynomial algorithm is very efficient on sparse graphs $(m \ll n^2)$, such as planar graphs for which $m \le 3n-6$.

It is especially used to find the shortest path in a road network.

2.2.4 A* algorithm

Dijkstra's algorithm processes the minimum value node at each stage without accounting for the remaining path (which is unknown).

The A* algorithm modifies Dijkstra's algorithm by using an estimate of the remaining cost in order to choose which node to process. This estimate is given by a function h called **heuristic**. The heuristic h associates to each node v_i an arbitrary estimate h_i of the remaining cost between the node v_i and the sink t.

The valuation of node v_i is the sum of the cost z_{si} (known) between s and v_i and the cost h_i (estimated) between v_i and t. This valuation is noted: $e_i = z_{si} + h_i$.

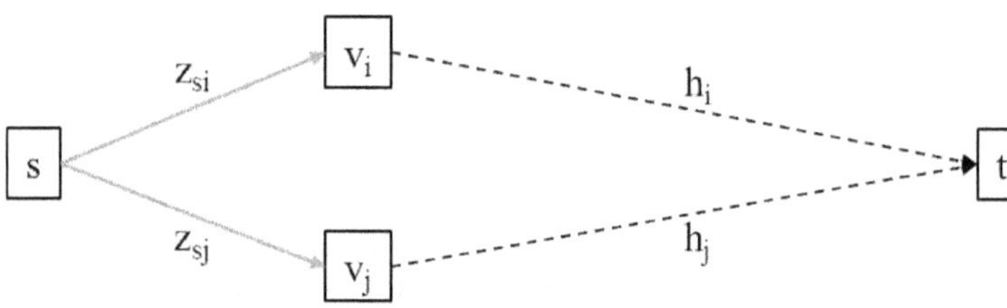

Figure 2-10: Valuation of a node with heuristic.

At each stage the A* algorithm processes the minimum valuation node, which is the most promising. Unlike Dijkstra's algorithm, each node can be processed several times. The mark z_{si} of the node v_i is the value of the best cost found between the source s and the node.

The stages of the algorithm are as follows.

Initialization
A value $z_s = 0$ is assigned to the source and values $z_{si} = \infty$ to the other nodes.
All nodes are initially closed, except for the source s.

Treatment of a node
A node is **open** when its cost is changed from a predecessor.
A node is **closed** when the costs of all its successors have been updated.

Among the open nodes we choose the node v_i of minimum valuation e_i.
The values of the nodes v_j successor of node v_i are updated.

If $z_{si} + c_{ij} < z_{sj}$,
passing through v_i improves the solution arriving at v_j
The value of v_j is updated: $z'_{sj} = z_{si} + c_{ij}$
and the node v_j is open.

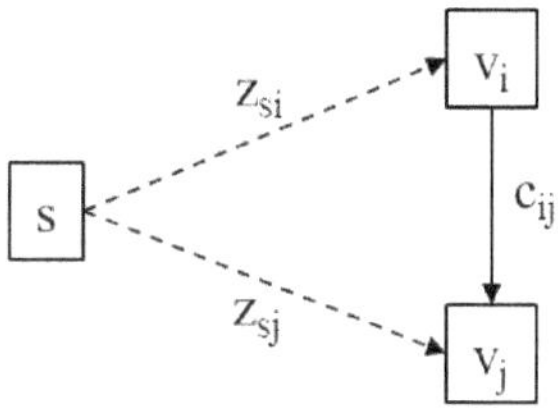

The node v_i is closed when all its successors have been processed.

End of the algorithm
We stop when the best open node is the sink t.
Each node is then marked with the value of the best path from the source.

The number of nodes processed and the path obtained depend on the heuristic chosen. The two extreme cases are the null heuristic and the perfect heuristic:

- the null heuristic $(h_i = 0)$ does not use an estimate of the remaining path. The chosen node is then the minimum cost node. This amounts to Dijkstra's algorithm and all nodes are processed once;
- the perfect heuristic $(h_i = z_{it})$ assumes that the cost of the remaining optimal path is known. The chosen node is then on the optimal path. The optimal path is obtained directly, because the evaluation gives the exact solution.

A poorly chosen heuristic may lead to more nodes being processed than Dijkstra's algorithm, or it may even result in a non-optimal path.

Before examining the properties of the heuristic, let us consider a first example of application of the A* algorithm where the heuristic is defined by the number of remaining arcs from the node to the sink.

Example 2-9: A* algorithm

Consider the 16-node graph in figure 2-11.

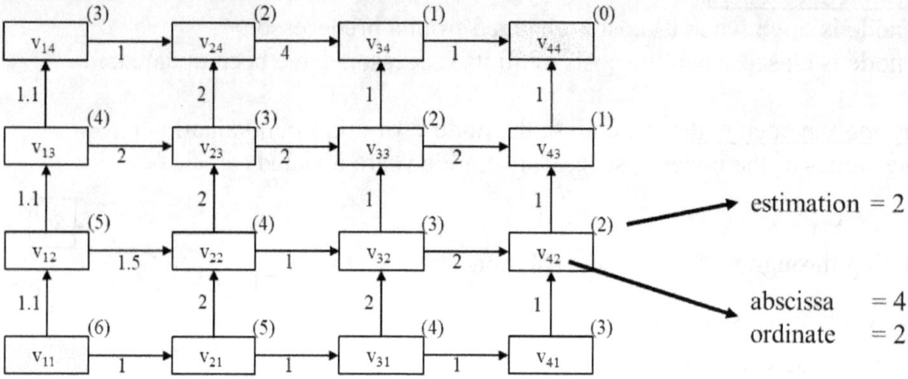

Figure 2-11: Heuristic with estimation by the number of remaining arcs.

The nodes are indicated by their abscissa and ordinate.
The node v_{ij} has coordinates (i ; j).
The start node (source) is v_{11}. The end node (sink) is v_{44}.
The valuations of the arcs are positive and indicated on the arrows.

The heuristic giving the remaining cost is the number of remaining arcs to v_{44}:
- for the starting node v_{11}, there are 6 arcs to go;
- for the node v_{43}, there is 1 arc left to go.

This estimate is indicated in brackets at the top right of each node.

Let us apply the A* algorithm to find the minimum cost paths from the source to each node. At each iteration, we indicate:
- the list of open nodes with their estimation;
- the node to be treated (having the lowest estimate);
- the best path found at this iteration;
- the total number of nodes evaluated since the beginning.

Nodes that have already been assessed are shown in dark grey, open nodes in light grey and the node to be treated is framed in thick lines.

<u>Iteration 1</u>

Open nodes:
v_{12} (6.1), v_{21} (6)
Node to be treated:
v_{21} (6)

Best path:
z = 1
$v_{11} - v_{21}$

Updated nodes: 3

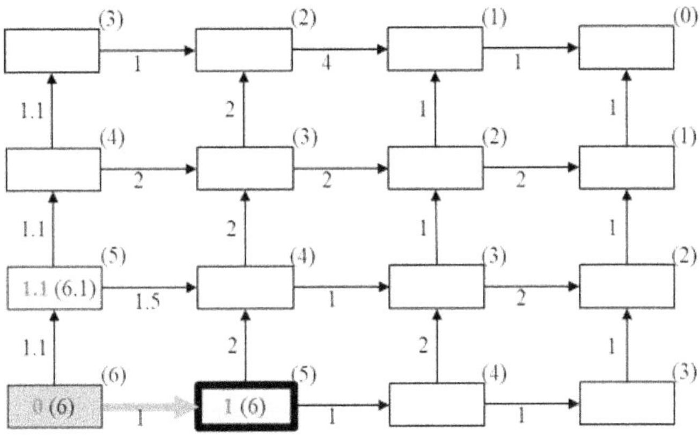

<u>Iteration 2</u>

Open nodes:
v_{12} (6.1), v_{22} (7),
v_{31} (6)
Node to be treated:
v_{31} (6)

Best path:
z = 2
$v_{11} - v_{21} - v_{31}$

Updated nodes: 5

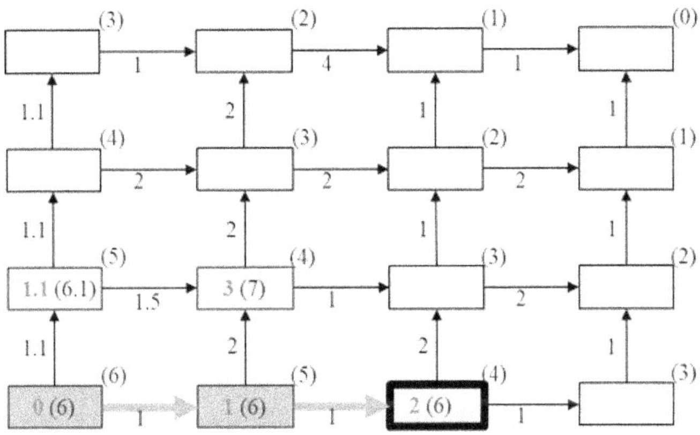

<u>Iteration 3</u>

Open nodes:
v_{12} (6.1), v_{22} (7),
v_{32} (7), v_{41} (6)
Node to be treated:
v_{41} (6)

Best path:
z = 3
$v_{11} - v_{21} - v_{31} - v_{41}$

Updated nodes: 7

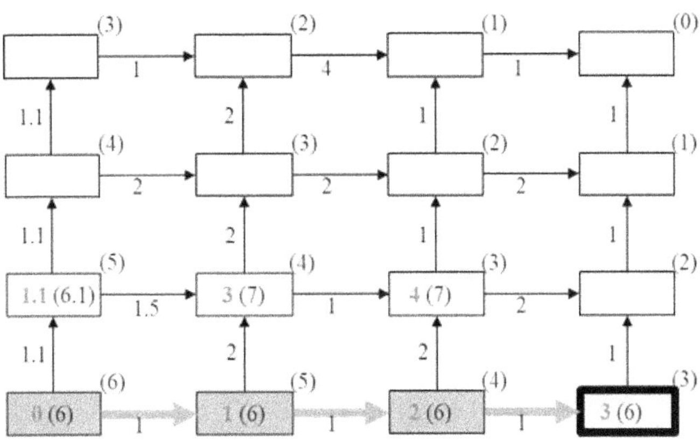

Iteration 4

Open nodes:
v_{12} (6.1), v_{22} (7),
v_{32} (7), v_{42} (6)
Node to be treated:
v_{42} (6)

Best path:
z = 4
$v_{11} - v_{21} - v_{31} - v_{41} - v_{42}$
Updated nodes: 8

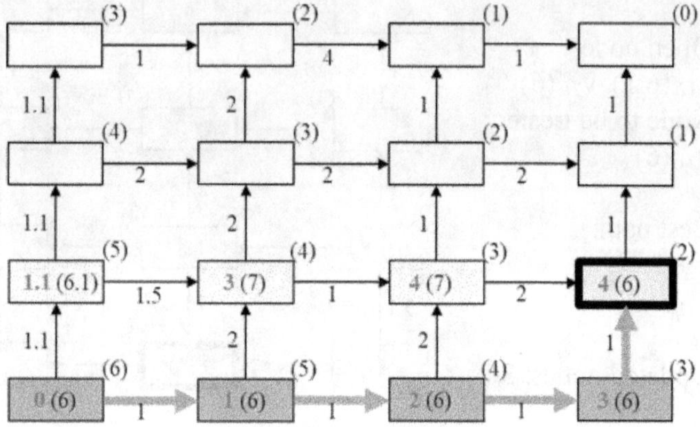

Iteration 5

Open nodes:
v_{12} (6.1), v_{22} (7),
v_{32} (7), v_{43} (6)
Node to be treated:
v_{43} (6)

Best path:
z = 5
$v_{11} - v_{21} - v_{31} - v_{41} - v_{42} - v_{43}$
Updated nodes: 9

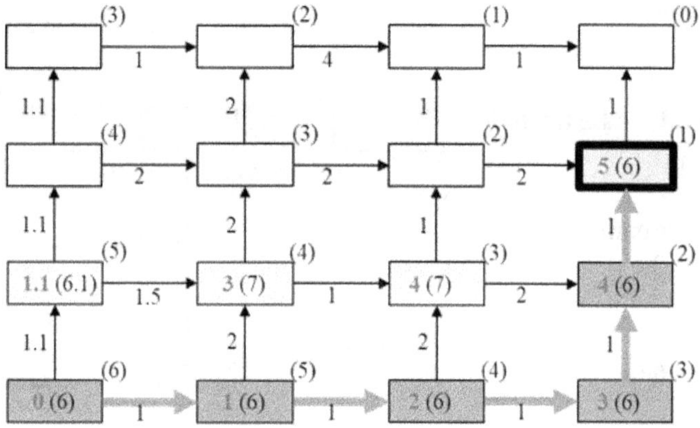

Iteration 6

Open nodes:
v_{12} (6.1), v_{22} (7),
v_{32} (7), v_{44} (6)
Node to be treated:
v_{44} (6)

Best path:
z = 6
$v_{11} - v_{21} - v_{31} - v_{41} - v_{42} - v_{43} - v_{44}$
Updated nodes: 10

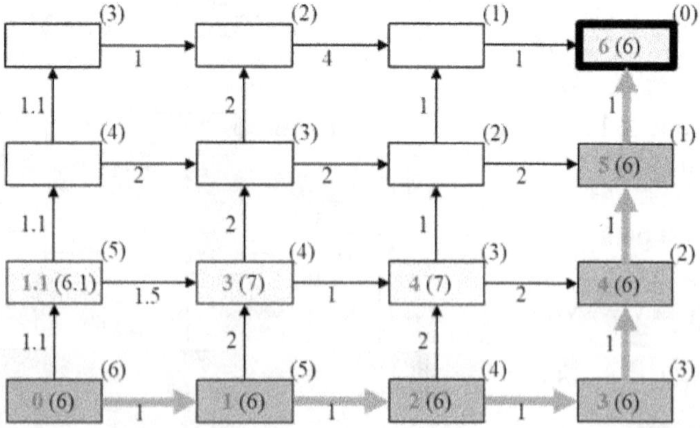

The algorithm is finished, as the node to be processed is the sink v_{44}.
The path obtained by A* algorithm is: $v_{11} - v_{21} - v_{31} - v_{41} - v_{42} - v_{43} - v_{44}$ of cost: z = 6.
The result is obtained in 6 iterations (= 6 processed nodes) and by updating a total of 10 nodes (nodes in grey that received an evaluation during the algorithm).

Let us compare this result to the one obtained by Dijkstra's algorithm, which processes each node only once by choosing the node of minimum evaluation. This is equivalent to applying the A* algorithm with the null heuristic as shown in figure 2-12 by the zero in brackets on each node. The nodes are then processed in the following order (with their evaluation in brackets).

v_{11} (0)
→ v_{21} (1) → v_{12} (1.1)
→ v_{31} (2) → v_{13} (2.2)
→ v_{22} (2.6) → v_{41} (3)
→ v_{14} (3.3) → v_{32} (3.6)
→ v_{42} (4) → v_{23} (4.2)
→ v_{24} (4.3) → v_{33} (4.6)
→ v_{43} (5) → v_{44} (6)

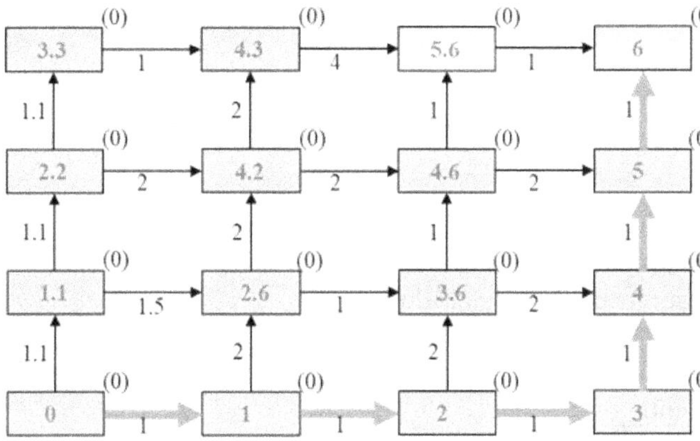

Figure 2-12: Optimal path with null heuristic (Dijkstra).

Dijkstra's algorithm gives the optimal path by processing 15 nodes and updating a total of 23 nodes.
A* algorithm gives the optimal path by processing only 6 nodes and updating a total of 10 nodes.

The performance of A* algorithm is much better in this example, but this is not always the case.
Let us take the graph from figure 2-11 and change the cost of arc v_{43}-v_{44}, which passes from 1 to 4.
The iterations of the A* algorithm remain the same until iteration 5, which gives the figure 2-13.

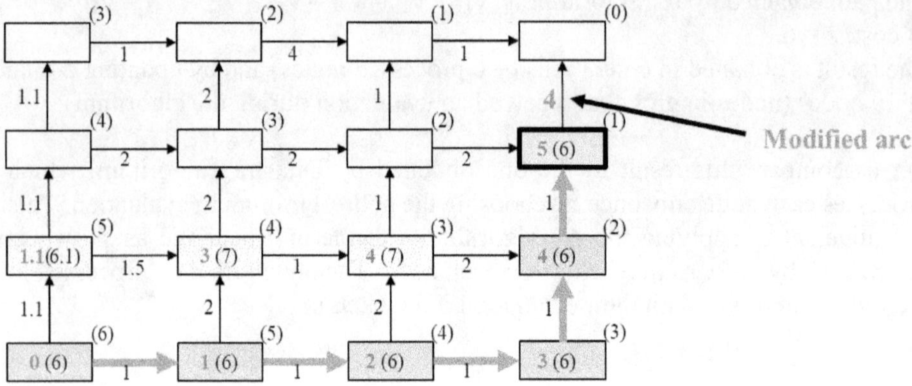

Figure 2-13: Change in the cost of arc v_{43}-v_{44}.

With the new cost of the arc v_{43}-v_{44}, the sink takes the value 9 and is no longer the minimum valuation node. In that case, A* algorithm is not finished and the iterations continue as follows.

Iteration 6

Open nodes:
v_{12} (6.1), v_{22} (7),
v_{32} (7), v_{44} (9)
Node to be treated:
v_{21} (6.1)

Best path:
z = 1.1
$v_{11} - v_{12}$

Updated nodes: 10

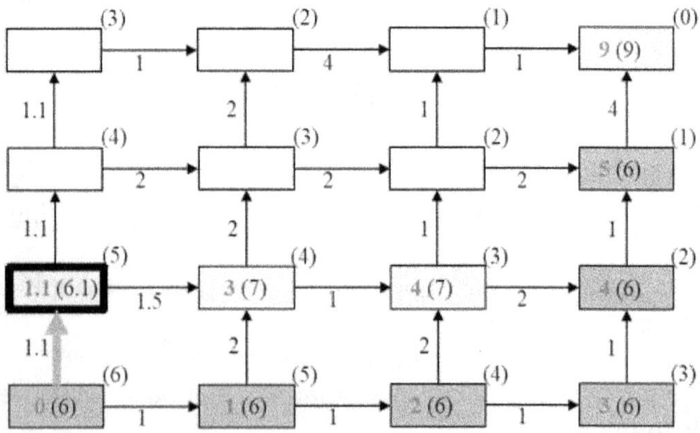

<u>Iteration 7</u>
Open nodes:
v_{13} (6.2), v_{22} (6.6),
v_{32} (7), v_{44} (9)
Node to be treated:
v_{13} (6.2)

Best path:
z = 2.2
$v_{11} - v_{12} - v_{13}$
Updated nodes: 12

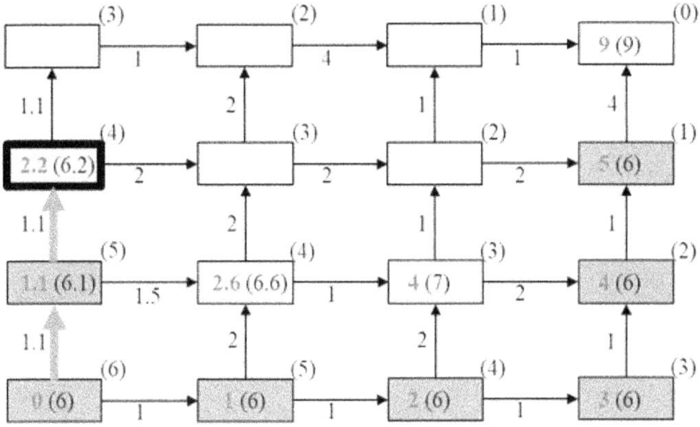

<u>Iteration 8</u>
Open nodes:
v_{14} (6.3), v_{22} (6.6),
v_{23} (7.2),
v_{32} (7), v_{44} (9)
Node to be treated:
v_{14} (6.3)

Best path:
z = 3.3
$v_{11} - v_{12} - v_{13} - v_{14}$
Updated nodes: 14

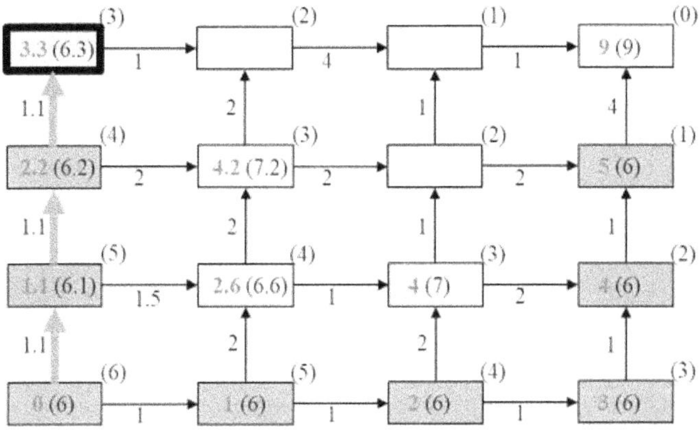

<u>Iteration 9</u>
Open nodes:
v_{22} (6.6), v_{23} (7.2),
v_{24} (6.3),
v_{32} (7), v_{44} (9)
Node to be treated:
v_{24} (6.3)

Best path:
z = 4.3
$v_{11} - v_{12} - v_{13} - v_{14} -$
v_{24}
Updated nodes: 15

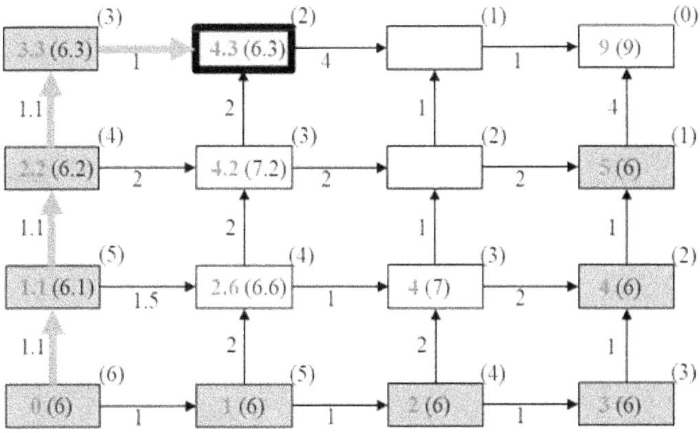

<u>Iteration 10</u>
Open nodes:
v_{22} (6.6), v_{23} (7.2),
v_{32} (7),
v_{34} (9.3), v_{44} (9)
Node to be treated:
v_{22} (6.6)

Best path:
z = 2.6
$v_{11} - v_{12} - v_{22}$
Updated nodes: 16

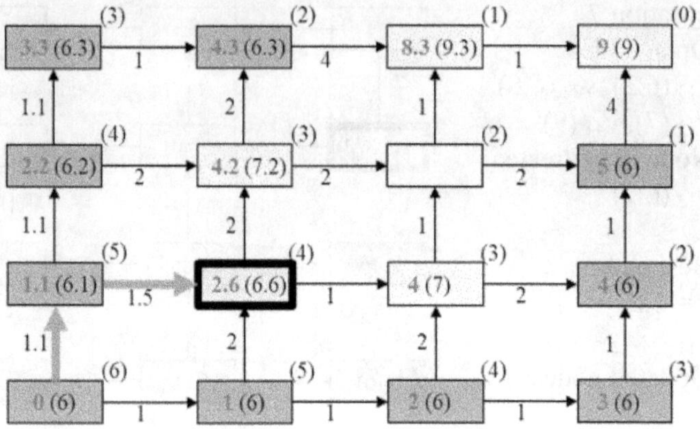

<u>Iteration 11</u>
Open nodes:
v_{23} (7.2), v_{32} (6.6),
v_{34} (9.3), v_{44} (9)
Node to be treated:
v_{32} (6.6)

Best path:
z = 3.6
$v_{11} - v_{12} - v_{22} - v_{32}$
Updated nodes: 18

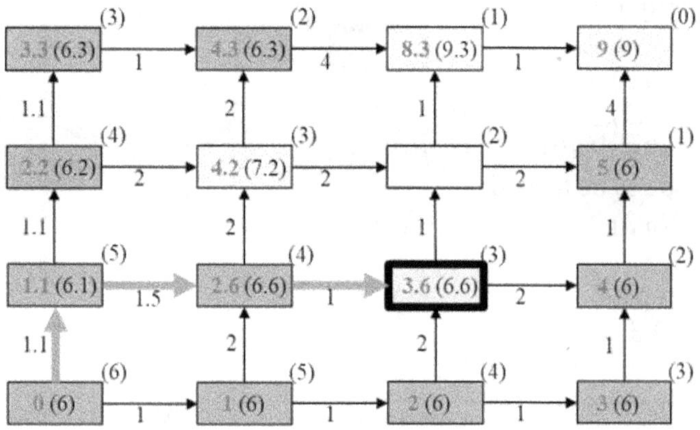

<u>Iteration 12</u>
Open nodes:
v_{23} (7.2), v_{33} (6.6),
v_{34} (9.3), v_{44} (9)
Node to be treated:
v_{33} (6.6)

Best path:
z = 4.6
$v_{11} - v_{12} - v_{22} - v_{32} - v_{33}$
Updated nodes: 20

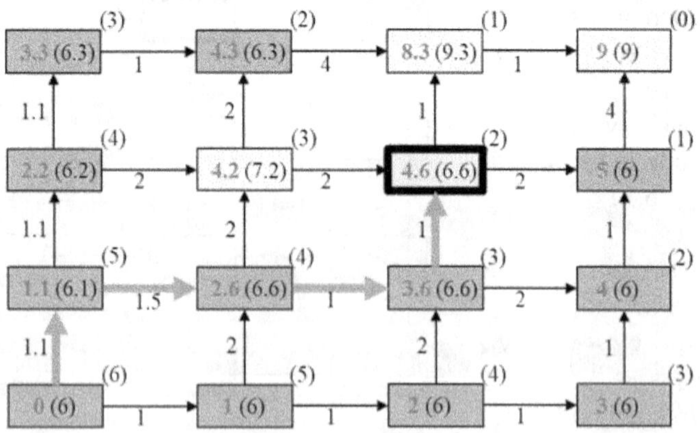

<u>Iteration 13</u>
Open nodes:
v_{23} (7.2), v_{34} (6.6),
v_{44} (9)
Node to be treated:
v_{34} (6.6)

Best path:
$z = 5.6$
$v_{11} - v_{12} - v_{22} - v_{32} - v_{33} - v_{34}$
Updated nodes: 22

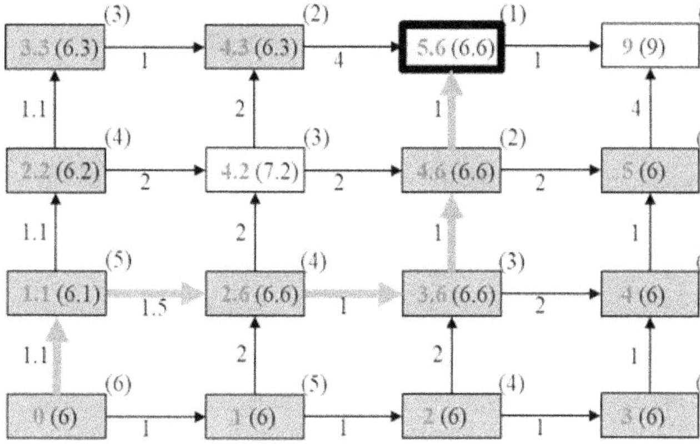

<u>Iteration 14</u>
Open nodes:
v_{23} (7.2), v_{44} (6.6)
Node to be treated:
v_{44} (6.6)

Best path:
$z = 6.6$
$v_{11} - v_{12} - v_{22} - v_{32} - v_{33} - v_{34} - v_{44}$
Updated nodes: 23

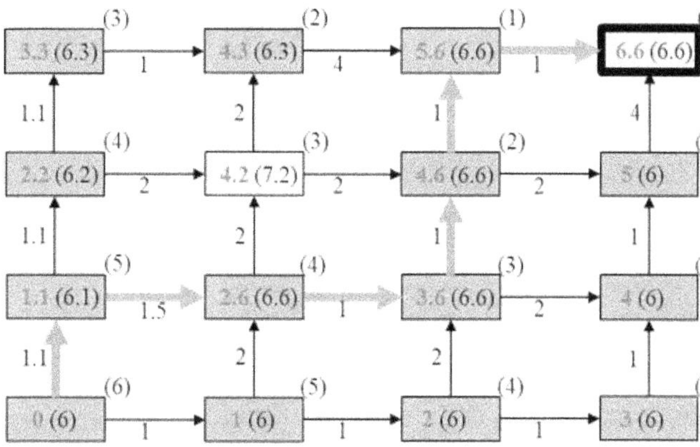

The algorithm is finished, as the node to be processed is the sink v_{44}.
The resulting path is: $v_{11} - v_{12} - v_{22} - v_{32} - v_{33} - v_{34} - v_{44}$ of cost: $z = 6.6$.
In this modified example (figure 2-13), the performance (14 nodes processed and 23 nodes updated) is strictly identical to that of Dijkstra's algorithm.

The behavior of the A* algorithm depends on the properties of the heuristic.

Let us go back to the notations of Figure 2-10 reproduced opposite.

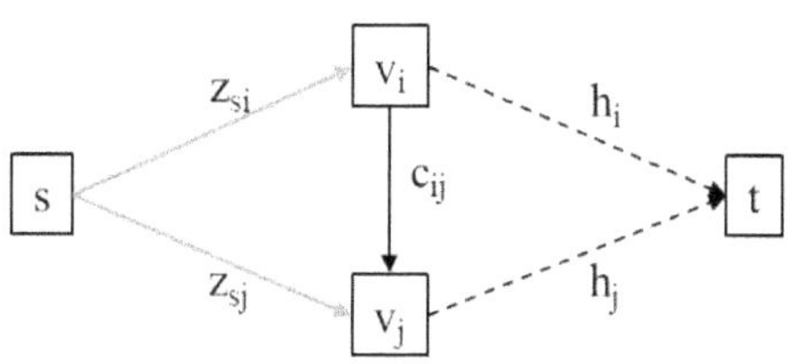

To analyze the properties of the heuristic, we construct the **revalued graph** in which the valuations of the arcs are replaced by

$$c'_{ij} = c_{ij} - (h_i - h_j) \tag{2.4}$$

Assume that these valuations are positive (assumption of consistency discussed below) and consider a node v_j at the current iteration. The best known path from the source s to node v_j is: $(s - v_{i1} - v_{i2} - \cdots - v_{ik} - v_j)$ and it has a cost z_{sj}.

The valuation of the node v_j with the heuristic h is given by

$$e_j = z_{sj} + h_j \tag{2.5}$$

Expressing the sum of the arc costs from s to v_j, and using (2.4), we obtain

$$\begin{aligned} e_j &= z_s + c_{s-i1} + c_{i1-i2} + \cdots + c_{ik-j} + h_j \\ &= z_s + c'_{s-i1} + c'_{i1-i2} + \cdots + c'_{ik-j} \\ &\quad + (h_s - h_{i1}) + (h_{i1} - h_{i2}) + \cdots + (h_{ik} - h_j) + h_j \\ &= z_s + c'_{s-i1} + c'_{i1-i2} + \cdots + c'_{ik-j} + h_s \\ &= z'_{sj} + h_s \end{aligned} \tag{2.6}$$

denoting z'_{sj} the cost of the path from s to v_j in the revalued graph.

Formula (2.6) shows that the node of best valuation e_j is also the node of best cost z'_{sj} in the revalued graph (within the constant h_s).

A* algorithm with heuristic h is thus equivalent to Dijkstra's algorithm in the revalued graph. Since A* algorithm stops when the best node is the sink, it examines at most the n nodes of the graph and gives the solution faster than Dijkstra's algorithm.

This assumes that the heuristic h satisfies certain conditions discussed below.

Let us first illustrate in example 2-9 the equivalence with Dijkstra's algorithm.

Example 2-10: Equivalence between A* and Dijkstra's algorithm

Let us retrieve the example 2-9 with the heuristic defined by the number of remaining arcs.
The solution obtained in iteration 6 is shown below.

Solution A*

Open nodes:
v_{12} (6.1), v_{22} (7),
v_{32} (7), v_{44} (6)

Node to be treated:
v_{44} (6)

Best path:
$z = 6$
$v_{11} - v_{21} - v_{31} - v_{41} - v_{42} - v_{43} - v_{44}$

Updated nodes: 10

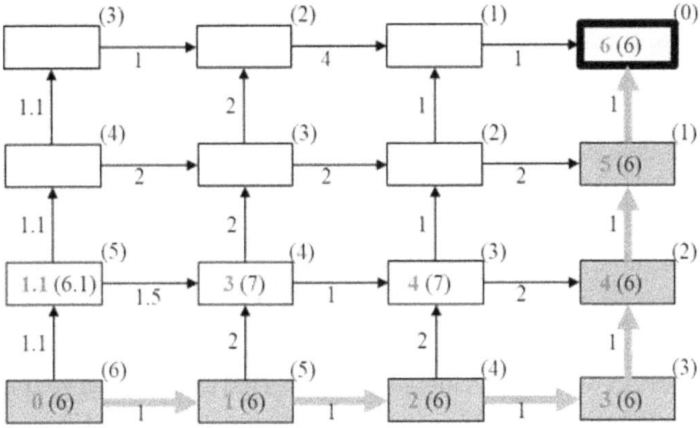

Figure 2-14 shows the revalued graph where the arc costs are changed to: $c'_{ij} = c_{ij} + (h_i - h_j)$. The heuristic being the number of remaining arcs, this amounts to decreasing the costs of the arcs by 1. The new costs are written to the right or above the arcs. Let us apply Dijkstra's algorithm to this revalued graph. This gives the solution in figure 2-14.

Dijkstra solution

Sequence of nodes processed:
v_{11} (0)
$\rightarrow v_{21}$ (0)
$\rightarrow v_{31}$ (0)
$\rightarrow v_{41}$ (0)
$\rightarrow v_{42}$ (0)
$\rightarrow v_{43}$ (0)
$\rightarrow v_{44}$ (0)

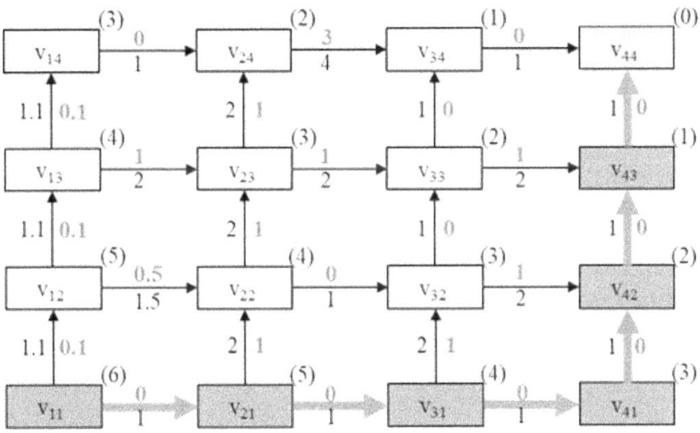

Figure 2-14: Dijkstra's algorithm in the revalued graph.

Dijkstra's algorithm in the revalued graph follows the same order of node processing as A* algorithm and it reaches the solution after 6 processed nodes. However, Dijkstra's algorithm continues to process the remaining 9 nodes, while A* algorithm stops.

The following properties guarantee that the heuristic h finds the optimal path, while performing fewer iterations than Dijkstra's algorithm.

Feasible heuristic

A heuristic is **feasible** if it underestimates the cost of the remaining path between each node and the sink: $h_i \leq z_{it}$.

A feasible heuristic ensures that the optimal path is found.

Let us assume that the optimal path has a cost z* and that a node v_i is located on the optimal path. Its valuation $e_i = z_{si} + h_i$ is lower than z*.

When all other paths have been traversed and show a cost greater than z*, the node v_i will be processed.

Conversely, a non-feasible heuristic may result in a non-optimal path as shown in the following example.

Example 2-11: Non-feasible heuristic

Consider the graph in figure 2-15a with the heuristic in parentheses on each node. This heuristic is not feasible, because it overestimates the path cost from v_2 to t:

$$\begin{cases} h_2 = 3 \\ z_{2t} = 1 \end{cases} \rightarrow h_2 > z_{2t}$$

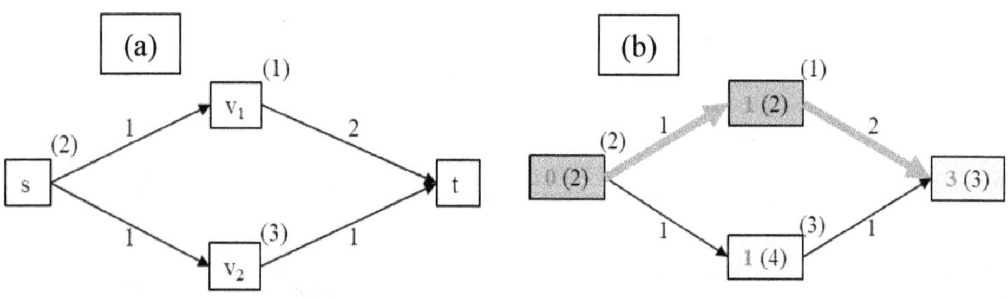

Figure 2-15: Non-feasible heuristic.

The optimal path is s – v_2 – t of cost $z^* = 2$.

A* algorithm results in the path: s – v_1 – t of cost $z = 3$ shown in figure 2-15b. The node v_2 remains open, as its valuation $e_2 = 4$ is higher than that of the sink $e_t = 3$. A* algorithm stops at a non-optimal solution.

Consistent heuristic

A heuristic is **consistent** if it underestimates the cost of arcs: $h_i \le h_j + c_{ij}$.

In this case, the valuations (2.4) are positive and A* algorithm is equivalent to Dijkstra's algorithm in the revalued graph. The optimal solution is then obtained by processing fewer (or as many) nodes as Dijkstra's algorithm.

A consistent heuristic is therefore feasible. It is also monotonous, which means that the evaluation decreases between a node and its successors: $e_i \le e_j$.

An inconsistent heuristic is detected by negative valuations (2.4). Such a heuristic can be feasible, but it may reopen nodes already processed as in example 2-12. In such cases, the A* algorithm performance may be strongly degraded.

Example 2-12: Inconsistent heuristic

Consider the graph in figure 2-16a with the heuristic in parentheses on each node. The heuristic is not consistent, as it overestimates the cost of the arc $v_2 - v_3$:

$$\begin{cases} h_2 = 2 \\ h_3 = 0,5 \quad \rightarrow h_2 > h_3 + c_{23} \\ c_{23} = 0 \end{cases}$$

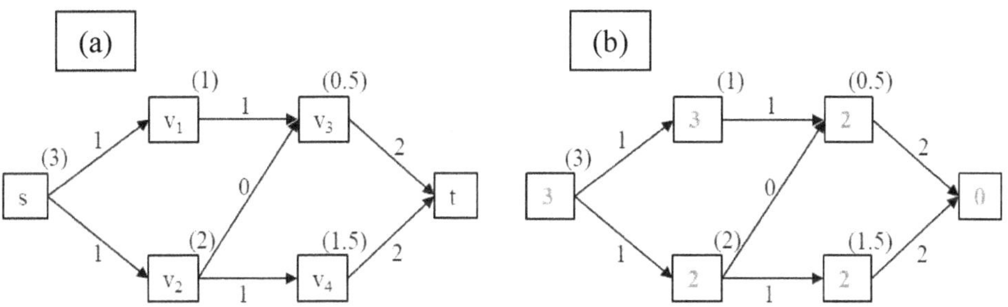

Figure 2-16: Inconsistent heuristic.

The heuristic is nevertheless feasible. Figure 2-16b shows for each node the optimal cost from that node to the sink. We can check that the heuristic underestimates each remaining cost.

$$\begin{cases} z_{4t} = 2 \ \ge \ h_4 = 1,5 \\ z_{3t} = 2 \ \ge \ h_3 = 0,5 \\ z_{2t} = 2 \ \ge \ h_2 = 2 \\ z_{1t} = 3 \ \ge \ h_1 = 1 \\ z_{st} = 3 \ \ge \ h_s = 3 \end{cases} \quad \rightarrow \text{feasible heuristic}$$

Let us apply A* algorithm with this feasible, but inconsistent, heuristic. The progress of the algorithm is shown in figure 2-17.

The table in figure 2-17 gives the iterations of the algorithm with the processed node in the first column. The node v_3 is opened a first time at iteration 3 corresponding to the top schema. This node v_3 is opened a second time at iteration 5 corresponding to the bottom schema.

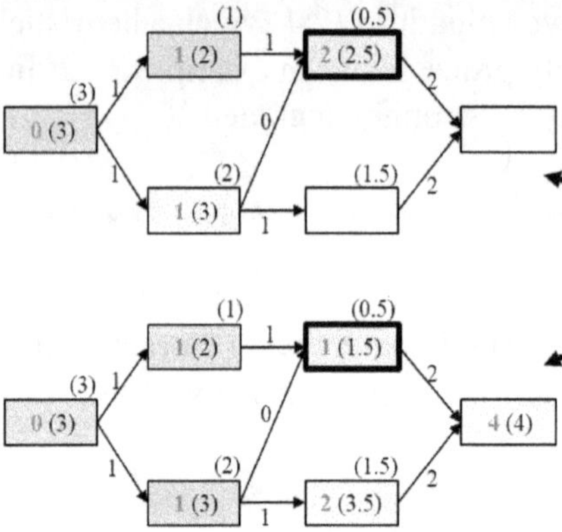

processed	open (evaluation)	closed
s	$v_1(2)$, $v_2(3)$	s , v_3 , v_4 , t
v_1	$v_2(3)$, $v_3(2.5)$	s , v_1 , v_4 , t
v_3	$v_2(3)$, t(4)	s , v_1 , v_3 , v_4
v_2	$v_3(1.5)$, $v_4(3.5)$, t(4)	s , v_1 , v_2
v_3	$v_4(3.5)$, t(3)	s , v_1 , v_2 , v_3

Optimal solution : $s - v_2 - v_3 - t$: $C = 3$

Figure 2-17: A* algorithm with inconsistent feasible heuristic.

The algorithm terminates at iteration 5, because the node to be processed is the sink of valuation 3. The optimal solution is well found, as the heuristic is feasible. The node v_3 has been processed twice, because the heuristic is not consistent (on the arc $v_2 - v_3$).

2.2.5 Democron and Floyd's algorithm

Ford's algorithm gives the minimum cost between the source and each node.

The algorithm of Demoucron and Floyd is an extension to obtain the minimum cost between any two nodes. The matrix of minimum costs between all pairs of nodes is called the distancer. This matrix D of size n×n is obtained in n iterations in a similar way to Ford's algorithm.

Initialization
The elements of the matrix D are filled with the value:
- $d_{ij} = c_{ij}$ if the nodes v_i and v_j are connected by an arc of cost c_{ij};
- $d_{ij} = \infty$ if the nodes v_i and v_j are not connected.

Iteration k $(1 \le k \le n)$
The matrix D gives the cost of the best path between each pair of nodes.
Iteration k consists of examining for each pair of nodes (v_i, v_j) the passage through the intermediate node v_k.

The best path from v_i to v_j has a cost d_{ij}.
Passing through v_k would have a cost $d_{ik} + d_{kj}$.

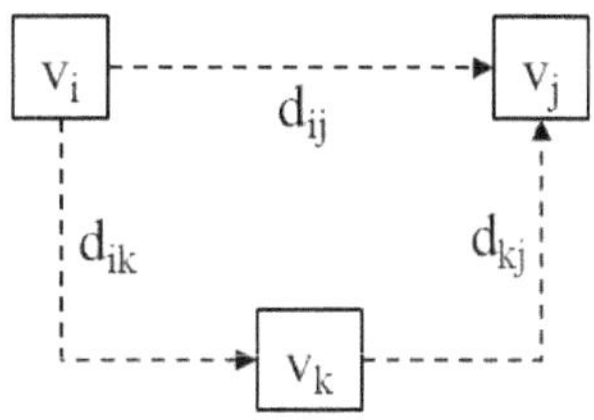

If $d_{ik} + d_{kj} < d_{ij}$,
passing through v_k improves the solution between v_i and v_j.
The value of d_{ij} is updated: $d'_{ij} = d_{ik} + d_{kj}$.

Each element of the matrix D is updated in this way during the iteration.

End of the algorithm
We stop when the n nodes have been considered as intermediate nodes.
The matrix D then contains the costs of the best paths between pairs of nodes.

Like Ford's algorithm, Demoucron and Floyd's algorithm is applicable in a graph with any valuations and assumes that the graph does not contain an absorbing circuit. The presence of an absorbing circuit is detected by the appearance of a negative element on the diagonal of D: $d_{ii} < 0$.

The algorithm consists of three nested loops:

From k=1 to n (iteration k = number of the intermediate node)
From i=1 to n (i = number of the starting node)
From j=1 to n (j = number of the arrival node)
Replace d_{ij} with $\min(d_{ij}, d_{ik} + d_{kj})$

Example 2-13 illustrates how Demoucron and Floyd's algorithm works.

Example 2-13: Demoucron and Floyd's algorithm (from [R6])

Consider the 6-node valued graph shown opposite.

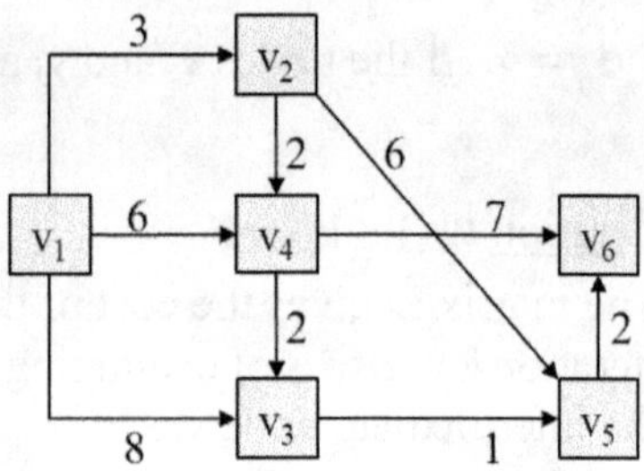

Let us apply the Demoucron and Floyd's algorithm to calculate the distancer.

Initialization

The matrix D is filled with:

$d_{ij} = c_{ij}$ if nodes v_i and v_j are connected.

$d_{ij} = \infty$ otherwise (shaded boxes).

	v_1	v_2	v_3	v_4	v_5	v_6
v_1		3	8	6		
v_2				2	6	
v_3					1	
v_4			2			7
v_5						2
v_6						

Iteration 1

We examine the passage through the node v_1.
The column v_1 is empty in the matrix.
The node v_1 does not receive any arcs.
v_1 cannot be an intermediate node. The matrix D is not changed.

Iteration 2

We examine the passage through the node v_2.
The column v_2 has a value from v_1.
We compare the costs d_{1j} and $d_{12} + d_{2j}$.

$d_{12} + d_{21} = \infty$

$d_{12} + d_{22} = \infty$

$d_{12} + d_{23} = \infty$

$d_{12} + d_{24} = 5 < 6 \quad \rightarrow$ improvement

$d_{12} + d_{25} = 9 < \infty \quad \rightarrow$ improvement

$d_{12} + d_{26} = \infty$

The matrix D is updated with the improved costs.

	v_1	v_2	v_3	v_4	v_5	v_6
v_1		3	8	**5**	**9**	
v_2				2	6	
v_3					1	
v_4			2			7
v_5						2
v_6						

Iteration 3

We examine the passage through the node v_3.
The column v_3 has values from v_1 and v_4.

We compare the costs d_{1j} and $d_{13} + d_{3j}$.

$d_{13} + d_{31} = \infty$, $d_{13} + d_{32} = \infty$
$d_{13} + d_{33} = \infty$, $d_{13} + d_{34} = \infty$
$d_{13} + d_{35} = 9 = d_{15}$ $\rightarrow$ identical
$d_{13} + d_{36} = \infty$

	v_1	v_2	v_3	v_4	v_5	v_6
v_1		3	8	5	9	
v_2				2	6	
v_3					1	
v_4			2		**3**	7
v_5						2
v_6						

We compare the costs d_{4j} and $d_{43} + d_{3j}$.

$d_{43} + d_{31} = \infty$, $d_{43} + d_{34} = \infty$
$d_{43} + d_{32} = \infty$, $d_{43} + d_{35} = 3 < \infty$ $\rightarrow$ improvement
$d_{43} + d_{33} = \infty$, $d_{43} + d_{36} = \infty$

The matrix D is updated with the improved costs.

For the following iterations, only the paths improved by the intermediate node are shown.

Iteration 4

We examine the passage through the node v_4.
The column v_4 has values from v_1 and v_2.

We compare the costs d_{1j} and $d_{14} + d_{4j}$.

$d_{14} + d_{43} = 7 < 8$ $\rightarrow$ improvement
$d_{14} + d_{45} = 8 < 9$ $\rightarrow$ improvement
$d_{14} + d_{46} = 12 < \infty$ $\rightarrow$ improvement

	v_1	v_2	v_3	v_4	v_5	v_6
v_1		3	**7**	5	**8**	**12**
v_2			**4**	2	**5**	**9**
v_3					1	
v_4			2		3	7
v_5						2
v_6						

We compare the costs d_{2j} and $d_{24} + d_{4j}$.

$d_{24} + d_{43} = 4 < \infty$ $\rightarrow$ improvement
$d_{24} + d_{45} = 5 < 6$ $\rightarrow$ improvement
$d_{24} + d_{46} = 9 < \infty$ $\rightarrow$ improvement

Iteration 5

We examine the passage through the node v_5.
The column v_5 has values from v_1, v_2, v_3, v_4.

We compare the costs d_{1j} and $d_{15}+d_{5j}$.

$d_{15}+d_{56}=10<12 \rightarrow$ improvement

We compare the costs d_{2j} and $d_{25}+d_{5j}$.

$d_{25}+d_{56}=7<9 \rightarrow$ improvement

We compare the costs d_{3j} and $d_{35}+d_{5j}$.

$d_{35}+d_{56}=3<\infty \rightarrow$ improvement

We compare the costs d_{4j} and $d_{45}+d_{5j}$.

$d_{45}+d_{56}=5<7 \rightarrow$ improvement

	v_1	v_2	v_3	v_4	v_5	v_6
v_1		3	7	5	8	**10**
v_2			4	2	5	**7**
v_3					1	**3**
v_4			2		3	**5**
v_5						2
v_6						

Iteration 6

We examine the passage through the node v_6. The row of v_6 is empty (no arc starts from v_6) and this node v_6 cannot be an intermediate node.

End of the algorithm

All 6 nodes were examined.
We obtain the distancer shown opposite.

	v_1	v_2	v_3	v_4	v_5	v_6
v_1		3	7	5	8	10
v_2			4	2	5	7
v_3					1	3
v_4			2		3	5
v_5						2
v_6						

2.3 Scheduling problem

Scheduling consists in planning a process made up of different tasks, while respecting certain time constraints between the tasks. The objective is to complete the process as quickly as possible.

The process consists of n tasks $T_1,\cdots,T_n$ with respective durations $d_1,\cdots,d_n$.
The start and end dates of the task T_i are noted $t_{i,d}$ and $t_{i,f}=t_{i,d}+d_i$.
The start dates of the tasks are the unknowns of the scheduling problem.
Fictitious tasks T_0 and T_∞ are also introduced, representing the beginning and end of the process, respectively.

There may be several types of constraints on task completion.

Potential constraints

- The task T_i cannot start before a given date: $t_{i,d} \geq t_{i\,min}$;
- the task T_i must be completed by a given date: $t_{i,f} \leq t_{i\,max}$;
- the task T_i cannot start before the task T_j is finished: $t_{i,d} \geq t_{j,f}$.

Disjunctive constraints

- The tasks T_i and T_j cannot occur simultaneous: $\left[t_{i,d}, t_{i,f}\right] \cap \left[t_{j,d}, t_{j,f}\right] = \varnothing$.

Cumulative constraints

These constraints reflect limitations related to the resources available (machines, personnel, cost, etc.) to perform the tasks. They take into account aspects other than the temporal order and are specific to each problem.

Standard scheduling algorithms only take into account potential constraints. The approximate GANTT method (bar chart) developed by K. Adamiecki (1896), then H. Gantt (1910) was for a long time the only one available. It is still useful for visualizing a schedule optimized by other methods, such as the PERT and MPM methods developed from 1958. These two methods based on a task graph are presented in the following sections.

2.3.1 PERT method

The PERT method (*Program Evaluation Research Task* or *Program Evaluation and Review Technique*) also known as the CPM (*Critical Task* Method) is one of the first operational research methods. It was developed in the USA in 1958 to plan the nuclear submarine program.

Each task T_i has a start event $T_{i,d}$ and an end event $T_{i,f}$.

These events form the nodes of a graph. The events T_0 and T_∞ are added to represent the beginning and the end of the process. The arcs connect events and have the value of the time between events:

- an arc between the start and end of a task has the value of the task duration;
- a potential constraint (between the end of a task T_i and the beginning of another task T_j) is taken into account by an arc of zero duration (or equal to the delay to be respected) between $T_{i,f}$ and $T_{j,d}$.

The PERT method consists in applying Bellman's algorithm (section 2.2.2) in this graph by looking for the path of maximum length between T_0 and T_∞ called the **critical path** of the process. The length of this path is the minimum time required to execute all tasks, while satisfying the potential constraints.

There may be several critical paths (of equal duration). A task on a critical path is a critical task. It cannot be delayed without increasing the total duration of the process. A non-critical task has a margin and can be delayed without affecting the total duration.

Example 2-14: PERT method (from [R6])

Let us consider a process with 9 tasks labelled A, B, C, D, E, F, G, H, I.
The tasks T_0 and T_∞ represent the beginning and end of the process. Table 2-3 shows the duration and prerequisite events for each task.

Task	A	B	C	D	E	F	G	H	I
Prerequisit	T_0+5	T_0	T_0+3	A,B	B	B,C	D,E,F	D,E,F	E,C_m
Duration	16	14	20	8	18	25	15	10	17

Table 2-3: Task durations and prerequisites.

Task I can start when task C is half completed. This intermediate event noted C_m is added to the set of tasks and gives table 2-4.

Task	A	B	C_m	C	D	E	F	G	H	I
Prerequisit	T_0+5	T_0	T_0+3	C_m	A,B	B	B,C	D,E,F	D,E,F	E,C_m
Duration	16	14	10	10	8	18	25	15	10	17

Table 2-4: Task durations and prerequisites with intermediate events.

The PERT graph is shown in figure 2-18. The nodes are the start and end events of the tasks given in table 2-4. The arcs connect successive events and have the value of the duration between the events. Dashed arcs mean zero duration. The point task C_m is represented by a single node for simplicity (it can be represented by its start and end linked by an arc of zero duration, but this arc does not interact with the rest of the graph).

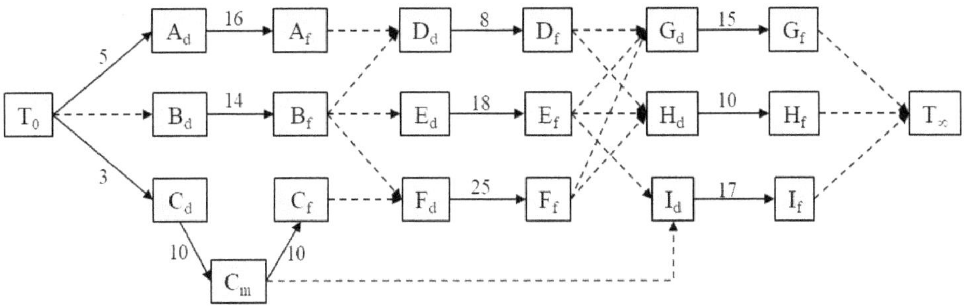

Figure 2-18: PERT graph.

Let us apply Bellman's algorithm to find the path of maximum duration from T_0 to T_∞. Figure 2-19 shows the costs of the nodes at the end of the algorithm.

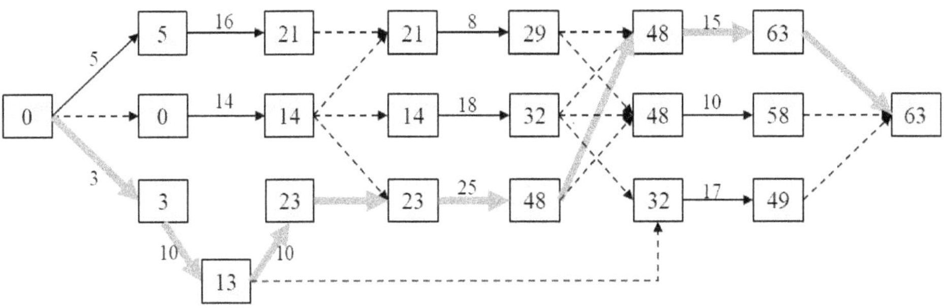

Figure 2-19: Bellman's algorithm on PERT graph.

The critical path (thick line) is: $C_d - C_m - C_f - F_d - F_f - G_d - G_f$ of duration 63.
The critical tasks are therefore C, F and G.
The valuations of the nodes represent the earliest start dates of each task. These dates are summarized in table 2-5.

Task	A	B	C_m	C	D	E	F	G	H	I
Prerequisit	T_0+5	T_0	T_0+3	C_m	A,B	B	B,C	D,E,F	D,E,F	E,C_m
Duration	16	14	10	10	8	18	25	15	10	17
Start min	5	0	3	13	21	14	23	48	48	32
End min	21	14	13	23	29	32	48	63	58	49

Table 2-5: Earliest start and end dates for each task.

2.3.2 MPM method

The MPM method (*Méthode des Potentiels Métra*) was developed by B. Roy of the Métra company to plan the construction of the liner "France" in 1958. Like the PERT method, it comes down to a path problem in a graph. Each node represents the start of a task. The arcs represent duration constraints. All arcs starting from a task $T_{i,d}$ have the same valuation d_i equal to the duration of the task T_i . This graph is similar to the PERT graph, except that it does not show "end of task" events.

The MPM method implicitly applies Bellman's algorithm in the form of a table indicating for each task the prior tasks and their durations. This table is filled in progressively from the beginning T_0 to find the earliest start and end dates of each task. The use of the MPM table is simple and avoids the explicit construction of the graph.

The MPM method is illustrated on the previous scheduling problem.

Example 2-15: MPM method (from [R6])

Let us retrieve the scheduling problem from example 2-14.
Table 2-4 is reproduced below with the intermediate task C_m representing half the completion of task C.

Task	A	B	C_m	C	D	E	F	G	H	I
Prerequisit	T_0+5	T_0	T_0+3	C_m	A,B	B	B,C	D,E,F	D,E,F	E,C_m
Duration	16	14	10	10	8	18	25	15	10	17

The MPM graph is shown in figure 2-20. The nodes are the beginnings of the tasks. The arcs originating from a node have the duration of the task as their value.

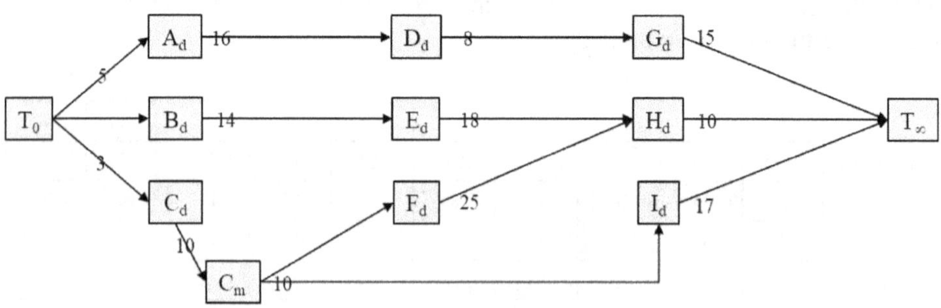

Figure 2-20: MPM graph.

Figure 2-21 shows the result obtained by Bellman's algorithm and the critical path C – F – G of duration 63.

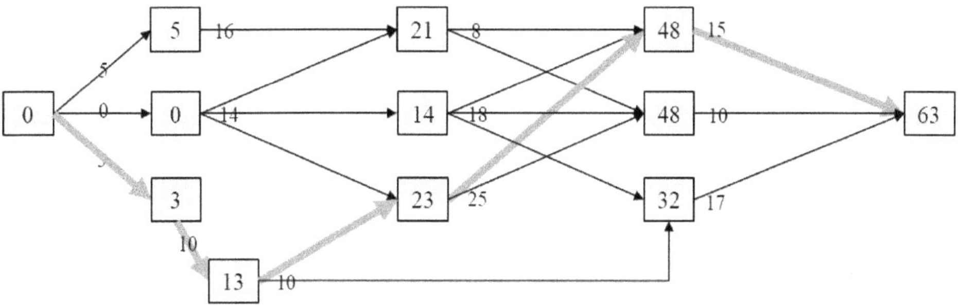

Figure 2-21: Bellman's algorithm on the MPM graph.

The MPM method consists of implicitly applying Bellman's algorithm in the form of an array. The initial MPM table is shown below.
For the following presentation, this table has been split into two parts: A-B-C-D then E-F-G-H-I-T_∞ .

The table is divided into columns associated with each task. For each task, the first row shows the name of the task and the rows below show all the previous tasks with their duration. For example, task G has as a prerequisite tasks D (duration 8), E (duration 18) and F (duration 25).
An initially empty column is added to the left of each task column.

These left-hand sub-columns will be filled in with the earliest dates.

Task		A		B		C_m		C		D
Prerequisit		T_0 5		T_0 0		T_0 3		C_m 10		A 16
										B 14

Task		E		F		G		H		I		T_∞
Prerequisit		B 14		B 14		D 8		D 8		E 18		G 15
				C 10		E 18		E 18		C_m 10		H 10
						F 25		F 25				I 17

Table 2-6: Initial MPM table.

First level

We start with the first task T_0 whose earliest date is known (= 0).

This date is entered to the left of T_0 in the sub-columns where T_0 appears, namely A, B, C_m. The columns A, B, C_m are then complete, because all their prior tasks have their date known. We can then calculate the earliest date of tasks A, B, C_m.

For example, the earliest date of A is:

0 (earliest date of T_0) + 5 (duration from T_0 to A).

This earliest date is entered in the top left-hand cell of column A (same procedure for B and C).

Task	5	A		0	B		3	C_m			C			D	
Prerequisit	0	T_0	5	0	T_0	0	0	T_0	3		C_m	10		A	16
														B	14

Task		E			F			G			H			I			T_∞	
Prerequisit		B	14		B	14		D	8		D	8		E	18		G	15
					C	10		E	18		E	18		C_m	10		H	10
								F	25		F	25					I	17

Second level

Columns A, B, C_m are complete. The earliest dates of tasks A, B, C_m are known. These dates are entered in the sub-columns where A, B, C_m appear, namely C, D, E, F.

Columns C, D, E are complete, as all their prior tasks have their date known. We can then calculate the earliest date of tasks C, D, E.

For example, the earliest date of D is the maximum between:

- 5 (earliest date of A) + 16 (duration from A to D);
- 0 (earliest date of B) + 14 (duration from B to D).

These earliest dates are entered in the top left-hand cell of columns C, D, E.

Task	5	A		0	B		3	C_m		13	C		21	D	
Prerequisit	0	T_0	5	0	T_0	0	0	T_0	3	3	C_m	10	5	A	16
													0	B	14

Task	14	E			F			G			H			I			T_∞	
Prerequisit	0	B	14	0	B	14		D	8		D	8		E	18		G	15
					C	10		E	18		E	18		C_m	10		H	10
								F	25		F	25					I	17

<u>Third level</u>

Columns C, D, E are complete. The earliest dates of tasks C, D, E are known. These dates are entered in the sub-columns where C, D, E appear, namely F, G, H, I.

The columns F, I are <u>complete</u>, because all their prior tasks have their date known. We can then calculate the earliest date of tasks F, I.

For example, the earliest date of F is the maximum between:

- 0 (earliest date of B) + 14 (duration from B to F);
- 13 (earliest date of C) + 10 (duration from C to F).

The earliest dates are entered in the top left-hand cell of columns F, I.

Task	5	A		0	B		3	C_m		13	C		21	D	
Prerequisit	0	T_0	5	0	T_0	0	0	T_0	3	3	C_m	10	5	A	16
													0	B	14

Task	14	E		23	F			G			H		32	I			T_∞	
Prerequisit	0	B	14	0	B	14	21	D	8	21	D	8	14	E	18		G	15
				13	C	10	14	E	18	14	E	18	3	C_m	10		H	10
								F	25		F	25					I	17

<u>Fourth level</u>

Columns F, I are complete. The earliest dates of tasks F, I are known.

These dates are entered in the sub-columns where F, I appear, namely G, H, T_∞.

Columns G, H are complete. The earliest date of tasks G, H can be calculated. The table can then be completed by calculating the earliest final date (column T_∞).

The final MPM table is presented in table 2-7.

Task	5	A		0	B		**3**	C_m		**13**	C		21	D	
Prerequisit	0	T_0	5	0	T_0	0	0	T_0	3	3	C_m	10	5	A	16
													0	B	14

Task	14	E		**23**	F		**48**	G		48	H		32	I		**63**	T_∞	
Prerequisit	0	B	14	0	B	14	21	D	8	21	D	8	14	E	18	48	G	15
				13	C	10	14	E	18	14	E	18	3	C_m	10	48	H	10
							23	F	25	23	F	25				32	I	17

Table 2-7: Final MPM table.

This final table shows the earliest dates of each task at the top left of the columns. The **critical path is** reconstructed from the end.
The critical operation in each column is the one that gave the earliest date.

$T_\infty \to G$	$G \to F$	$F \to C$	$C \to C_m$	$C_m \to T_0$
63 = 48+15	48 = 23+25	23 = 13+10	13 = 3+10	3 = 0+3

2.3.3 Margins

Bellman's algorithm in the form of PERT or MPM gives the critical path. Events on the critical path have an imperative date. For non-critical events, there may be a margin on the date of the event. A distinction is made between:
- the **free margin** which does not shift the dates of other events;
- the **total margin** that does not shift the final date of the process.

An event E_i has an earliest starting date $t_{i\,min}$ given by Bellman's algorithm and a latest starting date $t_{i\,max}$ (not known) which would not shift the end of the process. The durations d_{ij} separating consecutive events E_i and E_j are known. The earliest and latest dates as well as the margins are related by the following relations:

- the E_i event must wait for its predecessors: $t_{i\,min} = \max\limits_{j \in Pred(i)} (t_{j\,min} + d_{ji})$;
- the E_i event must not delay its successors: $t_{i\,max} = \min\limits_{j \in Succ(i)} (t_{j\,max} - d_{ij})$;
- if the event E_i is on the critical path: $t_{i\,max} = t_{i\,min}$;
- if the arc (or task) $E_i - E_j$ is on the critical path: $t_{j\,min} = t_{i\,min} + d_{ij}$.

The free margin on the task $E_i - E_j$ is: $m_{ij} = t_{j\,min} - t_{i\,min} - d_{ij}$.

The total margin on the task $E_i - E_j$ is: $M_{ij} = t_{j\,max} - t_{i\,min} - d_{ij}$.

The duration of the task $E_i - E_j$ can be increased:
- of m_{ij} without changing the dates of subsequent events;
- of M_{ij} without changing the final date, but this will postpone the later events to their latest date (and their free margins become then zero).

When Bellman's algorithm is complete, the latest dates are calculated backwards from the final date and the free and total margins are derived.

Example 2-16: Margins (from [R6])

Consider the 4-event problem A, B, C, D in figure 2-22.

The critical path:
$T_0 - B - D - T_\infty$
is of duration 6.

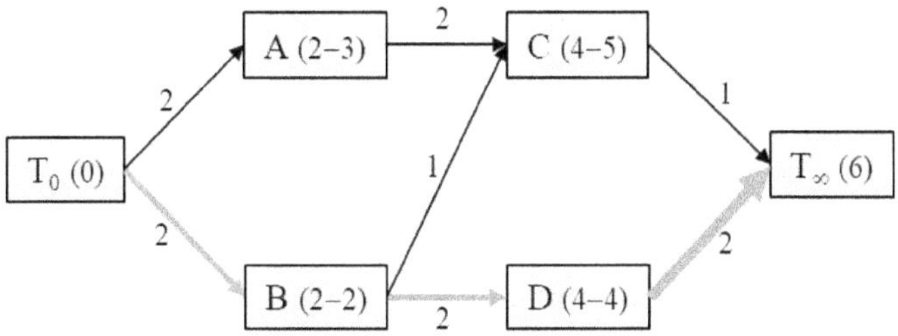

Figure 2-22: Critical path and margins.

Bellman's algorithm yields the event earliest dates (1st number in brackets in figure 2-22). The latest dates (2nd number in brackets) are calculated backwards from T_∞ by subtracting the durations of the arcs travelled. The latest date of an event depends on the latest dates of its successors: $t_{i\max} = \min_{j \in Succ(i)} (t_{j\max} - d_{ij})$.

These earliest and latest dates are summarized in table 2-8.

Event	T_0	A	B	C	D	T_∞
Earliest date	0	2	2	4	4	6
Latest date	0	5–2 = 3	4–2 = 2	6–1 = 5	6–2 = 4	6

Table 2-8: Earliest and latest dates.

The earliest and latest dates are used to determine for each task:

- its free margin: $m_{ij} = t_{j\min} - t_{i\min} - d_{ij}$;
- its total margin: $M_{ij} = t_{j\max} - t_{i\min} - d_{ij}$.

These margins are given in table 2-9 where the critical tasks are colored.

Task	$T_0 - A$	$T_0 - B$	A – C	B – C	B – D	$C - T_\infty$	$D - T_\infty$
Free margin	2–0–2 = 0	2–0–2 = 0	4–2–2 = 0	4–2–1 = 1	4–2–2 = 0	6–4–1 = 1	6–4–2 = 0
Total margin	3–0–2 = 1	2–0–2 = 0	5–2–2 = 1	5–2–1 = 2	4–2–2 = 0	6–4–1 = 1	6–4–2 = 0

Table 2-9: Free and total margins.

It can be seen that the tasks (T_0 – A) and (A – C) have a free margin of 0, although they are not critical. Therefore, these tasks cannot be delayed without shifting the subsequent tasks and the final date.

2.4 Flow problem

We consider a network (section 2.1.1) defined by its nodes and (oriented) arcs. The capacity and cost of the arc from node i to node j are noted f_{ij} and c_{ij}.

The nodes s and t are the entry (source) and exit (sink) points of the network.

The standard flow problems are:
- the search for the maximum flow that can pass through the network;
- the search for the maximum flow with minimum cost (if several solutions exist).

These problems can be solved by linear programming (section 1.4.4) or by the specific Ford-Fulkerson and Roy-Busacker-Gowen algorithms presented in the sections 2.4.1 and 2.4.2.

2.4.1 Ford-Fulkerson algorithm

Ford-Fulkerson algorithm determines the maximum flow that can traverse a network. It is an iterative method based on the notion of augmenting path.

<u>Augmenting path</u>
Suppose that an initial flow is known. Each arc v_i - v_j is traversed by a flow $x_{ij} \le f_{ij}$ and the conservation law is respected at each node.
An **elementary chain** is a sequence of <u>distinct</u> nodes connected by <u>(non-oriented)</u> edges. Figure 2-23 shows an elementary chain (v_1, v_2, v_3, v_4) with two direct arcs $(v_1 - v_2, v_3 - v_4)$ and one indirect dashed arc $(v_2 - v_3)$.

Figure 2-23: Elementary chain.

An elementary chain is said to be **augmenting** if:
- its direct arcs are not saturated: $x_{ij} < f_{ij}$;
- its indirect arcs are of non-zero flow: $x_{ij} > 0$.

The flow along the augmenting path can be increased of the quantity $\delta\varphi$:
- by increasing the flow of direct arcs: $x_{ij} < f_{ij} \rightarrow \delta\varphi \le f_{ij} - x_{ij}$;
- by decreasing the flow of indirect arcs: $x_{ij} > 0 \rightarrow \delta\varphi \le x_{ij}$.

This variation throughout the chain increases the total flow of $\delta\varphi$, while satisfying the conservation law (already satisfied by the initial flow). Figure 2-24 illustrates the effect of the variation $\delta\varphi$ along the augmenting path.

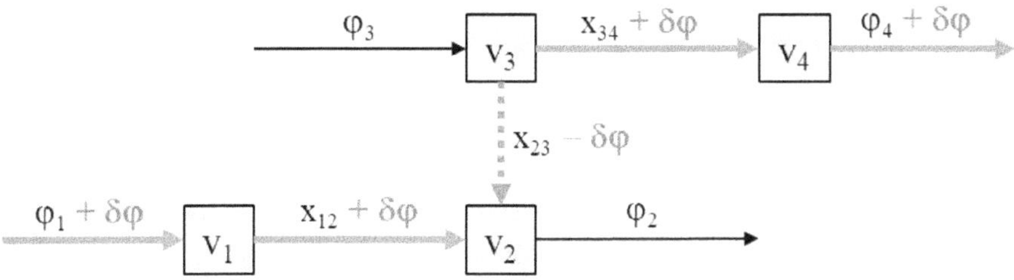

Figure 2-24: Augmenting path.

<u>Node marking</u>
The search for an augmenting path is carried out by marking the nodes, according to the following procedure:

- initialization:
  - the source is marked +;
  - the other nodes are unmarked.

- node marking:
  the set of arcs is traversed starting from the source and examining at each node the set of incoming and outgoing arcs. For an arc $[u-v]$:
  - if u marked and $[u-v]$ unsaturated, then v is marked + (direct arc);
  - if v marked and $[u-v]$ of non-zero flow, then u is marked – (indirect arc).

  For each node we store its predecessor in the order of marking.

- augmenting path:
  if the sink is marked, the flow is not maximum. The augmenting path is obtained by going back to the predecessors from the sink.
  The flow is increased by $\delta\varphi$ along the augmenting path:
  - the flow of direct arcs $[v_i - v_j]$ is increased by $\delta\varphi$ (with $\delta\varphi \le f_{ij} - x_{ij}$);
  - the flow of indirect arcs $[v_i - v_j]$ is decreased by $\delta\varphi$ (with $\delta\varphi \le x_{ij}$).

The procedure stops when the sink is unmarked. The flow is then maximum.

Example 2-17 illustrates the iterations of Ford-Fulkerson algorithm with the marking procedure.

Example 2-17: Ford-Fulkerson algorithm (from [R6])

Consider the network in figure 2-25 with the capacities in brackets.

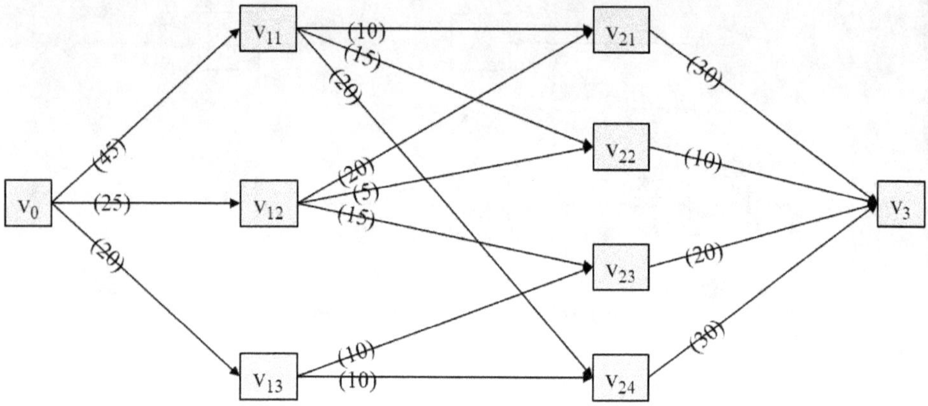

Figure 2-25: Transport network.

We construct an initial flow compliant with the arc capacities and satisfying the conservation law at each node. The flows are noted on each arc and the saturated arcs are in thick lines (figure 2-26). This flow has a value of 80.

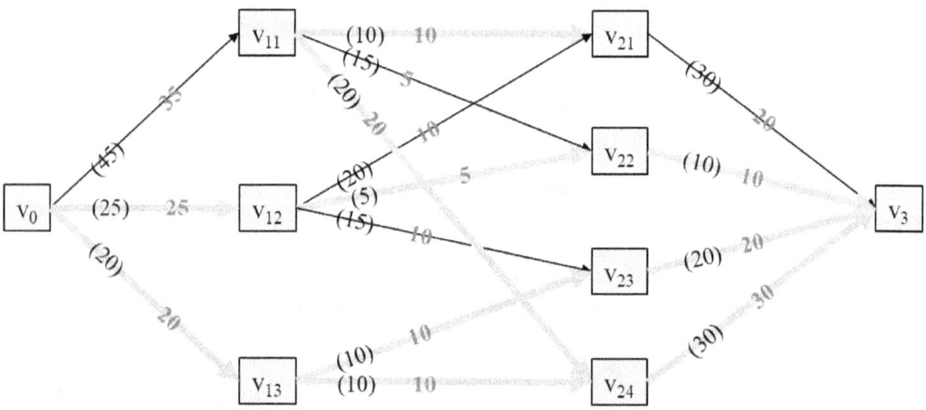

Figure 2-26: Initial flow.

First iteration

The nodes are marked from the source. Table 2-10 shows the order of processing of the arcs, the resulting marking and the predecessor of each node.

Arc	$[v_0-v_{11}]$	$[v_{11}-v_{22}]$	$[v_{12}-v_{22}]$	$[v_{12}-v_{21}]$	$[v_{12}-v_{23}]$	$[v_{21}-v_3]$	$[v_{13}-v_{23}]$
Marking	v_{11} (+)	v_{22} (+)	v_{12} (−)	v_{21} (+)	v_{23} (+)	v_3 (+)	v_{13} (−)
Predecessor	v_0	v_{11}	v_{22}	v_{12}	v_{12}	v_{21}	v_{23}

Table 2-10: Marking at first iteration.

Since the sink v_3 is marked, there is an augmenting path found by tracking back the predecessors from table 2-10: $v_3 \to v_{21} \to v_{12} \to v_{22} \to v_{11} \to v_0$. Table 2-11 shows the achievable variations on each arc of the augmenting path.
The maximum compatible variation of the arc capacities is: $\delta\varphi = 5$.

Node	v_3 (+)	v_{21} (+)	v_{12} (−)	v_{22} (+)	v_{11} (+)
Predecessor	v_{21}	v_{12}	v_{22}	v_{11}	v_0
Bound (+ −)	30	20	0	15	45
Flow on arc	20	10	5	5	35
Variation $\delta\varphi$	+10	+10	−5	+10	+10

Table 2-11: Variation achievable along the augmenting path.

The variation is applied to the augmenting path (thick lines in figure 2-27). The flow increases $(\delta\varphi = +5)$ on the direct arcs $[v_0 - v_{11}]$, $[v_{11} - v_{22}]$, $[v_{12} - v_{21}]$, $[v_{21} - v_3]$ and decreases $(\delta\varphi = -5)$ on the indirect arc $[v_{12} - v_{22}]$.

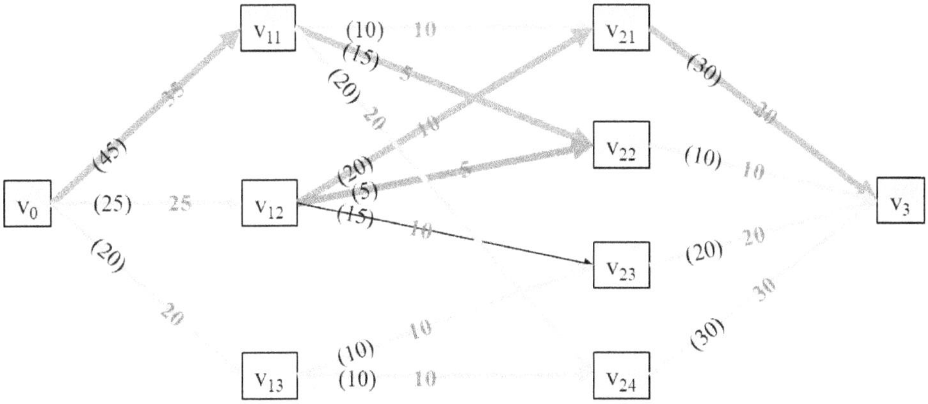

Figure 2-27: Augmenting path.

The new flow of value 85 is shown in figure 2-28. The saturated arcs are in thick lines.

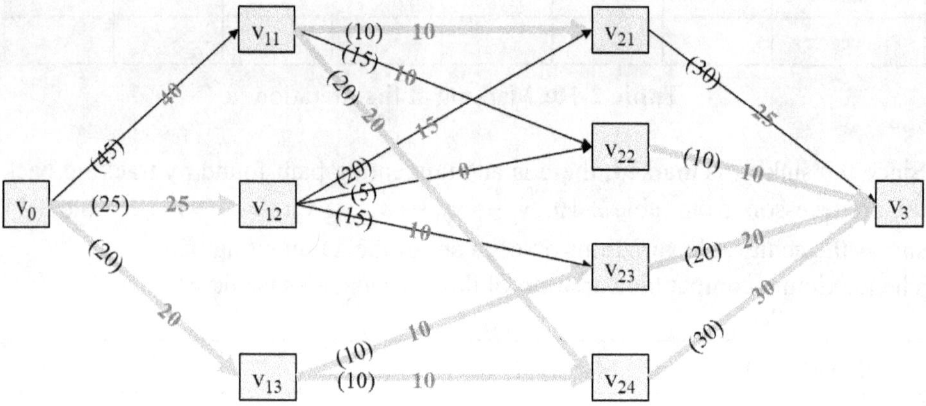

Figure 2-28: Flow after the first iteration.

<u>Second iteration</u>

The marking of the nodes starting from the source is given in table 2-12.

Arc	$[v_0 - v_{11}]$	$[v_{11} - v_{22}]$
Marking	v_{11} (+)	v_{22} (+)
Predecessor	v_0	v_{11}

Table 2-12: Marking at second iteration.

The marking stops at the node v_{22}, whose arcs are all zero or saturated. As the sink v_3 is not marked, the flow of figure 2-28 is maximum and has the value 85.

<u>Cut</u>

The optimality of the flow obtained by Ford-Fulkerson algorithm is based on the notion of cut. The **cut** associated with a flow is the set W of marked nodes as shown in figure 2-29. Each node belongs either to W or to its complementary $\bar{W}$:

- an arc $[v_i - v_j]$ going out of W is saturated.
  Otherwise v_j would be marked (+) and would be in W;
- an arc $[v_k - v_l]$ entering W has a zero flow.
  Otherwise v_k would be marked (–) and would be in W.

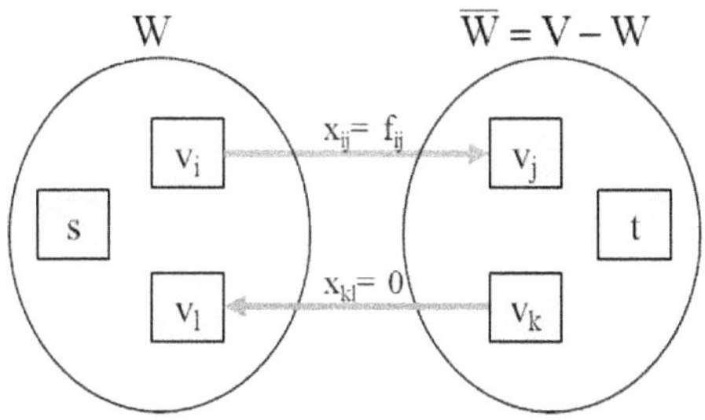

Figure 2-29: Cut associated with the flow.

The flow inside the cut W can eventually increase, but the total flow out of W cannot be increased, nor can the total flow into W be decreased.
The source is always in the cut. If the sink is not in the cup (unmarked), the flow arriving at it cannot be increased. This flow is then maximum.

Example 2-18: Cut

Let us retrieve the maximum flow example 2-17 shown in figure 2-28.
The nodes marked v_0, v_{11}, v_{22} (table 2-12) form the set W circled in figure 2-30. The arcs leaving W in solid lines are all saturated. The arcs entering W in dotted lines are all of zero flow. The sink v_3 is outside W and we cannot increase the flow out of W. This flow is maximum.

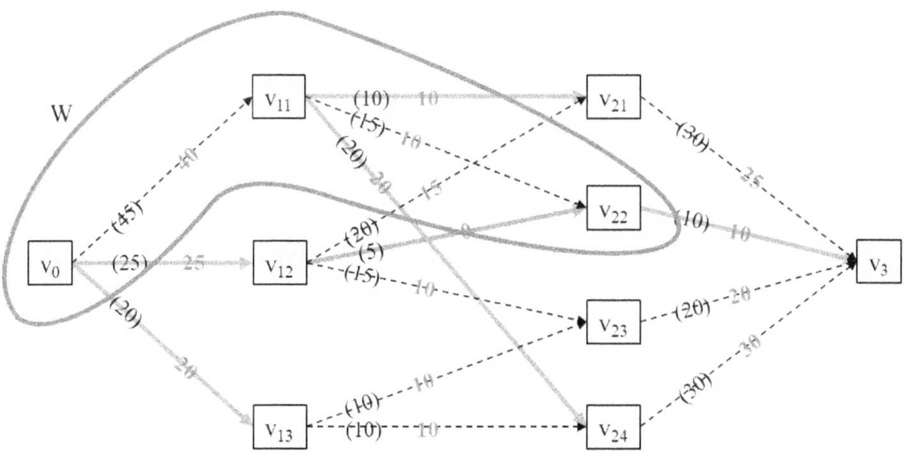

Figure 2-30: Cut of the maximum flow.

2.4.2 Roy-Busacker-Gowen algorithm

There are usually several solutions for the maximum flow. These solutions do not have the same cost depending on the arcs selected for the flow. Roy-Busacker-Gowen algorithm determines the maximum flow of minimum cost that can traverse a network. This method uses the deviation graph associated with the flow.

Deviation graph

The graph G is assumed to be antisymmetric. There is at most one arc between two nodes: if the arc $[v_i - v_j]$ exists, then the arc $[v_j - v_i]$ does not exist.

For a given flow Φ, the **deviation graph** $G_e(\Phi)$ associated with the flow Φ is a graph with the same nodes as G and whose arcs are defined as follows:

- if the arc $[v_i - v_j]$ of G is saturated $(x_{ij} = f_{ij})$,
  G_e does not include the arc $[v_i - v_j]$
  and includes the arc $[v_j - v_i]$ of capacity f_{ij};

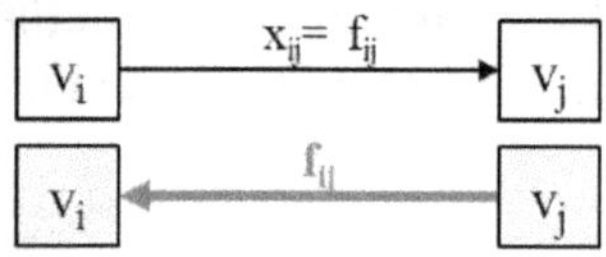

- if the arc $[v_i - v_j]$ of G has a zero flow $(x_{ij} = 0)$,
  G_e includes the arc $[v_i - v_j]$ of capacity f_{ij}
  and does not include the arc $[v_j - v_i]$;

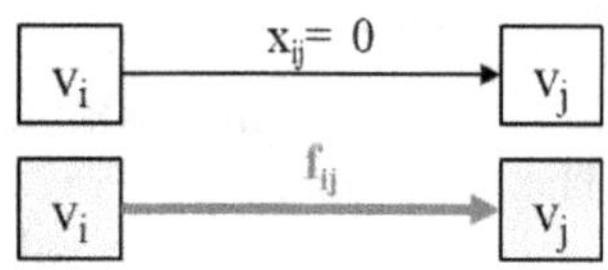

- if the arc $[v_i - v_j]$ of G is of intermediate flow $(0 < x_{ij} < f_{ij})$,
  G_e includes the arc $[v_i - v_j]$ of capacity $f_{ij} - x_{ij}$
  and includes the arc $[v_j - v_i]$ of capacity x_{ij}.

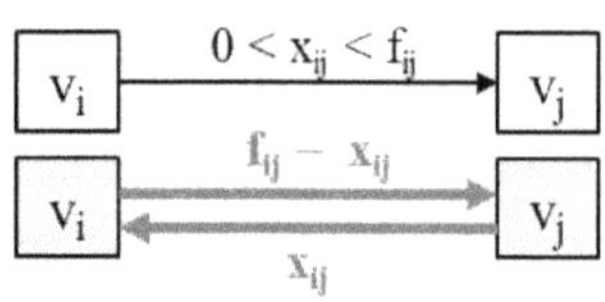

A path from the source to the sink in the deviation graph G_e corresponds to an augmenting path in the graph G. Indeed, this path will either take direct unsaturated arcs or indirect arcs of non-zero flow. Example 2-19 illustrates the use of the deviation graph to find an augmenting path.

Example 2-19: Deviation graph

Consider an initial flow Φ_0 in the graph G in figure 2-31a. The flow is indicated on each arc with the capacity in brackets. Saturated and zero flow arcs are in thick lines. The associated deviation graph $G_e(\Phi_0)$ is plotted in figure 2-31b.

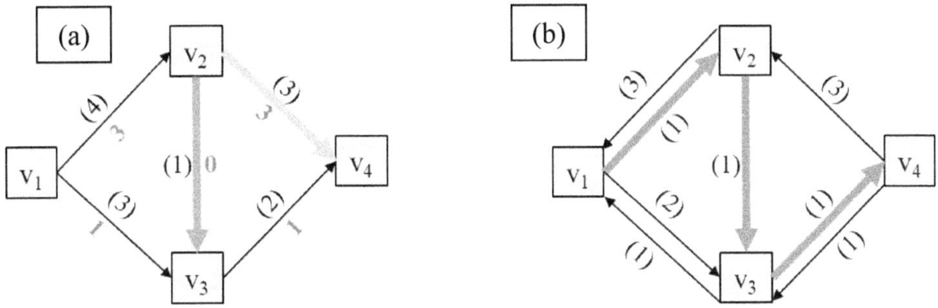

Graph G with flow Φ_0 Deviation graph $G_e(\Phi_0)$

Figure 2-31: Deviation graph associated with the initial flow.

The table 2-13 details each arc of $G_e(\Phi_0)$ according to the flow through the arc.

Arc of G	$[v_1-v_2]$ (4)	$[v_1-v_3]$ (3)	$[v_2-v_3]$ (1)	$[v_2-v_4]$ (3)	$[v_3-v_4]$ (2)
Flow through arc	unsaturated (3)	unsaturated (1)	null flow (0)	saturated (3)	unsaturated (1)
Direct arc of G_e	$[v_1-v_2]$ (1)	$[v_1-v_3]$ (2)	$[v_2-v_3]$ (1)		$[v_3-v_4]$ (1)
Indirect arc of G_e	$[v_2-v_1]$ (3)	$[v_3-v_1]$ (1)		$[v_4-v_2]$ (3)	$[v_4-v_3]$ (1)

Table 2-13: Arcs of the initial deviation graph.

There is a path from source v_1 to the sink v_4 in the deviation graph $G_e(\Phi_0)$.
This path shown in thick lines in figure 2-31b corresponds to the augmenting path $v_1 \rightarrow v_2 \rightarrow v_3 \rightarrow v_4$ in the graph G. The flow can be increased by $\delta\varphi = 1$ along this augmenting path.
Figure 2-32a shows the new flow Φ_1 with the corresponding deviation graph $G_e(\Phi_1)$ in figure 2-32b.

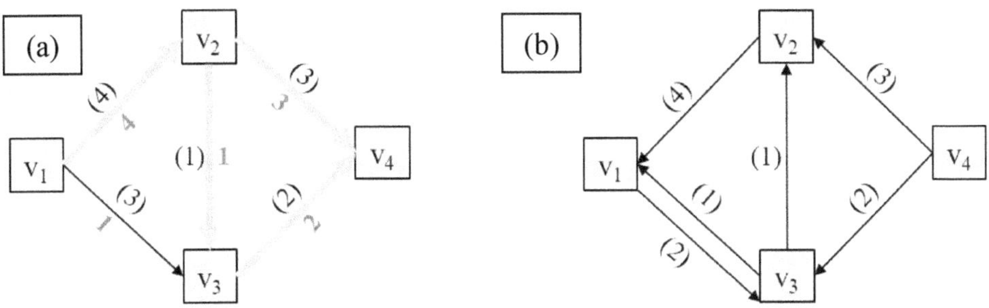

Graph G with flow Φ_1 Deviation graph $G_e(\Phi_1)$

Figure 2-32: Deviation graph associated with the new flow.

Table 2-14 details each arc of $G_e(\Phi_1)$ according to the flow through the arc.

Arc of G	$[v_1-v_2]$ (4)	$[v_1-v_3]$ (3)	$[v_2-v_3]$ (1)	$[v_2-v_4]$ (3)	$[v_3-v_4]$ (2)
Flow through arc	saturated (4)	unsaturated (1)	saturated (1)	saturated (3)	saturated (2)
Direct arc of G_e		$[v_1-v_3]$ (2)			
Indirect arc of G_e	$[v_2-v_1]$ (4)	$[v_3-v_1]$ (1)	$[v_3-v_2]$ (1)	$[v_4-v_2]$ (3)	$[v_4-v_3]$ (2)

Table 2-14: Arcs of the new deviation graph.

There is no path from v_1 to v_4 in the deviation graph $G_e(\Phi_1)$.
The flow Φ_1 is therefore maximum.

Let us now turn to the problem of the maximum flow with minimum cost.
The cost of a flow Φ is: $C = \sum c_{ij}x_{ij}$, where x_{ij} is the flow through the arc $[v_i - v_j]$ and c_{ij} is the unit cost of the arc.

The costs c_{ij} are marked on the deviation graph $G_e(\Phi)$ as follows:

- the direct arcs of the deviation graph $G_e(\Phi)$ are assigned the costs $+c_{ij}$;
- the indirect arcs of the deviation graph $G_e(\Phi)$ are assigned the costs $-c_{ij}$.

In the deviation graph valuated with these costs, we make the following diagnoses:

- the flow Φ is <u>of maximum value</u> if there is no path from the source to the sink in the deviation graph;
- the flow Φ is <u>of minimum cost</u> if there are no absorbing cost circuits (negative cost circuits) in the deviation graph (Roy's theorem).

Example 2-20: Optimality diagnosis in the deviation graph

Let us retrieve the example 2-19 and the deviation graph associated with the maximum flow Φ_1 (figure 2-32). This graph is shown in figure 2-33, indicating on each arc its capacity and its cost in the form: (f_{ij}, c_{ij}). The saturated arcs are in thick lines. The flow Φ_1 has a value of $\varphi = 5$ and a cost of $c = 20$.

The associated deviation graph $G_e(\Phi_1)$ is shown in figure 2-33b. The flow is maximum, because there is no path from the source v_1 to the sink v_4. On the other hand, this flow is not of minimum cost, since there exists in $G_e(\Phi_1)$ an absorbing circuit $v_1 \to v_3 \to v_2 \to v_1$ of cost –5 (thick lines). It is therefore possible to find a flow with the same value $\varphi = 5$ and a lower cost (calculated in example 2-21).

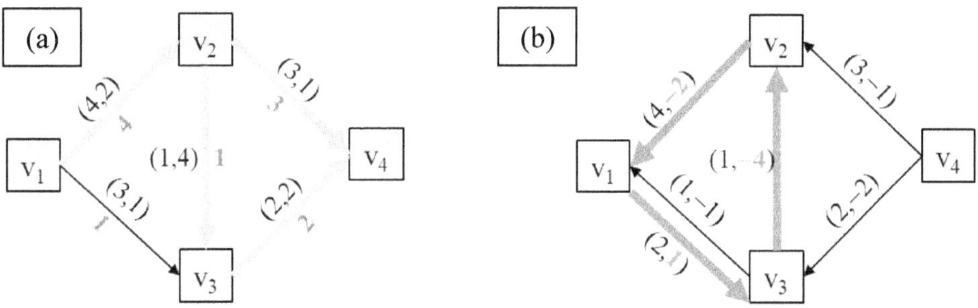

Graph G with flow Φ_1 Deviation graph $G_e(\Phi_1)$

Figure 2-33: Deviation graph with arc costs.

The search for an augmenting path is equivalent to the search for a path between the source and the sink in the deviation graph. Roy-Busacker-Gowen algorithm consists in choosing at each iteration the augmenting path <u>of minimum cost</u>.

The minimum cost chain is obtained by applying Ford's algorithm (section 2.2.1) in the deviation graph. The flow along the augmenting path is then increased and the gap graph associated with the new flow is updated. The algorithm ends when there is no more path from the source to the sink in the gap graph. The resulting flow is then of maximum value and minimum cost.

Ford's algorithm (section 2.2.1) applies for valuations of any sign and detects an absorbing circuit. Such a circuit leads to an infinite loop of Ford's algorithm and a blocking of Roy-Busacker-Gowen algorithm. By selecting the minimum cost augmenting path at each iteration, the updated deviation graph does not contain an absorbing circuit and the algorithm terminates in a finite number of iterations.

Example 2-21 illustrates Roy-Busacker-Gowen algorithm.

Example 2-21: Roy-Busacker-Gowen algorithm

Let us retrieve the graph from figure 2-33 and apply Roy-Busacker-Gowen algorithm starting from a zero flow. This flow Φ_0 and the associated deviation graph are shown in figure 2-34.
For the zero flow, the deviation graph G_e is identical to the graph G.

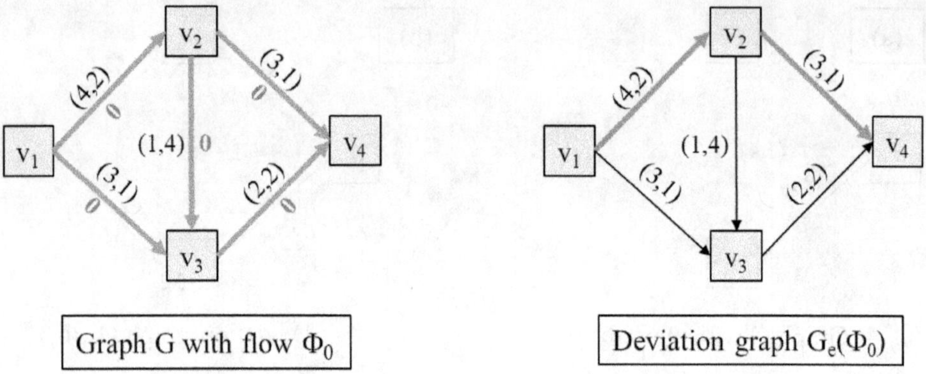

Figure 2-34: Deviation graph for the initial zero flow.

There are three paths from source v_1 to sink v_4 in the deviation graph $G_e(\Phi_0)$:
- the path $v_1 \to v_2 \to v_4$ has a cost of 3;
- the path $v_1 \to v_2 \to v_3 \to v_4$ has a cost of 8;
- the path $v_1 \to v_3 \to v_4$ has a cost of 3.

Let us choose the minimum cost path $v_1 \to v_2 \to v_4$ as the augmenting path.
The flow can be increased by $\delta\varphi = 3$ along this path.
The new flow Φ_1 with value $\varphi = 3$ and cost $c = 8$ is shown in figure 2-35.

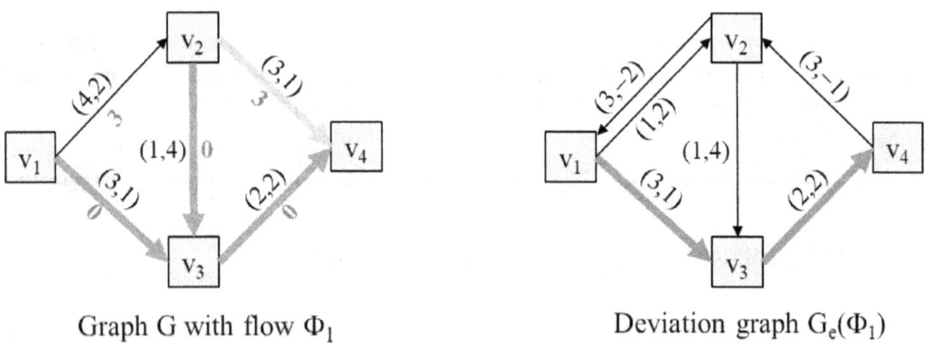

Figure 2-35: Deviation graph for the first iteration flow.

There are two paths from source v_1 to sink v_4 in the deviation graph $G_e(\Phi_1)$:
- the path $v_1 \to v_2 \to v_3 \to v_4$ has a cost of 8;
- the path $v_1 \to v_3 \to v_4$ has a cost of 3.

Let us choose the minimum cost path $v_1 \to v_3 \to v_4$ as the augmenting path.

The flow can be increased by $\delta\varphi = 2$ along this path.
The new flow Φ_2 with value $\varphi = 5$ and cost $c = 15$ is shown in figure 2-36.

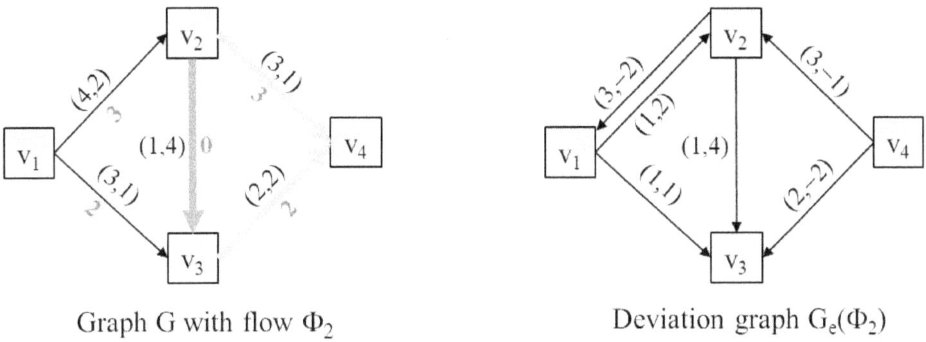

Figure 2-36: Deviation graph for the second iteration flow.

Let us check the optimality of the flow Φ_2 of value $\varphi = 5$ and cost $c = 15$.
This flow is of maximum value, because there is no path from v_1 to v_4 in $G_e(\Phi_2)$.
This flow is of minimum cost, because there is no absorbing circuit in $G_e(\Phi_2)$.
This flow Φ_2 is of lower cost than the one in figure 2-33.

2.5 Assignment problem

The objective is to assign n resources to n tasks, while minimizing the total cost.
The cost of assigning resource i to task j is noted c_{ij}.
This problem can be solved by mixed linear programming (section 1.4.2), by Roy-Busacker-Gowen algorithm or by the Hungarian method.

2.5.1 Equivalent flow problem

The assignment problem can be formulated as a flow problem.
The nodes of the graph are the resources $(r_i)_{i=1 \text{ to } n}$ and the tasks $(t_i)_{i=1 \text{ to } n}$.

This graph is **bipartite**, because resources on the one hand and tasks on the other form two subsets of disjoint nodes (unlinked nodes within the subsets):
- each resource r_i is linked to a task t_j by an arc of capacity 1 and cost c_{ij};
- all resources are linked to a source r_0 by arcs of capacity 1 and cost 0;
- all tasks are linked to a sink t_0 by arcs of capacity 1 and cost 0.

We are looking for the maximum flow of minimum cost between r_0 and t_0.

The value of the flow will be equal to n. The graph of the equivalent flow problem is plotted in figure 2-37.

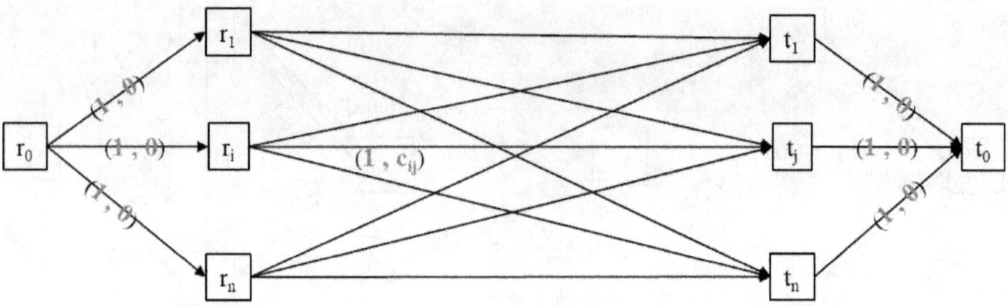

Figure 2-37: Flow problem equivalent to the assignment problem.

Example 2-22: Equivalent flow problem

Consider an assignment problem with 4 resources and 4 tasks.
The cost matrix is shown opposite.
(rows = resources, columns = tasks)

24	10	21	11
14	22	10	15
15	17	20	19
11	19	14	13

The graph of the equivalent flow problem is shown in figure 2-38.

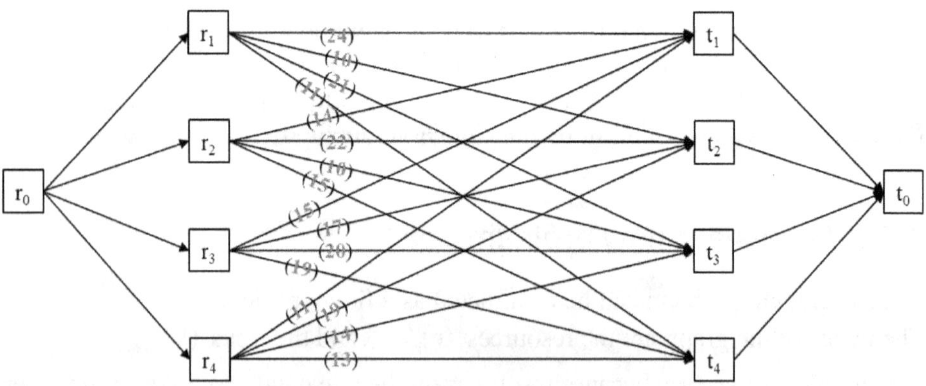

Figure 2-38: Equivalent flow graph.

An initial flow is constructed by a greedy method. The resources are processed one by one. For each resource (line), the task with the lowest cost is chosen among those still available. Figure 2-39 shows this initial assignment. The rows and columns chosen are progressively shaded as the assignments are made.

$(r_1 - t_2)$

24	10	21	11
14	22	10	15
15	17	20	19
11	19	14	13

$(r_2 - t_3)$

24	10	21	11
14	22	10	15
15	17	20	19
11	19	14	13

$(r_4 - t_1)$

24	10	21	11
14	22	10	15
15	17	20	19
11	19	14	13

$(r_3 - t_4)$

24	10	21	11
14	22	10	15
15	17	20	19
11	19	14	13

Figure 2-39: Initial greedy assignment.

This assignment corresponds to a flow Φ_1 with a value of 4 and a cost of 50. Figure 2-40a shows the flow graph (only saturated arcs are shown with their cost) and figure 2-40b shows the associated deviation graph. The flow Φ_1 is maximum, as there is no path from r_0 to t_0. This flow is not of minimum cost, because there is an absorbing circuit $r_3 \to t_1 \to r_4 \to t_4 \to r_3$ drawn in thick lines.

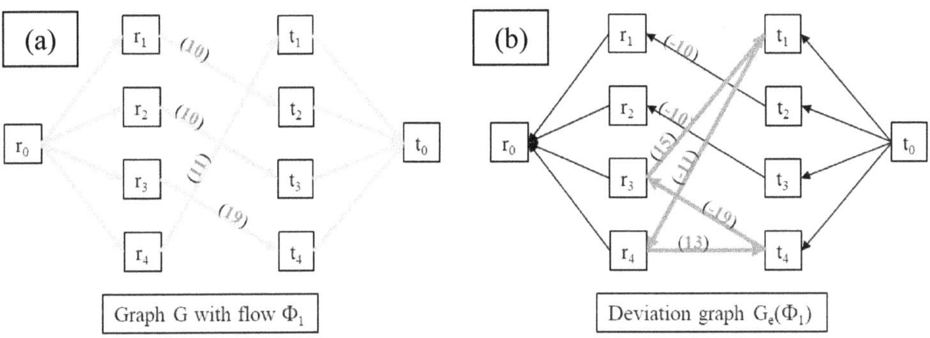

Figure 2-40: Flow associated with the initial assignment.

The absorbing circuit indicates that the assignments of r_3 to t_4 and of r_4 to t_1 should be swapped. The resource r_3 is assigned to t_1 and the resource r_4 is assigned to t_4. Figure 2-41 shows the graph of the new flow Φ_2. This flow has a maximum value of 4 and a minimum cost, as there is no absorbing circuit in the deviation graph.

The optimal assignment is shown opposite.
The cost is 48.

24	10	21	11
14	22	10	15
15	17	20	19
11	19	14	13

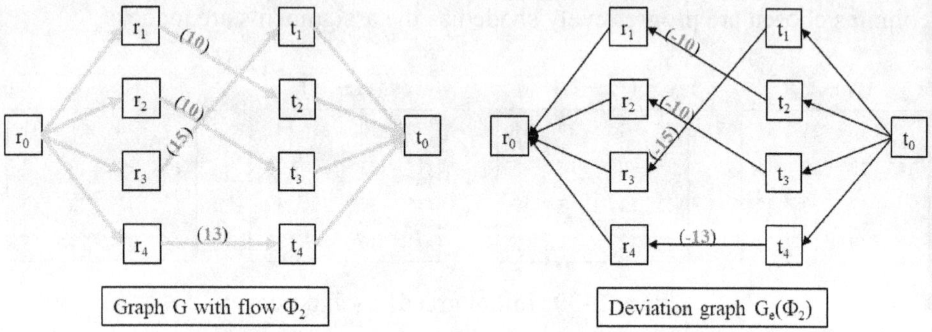

Figure 2-41: Optimal flow deviation graph.

2.5.2 Hungarian method

The Hungarian method (Kühn, Evergary, König, 1955) directly uses the cost matrix $(c_{ij})_{i,j=1 \text{ to } n}$ to determine the optimal assignment.

This iterative method consists of the following stages:

- stage 1: row reduction
  Each row is subtracted from its smallest element. This yields a zero per line;

- stage 2: column reduction
  Each column is subtracted from its smallest element. This yields at least one zero per column;

- stage 3: covering the zeros
  The minimum number p of rows and columns covering all zeros is sought.
  This search using Ford-Fulkerson algorithm is detailed below (zero covering):
  - if $p < n$, the solution is not optimal. We proceed to stage 4;
  - if $p = n$, the solution is optimal. We proceed to stage 5;

- stage 4: increase in cost
  We look for the smallest cost among the uncovered rows or columns.
  It is subtracted from the elements not covered.
  It is added to the elements covered twice. We come back to stage 3;

- stage 5: optimal solution
  One zero is selected per row and per column, corresponding to the assignment.

Example 2-23: Hungarian method

Let us retrieve the problem of example 2-22 whose matrix is given opposite.

24	10	21	11
14	22	10	15
15	17	20	19
11	19	14	13

Let us apply the stages of the Hungarian method.

Stage 1: row reduction
The smallest element in each row is subtracted.
$\rightarrow C = 10 + 10 + 15 + 11 = 46$

24	**10**	21	11	→ 10
14	22	**10**	15	→ 10
15	17	20	19	→ 15
11	19	14	13	→ 11

→

14	0	11	1
4	12	0	5
0	2	5	4
0	8	3	2

Stage 2: column reduction
The smallest element in each column is subtracted.
$\rightarrow C = 46 + 0 + 0 + 0 + 1 = 47$

14	**0**	11	**1**
4	12	**0**	5
0	2	5	4
0	8	3	2
↓	↓	↓	↓
0	0	0	1

→

14	0	11	0
4	12	0	4
0	2	5	3
0	8	3	1

Stage 3: covering the zeros
The method is detailed below. It is found that 1 row and 2 columns are needed to cover all zeros.
$p = 1 + 2 < n = 4 \rightarrow$ not optimal

14	0	11	0
4	12	0	4
0	2	5	3
0	8	3	1

Stage 4
Smallest element not covered: $c_{min} = 1$
It is subtracted from the elements not covered. $\rightarrow -1$
It is added to the elements covered twice. $\rightarrow +1$

14	0	11	0
4	12	0	4
0	2	5	3
0	8	3	1

14	0	11	0
4	11	0	3
0	1	5	2
0	7	3	0

15	0	12	0
4	11	0	3
0	1	5	2
0	7	3	0

$\rightarrow C = 47 + 1 = 48$

Stage 3: covering the zeros
The method is detailed below. It is found that 2 rows and 2 columns are needed to cover all zeros.
$p = 2 + 2 = n \rightarrow$ optimal

15	0	12	0
4	11	0	3
0	1	5	2
0	7	3	0

Stage 5: selecting a zero per row and column
We return to the original cost matrix.
The chosen zeros define the assignments.
The corresponding elements are highlighted.

24	10	21	11
14	22	10	15
15	17	20	19
11	19	14	13

The optimal assignment $(r_1 - t_2)$; $(r_2 - t_3)$; $(r_3 - t_1)$; $(r_4 - t_4)$ has a cost of 48.

Zero covering

The third stage of the Hungarian method requires finding the minimum number of rows and columns covering all zeros of the matrix.
This search is performed in a graph defined as follows:
- a zero in row i and column j gives an arc between r_i and t_j of capacity 1;
- the arcs from the source r_0 and the arcs to the sink t_0 are of capacity 1.

We apply Ford-Fulkerson algorithm starting from the zero flow to determine the maximum flow from r_0 to t_0 . If the maximum flow is of value n, all resources and tasks are assigned. The optimal assignments correspond to the arcs $(r_i - t_j)$ of flow 1 (= selected zeros).

Example 2-24: Zero covering

Let us go back to stage 3 of example 2-23 with the matrix given opposite.

14	0	11	0
4	12	0	4
0	2	5	3
0	8	3	1

The graph associated with the zeros of the matrix is shown below. The initial flow Φ_0 is zero.

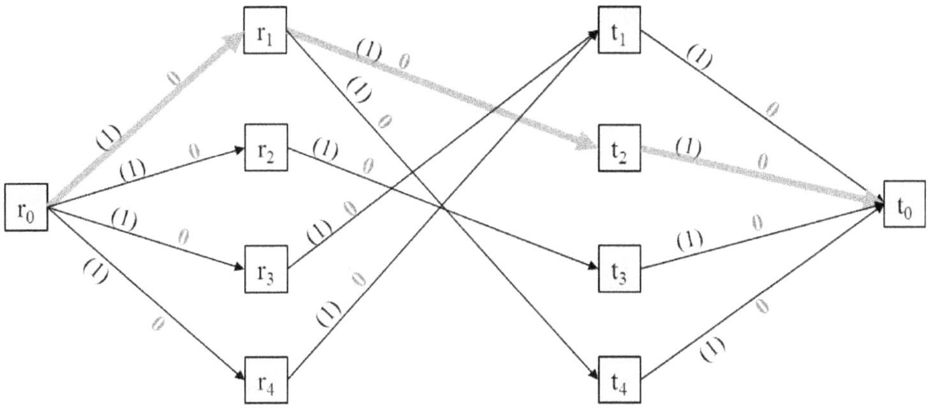

Marking the nodes (section 2.4.1) from r_0 yields the augmenting path:
$r_0\ (+) \rightarrow r_1\ (+) \rightarrow t_2\ (+) \rightarrow t_0\ (+)$ (plotted in thick green lines).
With the sink t_0 marked, the flow can increase by $\delta\varphi=1$ along the augmenting path: $(r_0 - r_1 - t_2 - t_0)$. The new flow Φ_1 of value 1 is shown below.

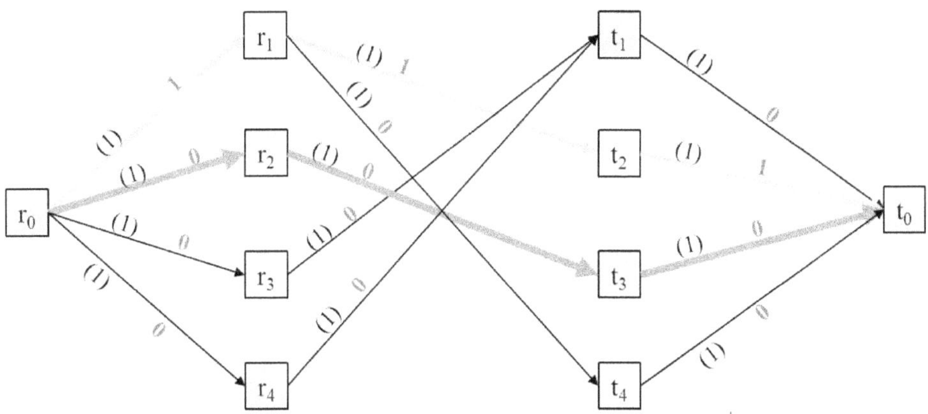

Marking the nodes from r_0 yields the augmenting path:
$r_0\ (+) \rightarrow r_2\ (+) \rightarrow t_3\ (+) \rightarrow t_0\ (+)$ (plotted in thick green lines).
With the sink t_0 marked, the flow can increase by $\delta\varphi=1$ along the augmenting path: $(r_0 - r_2 - t_3 - t_0)$. The new flow Φ_2 of value 2 is shown on the next page.

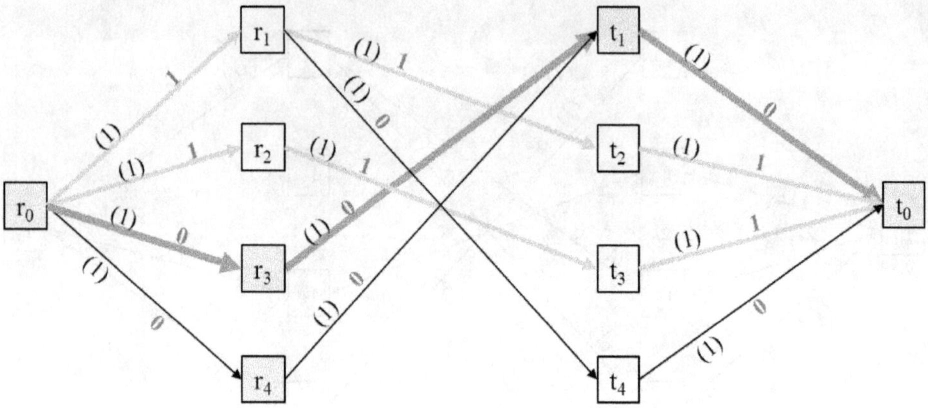

Marking the nodes from r_0 yields the augmenting path:
$r_0 (+) \rightarrow r_3 (+) \rightarrow t_1 (+) \rightarrow t_0 (+)$ (plotted in thick green lines).
With the sink t_0 marked, the flow can increase by $\delta\varphi=1$ along the augmenting path: $(r_0 - r_3 - t_1 - t_0)$. The new flow Φ_3 of value 3 is shown below.

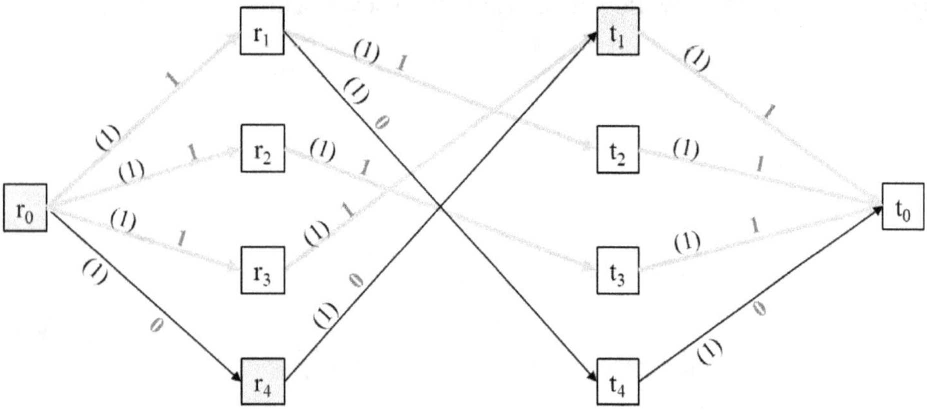

Marking the nodes from r_0 yields the chain: $r_0 (+) \rightarrow r_3 (+) \rightarrow t_1 (+)$.
As the sink t_0 is not marked, the flow is maximum. The minimum number of rows and columns covering the zeros is 3, less than the assignments to be made (4).

This brings us to stage 4 of the Hungarian method.

After stage 4, the cost matrix (opposite) has an additional zero in row 4 column 4.
This zero adds an arc (r_4-t_4) in the previous graph and allows the sink to be marked by the chain $(r_0 - r_4 - t_4 - t_0)$.

15	0	12	0
4	11	0	3
0	1	5	2
0	7	3	0

The flow Φ_4 of value 4 is shown below with the saturated arcs in thick lines.

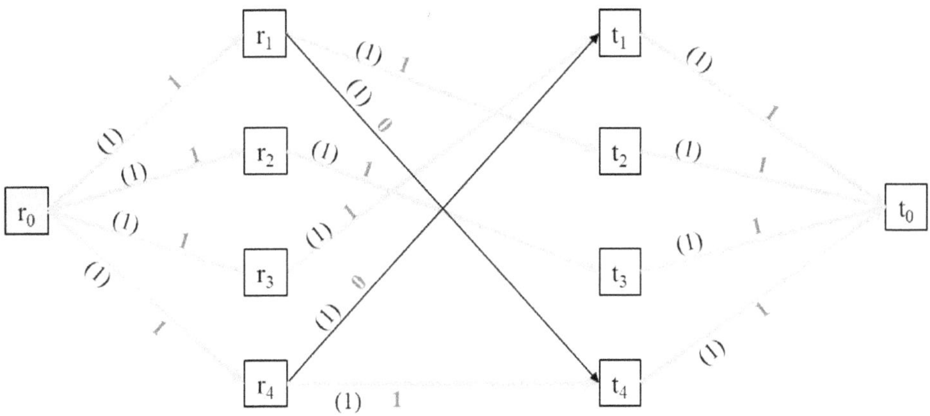

This flow is maximum. The optimal assignment corresponds to the saturated arcs: $(r_1 - t_2)$; $(r_2 - t_3)$; $(r_3 - t_1)$; $(r_4 - t_4)$

Extensions

The Hungarian method applies also to maximization problems (e.g. to maximize satisfaction on a preference matrix). To do this, we simply take the largest element c_{max} on each row of the matrix C, and replace each element c_{ij} in the row with $c_{max} - c_{ij}$. The Hungarian method applies then as above to the matrix $C_{max} - C$.

The Hungarian method applies also if the number m of resources m is different from the number p of tasks. In this case, stage 3 of zero covering is carried out by Roy-Busacker-Gowen algorithm giving the maximum flow of minimum cost. If the maximum flow is $\min(m, p)$, the solution is optimal. Otherwise, the flow is increased along the minimum cost augmenting path. Example 2-25 illustrates the case of a number of resources different from the number of tasks.

Example 2-25: Different numbers of resources and tasks

Consider the problem with 2 resources and 3 tasks with the cost matrix opposite.

1	2	3
1	3	3

After the row and column reduction, we obtain the matrices shown opposite.

0	1	2
0	2	2

→

0	0	0
0	1	0

We are looking for the minimum cost covering of the zeros. In this simple case, we can list the 4 possible covering and compare their cost.
$(r_1 - t_1)$ and $(r_2 - t_3)$ $\rightarrow C = 1 + 3 = 4$
$(r_1 - t_2)$ and $(r_2 - t_1)$ $\rightarrow C = 2 + 1 = 3$ $\rightarrow$ minimum cost
$(r_1 - t_2)$ and $(r_2 - t_3)$ $\rightarrow C = 2 + 3 = 5$
$(r_1 - t_3)$ and $(r_2 - t_1)$ $\rightarrow C = 3 + 1 = 4$

In practice, the Roy-Busacker-Gowen algorithm is used in the graph of the zeros shown below. The initial flow Φ_0 is zero.

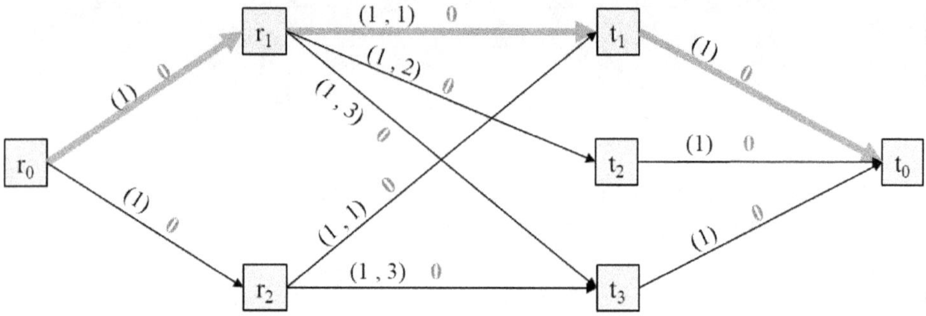

There are two augmenting paths of cost +1: $(r_0 - r_1 - t_1 - t_0)$ and $(r_0 - r_2 - t_1 - t_0)$.
The other augmenting paths are of higher cost.
The flow is increased by 1 on the chain $(r_0 - r_1 - t_1 - t_0)$.
The new flow Φ_1 of value 1 is shown below.

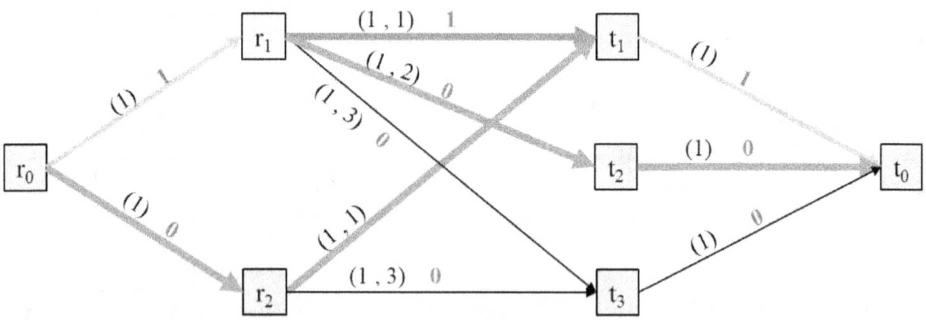

There are 3 augmenting paths:
- $(r_0 - r_2 - t_3 - t_0)$ of cost +3;
- $(r_0 - r_2 - t_1 - r_1 - t_2 - t_0)$ of cost $+ 1 - 1 + 2 = +2$;
- $(r_0 - r_2 - t_1 - r_1 - t_3 - t_0)$ of cost $+ 1 - 1 + 3 = +3$.

The flow is increased by 1 on the minimum cost chain $(r_0 - r_2 - t_1 - r_1 - t_2 - t_0)$ (thick lines). The new flow Φ_2 of value 2 is shown below.

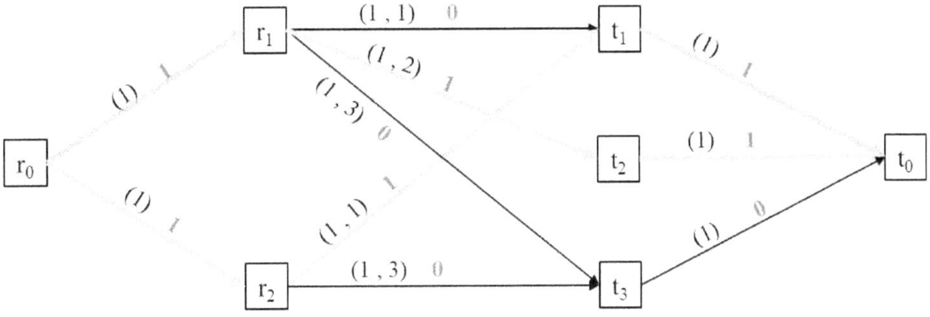

This flow is maximum, as all resources are assigned.
The total cost is 1 + 2 = 3.
The optimal assignment is: $(r_1 - t_2) \rightarrow$ cost = 2;
$(r_2 - t_1) \rightarrow$ cost = 1.

1	2	3
1	3	3

2.5.3 Theoretical justification

The assignment problem can be formulated as a linear problem in binary variables (section 1.4.2). The Hungarian method is then interpreted as an iterative solution of the dual problem.

Justification of the Hungarian method

The assignment problem is formulated as a linear problem denoted (P).

$$\min_{x_{ij}} z = \sum_{i=1}^{n}\sum_{j=1}^{n} c_{ij}x_{ij} \quad \text{s.t.} \begin{cases} \sum_{i=1}^{n} x_{ij} = 1 \;,\; j = 1 \text{ to } n & \rightarrow \text{multipliers u} \\ \sum_{j=1}^{n} x_{ij} = 1 \;,\; i = 1 \text{ to } n & \rightarrow \text{multipliers v} \end{cases}$$

The binary variable x_{ij} is 1 if resource i is assigned to task j, 0 otherwise.

The constraints express that each resource (i = 1 to n) and each task (j = 1 to n) is assigned once and only once.
The Lagrange multipliers associated with the constraints are noted as:
- $u = (u_1, \cdots, u_n)$ for n constraints on the resources;
- $v = (v_1, \cdots, v_n)$ for the n constraints on the tasks.

Let us rewrite this problem (P) as: $\min\limits_{x\in\{0;1\}} z = c^T x$ s.t. $Ax = b$

with the matrix A and the vectors b, c, x defined by

$$
\begin{array}{c}
\begin{array}{cccccc|cccccc|c|cccccc}
 & & x_{1j} & & & & & & x_{2j} & & & & & & & x_{nj} & & & \\
\end{array}\\
A = \left(\begin{array}{cccccc|cccccc|c|cccccc}
1 & 1 & \cdots & \cdots & 1 & 1 & 0 & 0 & \cdots & \cdots & 0 & 0 & & 0 & 0 & \cdots & \cdots & 0 & 0 \\
0 & 0 & \cdots & \cdots & 0 & 0 & 1 & 1 & \cdots & \cdots & 1 & 1 & & 0 & 0 & \cdots & \cdots & 0 & 0 \\
\vdots & \vdots & & & \vdots & \vdots & \vdots & \vdots & & & \vdots & \vdots & & \vdots & \vdots & & & \vdots & \vdots \\
\vdots & \vdots & & & \vdots & \vdots & \vdots & \vdots & & & \vdots & \vdots & & \vdots & \vdots & & & \vdots & \vdots \\
0 & 0 & \cdots & \cdots & 0 & 0 & 0 & 0 & \cdots & \cdots & 0 & 0 & & 1 & 1 & \cdots & \cdots & 1 & 1 \\
\hline
1 & 0 & \cdots & \cdots & 0 & 0 & 1 & 0 & \cdots & \cdots & 0 & 0 & & 1 & 0 & \cdots & \cdots & 0 & 0 \\
0 & 1 & \cdots & \cdots & 0 & 0 & 0 & 1 & \cdots & \cdots & 0 & 0 & & 0 & 1 & \cdots & \cdots & 0 & 0 \\
\vdots & \vdots & & & \vdots & \vdots & \vdots & \vdots & & & \vdots & \vdots & & \vdots & \vdots & & & \vdots & \vdots \\
\vdots & \vdots & & & \vdots & \vdots & \vdots & \vdots & & & \vdots & \vdots & & \vdots & \vdots & & & \vdots & \vdots \\
0 & 0 & \cdots & \cdots & 0 & 1 & 0 & 0 & \cdots & \cdots & 0 & 1 & & 0 & 0 & \cdots & \cdots & 0 & 1
\end{array}\right)
\begin{array}{c} u_1 \\ u_2 \\ \vdots \\ \vdots \\ u_n \\ v_1 \\ v_2 \\ \vdots \\ \vdots \\ v_n \end{array}
, \quad b = \begin{pmatrix} 1 \\ 1 \\ \vdots \\ \vdots \\ 1 \\ 1 \\ 1 \\ \vdots \\ \vdots \\ 1 \end{pmatrix}
\end{array}
$$

$$c^T = (c_{11}\ c_{12}\ \cdots\ c_{1n}\ c_{21}\ c_{22}\ \cdots\ c_{2n}\ \cdots\ c_{n1}\ c_{n2}\ \cdots\ c_{nn})$$
$$x^T = (x_{11}\ x_{12}\ \cdots\ x_{1n}\ x_{21}\ x_{22}\ \cdots\ x_{2n}\ \cdots\ x_{n1}\ x_{n2}\ \cdots\ x_{nn})$$

The solution of the primal problem (P) is naturally binary, because the matrix A is **unimodular**. The dual problem (D) is formulated as

$$\max_{\lambda} z' = b^T\lambda \quad \text{s.t.} \quad A^T\lambda \le c \quad \text{with } \lambda = \begin{pmatrix} u \\ v \end{pmatrix} \begin{matrix} \to n \\ \to n \end{matrix}$$

or by making the variables and constraints explicit

$$\max_{u_i, v_j} z' = \sum_{i=1}^{n} u_i + \sum_{j=1}^{n} v_j \quad \text{s.t.} \quad u_i + v_j \le c_{ij} \quad , \quad i = 1 \text{ to } n \ , \ j = 1 \text{ to } n$$

The dual problem has 2n variables and n^2 constraints.
The dual cost z' is the sum of the dual variables u and v. To maximize the dual cost, we will try to saturate the constraints inequalities. The second members of the inequalities correspond to the cost matrix shown below.

$$
\begin{array}{c}
\begin{array}{ccccc} v_1 & \cdots & v_j & \cdots & v_n \end{array} \\
\begin{array}{c} u_1 \\ \vdots \\ u_i \\ \vdots \\ u_n \end{array}
\begin{pmatrix}
c_{11} & \cdots & c_{1j} & \cdots & c_{1n} \\
\vdots & & \vdots & & \vdots \\
c_{i1} & \cdots & c_{ij} & \cdots & c_{in} \\
\vdots & & \vdots & & \vdots \\
c_{n1} & \cdots & c_{nj} & \cdots & c_{nn}
\end{pmatrix}
\end{array}
$$

Each variable u_i is bounded by the minimum cost on the row:

$$u_i \le \min_j c_{ij} \underset{\text{noted}}{=} u_{i\max} \ , \ i = 1 \text{ to } n .$$

Each variable v_j is bounded by the minimum cost on the column:

$$v_j \le \min_i c_{ij} \underset{\text{noted}}{=} v_{j\max} \ , \ j = 1 \text{ to } n .$$

Let us show that the Hungarian method is an iterative solution of the dual problem.

Stage 1: row reduction
Subtract from each row of the matrix (c_{ij}) the smallest element noted $u_{i\max}$.

This operation is equivalent to changing the variables by: $\begin{cases} u'_i = u_i - u_{i\max} \\ c'_{ij} = c_{ij} - u_{i\max} \end{cases}$.

It makes at least one zero appear in each row of the matrix (c'_{ij}).

Stage 2: column reduction
Subtract from each column of the matrix (c'_{ij}) the smallest element noted $v_{j\max}$.

This operation is equivalent to changing the variables by: $\begin{cases} v'_j = v_j - v_{j\max} \\ c''_{ij} = c'_{ij} - v_{j\max} \end{cases}$.

It makes at least one zero appear in each column of the matrix (c''_{ij}).
With these changes of variables, the dual problem becomes

$$\max_{u'_i, v'_j} z' = \sum_{i=1}^{n} u'_i + \sum_{j=1}^{n} v'_j + z_{\max} \quad \text{s.t.} \quad u'_i + v'_j \le c''_{ij}, \; i,j = 1 \text{ to } n$$

The value $z_{\max} = \sum_{i=1}^{n} u_{i\max} + \sum_{j=1}^{n} v_{j\max}$ gives a lower bound on the cost.

The matrix $(c''_{ij} = c_{ij} - u_{i\max} - v_{j\max})$ has at least one zero per row and per column.

Stage 3→ 5: covering zeros with n rows and columns
Suppose we can choose n zero elements c''_{ij} so as to cover each row and each column.
For each zero c''_{ij}, we group the corresponding dual variables u'_i and v'_j.
We obtain n pairs (u'_i, v'_j) satisfying the constraints: $u'_i + v'_j \le 0$.
We make these couples appear in the cost z' by noting $(i_k, j_k)_{k=1 \text{ to } n}$ their indices.

$$z' = \sum_{i=1}^{n} u'_i + \sum_{j=1}^{n} v'_j + z_{\max} = \sum_{k=1}^{n} (u'_{i_k} + v'_{j_k}) + z_{\max} \le z_{\max} \quad \text{because} \quad u'_{i_k} + v'_{j_k} \le 0$$

We deduce that the dual cost z' is at most $z_{\max}$. This value is reached for the zero solution $(u'_i = 0, v'_j = 0)$ which is therefore optimal of cost $z_{\max}$.
The optimal assignments $(i_k, j_k)_{k=1 \text{ to } n}$ correspond to the n zero elements c''_{ij} chosen to cover all rows and columns.

Stage 3→ 4: covering zeros with less than n rows and columns

Suppose that we can cover each row and each column of the matrix with less than n zero elements c''_{ij}.

The smallest non-zero c_{ij} on the rows with a single zero is chosen.

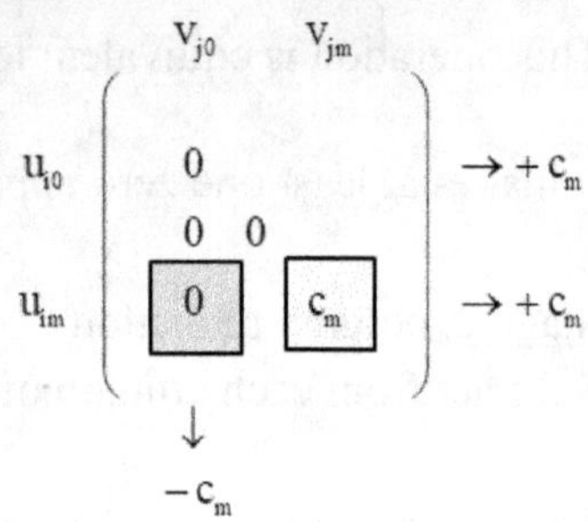

This element noted c_m is located in row i_m and in column j_m.

$\rightarrow c_m = c_{i_m j_m}$

The row i_m has a single zero in the column j_0.

$\rightarrow c_{i_m j_0} = 0$

Consider then the zero solution $(u_i = 0 \,,\, v_j = 0)$ except for:

- the variable v_{j0} : $v_{j0} = -c_m$;
- the variables u_{i0} : $u_{i0} = +c_m$ $\rightarrow$ rows i_0 with a single zero in the column j_0.

This solution is feasible:

$$\begin{cases} u_{i0} + v_j = c_m + 0 = c_m & \le c_{i_0 j} \text{ if } j \ne j_0 \quad \rightarrow \text{by choice of } c_m \\ u_{i0} + v_{j0} = c_m - c_m = 0 & = c_{i_0 j_0} \text{ if } j = j_0 \\ u_i + v_{j0} = 0 - c_m = -c_m & < 0 \le c_{ij_0} \text{ if } i \ne i_0 \end{cases}$$

The cost is greater than z_{max}:

$$z = \sum_{i_0 \ne i_m} \underset{\substack{\downarrow \\ +c_m}}{u_i} + \underset{\substack{\downarrow \\ +c_m}}{u_{i_m}} + \underset{\substack{\downarrow \\ -c_m}}{v_{j_0}} + z_{max} = \sum_{i_0 \ne i_m} c_m + z_{max} \ge z_{max}$$

The reduction in stage 4 corresponds to the following change of variables to come back to the zero solution.

$$\begin{cases} u'_{i_0} = u_{i_0} + c_m \\ c'_{i_0 j} = c_{i_0 j} - c_m \end{cases} \quad \rightarrow \text{on all rows } i_0 \text{ with a single zero in the column } j_0$$

$$\begin{cases} v'_{j_0} = v_{j_0} - c_m \\ c'_{ij_0} = c_{ij_0} + c_m \end{cases} \quad \rightarrow \text{on the column } j_0$$

The effect of this change of variable on the cost matrix is shown below.

$$
\begin{array}{c} \\ u_{i0} \\ \\ u_{im} \\ \\ \end{array}
\begin{array}{c}
\begin{array}{cc} v_{j0} & v_{jm} \end{array} \\
\left(\begin{array}{ccc} 0 & & \\ 0 & 0 & \\ 0 & & c_m \end{array}\right) \\
\downarrow \\
+c_m
\end{array}
\begin{array}{c} \\ \to -c_m \\ \\ \to -c_m \\ \\ \end{array}
\quad \longrightarrow \quad
\begin{array}{c} \\ u'_{i0} \\ \\ u'_{im} \end{array}
\begin{array}{c}
\begin{array}{cc} v'_{j0} & v'_{jm} \end{array} \\
\left(\begin{array}{ccc} 0 & & \\ c_m & 0 & \\ 0 & & 0 \end{array}\right)
\end{array}
$$

The cost matrix after reduction becomes (c'_{ij}):

→ it makes a zero appear in row i_m and column j_m instead of c_m (smallest non-zero element of single-zero rows);

→ it keeps the zeros in the column j_0 on the single-zero rows.
It keeps at least one zero per row and per column.

The zero solution $(u'_i = 0, v'_j = 0)$ is feasible for the new reduced problem.
This solution is optimal if all rows and all columns can be covered by choosing n elements c'_{ij}.

2.6 Heuristics

It is usually not possible to solve large combinatorial problems exactly. One can then resort to **heuristics**, which are empirical methods adapted to a particular problem. They are mostly based on so-called greedy approaches that make the most optimal immediate choice at each stage. These methods are simple and fast, as the solution is developed without iterations, but there is no guarantee on the quality of the result. The result of a heuristic can be used to initialize a metaheuristic (Volume I, chapter 2) or a tree-based method (section 1.3).

This section presents some specific heuristics adapted to classical combinatorial problems.

2.6.1 Stacking problem

Statement

We have a set of n objects and a box of size b.

Object number i has size a_i and value c_i. We have to choose the objects to stack in the box to maximize the total value. This stacking problem, also known as knapsack, is NP-complete.

Heuristics

The principle is to choose as a priority the objects with the best value/size ratio, called utility. The objects are sorted in decreasing utility

$$\frac{c_1}{a_1} \geq \frac{c_2}{a_2} \geq \cdots \geq \frac{c_n}{a_n}$$

and then stacked in that order as long as the size b of the box allows.

$$\rightarrow \begin{array}{ll} a_1 + a_2 + \cdots + a_p & \leq b \\ a_1 + a_2 + \cdots + a_p + a_{p+1} & > b \end{array}$$

The same principle applies in the case where several copies of each object can be selected. Example 2-26 shows that this intuitive approach can lead to poor results, even in very simple cases.

Example 2-26: Stacking by greedy method

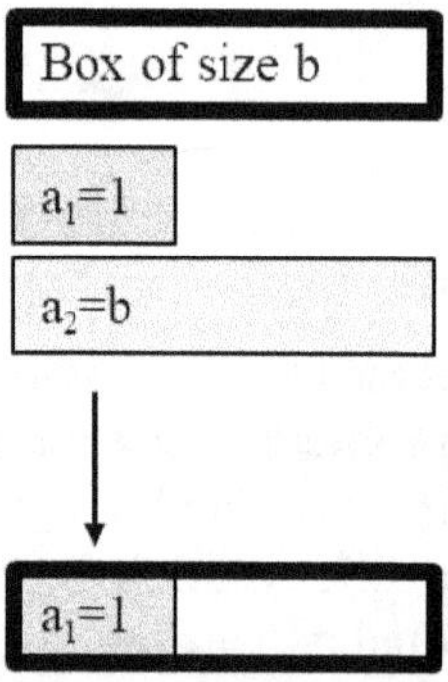

Problem 1: choice of two objects

The size of the box is: $b > 2$.

The values of the objects are: $c = (1, b-1)$.

The sizes of the objects are: $a = (1, b)$.

The objects are sorted in decreasing order of utility:

$$\frac{c_1}{a_1} = 1 \geq \frac{c_2}{a_2} = \frac{b-1}{b}$$

The greedy solution is to first choose object 1 of size $a_1 = 1$.

Object 2 of size $a_2 = b$ no longer fits. The result is a stack of value $c = 1$.

The optimal solution is to take object 2, which gives a stack of value $c = b - 1$.

The greedy solution is poor if b is large, because the box is little filled.

The greedy solution is better when the objects are small compared to the box $(a_i \ll b)$, as it is possible to approach a maximum filling.

Problem 2: cutting a cable
A seller has a cable of length: b = 6 790.
Ten customers place orders for the following lengths:
a = (2597 , 2463 , 2292 , 1625 , 1283 , 1089 , 759 , 599 , 315 , 211):

- the greedy solution is to satisfy the largest orders first. The length sold is: 2597 + 2463 + 1625 = 6685 and there remains 105 unsold;

- the optimal solution is to satisfy orders 1, 4, 5, 7, 9, 10. The length sold is: 2597 + 1625 + 1283 + 759 + 315 + 211 = 6790, which is the entire length available.

2.6.2 Bin packing problem

Statement
We have a set of n objects and several boxes of identical sizes b. Object number i is of size a_i. All objects must be packaged using as few boxes as possible. This bin packing problem is NP-complete.

Heuristics
The principle is to place the largest objects first, trying as much as possible to fill the boxes already used.
The objects are sorted in descending order of size: $a_1 \geq a_2 \geq \cdots \geq a_n$, and then placed in that order, if possible in a box already in use. If the object fits in more than one box, the one that the object fills the most is chosen.

This heuristic works well when the objects are small compared to the boxes $(a \ll b)$.
Worst cases occur for problems with few objects like the one in example 2-27.

Example 2-27: Bin packing by greedy method

We are trying to place the 6 objects opposite
of sizes a = (4, 3, 3, 2, 2, 2)
in boxes of size b = 8.

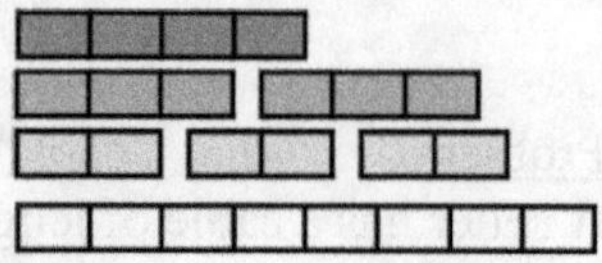

The **greedy** solution consists of choosing to place the objects by decreasing size, adding a new box when necessary. It requires 3 boxes.

Object	a_1=4	a_2=3	a_3=3	a_4=2	a_5=2	a_6=2	Total
Box 1	4	7					7
Box 2			3	5	7		7
Box 3						2	2

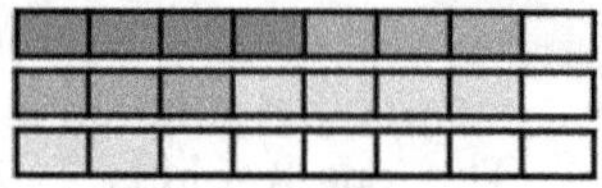

The optimal solution shown here uses only 2 boxes.

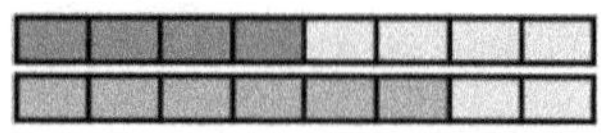

The relative performance of the heuristic on this case is 50% compared to the optimum. This is a worst case for this type of problem.

2.6.3 Set covering problem

Statement

There is a set of n resources to perform m tasks. Resource number j has a cost c_j and can perform N_j tasks. Resources must be chosen to perform all tasks, while minimizing the total cost. Unlike the assignment problem (section 2.5), each resource can perform multiple tasks.
This set covering problem is NP-complete.

Heuristics

The principle is to select resources according to their cost wrt their feasible tasks.

The average cost of resource j is $\frac{c_j}{N_j}$. At each stage, the resource with the lowest average cost is chosen and assigned all the tasks it can perform.
These tasks are removed from the list and the average resource costs on the remaining tasks are recalculated.

To facilitate the process, the resource/task compatibility matrix A is defined with elements $a_{ij} = 1$ if resource j can perform task i and 0 otherwise. After each assignment the corresponding rows and columns are deleted until all tasks are covered.

Example 2-28: Set covering by greedy method

The aim is to achieve 4 tasks with 5 resources of costs $c = \begin{pmatrix} 7 & 3 & 7 & 12 & 6 \end{pmatrix}$.
The resource/task compatibility matrix is

$$A = \begin{matrix} & r_1 & r_2 & r_3 & r_4 & r_5 \\ t_1 & 1 & 0 & 1 & 1 & 0 \\ t_2 & 0 & 1 & 0 & 1 & 0 \\ t_3 & 1 & 0 & 0 & 0 & 1 \\ t_4 & 0 & 0 & 1 & 0 & 1 \end{matrix}$$

Figure 2-42 shows the sequence of successive assignments. At each stage, the average cost of the remaining resources (columns) on the remaining tasks they can perform is calculated. This average cost shown in the last row is used to select the least expensive resource.
After the 3 stages, we obtain the assignments: ($r_1 \rightarrow t_1$), ($r_2 \rightarrow t_2$), ($r_5 \rightarrow t_3, t_4$).
This choice of cost 7 + 3 + 6 = 16 is the optimal solution.

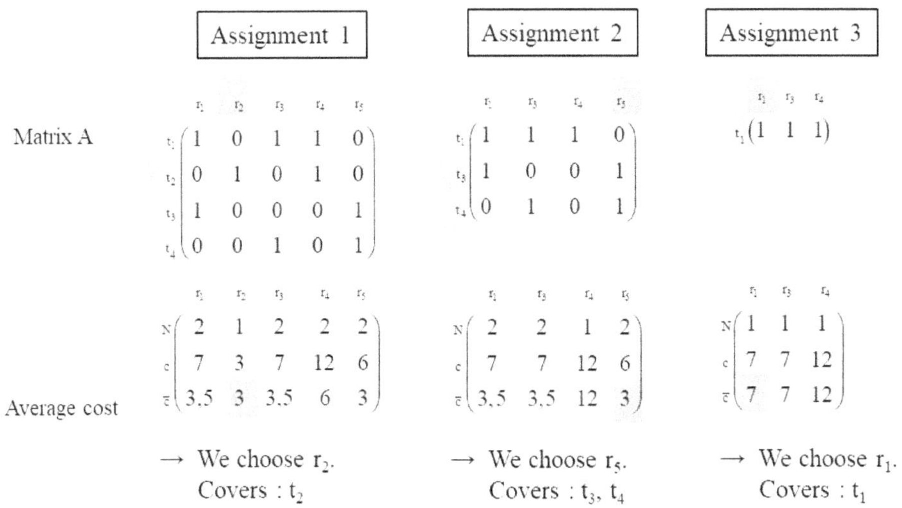

Figure 2-42: Set covering by greedy method.

2.6.4 Coloring problem

Statement

The nodes of a graph should be colored with as few colors as possible. Adjacent (or neighboring) nodes must be of different colors.
This coloring problem is NP-complete.

Heuristics

The DSATUR algorithm of Brelaz (1979) consists in coloring the nodes one by one, choosing at each stage the most "difficult" node to color. We define:
- the degree δ_i of the node v_i which is its number of neighbors;
- the saturation σ_i of the node v_i which is the number of colors of the neighbors.

At each iteration the node with the maximum saturation (and maximum degree if several nodes have the same saturation) is selected. The selected node is colored if possible with an existing color, otherwise with a new color.

Example 2-29: Coloring by greedy method

We want to color a graph with 8 nodes. The following figures show the successive iterations with the state of the graph on the left and the table giving the degrees, the saturations and the colors of the nodes.

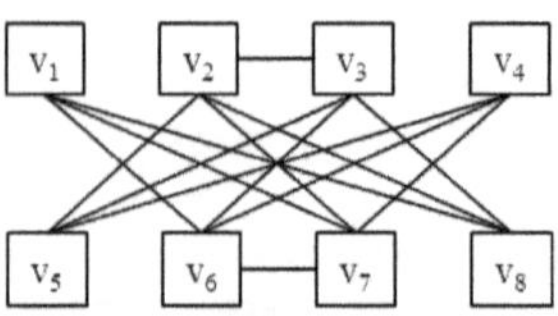

$$\begin{matrix} & v_1 & v_2 & v_3 & v_4 & v_5 & v_6 & v_7 & v_8 \\ \delta & 3 & 4 & 4 & 3 & 3 & 4 & 4 & 3 \\ \sigma & 0 & 0 & 0 & 0 & 0 & 0 & 0 & 0 \\ C & 0 & 0 & 0 & 0 & 0 & 0 & 0 & 0 \end{matrix}$$

→ We color v_2 with color 1.

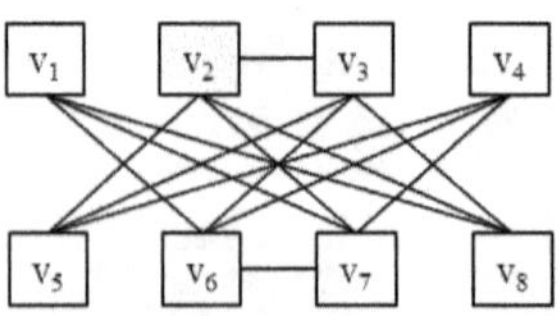

$$\begin{matrix} & v_1 & v_2 & v_3 & v_4 & v_5 & v_6 & v_7 & v_8 \\ \delta & 3 & 4 & 4 & 3 & 3 & 4 & 4 & 3 \\ \sigma & 0 & - & 1 & 0 & 1 & 0 & 1 & 1 \\ C & 0 & 1 & 0 & 0 & 0 & 0 & 0 & 0 \end{matrix}$$

→ We color v_3 with color 2.

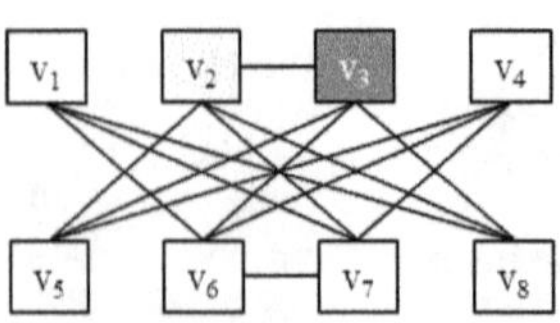

$$\begin{matrix} & v_1 & v_2 & v_3 & v_4 & v_5 & v_6 & v_7 & v_8 \\ \delta & 3 & 4 & 4 & 3 & 3 & 4 & 4 & 3 \\ \sigma & 0 & - & - & 0 & 2 & 1 & 1 & 2 \\ C & 0 & 1 & 2 & 0 & 0 & 0 & 0 & 0 \end{matrix}$$

→ We color v_5 with color3.

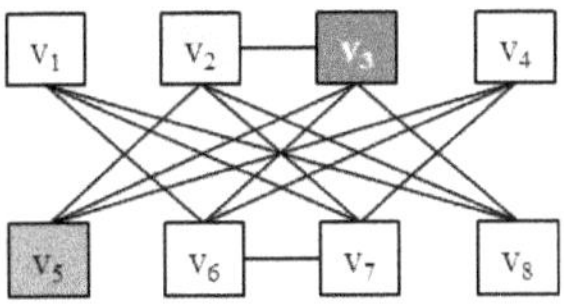

$$\begin{array}{c|cccccccc} & v_1 & v_2 & v_3 & v_4 & v_5 & v_6 & v_7 & v_8 \\ \hline \delta & 3 & 4 & 4 & 3 & 3 & 4 & 4 & 3 \\ \sigma & 0 & - & - & 1 & - & 1 & 1 & 2 \\ C & 0 & 1 & 2 & 0 & 3 & 0 & 0 & 0 \end{array}$$

→ We color v_8 with color 3.

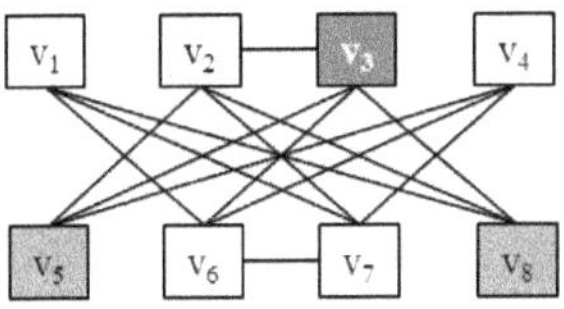

$$\begin{array}{c|cccccccc} & v_1 & v_2 & v_3 & v_4 & v_5 & v_6 & v_7 & v_8 \\ \hline \delta & 3 & 4 & 4 & 3 & 3 & 4 & 4 & 3 \\ \sigma & 1 & - & - & 1 & - & 1 & 1 & - \\ C & 0 & 1 & 2 & 0 & 3 & 0 & 0 & 3 \end{array}$$

→ We color v_6 with color 1.

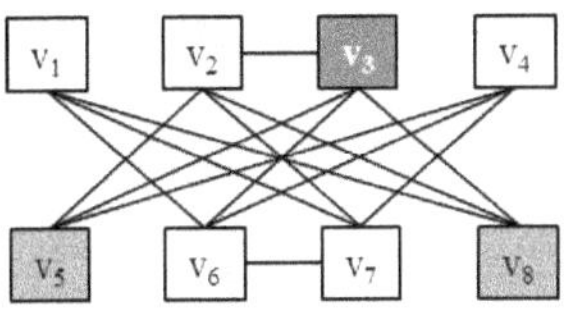

$$\begin{array}{c|cccccccc} & v_1 & v_2 & v_3 & v_4 & v_5 & v_6 & v_7 & v_8 \\ \hline \delta & 3 & 4 & 4 & 3 & 3 & 4 & 4 & 3 \\ \sigma & 2 & - & - & 2 & - & - & 2 & - \\ C & 0 & 1 & 2 & 0 & 3 & 1 & 0 & 3 \end{array}$$

→ We color v_7 with color 2.

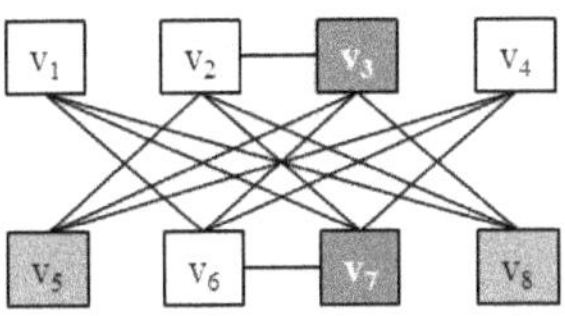

$$\begin{array}{c|cccccccc} & v_1 & v_2 & v_3 & v_4 & v_5 & v_6 & v_7 & v_8 \\ \hline \delta & 3 & 4 & 4 & 3 & 3 & 4 & 4 & 3 \\ \sigma & 3 & - & - & 3 & - & - & - & - \\ C & 0 & 1 & 2 & 0 & 3 & 1 & 2 & 3 \end{array}$$

→ We color v_1 with color 4.

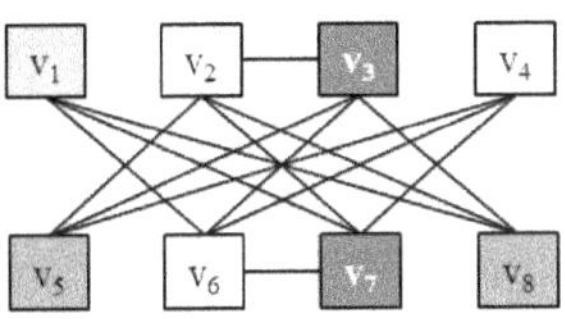

$$\begin{array}{c|cccccccc} & v_1 & v_2 & v_3 & v_4 & v_5 & v_6 & v_7 & v_8 \\ \hline \delta & 3 & 4 & 4 & 3 & 3 & 4 & 4 & 3 \\ \sigma & - & - & - & 3 & - & - & - & - \\ C & 4 & 1 & 2 & 0 & 3 & 1 & 2 & 3 \end{array}$$

→ We color v_4 with color 4.

The result is a solution using 4 colors.

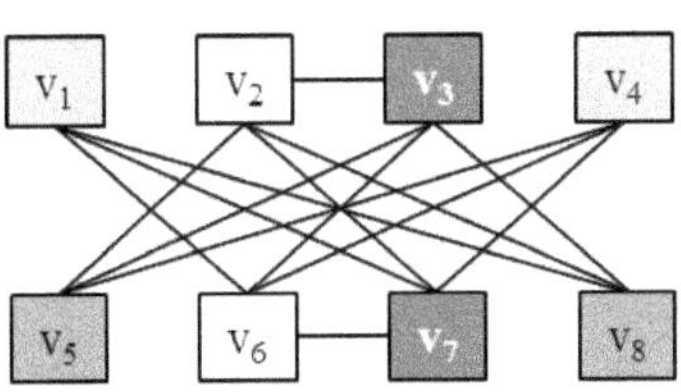

Color 1: v_2 , v_6
Color 2: v_3 , v_7
Color 3: v_5 , v_8
Color 4: v_1 , v_4

The coloring algorithm can be used for the numerical calculation of derivatives. Consider a simulator with n inputs and m outputs. To calculate the Jacobian matrix (partial derivatives) by finite differences, the simulator must be called n times by successively varying $x_1, x_2, \cdots, x_n$. This operation is costly in terms of computing time in an optimization algorithm.

If the outputs y_j do not depend on all the inputs x_i (sparse Jacobian matrix), several inputs can be grouped together to vary them simultaneously. The structure of the Jacobian matrix must be known (theoretically or experimentally).

The grouping of variables is obtained by forming the input incompatibility graph (inputs that cannot be varied simultaneously) and applying a coloring algorithm to group compatible inputs. The inputs with the same color are those that can be varied simultaneously.

Example 2-30: Coloring for numerical derivation

Consider a simulator with 5 inputs and 3 outputs of equations: $\begin{cases} y_1 = x_1 x_4 \\ y_2 = x_1 x_2 x_5 \\ y_3 = x_3 x_5 \end{cases}$.

The structure of the Jacobian matrix is: $\dfrac{\partial y}{\partial x} = \begin{pmatrix} \times & 0 & 0 & \times & 0 \\ \times & \times & 0 & 0 & \times \\ 0 & 0 & \times & 0 & \times \end{pmatrix}$.

The input incompatibility graph is constructed.
The nodes of this graph correspond to the input variables $(x_1, x_2, x_3, x_4, x_5)$.
If an output depends on several inputs, these inputs are connected by an edge:

- output y_1 $\rightarrow$ (x_1, x_4);
- output y_2 $\rightarrow$ (x_1, x_2), (x_1, x_5), (x_2, x_5);
- output y_3 $\rightarrow$ (x_3, x_5).

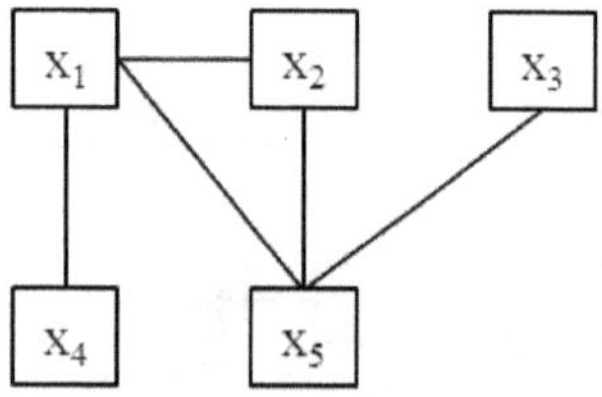

We apply the greedy coloring algorithm in this graph.

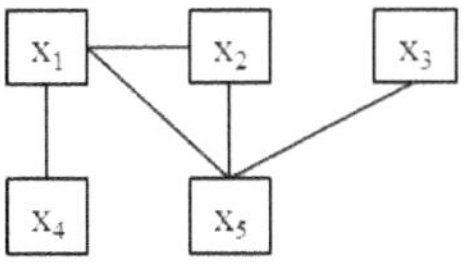

$$\begin{array}{c|ccccc} & x_1 & x_2 & x_3 & x_4 & x_5 \\ \hline \delta & 3 & 2 & 1 & 1 & 3 \\ \sigma & 0 & 0 & 0 & 0 & 0 \\ C & 0 & 0 & 0 & 0 & 0 \end{array}$$

→ We color x_1 with color 1.

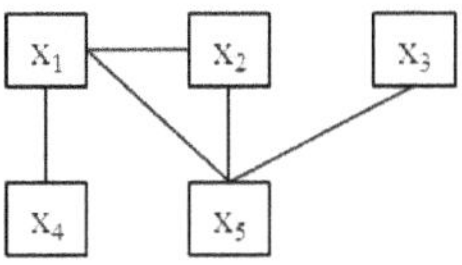

$$\begin{array}{c|ccccc} & x_1 & x_2 & x_3 & x_4 & x_5 \\ \hline \delta & 3 & 2 & 1 & 1 & 3 \\ \sigma & - & 1 & 0 & 1 & 1 \\ C & 1 & 0 & 0 & 0 & 0 \end{array}$$

→ We color x_5 with color 2.

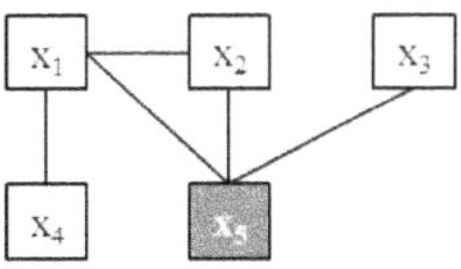

$$\begin{array}{c|ccccc} & x_1 & x_2 & x_3 & x_4 & x_5 \\ \hline \delta & 3 & 2 & 1 & 1 & 3 \\ \sigma & - & 2 & 1 & 1 & - \\ C & 1 & 0 & 0 & 0 & 2 \end{array}$$

→ We color x_2 with color 3.

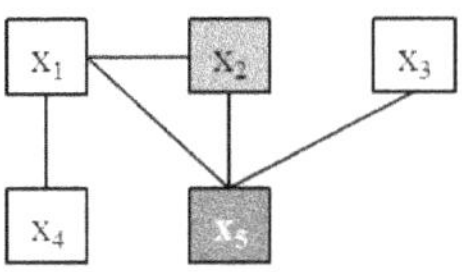

$$\begin{array}{c|ccccc} & x_1 & x_2 & x_3 & x_4 & x_5 \\ \hline \delta & 3 & 2 & 1 & 1 & 3 \\ \sigma & - & - & 1 & 1 & - \\ C & 1 & 3 & 1 & 0 & 2 \end{array}$$

→ We color x_3 with color 1.

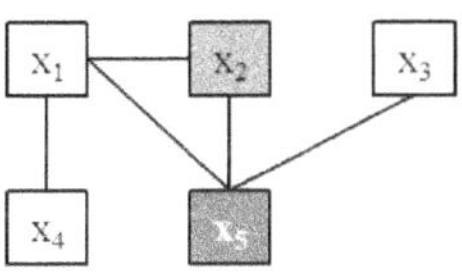

$$\begin{array}{c|ccccc} & x_1 & x_2 & x_3 & x_4 & x_5 \\ \hline \delta & 3 & 2 & 1 & 1 & 3 \\ \sigma & - & - & - & 1 & - \\ C & 1 & 3 & 1 & 0 & 2 \end{array}$$

→ We color x_4 with color 2.

The result is a solution using 3 colors.

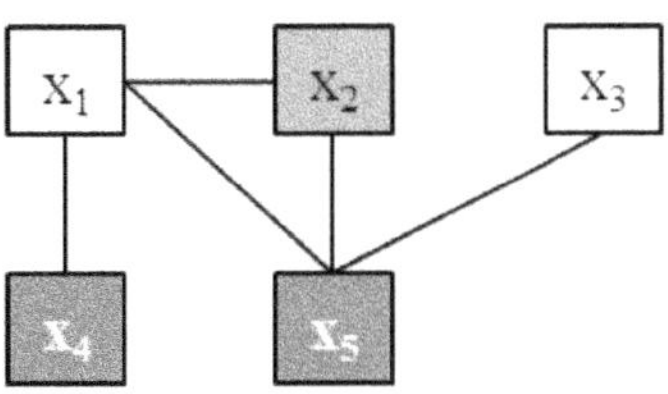

Color 1: x_1 , x_3

Color 2: x_4 , x_5

Color 3: x_2

The Jacobian matrix can be calculated in 3 calls (instead of 5) by successively varying (x_1 and x_3), then (x_4 and x_5), then x_2.

2.6.5 Travelling salesman problem

Statement

The travelling salesman must visit n cities travelling the shortest distance possible. Each city must be visited once and only once, whilst returning to the starting city at the end. The travelling salesman problem (denoted TSP) is NP-complete.

Heuristics

There are many heuristics suitable for the travelling salesman problem and its variants, including the following three:

- minimum cost arc heuristic:
  at each stage the path is extended by the minimum cost arc;
- maximum regret arc heuristic:
  at each stage the maximum regret arc (= non-selection cost) is chosen;
- minimum cost insertion heuristic:
  at each stage a new node is inserted at the position that minimizes the cost.

These three heuristics are illustrated in example 2-31.

Example 2-31: Heuristics for TSP

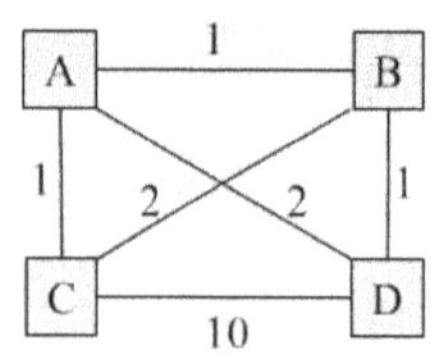

Consider a 4-city problem A, B, C, D opposite.
The optimal solution is the path (ADBCA) of length 6.

The performances of the 3 heuristics are compared.

Minimum cost arc heuristic

We start from A and extend at each stage through the arc of minimum cost. The stages are illustrated in figure 2-43. This gives the path (ABDCA) of length 13.

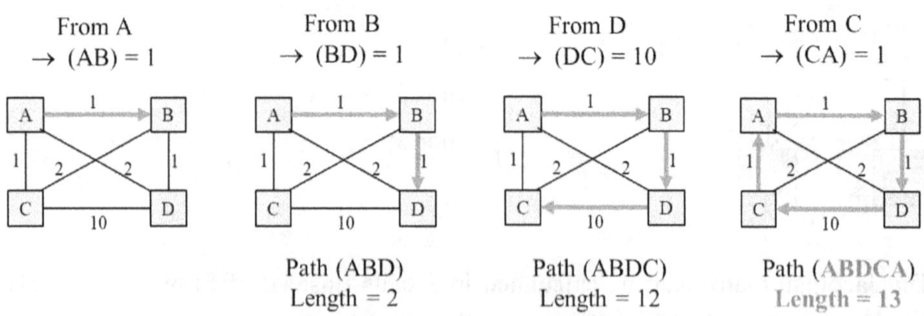

Figure 2-43: Minimum cost arc heuristics.

Maximum regret arc heuristic

The travelling salesman must pass through all the nodes. The consequences of not selecting an arc from node v to node w are examined.

If the arc (v,w) is not selected, we will have to select:

- another arc starting from v and generating a cost at least equal to r_v;

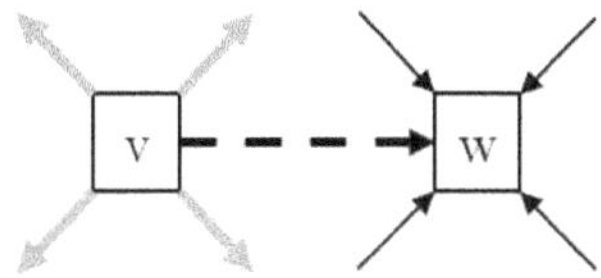

- another arc arriving at w and generating a cost at least equal to r_w.

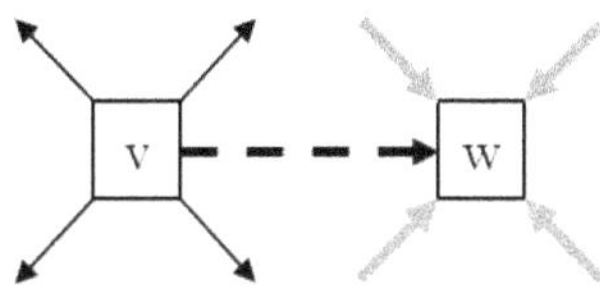

The **regret** associated with arc (v,w) is the minimum cost of not selecting arc (v,w). Its value is $r_{vw} = \max(r_v, r_w)$. The regret matrix is deduced from the distance matrix by taking the value of r_v (on the row) and r_w (on the column). For the arc CA, we have for example $r_v = 2$ (arc CB) and $r_w = 1$ (arc BA). Therefore, the value 2 is entered for the CA arc in the regret matrix.

$$\begin{array}{c} \\ A \\ B \\ C \\ D \end{array} \begin{array}{c} \begin{array}{cccc} A & B & C & D \end{array} \\ \begin{pmatrix} - & 1 & 1 & 2 \\ 1 & - & 2 & 1 \\ 1 & 2 & - & 10 \\ 2 & 1 & 10 & - \end{pmatrix} \end{array} \qquad \begin{array}{c} \\ A \\ B \\ C \\ D \end{array} \begin{array}{c} \begin{array}{cccc} A & B & C & D \end{array} \\ \begin{pmatrix} - & 1 & 2 & 1 \\ 1 & - & 1 & 2 \\ 2 & 1 & - & 1 \\ 1 & 2 & 1 & - \end{pmatrix} \end{array}$$

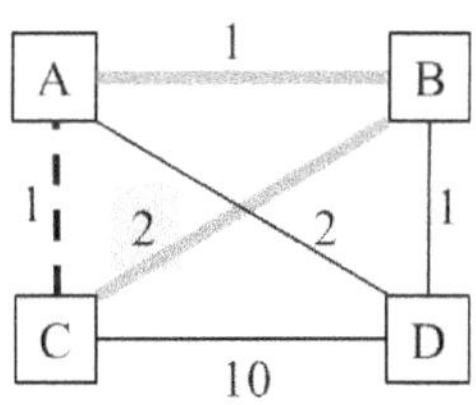

Figure 2-44: Regret matrix.

Let us choose among the arcs of maximum regret (= 2) the arc CA. We delete its row and column in the distance matrix, and look for the next arc among the arcs still available.

The stages 1, 2 and 3 are illustrated, with the matrix of available arcs on the left and the matrix of regrets in the center. The successively selected arcs are in thick lines. This gives the path (CADBC) of length 6.

Costs of available arcs — Regret matrix — Maximum regret arc

Stage 1

Costs of available arcs:

	A	B	C	D
A	–	1	1	2
B	1	–	2	1
C	1	2	–	10
D	2	1	10	–

Regret matrix:

	A	B	C	D
A	–	1	2	1
B	1	–	1	2
C	2	1	–	1
D	1	2	1	–

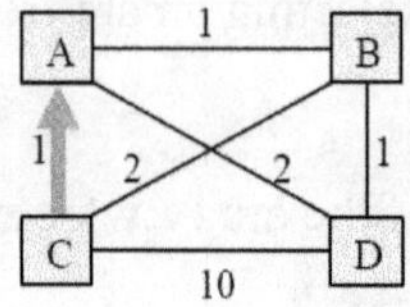

Stage 2

Costs of available arcs:

	A	B	C	D
A		1		2
B		–	2	1
C				
D		1	10	–

Regret matrix:

	A	B	C	D
A	–	2	–	1
B	–	–	10	2
C	–	–	–	–
D	–	10	2	–

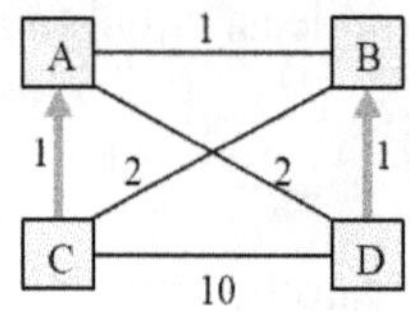

Stage 3

Costs of available arcs:

	A	B	C	D
A				2
B			2	
C				
D				

→ Arcs (BC) and (AD)

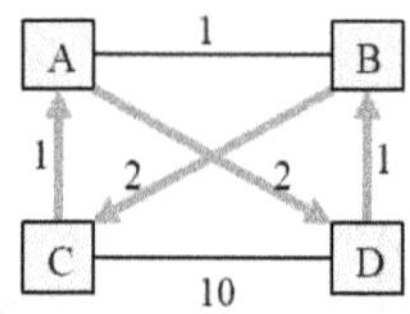

Minimum cost insertion heuristic

At each iteration a new node is inserted at the position that minimizes the cost:

- initialization
  We choose an arc with minimum cost → (AB) = 1;

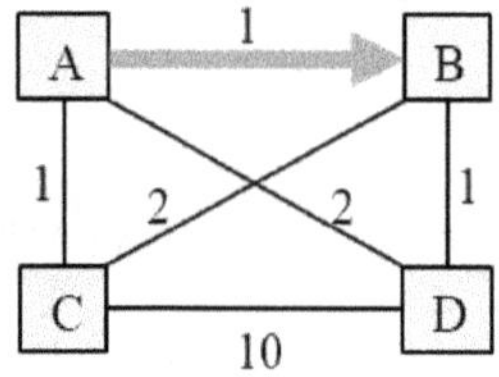

- inserting a new node on (AB)
  Node C: best insertion → (CAB) = 2
  Node D: best insertion → (ABD) = 2
  The path (CAB) of length = 2 is chosen;

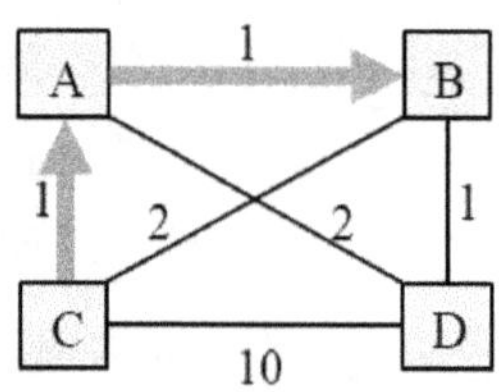

- inserting a new node on (CAB)
  Node D: best insertion → (CADBC) = 6
  (taking into account the return to the initial node)
  This gives the path (CADBC) of length = 6.

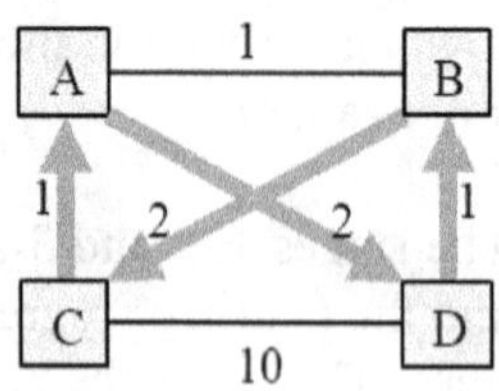

2.7 Conclusion

2.7.1 The key points

- Many combinatorial problems can be reduced to route problems in graphs;
- path problems are solved by Ford's, Bellman's, Dijkstra's or the A* algorithm;
- flow problems are solved by Ford-Fulkerson and Roy-Busacker-Gowen algorithms;
- assignment problems are solved by the Hungarian method;
- greedy heuristics quickly give approximate solutions.

2.7.2 To go further

- *Algorithmes de graphes (P. Lacomme, C. Prins, M. Sevaux, Eyrolles 2007)*

  The book is devoted to optimization problems in graphs. Chapters 1, 2 and 3 introduce theoretical notions on graphs, complexity and hard problems. Chapter 4 introduces structures for the implementation of algorithms in the Delphi language. Classical combinatorial optimization problems are addressed in chapter 5 with the description and implementation of dedicated algorithms. The book is accompanied by a CD-Rom with all the software presented.

- *Précis de recherche opérationnelle (R. Faure, B. Lemaire, C. Picouleau, Dunod 2009)*

  The book deals with operations research problems including combinatorial optimization and stochastic problems. Chapters 1 to 3 introduce the notions of graphs and complexity. Chapter 4 presents algorithms adapted to path problems. Numerous examples and exercises allow the reader to become familiar with the different methods.

- *Stochastic optimization (J.J. Schneider, S. Kirkpatrick, Springer 2006)*

  This book presents many heuristics (a term preferred to "metaheuristics") and analyzes more particularly the simulated annealing method, of which the authors are specialists. The different methods are applied and compared on the travelling salesman problem.

3. Functional optimization

Functional optimization, also called calculus of variations, concerns problems where the unknown is a function. Functional optimization is said to be in infinite dimension, as opposed to continuous or discrete optimization which deals with a finite number of variables. The unknown function can indeed be conceived as an infinite number of unknown variables which would be the values of the function.

Section 1 introduces the notions of functional (function of functions), neighborhood of functions and variation of a functional. These notions allow to define a strong or weak minimum of a functional depending on whether we consider neighborhood in value or in value and derivative. The standard problem of the calculus of variations consists in minimizing an integral type functional.

Section 2 establishes various optimality conditions. The necessary conditions of weak minimum lead to the Euler-Lagrange equation, whose solutions form the extremals of the functional. The sufficient conditions are based on the study of the conjugate points associated with the Jacobi equation. The extension to piecewise derivable functions leads to the Weierstrass-Erdmann corner conditions. The necessary strong minimum conditions based on the Weierstrass function are the equivalent of the maximum principle in optimal control.

Section 3 introduces the consideration of constraints. Constraints can be on the endpoint conditions (final constraints), on the fixed value of an integral (integral constraint) or on each point of the unknown function (current constraint). These constraints add optimality conditions in the form of transversality conditions for final constraints or in the form of unknown multipliers for integral or current constraints.

Section 4 presents the canonical form of a calculus of variations problem. The canonical variables form a Hamiltonian system with special properties, useful especially for applications in mechanics. The canonical form allows to derive the Hamilton-Jacobi-Bellman equation satisfied by the value function over the whole extremal field of the functional.

Section 5 introduces the essential notions of dynamic systems. This modelling allows us to move from a problem of calculating variations to a much more general problem of optimal control which is the subject of chapter 4.

3.1 Formulation

3.1.1 Functional

We consider real functions y defined on an interval $[a;b]$.

$$y : x \in [a;b] \mapsto y(x) \in \mathbb{R} \tag{3.1}$$

These functions belong to a space noted Y which can be, according to the case:
- the space of continuous or differentiable functions on $[a;b]$;
- the space of continuous or piecewise derivable functions on $[a;b]$.

A functional J is a function of Y in $\mathbb{R}$.

$$J : y \in Y \mapsto J(y) \in \mathbb{R} \tag{3.2}$$

A functional is therefore a "function of functions".
The objective is to find the functions $y \in Y$ minimizing J. This leads to defining the notions of neighborhood and variation of the functional J.

3.1.2 Neighborhood

The space of functions Y is provided with a norm measuring the distance between two functions. A norm on Y is a function of Y in $\mathbb{R}^+$ denoted $\| \; \|$ and satisfying

$$\forall y,z \in Y \,,\; \forall \lambda \in \mathbb{R} \,,\; \begin{cases} \|\lambda y\| = |\lambda| . \|y\| \\ \|y+z\| \leq \|y\| + \|z\| \end{cases} \tag{3.3}$$

The second property is the triangular inequality.

The usual norms on function spaces are:

- the 0-norm (for continuous functions): $\|y\|_0 = \max\limits_{x \in [a;b]} |y(x)|$;
- the 1-norm (for differentiable functions): $\|y\|_1 = \max\limits_{x \in [a;b]} |y(x)| + \max\limits_{x \in [a;b]} |y'(x)|$;
- the L_p norm (for integrable functions): $\|y\|_p = \left(\int_a^b |y(x)|^p \, dx \right)^{1/p}$.

  The L_∞ norm is identical to the 0-norm: $\|y\|_\infty = \|y\|_0$.

The neighborhood of a function y* is the set of functions of Y such that

$$\|y - y^*\| < \varepsilon \tag{3.4}$$

for a given positive real ε. The neighborhood depends on the norm considered.

The neighborhood in 0-norm is the set of continuous functions close to y* in value. These functions are not necessarily derivable.

$$|y(x) - y^*(x)| < \varepsilon \quad \text{on} \quad [a;b] \tag{3.5}$$

The neighborhood in -norm is the set of functions that are derivable and close to y* in value and derivative. This neighborhood is smaller (because it contains fewer functions) than the 0-norm neighborhood.

$$|y(x) - y^*(x)| + |y'(x) - y^{*\prime}(x)| < \varepsilon \quad \text{on} \quad [a;b] \tag{3.6}$$

Figure 3-1 shows functions y_0 and y_1 close to y for norms 0 and 1.

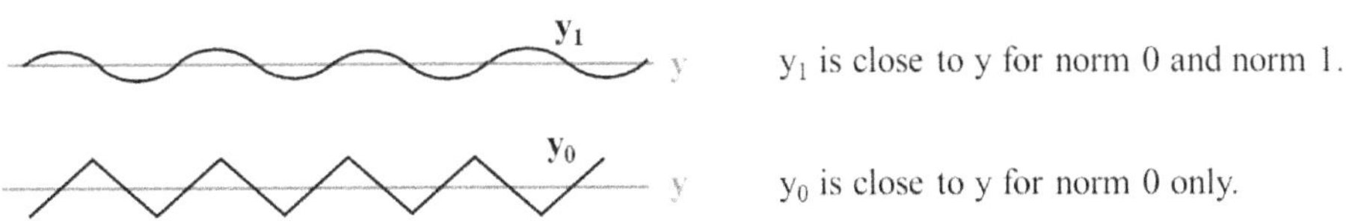

Figure 3-1: Neighborhood in norm 0 and 1.

3.1.3 Variation

To study the variation of a functional J in the neighborhood of y, we consider functions of the form $y + \varepsilon\eta$ where η is a function and ε is a "small" real. The function η (called perturbation) is feasible if $y + \varepsilon\eta$ belongs to Y. The first and second variations of J in the neighborhood of y are defined as follows.

<u>First variation</u>
The first variation of J in y is the linear functional δJ_y such that

$$\forall \eta, \forall \varepsilon \in \mathbb{R}, \quad J(y + \varepsilon\eta) = J(y) + \delta J_y(\eta)\varepsilon + o(\varepsilon) \tag{3.7}$$

By defining the real-variable function g_η associated with the function η

$$g_\eta : \varepsilon \in \mathbb{R} \mapsto g_\eta(\varepsilon) = J(y + \varepsilon\eta) \tag{3.8}$$

the first variation of J in y is expressed as

$$\delta J_y(\eta) = \lim_{\varepsilon\to 0}\frac{J(y+\varepsilon\eta)-J(y)}{\varepsilon} = \lim_{\varepsilon\to 0}\frac{g_\eta(\varepsilon)-g_\eta(0)}{\varepsilon} = g'_\eta(0) \tag{3.9}$$

The first variation is a linear functional, which means that it satisfies

$$\delta J_y(\alpha_1\eta_1+\alpha_2\eta_2) = \alpha_1\delta J_y(\eta_1)+\alpha_2\delta J_y(\eta_2) \tag{3.10}$$

Second variation

The second variation of J in y is the quadratic functional $\delta^2 J_y$ such that

$$\forall\eta\,,\ \forall\varepsilon\in\mathbb{R}\,,\ J(y+\varepsilon\eta) = J(y)+\delta J_y(\eta)\varepsilon+\delta^2 J_y(\eta)\varepsilon^2+o(\varepsilon^2) \tag{3.11}$$

With the function g_η defined in (3.8), the second variation is expressed as

$$\begin{aligned}\delta^2 J_y(\eta) &= \lim_{\varepsilon\to 0}\frac{J(y+\varepsilon\eta)-J(y)-\delta J_y(\eta)\varepsilon}{\varepsilon^2}\\ &= \lim_{\varepsilon\to 0}\frac{g_\eta(\varepsilon)-g_\eta(0)-g'_\eta(0)\varepsilon}{\varepsilon^2} = \frac{1}{2}g''_\eta(0)\end{aligned} \tag{3.12}$$

The second variation is a quadratic functional, which means that it is of the form $\delta^2 J_y(\eta) = K(\eta,\eta)$ where $K(\eta_1,\eta_2)$ is a bilinear functional (linear functional with respect to each of the two functions η_1 and η_2).

3.1.4 Minimum

A function y* is a local minimum of the functional J if the neighboring functions of y* give higher values for J. For any small real $\delta>0$, we have

$$\forall y\in Y\,,\ \|y-y^*\|<\delta \Rightarrow J(y^*)\le J(y) \tag{3.13}$$

The minimum is said to be weak or strong depending on the norm considered for the neighborhood:

- a weak minimum corresponds to a neighborhood in 1-norm. The condition (3.13) concerns only differentiable functions;
- a strong minimum corresponds to a neighborhood in 0-norm. The condition (3.13) concerns all continuous functions. The minimum is said to be strong, because this neighborhood contains more functions than the neighborhood in norm 1.

Using the second-order expansion (3.11), the minimum condition gives:

$$\begin{aligned} J(y) = J(y^* + \varepsilon\eta) = J(y^*) + \delta J_{y^*}(\eta)\varepsilon + \delta^2 J_{y^*}(\eta)\varepsilon^2 + o(\varepsilon^2) \geq J(y^*) \\ \Rightarrow \delta J_{y^*}(\eta)\varepsilon + \delta^2 J_{y^*}(\eta)\varepsilon^2 \geq 0 \end{aligned} \quad (3.14)$$

This inequality must be satisfied for any feasible perturbation η and any small real ε. It leads to conditions of order 1 (on δJ_y) and order 2 (on $\delta^2 J_y$) as follows. The necessary conditions of minimum are obtained for a broad inequality.

$$y^* \text{ minimum of } J \Rightarrow \begin{cases} \delta J_{y^*}(\eta) = 0 \\ \delta^2 J_{y^*}(\eta) \geq 0 \end{cases}, \quad \forall \eta \text{ feasible} \quad (3.15)$$

The sufficient conditions of minimum are obtained for a strict inequality.

$$\begin{cases} \delta J_{y^*}(\eta) = 0 \\ \delta^2 J_{y^*}(\eta) > 0 \end{cases}, \quad \forall \eta \text{ feasible} \Rightarrow y^* \text{ minimum of } J \quad (3.16)$$

The strictly positive second order term allows to neglect the third order terms.

The section 3.1.5 applies these conditions to integral functions.

3.1.5 Standard problem

The standard problem of the calculus of variations is of the form

$$\min_{y \in Y} J(y) = \int_a^b L[x, y(x), y'(x)] dx \quad (3.17)$$

The function $L : \mathbb{R}^3 \to \mathbb{R}$ in the integral is called the **Lagrangian** of the problem.

This function depends on three real variables (x,y,z) and it is assumed to be "sufficiently derivable".

To simplify the equations, the following notations are used:

$$\begin{aligned} & L[x, y(x), y'(x)] \underset{\text{noted}}{=} L(x, y, y') \\ & \frac{\partial L}{\partial y} \underset{\text{noted}}{=} L_y \;,\; \frac{\partial L}{\partial z} \underset{\text{noted}}{=} L_z \text{ or } L_{y'} \;,\; \frac{d}{dx}\left(\frac{\partial L}{\partial z}\right) \underset{\text{noted}}{=} L'_z \end{aligned} \quad (3.18)$$

The second partial derivatives are similarly noted L_{yy}, L_{yz}, L_{zz}.

The notations y' and L'_z refer to the total derivatives of the functions y and L_z with respect to the variable x.

Y is the space of functions of class C^1 on $[a;b]$ and with fixed values in a and b.

$$y \in Y \quad \Leftrightarrow \quad y \in C^1([a;b]) \text{ and } \begin{cases} y(a) = \alpha \\ y(b) = \beta \end{cases} \tag{3.19}$$

Figure 3-2 shows three functions y_1, y_2 and y^* belonging to Y (functions of class C^1 with fixed values at the ends of the interval).

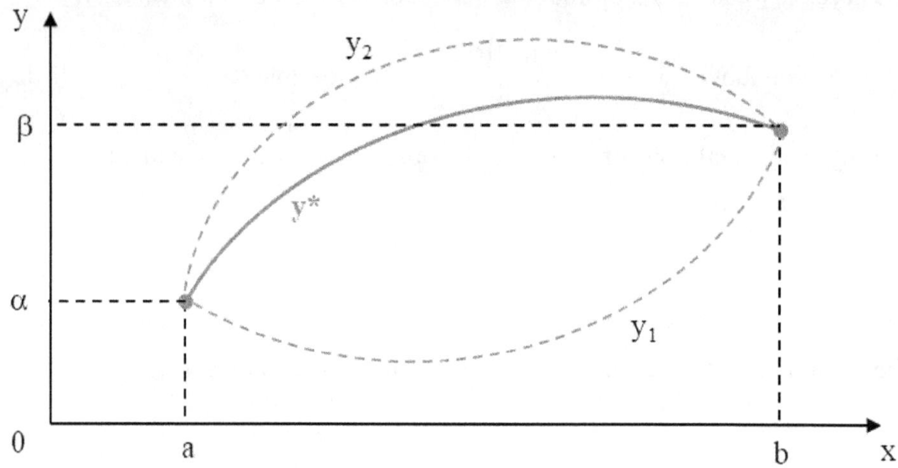

Figure 3-2: Functions with fixed endpoint values.

The neighborhood of the function y^* is the set of functions of the form $y = y^* + \varepsilon\eta$ belonging to Y. The perturbation η must therefore satisfy

$$\eta \in C^1([a;b]) \text{ and } \begin{cases} y(a) = \alpha \Rightarrow \eta(a) = 0 \\ y(b) = \beta \Rightarrow \eta(b) = 0 \end{cases} \tag{3.20}$$

The first and second variations of the functional J defined by (3.17) have then the following form:

$$\begin{aligned} \delta J_{y^*}(\eta) &= \int_a^b \eta\left(L_y - L'_z\right)dx \\ \delta^2 J_{y^*}(\eta) &= \frac{1}{2}\int_a^b \left(\eta^2(L_{yy} - L'_{yz}) + \eta'^2 L_{zz}\right)dx \end{aligned} \tag{3.21}$$

These formulas demonstrated below will be used in the section 3.2 to establish the necessary and sufficient conditions for local minimum.

Demonstration

The variations of J are obtained from definition (3.11) recalled here:

$J(y^*+\varepsilon\eta) = J(y^*) + \delta J_{y^*}(\eta)\varepsilon + \delta^2 J_{y^*}(\eta)\varepsilon^2 + o(\varepsilon^2)$.

Let us calculate $J(y^*+\varepsilon\eta)$ from (3.17): $J(y^*+\varepsilon\eta) = \int_a^b L\left[x, y^*+\varepsilon\eta, (y^*+\varepsilon\eta)'\right]dx$.

The Lagrangian is expanded to order 2 with respect to ε.

$$L\left[x, y^*+\varepsilon\eta, (y^*+\varepsilon\eta)'\right] = L\left[x, y^*, y^{*\prime}\right] + \varepsilon\eta L_y + \varepsilon\eta' L_z + \frac{1}{2}\varepsilon^2\eta^2 L_{yy} + \varepsilon^2\eta\eta' L_{yz} + \frac{1}{2}\varepsilon^2\eta'^2 L_{zz} + o(\varepsilon^2)$$

All partial derivatives of L are evaluated in $(x, y^*, y^{*\prime})$. By replacing in the expression of $J(y^*+\varepsilon\eta)$, and then identifying the terms in ε and ε^2, we obtain

$$\begin{cases} \delta J_{y^*}(\eta) = \int_a^b (\eta L_y + \eta' L_z)dx \\ \delta^2 J_{y^*}(\eta) = \frac{1}{2}\int_a^b (\eta^2 L_{yy} + 2\eta\eta' L_{yz} + \eta'^2 L_{zz})dx \end{cases}$$

The first equation is integrated by parts, noting $L_z' = \frac{d}{dx}\left(\frac{\partial L}{\partial z}(x, y^*, y^{*\prime})\right)$.

$$\delta J_{y^*}(\eta) = \int_a^b (\eta L_y + \eta' L_z)dx = \int_a^b \eta L_y dx + \left[\eta L_z\right]_a^b - \int_a^b \eta L_z' dx$$

$$\Rightarrow \delta J_{y^*}(\eta) = \int_a^b \eta(L_y - L_z')dx \quad \text{because } \eta(a) = \eta(b) = 0 \text{ (feasible perturbation } \eta\text{)}$$

The second equation is also integrated by parts.

$$\delta^2 J_{y^*}(\eta) = \frac{1}{2}\int_a^b (\eta^2 L_{yy} + 2\eta\eta' L_{yz} + \eta'^2 L_{zz})dx$$

with $\int_a^b 2\eta\eta' L_{yz} dx = \int_a^b (\eta^2)' L_{yz} dx = \left[\eta^2 L_{yz}\right]_a^b - \int_a^b \eta^2 L_{yz}' dx = -\int_a^b \eta^2 L_{yz}' dx$

$$\Rightarrow \delta^2 J_{y^*}(\eta) = \frac{1}{2}\int_a^b \left(\eta^2 (L_{yy} - L_{yz}') + \eta'^2 L_{zz}\right)dx$$

This gives the expressions (3.21).

3.2 Optimality conditions

In this section we establish optimality conditions for the standard problem of calculus of variations (3.17) recalled below.

$$\min_y J(y) = \int_a^b L(x, y, y')dx \quad \text{with} \begin{cases} y(a) = \alpha \\ y(b) = \beta \end{cases} \tag{3.22}$$

For this type of problem, we consider derivable functions on $[a;b]$.
A distinction is made between a weak and a strong minimum:

- a weak minimum concerns class C^1 functions with the 1-norm neighborhood. Weak minimum conditions are discussed in sections 3.2.1 and 3.2.2;
- a strong minimum concerns piecewise class C^1 functions with the 0-norm neighborhood.
  Strong minimum conditions are discussed in sections 3.2.3 and 3.2.4.

3.2.1 Weak minimum necessary conditions

The functions considered in this section are of class C^1 on $[a;b]$.
The theorem 3.1 gives necessary conditions that a function y* must satisfy to be a weak local minimum of the functional J.

Theorem 3-1: Necessary conditions for a weak minimum

If y* is a weak local minimum of problem (3.22), then y* satisfies:

- the Euler-Lagrange equation (first order condition):

$$L_y - L'_z = 0. \tag{3.23}$$

- the Legendre condition (condition of order 2):

$$L_{zz} \geq 0. \tag{3.24}$$

Demonstration (see [R14])

The Euler-Lagrange equation (1755) and the Legendre condition (1786) are established from the first and second variations of J in y*.

Condition of order 1

Assume that y* is a weak local minimum of J.
According to (3.15), the first variation of J must be zero for any feasible perturbation η. Using the expression (3.21), we must have

$$\delta J_{y*}(\eta) = \int_a^b \eta(L_y - L_z')dx = 0 \quad , \quad \forall \eta$$

For a weak minimum, we consider class C^1 functions η such that: $\eta(a)=\eta(b)=0$.

Let us define the function: $P(x) = L_y - L_z' \quad \Rightarrow \quad \delta J_{y*}(\eta) = \int_a^b \eta P dx$.

Assume by contradiction that the function P is not identically zero on $[a;b]$.
Then there exists $c \in [a;b]$ such that $P(c) = 2\delta > 0$ (the reasoning is similar if $P(c) = 2\delta < 0$). Since the function P is continuous, there is a non-zero interval $[c_1;c_2]$ around c on which $P(x) > \delta$.

Let us define the perturbation η as

$$\eta(x) = \begin{cases} (x-c_1)^2(c_2-x)^2 & \text{if } x \in [c_1;c_2] \\ 0 & \text{if } x \notin [c_1;c_2] \end{cases}$$

This perturbation is feasible
(of class C^1 and taking zero value in a and b).

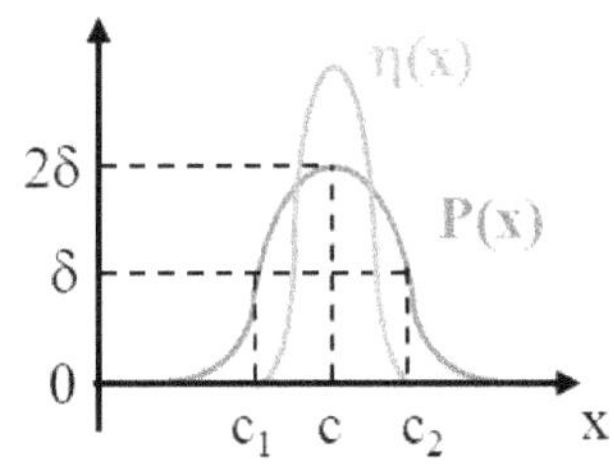

Let us calculate the first variation of J for this perturbation η.

$$\delta J_{y*}(\eta) = \int_a^b \eta P dx = \int_{c_1}^{c_2} P(x-c_1)^2(c_2-x)^2 dx \;\geq\; \int_{c_1}^{c_2} \delta(x-c_1)^2(c_2-x)^2 dx$$

This integral is strictly positive: $\delta > 0$ and $(x-c_1)^2(c_2-x)^2 > 0$ on $]c_1;c_2[$.
This would yield $\delta J_{y*}(\eta) > 0$, which would contradict the necessary condition of order 1 (3.15). The initial assumption is therefore false: the function $P(x)=L_y - L_z'$ is identically zero on $[a;b]$. This establishes the Euler-Lagrange equation.

Condition of order 2

Assume that y* is a weak local minimum of J.
According to (3.15), the second variation of J must be positive for any feasible perturbation η. Using the expression (3.21), we must have

$$\delta^2 J_{y*}(\eta) = \frac{1}{2}\int_a^b \left(\eta^2(L_{yy} - L_{yz}') + \eta'^2 L_{zz}\right)dx \geq 0 \quad , \quad \forall \eta$$

Let us note: $\begin{cases} P(x) = L_{zz} \\ Q(x) = L_{yy} - L'_{yz} \end{cases} \Rightarrow \delta^2 J_{y*}(\eta) = \frac{1}{2}\int_a^b \left(\eta'^2 P + \eta^2 Q\right) dx$.

Assume by contradiction that the function P is not positive or zero on $[a;b]$. Then there exists $c \in [a;b]$ such that $P(c) = -2\delta < 0$. Since the function P is continuous, there is an interval $[c-\varepsilon; c+\varepsilon]$ around c on which $P(x) < -\delta$.

Let us define the perturbation η as

$$\eta(x) = \begin{cases} \sin^2 \dfrac{\pi(x-c)}{\varepsilon} & \text{if } x \in [c-\varepsilon; c+\varepsilon] \\ 0 & \text{else} \end{cases}$$

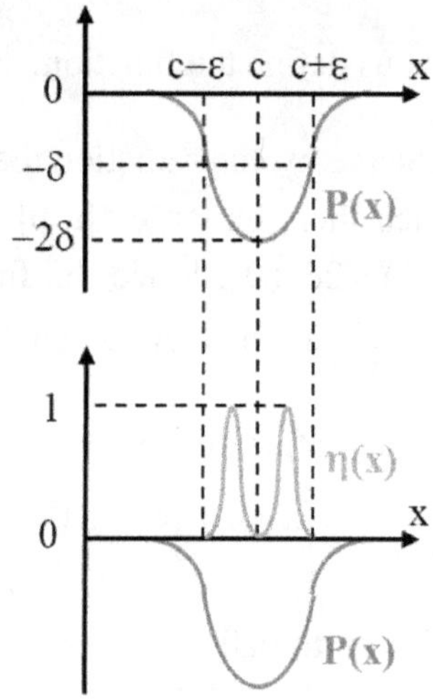

This perturbation is feasible
(of class C^1 and taking zero value in a and b).

The figure opposite shows the shape of the functions $P(x)$ and $\eta(x)$ in the vicinity of c.

Let us calculate the second variation of J for this perturbation η.

$$\delta J_{y*}(\eta) = \frac{1}{2}\int_a^b \left(\eta'^2 P + \eta^2 Q\right) dx = \frac{1}{2}\int_{c-\varepsilon}^{c+\varepsilon} \frac{\pi^2}{\varepsilon^2} P \sin^2 \frac{2\pi(x-c)}{\varepsilon} dx + \frac{1}{2}\int_{c-\varepsilon}^{c+\varepsilon} Q \sin^4 \frac{\pi(x-c)}{\varepsilon} dx$$

$$\leq \varepsilon \frac{\pi^2}{\varepsilon^2} \max_{c-\varepsilon \leq x \leq c+\varepsilon} P + \varepsilon \max_{c-\varepsilon \leq x \leq c+\varepsilon} Q$$

by taking a majorant under each integral.

By assumption we have: $P(x) < -\delta < 0$ on $[c-\varepsilon; c+\varepsilon]$

so that the inequality yields: $\delta J_{y*}(\eta) \leq -\dfrac{\pi^2}{\varepsilon}\delta + \varepsilon \max\limits_{c-\varepsilon \leq x \leq c+\varepsilon} Q$.

For ε sufficiently small, the second member becomes negative.
This would yield $\delta^2 J_{y*}(\eta) < 0$, which would contradict the necessary condition of order 2 (3.15).
The initial assumption is therefore false: the function $P(x) = L_{zz}$ is positive or zero on $[a;b]$. This establishes the Legendre condition.

The Euler-Lagrange equation (denoted EL) is a differential equation of order 2. The integration constants are determined by the boundary conditions $y(a)=\alpha$ and $y(b)=\beta$. In detailed form, equation (EL) is written as

$$\frac{\partial L}{\partial y}(x,y,y')-\frac{d}{dx}\left(\frac{\partial L}{\partial z}\right)(x,y,y')=0 \tag{3.25}$$

or by developing the composed derivative of the second term

$$\frac{\partial L}{\partial y}\left[x,y(x),y'(x)\right]=\left(\frac{\partial L_z}{\partial x}+y'\frac{\partial L_z}{\partial y}+y''\frac{\partial L_z}{\partial z}\right)\left[x,y(x),y'(x)\right] \tag{3.26}$$

Extremal and first integral

A function y satisfying equation (EL) is called an **extremal** of the functional (3.22). The extremal field of the functional is the set of solutions of equation (EL). These solutions are parameterized by two integration constants. An extremal is not necessarily a minimum of the functional, because equation (EL) is only a necessary condition. A **first integral** of J is a function $\Phi(x,y)$ that keeps a constant value along any extremal y*.

$$\Phi\left[x,y^*(x)\right]=C^{te} \tag{3.27}$$

Special cases:

- if the function L does not depend explicitly on y, then $L_y=0$.

The Euler-Lagrange equation in its form (3.25) reduces to

$$\frac{d}{dx}\left(\frac{\partial L}{\partial y'}\right)=0 \;\Rightarrow\; L_{y'}=C^{te} \tag{3.28}$$

The function $L_{y'}$ is in this case a first integral (corresponding to the impulse or the momentum in mechanics);

- if the function L does not depend explicitly on x, then $L_{zx}=0$.

The Euler-Lagrange equation in its form (3.26) reduces to

$$L_y=y'L_{zy}+y''L_{zz} \tag{3.29}$$

It can be deduced that (the verification is presented afterwards):

$$L_zy'-L=C^{te} \tag{3.30}$$

The function $H = L_{y'}y' - L$ is in this case a first integral (corresponding to the Hamiltonian or energy in mechanics).

Verification

Let us calculate the total derivative of $L_z y' - L$ with respect to x.

$$\frac{d}{dx}(L_z y' - L) = (L_{zx} + L_{zy}y' + L_{zz}y'')y' + L_z y'' - (L_x + L_y y' + L_z y'')$$
$$= (L_{zy}y' + L_{zz}y'' - L_y)y' + L_{zx}y' - L_x$$

The first term is zero according to the Euler-Lagrange equation (3.29). The other terms are zero because L and therefore L_z are not explicit in x by assumption. The derivative of $L_z y' - L$ is therefore zero and this quantity is a first integral.

Example 3-1 shows the application of the weak minimum conditions to the calculation of the shortest distance between two points in the plane.

Example 3-1: Minimum length curve in the plane

We seek the curve $y(x)$ minimizing the distance between A and B in the plane.

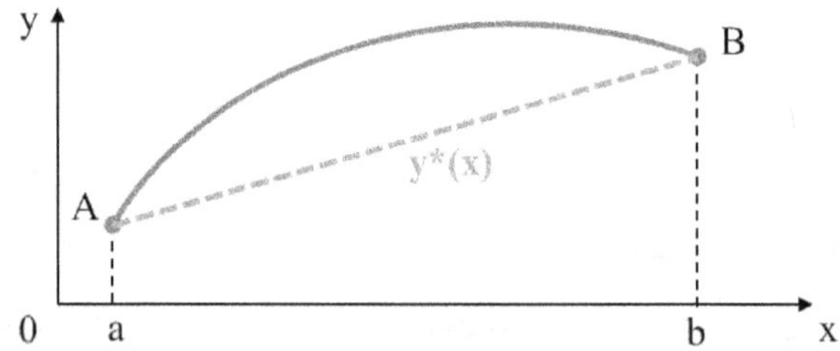

The curvilinear distance is: $D_{AB} = \int_A^B ds$ with $ds^2 = dx^2 + dy^2 = dx^2\left(1 + y'(x)^2\right)$.

We are looking for the curve $y^*(x)$ solution of

$$\min_y \int_a^b \sqrt{1 + y'(x)^2}\,dx \quad \Leftrightarrow \quad \min_y \int_a^b L(x, y, y')dx \quad \text{with } L(x, y, z) = \sqrt{1 + z^2}$$

The Lagrangian is not explicit in y. The Euler-Lagrange equation yields

$$\frac{\partial L}{\partial y}=\frac{d}{dx}\left(\frac{\partial L}{\partial y'}\right) \Rightarrow \frac{\partial L}{\partial y'}=\frac{y'}{\sqrt{1+y'^2}}=C^{te} \Rightarrow y'=C^{te} \quad \text{(first integral)}$$

The extremals of this problem are curves of constant slope, hence straight lines. The particular solution associated with two points A and B is the segment joining these two points.

Legendre's necessary condition is also satisfied: $L_{zz}=\left(1+y'^2\right)^{\frac{3}{2}} \geq 0$.

All straight lines do satisfy the necessary weak minimum conditions.

We will see in example 3-4 that the sufficient conditions are satisfied.

Example 3-2 is the historical brachistochrone problem formulated by Bernoulli in 1696. This problem led to the development of the calculus of variations.

Example 3-2: Brachistochrone

We are looking for the shape of the curve AB that minimizes the travel time of a particle sliding under the effect of gravity (without friction), in other words the shape of the optimal toboggan from A to B.

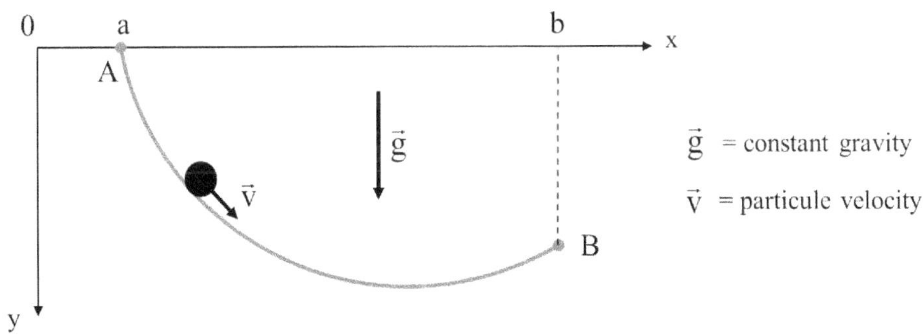

Figure 3-3: Brachistochrone problem.

The term "brachistochrone" is of Greek origin:
"βραχιστοσ" = shortest, "χρονοσ" = time.

Formulation of the problem

The first stage is to write the problem in the form (3.22).

The particle starts from the point $A(a;\alpha=0)$ with a zero velocity and arrives at the point $B(b;\beta)$.

The energy of the particle is constant because the gravity field is conservative. The energy at A is zero (taking the y-intercept at A), which determines its velocity as a function of the height: $E=\frac{1}{2}mv^2-mgy=0 \Rightarrow v=\sqrt{2gy}$.

The curve from A to B has the equation $y(x)$. The travel time from A to B is

$$T_{AB}=\int_A^B \frac{ds}{v} \quad \text{with} \quad ds^2=dx^2+dy^2=dx^2\left(1+y'(x)^2\right)$$

The problem is formulated as: $\min\limits_y \int_a^b \sqrt{\frac{1+y'(x)^2}{y(x)}}dx \quad \text{with} \quad \begin{cases} y(a)=0 \\ y(b)=\beta \end{cases}$.

Resolution of the problem

The Lagrangian: $L(x,y,z)=\sqrt{\frac{1+z^2}{y}}$ is not explicit in x.

According to (3.30), the Euler-Lagrange equation gives the first integral

$$L_{y'}y'-L=H=C^{te} \Rightarrow L_{y'}y'-L=\frac{y'^2}{\sqrt{y(1+y'^2)}}-\sqrt{\frac{1+y'^2}{y}}=H \Rightarrow y(1+y'^2)=\frac{1}{H^2}$$

The differential equation to solve is: $y(1+y'^2)=2c \quad \text{with} \quad c=\frac{1}{2H^2}$.

Let us perform the function change: $y'=-\tan\frac{\theta}{2}$ (**equation E_1**)

and replace in the differential equation to express y.

$$y=\frac{2c}{1+y'^2}=2c\cos^2\frac{\theta}{2} \Rightarrow y=c(1+\cos\theta)$$

Deriving this expression yields: $y'=-c\theta'\sin\theta=-2c\theta'\sin\frac{\theta}{2}\cos\frac{\theta}{2}$ (**equation E_2**).

By identifying the expressions (E_1) and (E_2) of y', we obtain an equation in θ.

$$y'=-\tan\frac{\theta}{2}=-2c\theta'\sin\frac{\theta}{2}\cos\frac{\theta}{2} \Rightarrow 2c\theta'\cos^2\frac{\theta}{2}=1 \Rightarrow c(1+\cos\theta)d\theta=dx$$

This differential equation is integrated into: $x=c(\theta-\theta_0+\sin\theta)$.

We obtain expressions for x and y parametrized in terms of θ.

The solutions are the curves of equation: $\begin{cases} x(\theta) = c(\theta - \theta_0 + \sin\theta) = x_c + c\sin\theta \\ y(\theta) = c(1+\cos\theta) \qquad = y_c + c\cos\theta \end{cases}$.

These extremals are cycloids with center $(x_c;y_c)$ and radius c (figure 3-4).
The conditions in A and B give the 2 integration constants c and θ_0.

A cycloid results from a uniform circular motion around a uniformly translating center (e.g. the motion relative to the ground of a point on a bicycle wheel).

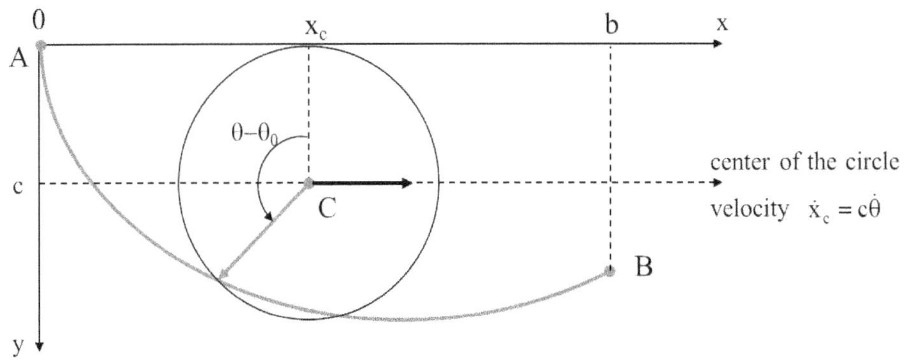

Figure 3-4: Cycloid solution of the brachistochrone.

The conditions of theorem 3-1 can be generalized to different problems.

Problem with several variables

Consider the following functional optimization problem.

$$\min_z J(z) = \iint_{(x,y)\in D\subset \mathbb{R}^2} L\left(x,y,z,z_x,z_y\right)dxdy \tag{3.31}$$

The Lagrangian here is a function of $\mathbb{R}^5$ in $\mathbb{R}$.

The functional J deals with a function z of two variables x and y, defined over a region D of $\mathbb{R}^2$. A necessary condition for a weak minimum is then

$$\frac{\partial L}{\partial z} = \frac{\partial}{\partial x}\left(\frac{\partial L}{\partial z_x}\right) + \frac{\partial}{\partial y}\left(\frac{\partial L}{\partial z_y}\right) \tag{3.32}$$

This second-order partial differential equation is the generalization to two variables of the second-order Euler-Lagrange differential equation (3.23). It can be proved in a similar way from the first variation of the functional J (3.31). Equation (3.32) generalizes to a function z of n variables.

Problem with several functions

Consider the functional optimization problem:

$$\min_{y_1,\cdots,y_n \in Y} J(y_1,\cdots,y_n) = \int_a^b L\left(x, y_1,\cdots,y_n, y_1',\cdots,y_n'\right)dx \tag{3.33}$$

The Lagrangian here is a function of $\mathbb{R}^{2n+1}$ in $\mathbb{R}$.

The functional J depends on n functions $y_1, \ldots, y_n$ of the variable x. These functions of class C^1 have their values fixed in a and b: $\begin{cases} y_j(a) = \alpha_j \\ y_j(b) = \beta_j \end{cases}$, $j = 1$ to n.

A necessary condition for a weak minimum is then

$$\frac{\partial L}{\partial y_j} = \frac{d}{dx}\left(\frac{\partial L}{\partial y_j'}\right) \quad , \ j = 1 \text{ to } n \tag{3.34}$$

Each function $(y_j)_{j=1 \text{ to } n}$ satisfies the Euler-Lagrange equation: $L_{y_j} = L'_{z_j}$.

The demonstration is similar to the case of a single function (by applying n independent perturbations η_j). Legendre's condition generalizes in the same way.

Parameterized curve

Let us retrieve the functional optimization problem in its standard form.

$$\min_y J(y) = \int_a^b L(x, y, y')dx \quad \text{with} \quad \begin{cases} y \in C^1[a;b] \\ y(a) = \alpha \ , \ y(b) = \beta \end{cases} \tag{3.35}$$

This formulation restricts the solutions to functions $y : x \in \mathbb{R} \mapsto y(x) \in \mathbb{R}$.
It does not allow to find general curves in the plane (x, y).

To enlarge the set of solutions, we consider curves parametrized by a real variable $t \in [t_a ; t_b]$.

$$\begin{cases} x = x(t) \\ y = y(t) \end{cases} \quad \text{with} \quad \begin{cases} x(t_a) = a \\ x(t_b) = b \end{cases} \quad \text{and} \quad \begin{cases} y(t_a) = \alpha \\ y(t_b) = \beta \end{cases} \tag{3.36}$$

By introducing the notations

$$\frac{dx}{dt} \underset{\text{noted}}{=} \dot{x} \ , \ \frac{dy}{dt} \underset{\text{noted}}{=} \dot{y} \ , \ \frac{dy}{dx} \underset{\text{noted}}{=} y' \ \Rightarrow \ y' = \frac{\dot{y}}{\dot{x}} \tag{3.37}$$

the term under the integral (3.35) is expressed as

$$L(x,y,y')dx = L\left(x(t),y(t),\frac{\dot{y}(t)}{\dot{x}(t)}\right)\dot{x}(t)dt \tag{3.38}$$

Problem (3.35) can then be written as

$$\begin{aligned} &\min_{x,y} K(x,y) = \int_{t_a}^{t_b} M(t,x,y,\dot{x},\dot{y})dt \\ &\quad \text{with } M(t,x,y,\dot{x},\dot{y}) = L\left(x(t),y(t),\frac{\dot{y}(t)}{\dot{x}(t)}\right)\dot{x}(t) \end{aligned} \tag{3.39}$$

This is a problem of calculus of variations of the form (3.33) with two unknown functions $x(t)$ and $y(t)$, with end conditions in t_a and t_b.
Each unknown function satisfies the Euler-Lagrange equation (3.34).

$$\begin{cases} M_x = \dot{M}_{\dot{x}} & \Leftrightarrow \dfrac{\partial M}{\partial x} = \dfrac{d}{dt}\left(\dfrac{\partial M}{\partial \dot{x}}\right) \\ M_y = \dot{M}_{\dot{y}} & \Leftrightarrow \dfrac{\partial M}{\partial y} = \dfrac{d}{dt}\left(\dfrac{\partial M}{\partial \dot{y}}\right) \end{cases} \tag{3.40}$$

This differential system on $x(t)$ and $y(t)$ is of dimension 2 and of order 2.
The two differential equations are in fact not independent. Each of them is equivalent to the Euler-Lagrange equation of the original problem (3.35).

Verification

Let us first consider the first equation (3.40) associated with $x(t)$: $M_x = \dot{M}_{\dot{x}}$.

We express the derivatives M_x and $\dot{M}_{\dot{x}}$ of: $M(t,x,y,\dot{x},\dot{y}) = L\left(x,y,\frac{\dot{y}}{\dot{x}}\right)\dot{x}$.

$\rightarrow \ M_x = L_x\dot{x} \ , \ M_{\dot{x}} = L_z\frac{-\dot{y}}{\dot{x}^2}\dot{x} + L = L - L_z\frac{\dot{y}}{\dot{x}} = L - L_z y'$ with the notations (3.37)

then replace in $M_x = \dot{M}_{\dot{x}}$.

$$L_x\dot{x} = \frac{d}{dt}(L - L_z y') = \frac{d}{dx}(L - L_z y')\dot{x} \quad \textbf{(equation E)}$$

Let us develop the derivatives $\frac{d}{dx}$: $\begin{cases} \frac{d}{dx}(L) &= L_x + L_y y' + L_z y'' \\ \frac{d}{dx}(L_z y') &= \frac{d}{dx}(L_z) y' + L_z y'' \end{cases}$

By replacing in equation (E), the terms in L_x and y'' are eliminated.

This yields: $L_y = \frac{d}{dx}(L_z)$ which is Euler-Lagrange equation of problem (3.35).

Then consider the second equation (3.40) associated with $y(t)$: $M_y = \dot{M}_{\dot{y}}$.

Let us express the derivatives M_y and $\dot{M}_{\dot{y}}$ of: $M(t,x,y,\dot{x},\dot{y}) = L\left(x,y,\frac{\dot{y}}{\dot{x}}\right)\dot{x}$

$\rightarrow M_y = L_y \dot{x}$, $M_{\dot{y}} = L_z \frac{1}{\dot{x}} \dot{x} = L_z$ with the notations (3.37)

then replace in $M_y = \dot{M}_{\dot{y}}$: $L_y \dot{x} = \frac{d}{dt}(L_z) = \frac{d}{dx}(L_z)\dot{x} \Rightarrow L_y = \frac{d}{dx}(L_z)$

The Euler-Lagrange equation of the original problem (3.35) is again found.

The formulation (3.39) allows any curves in the plane (x,y) as a solution. This formulation applies especially to the search for geodesics.

Geodesics

A geodesic is a curve minimizing the distance between two points on a given surface. Consider a surface defined in a general way by a position vector $\vec{r}$ depending on two real variables (u,v). A geodesic on the surface is a parametric curve of the form: $\vec{r}(t) = r(u(t), v(t))$.

The length element on the surface is expressed as

$$ds^2 = \|d\vec{r}\|^2 = \|\vec{r}_u du + \vec{r}_v dv\|^2 = r_u^2 du^2 + 2\vec{r}_u \cdot \vec{r}_v du dv + r_v^2 dv^2 \tag{3.41}$$

or by posing: $E = r_u^2$, $F = \vec{r}_u \cdot \vec{r}_v$, $G = r_v^2$

$$ds = \sqrt{Eu'^2 + 2Fu'v' + Gv'^2}\, dt \tag{3.42}$$

A geodesic between two points A and B on the surface is solution to the problem

$$\min_{u,v} J(u,v) = \int_A^B ds = \int_{t_a}^{t_b} \sqrt{Eu'^2 + 2Fu'v' + Gv'^2}\, dt \tag{3.43}$$

This is a problem with two unknown functions $u(t), v(t)$ with end conditions in t_a and t_b. The Euler-Lagrange equations (3.40) for u and v yield

$$\begin{cases} \dfrac{E_u u'^2 + 2F_u u'v' + G_u v'^2}{\sqrt{Eu'^2 + 2Fu'v' + Gv'^2}} = \dfrac{d}{dt}\left(\dfrac{2(Eu' + Fv')}{\sqrt{Eu'^2 + 2Fu'v' + Gv'^2}}\right) & \rightarrow \text{for u} \\ \dfrac{E_v u'^2 + 2F_v u'v' + G_v v'^2}{\sqrt{Eu'^2 + 2Fu'v' + Gv'^2}} = \dfrac{d}{dt}\left(\dfrac{2(Fu' + Gv')}{\sqrt{Eu'^2 + 2Fu'v' + Gv'^2}}\right) & \rightarrow \text{for v} \end{cases} \tag{3.44}$$

The functions $E = \vec{r}_u^{\,2}$, $F = \vec{r}_u \cdot \vec{r}_v$, $G = \vec{r}_v^{\,2}$ are called coefficients of the fundamental quadratic form of the surface. The geodesics in the plane (straight lines) were determined in example 3-1. Example 3-3 illustrates the search for geodesics on a cylinder using equations (3.44).

Example 3-3: Geodesics on a cylinder

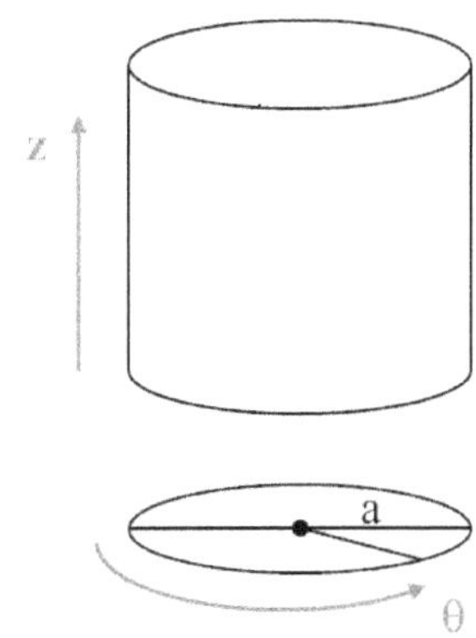

Let us consider a right cylinder with a circular base and a radius a.
The position on the cylinder is defined by an angle θ and a height z.

$$\vec{r} = \begin{pmatrix} a\cos\theta \\ a\sin\theta \\ z \end{pmatrix} \Rightarrow \vec{r}_\theta = \begin{pmatrix} -a\sin\theta \\ a\cos\theta \\ 0 \end{pmatrix} , \quad \vec{r}_z = \begin{pmatrix} 0 \\ 0 \\ 1 \end{pmatrix}$$

The geodesic is sought in parametric form: $\vec{r}(t) = \vec{r}\big(\theta(t), z(t)\big)$.

We calculate the coefficients E, F, G: $E = \vec{r}_\theta^{\,2} = a^2$, $F = \vec{r}_\theta \cdot \vec{r}_z = 0$, $G = \vec{r}_z^{\,2} = 1$.

Equations (3.44) give then:

$$\begin{cases} \dfrac{d}{dt}\left(\dfrac{a^2\theta'}{\sqrt{a^2\theta'^2 + z'^2}}\right) = 0 \\ \dfrac{d}{dt}\left(\dfrac{z'}{\sqrt{a^2\theta'^2 + z'^2}}\right) = 0 \end{cases} \Rightarrow \begin{cases} \dfrac{a^2\theta'}{\sqrt{a^2\theta'^2 + z'^2}} = c_1 \\ \dfrac{z'}{\sqrt{a^2\theta'^2 + z'^2}} = c_2 \end{cases} \Rightarrow c_1 z' = c_2 a^2 \theta' \Rightarrow \frac{dz}{d\theta} = c$$

The solution is of the form: $z = z_0 + c(\theta - \theta_0)$ which is the equation of a helix.

3.2.2 Weak minimum sufficient conditions

The functions considered in this section are of class C^1 on $[a;b]$.

The sufficient minimum conditions are based on the study of neighboring extremals and conjugate points. These two notions are introduced below.

Neighboring extremals

Recall that an extremal of the functional J (3.22) is a solution function of the Euler-Lagrange equation (3.23). An extremal does not necessarily satisfy the endpoint conditions and it is not necessarily a minimum of J.

Two extremals y_1 and y_2 are called neighboring extremals if their distance in norm is small: $\|y_2 - y_1\| < \varepsilon$.

The neighboring extremals satisfy the property 3-2.

Property 3-2: Jacobi equation

If y^* and y are two neighboring extremals of the J-functional (3.22), then their difference $v = y - y^*$ satisfy the Jacobi equation associated with the J functional.

$$Qv = (Pv')' \quad \text{with} \quad \begin{cases} P = L_{zz} \\ Q = L_{yy} - L'_{yz} \end{cases} \tag{3.45}$$

Demonstration

Each extremal satisfies the Euler-Lagrange equation, which gives the system

$$\begin{cases} L_y(x,y^*,y^{*\prime}) - L'_z(x,y^*,y^{*\prime}) = 0 & \rightarrow \text{ for } y^* \\ L_y(x,y,y') \quad - L'_z(x,y,y') \quad = 0 & \rightarrow \text{ for } y = y^* + v \end{cases} \quad \textbf{(system S)}$$

Let us expand the terms of the second equation to order 1 with respect to v.

$$\begin{aligned} L_y(x,y,y') &= L_y(x,y^*+v,(y^*+v)') \\ &= L_y(x,y^*,y^{*\prime}) + L_{yy}(x,y^*,y^{*\prime})v + L_{yz}(x,y^*,y^{*\prime})v' \\ &\underset{\text{noted}}{=} L_y + L_{yy}v + L_{yz}v' \end{aligned}$$

$$L_z(x,y,y') \underset{\text{noted}}{=} L_z + L_{zy}v + L_{zz}v'$$

$$L'_z(x,y,y') = (L_z + L_{zy}v + L_{zz}v')' = L'_z + L'_{zy}v + L_{zy}v' + (L_{zz}v')'$$

Let us replace in the system S the second equation by its development.

$$\begin{cases} L_y \qquad - L'_z \qquad = 0 & \rightarrow \text{for } y^* \\ \left(L_y + L_{yy}v + L_{yz}v'\right) - \left(L'_z + L'_{zy}v + L_{zy}v' + (L_{zz}v')'\right) = 0 & \rightarrow \text{for } y = y^* + v \end{cases}$$

Using the first equation, the second equation reduces to

$L_{yy}v - L'_{zy}v - (L_{zz}v')' = 0$ which is the Jacobi equation (3.45).

The Jacobi equation is obtained by expanding the Euler-Lagrange equation to order 1 in the vicinity of a solution y*. The Jacobi equation is said to be the variational equation associated with the Euler-Lagrange equation.
The neighboring extremals can also be determined from the accessory minimum problem stated in property 3-3.

Property 3-3: Accessory minimum problem

If y* and y are two neighboring extremals of the functional J (3.22), then their difference $v = y - y^*$ is a solution of the problem

$$\min_v K(v) = \int_a^b \left(Pv'^2 + Qv^2\right)dx \quad \text{with} \quad \begin{cases} P = L_{zz} \\ Q = L_{yy} - L'_{yz} \end{cases} \tag{3.46}$$

This problem is equivalent to minimizing the second-order expansion of the functional J in the neighborhood of the extremal y*.

Demonstration

Let us first show the connection between problems (3.45) and (3.46).
The functional K is of the form

$$K(v) = \int_a^b M(x, v, v')dx \quad \text{with} \quad M(x, v, v') = Pv'^2 + Qv^2$$

Its Euler-Lagrange equation is: $M_v = M'_{v'} \Rightarrow Qv = (Pv')'$ which is the Jacobi equation (3.45). The Jacobi equation represents the Euler-Lagrange equation of the accessory minimum problem.

Let us then establish the problem (3.46) from the expansion of J to order 2 in the neighborhood of y*. This expansion is given by (3.11).

$$J(y^* + \varepsilon v) = J(y^*) + \delta J_{y^*}(v)\varepsilon + \delta^2 J_{y^*}(v)\varepsilon^2 + o(\varepsilon^2)$$

The first and second variations are given by (3.21).

$$\delta J_{y^*}(v) = \int_a^b v\left(L_y - L'_z\right)dx = 0 \text{ since } y^* \text{ (extremal) satisfies equation (EL): } L_y = L'_z$$

$$\delta^2 J_{y^*}(v) = \frac{1}{2}\int_a^b \left(v^2(L_{yy} - L'_{yz}) + v'^2 L_{zz}\right)dx = \frac{1}{2}\int_a^b \left(Pv'^2 + Qv^2\right)dx$$

Let us replace J by its second-order expansion in the formulation (3.22).

$$\min_y J(y) \quad \rightarrow \quad \min_v J(y^* + \varepsilon v) = J(y^*) + \frac{1}{2}\varepsilon^2 \int_a^b \left(Pv'^2 + Qv^2\right)dx$$

This gives the formulation of the accessory problem.

Conjugate points

The Jacobi equation allows us to determine the extremals close to y* without accounting for the endpoint conditions: $y(a) = \alpha$, $y(b) = \beta$.
The points $x = a$ and $x = c \in]a;b]$ are called conjugate points of the functional J if there is a neighboring extremal $y = y^* + v$ distinct from y* with $v(a)=v(c)=0$.
The existence of a conjugate point means that there exist two distinct extremals starting from the same point $A(a;\alpha)$ and crossing at a point $C(c;\gamma)$ with $c \in]a;b]$.

Figure 3-5a shows points A and C conjugate w.r.t an extremal AB and figure 3-5b shows points A and B conjugate for the minimum distance on a sphere. There are indeed an infinite number of geodesics between two diametrically opposed points. The study of the geodesics on a sphere is detailed in example 3-5.

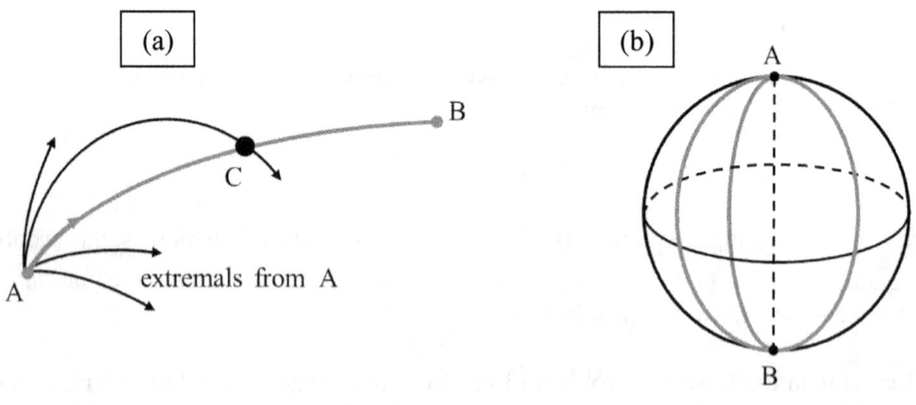

Figure 3-5: Conjugate points.

Example 3-4: Conjugate points for the geodesics in the plane

The geodesics in the plane were determined in example 3-1.

Recall the expression of the Lagrangian for this problem: $L(x,y,z)=\sqrt{1+z^2}$.

The extremals (solutions of Euler-Lagrange equation) are linear functions: $y'=C^{te}$

They satisfy the Legendre condition: $L_{zz}=\left(1+y'^2\right)^{\frac{3}{2}}>0$.

Neighboring extremals

The neighboring extremals of y* are found by solving the Jacobi equation.

$$Qv=(Pv')' \quad \text{with} \quad \begin{cases} P(x)=L_{zz}=\left(1+y^{*\prime 2}\right)^{-\frac{3}{2}}=C^{te} & \text{because } y^{*\prime}=C^{te} \\ Q(x)=L_{yy}-L'_{yz}=0 & \text{because } L_y=0 \end{cases}$$

The equation reduces to $v''=0$ whose solution is: $v(x)=v(a)+v'(a)(x-a)$.

We find that the distance between two neighboring extremals is a linear function (consistent with the fact that the extremals are all linear functions).

Conjugate points

Consider a neighboring extremal to y* starting from the same point A: $v(a)=0$.

If this extremal is distinct from y*, then v is not identically zero and $v'(a)\neq 0$.

The linear function $v(x)$, which is zero at a and has a non-zero slope, cannot cancel on the interval $]a;b]$.

There is therefore no conjugate point for the plane geodesics problem.

It is established below that the absence of conjugate points is a sufficient condition for a minimum.

Example 3-4 fulfils special conditions that allow to establish the absence of a conjugate point. Let us examine these special conditions.

Suppose on the one hand that the Lagrangian L is not explicit in y: $L_y=0$.

Jacobi's equation (3.45) simplifies, because $Q=L_{yy}-L'_{yz}=0$.

$$(Pv')'=Qv=0 \Rightarrow Pv'=C^{te} \tag{3.47}$$

Assume moreover that y* satisfies the strong Legendre condition: $P=L_{zz}>0$.

Equation (3.47) then shows that the derivative of v has a constant sign.

If $v(a)=0$, then the function v (not identically zero) no longer cancels on $]a;b]$ and there is no conjugate point for the extremal y*.

Jacobi's condition (1837) concerns the absence of a conjugate point. It establishes sufficient conditions that a function y* must satisfy to be a weak local minimum of the functional J.

Theorem 3-4: Sufficient conditions for a weak minimum

Assume that y* satisfies:

- the Euler-Lagrange equation (first order condition):

$$L_y - L_z' = 0 . \tag{3.48}$$

- the strong condition of Legendre (second order condition):

$$L_{zz} > 0 . \tag{3.49}$$

- the condition of Jacobi:

$$x = a \text{ does not have any conjugate point in }]a;b] . \tag{3.50}$$

Then y* is a weak local minimum of the problem (3.22).

Demonstration (see [R14])

The use of the three assumptions is highlighted in the demonstration.
We start from the expansion (3.11) of the functional J to order 2 in the neighborhood of y*, with the first and second variations given by (3.21).

$$J(y^*+\varepsilon\eta) = J(y^*) + \delta J_{y^*}(\eta)\varepsilon + \delta^2 J_{y^*}(\eta)\varepsilon^2 + o(\varepsilon^2)$$

$$\text{with } \begin{cases} \delta J_{y^*}(\eta) = \displaystyle\int_a^b \eta\left(L_y - L_z'\right)dx \\ \delta^2 J_{y^*}(\eta) = \displaystyle\frac{1}{2}\int_a^b \left(\eta^2 (L_{yy} - L_{yz}') + \eta'^2 L_{zz}\right)dx \end{cases}$$

The first variation is zero, because y* satisfies (by assumption) the Euler-Lagrange equation: $L_y = L_z'$.

The second variation takes the form

$$\delta^2 J_{y*}(\eta) = \frac{1}{2}\int_a^b \left(P\eta'^2 + Q\eta^2\right)dx \quad \text{with} \quad \begin{cases} P = L_{zz} \\ Q = L_{yy} - L'_{yz} \end{cases}$$

We aim to show that, under the assumptions of the theorem, we have

$\delta^2 J_{y*}(\eta) > 0 \ , \ \forall \eta$ feasible , $\eta \neq 0$

The feasible perturbations η are of class C^1 and such as: $\eta(a) = \eta(b) = 0$.

The demonstration consists in transforming $\delta^2 J_{y*}(\eta)$ to make a square appear under the integral.

Consider a term of the form: $I = \int_a^b d(w\eta^2)$ where w is a function of class C^1.

This term is always zero: $I = \int_a^b d(w\eta^2) = \left[w\eta^2\right]_a^b = 0$ because $\eta(a) = \eta(b) = 0$,

so that it can be added to $\delta^2 J_{y*}(\eta)$ without changing its value.

$$\delta^2 J_{y*}(\eta) = \delta^2 J_{y*}(\eta) + I = \frac{1}{2}\int_a^b \left(\eta'^2 P + 2\eta\eta' w + \eta^2 (Q + w')\right)dx$$

By Legendre's strong condition: $P > 0$. The function P does not cancel on $[a;b]$.

Suppose there is a function w such that: $P(Q + w') = w^2$

(the existence of such a function is demonstrated below).

Factoring by P under the integral, and with the equation satisfied by w, we get

$$\delta^2 J_{y*}(\eta) = \frac{1}{2}\int_a^b P\left(\eta'^2 + 2\eta\eta'\frac{w}{P} + \eta^2\frac{w^2}{P^2}\right)dx = \frac{1}{2}\int_a^b P\left(\eta' + \frac{w}{P}\eta\right)^2 dx \ \geq \ 0$$

This integral can cancel only if: $\eta' + \frac{w}{P}\eta = 0 \ \Rightarrow \ \eta = 0$ because $\eta(a) = 0$.

For a non-zero feasible perturbation, we have: $\delta^2 J_{y*}(\eta) > 0$.

It remains to show the existence of w of class C^1 such that: $P(Q + w') = w^2$.

Consider a solution v of the Jacobi equation: $Qv - \frac{d}{dx}(Pv') = 0$.

This solution depends continuously on the initial value $v(a)$.

By Jacobi's condition, there is no point on the interval $]a;b]$ that is conjugate to the point $x=a$. If $v(a)=0$, then v does not cancel on the interval $]a;b]$.
By continuity, for $v(a)=\varepsilon$ with ε small, the solution v does not cancel on the interval $]a;b]$. Therefore, there is a solution v that does not cancel on $]a;b]$.

Let us then define the function w as: $w=-\frac{v'}{v}P$.

This function of class C^1 satisfies: $vw=-Pv' \Rightarrow \frac{d}{dx}(vw)=-\frac{d}{dx}(Pv')=-Qv$

(because v is a solution of Jacobi's equation). Deriving (vw), we obtain

$v'w+vw'=-Qv \Rightarrow Q+w'=-\frac{v'}{v}w=\frac{w^2}{P}$ (since $\frac{v'}{v}=-\frac{w}{P}$ by definition of w).

Thus, we have shown the existence of a function w such that: $P(Q+w')=w^2$.

Notes on Legendre's strong condition: $P=L_{zz}>0$

- *It can be shown that a weak minimum satisfying the strong Legendre condition does not admit a conjugate point. The absence of a conjugate point is therefore a necessary condition of minimum when* $P=L_{zz}>0$.
- *In the case where L is not explicit in y:* $L_y=0$, *we have established above (3.47) that the strong Legendre condition leads to the absence of a conjugate point. This condition alone is then sufficient to guarantee a weak minimum.*

Example 3-5 illustrates the application of sufficient conditions to determine geodesics on a sphere.

Example 3-5: Geodesics on a sphere

We seek minimum distance curves between two points A and B on the unit sphere. The position on the sphere is defined by the latitude φ and the longitude λ. To facilitate the calculations, the axes are chosen so that points A and B are on the meridian of zero longitude (as in figure 3-6a).

With this choice of axes, points A and B have the coordinates:

$$A\begin{pmatrix}\varphi_A\\ \lambda_A=0\end{pmatrix},\quad B\begin{pmatrix}\varphi_B\\ \lambda_B=0\end{pmatrix}$$

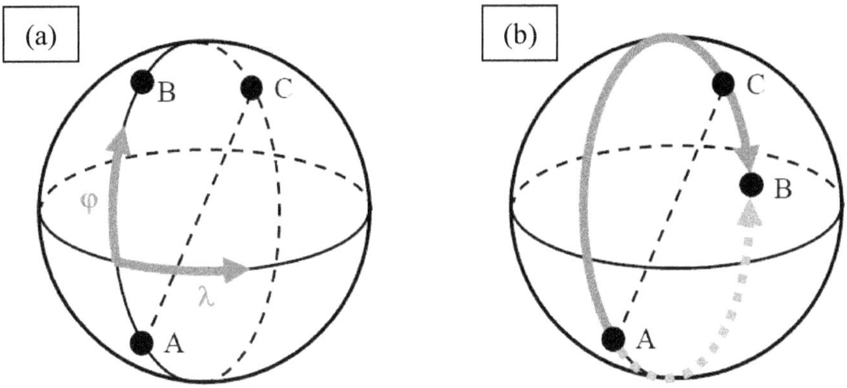

Figure 3-6: Geodesics and conjugate points on a sphere.

Formulation of the problem

The distance element ds on the sphere is expressed in spherical coordinates.

$$\begin{cases} x = \cos\varphi\cos\lambda \\ y = \cos\varphi\sin\lambda \\ z = \sin\varphi \end{cases} \rightarrow ds^2 = dx^2 + dy^2 + dz^2 = d\varphi^2 + d\lambda^2\cos^2\varphi .$$

We are looking for the function $\lambda(\varphi)$ which minimizes the distance from A to B.

The problem is formulated as: $\min_{\lambda} D_{AB} = \int_A^B ds = \int_{\varphi_A}^{\varphi_B} \sqrt{1+\lambda'(\varphi)^2\cos^2\varphi}\, d\varphi$

with the Lagrangian defined by: $L(\varphi,\lambda,\lambda') = \sqrt{1+\lambda'^2\cos^2\varphi}$.

Searching for extremals

The extremals are the solutions of the Euler-Lagrange equation: $\dfrac{\partial L}{\partial \lambda} = \dfrac{d}{d\varphi}\left(\dfrac{\partial L}{\partial \lambda'}\right)$.

with L not explicit in λ : $\dfrac{d}{d\varphi}\left(\dfrac{\partial L}{\partial \lambda'}\right) = 0 \Rightarrow \dfrac{\partial L}{\partial \lambda'} = c \Rightarrow \dfrac{\lambda'\cos^2\varphi}{\sqrt{1+\lambda'^2\cos^2\varphi}} = c$.

This gives the differential equation: $\dfrac{d\lambda}{d\varphi} = \dfrac{c}{\cos\varphi\sqrt{\cos^2\varphi - c^2}}$.

A (non-obvious) solution is: $\tan(\lambda - \lambda_0) = \dfrac{c\sin\varphi}{\sqrt{\cos^2\varphi - c^2}}$ (checked by derivation).

The integration constants λ_0, c are determined by the conditions in A and B.

$$\left.\begin{array}{ll} \text{In } A(\varphi_A;\lambda_A=0): & -\tan\lambda_0=\dfrac{c\sin\varphi_A}{\sqrt{\cos^2\varphi_A-c^2}} \\ \text{In } B(\varphi_B;\lambda_B=0): & -\tan\lambda_0=\dfrac{c\sin\varphi_B}{\sqrt{\cos^2\varphi_B-c^2}} \end{array}\right\} \Rightarrow c=0 \ , \ \lambda_0=0 \text{ or } \pi .$$

The extremal from A to B therefore has the equation: $\tan\lambda=0$.
This extremal is a great circle arc following the meridian through A and B.
Let us see if this extremal satisfies the sufficient conditions of theorem 3-4.

Legendre's strong condition

The function P is defined by: $P(\varphi)=\dfrac{\partial}{\partial\lambda'}\left(\dfrac{\partial L}{\partial\lambda'}\right)$ with

$$L(\varphi,\lambda,\lambda')=\sqrt{1+\lambda'^2\cos^2\varphi} \Rightarrow \frac{\partial L}{\partial\lambda'}=\frac{\lambda'\cos^2\varphi}{L} \Rightarrow \frac{\partial}{\partial\lambda'}\left(\frac{\partial L}{\partial\lambda'}\right)=\frac{\cos^2\varphi}{L}-\frac{\lambda' L_{\lambda'}\cos^2\varphi}{L^2}$$

The extremal has the equation: $\tan\lambda=0$, which leads to
$\lambda'=0$, $L(\varphi,\lambda,\lambda')=1$ and $P(\varphi)=\cos^2\varphi$.
Legendre's strong condition: $P(\varphi)>0$ is satisfied if the extremal does not pass through a pole.

Jacobi's condition

The Jacobi equation is: $(Pv')'=Qv$ with $Q=0$, because L is not explicit in λ.

The solutions are: $\quad Pv'=k \Rightarrow v'=\dfrac{k}{\cos^2\varphi} \Rightarrow v=v_0+k\tan\varphi$

If $v(\varphi_A)=0$, then v cancels again for $\varphi_C=\varphi_A+\pi$.
The point A has as for conjugate point the diametrically opposite point C.

Let us summarize the sufficient conditions for the distance AB to be a minimum:

- the extremal (AB) is a great circle meridian arc of equation $\lambda=0 \Rightarrow L=1$, $D_{AB}=\varphi_B-\varphi_A$;
- Legendre's strong condition holds if the extremal does not pass through a pole $\left(\varphi=\pm\dfrac{\pi}{2}\right)$;
- Jacobi's condition (no conjugate point) is satisfied if the length of (AB) is less than the length of a half great circle.

Note 1
If A and B are not diametrically opposed, one can choose the reference frame so that $-\frac{\pi}{2} < \varphi_A < \varphi_B < \frac{\pi}{2}$ and the sufficient minimum conditions are then satisfied.

Note 2
The formulation in $\lambda(\varphi)$ assumes the variable φ is increasing. If the latitudes are oriented in the "wrong direction" (*i.e.* $\varphi_B > \varphi_A + \pi$), the length of the arc (AB) is greater than the length of a half great circle and Jacobi's condition is not satisfied. In this case, there is a shorter path by traversing the arc in a retrograde direction (with φ decreasing) as shown in figure 3-6b.

Note 3
If A and B are diametrically opposed, all semicircles (AB) are solutions.

3.2.3 Corner necessary conditions

Theorem 3-1 gives necessary conditions for a weak minimum on the space of functions of class C^1 on $[a;b]$.

This section extends the space to functions of class C^1 by pieces on $[a;b]$.

The function y(x) may have a derivative discontinuity at a point c as in figure 3-7. Such a point is called a "corner" or an "edge".

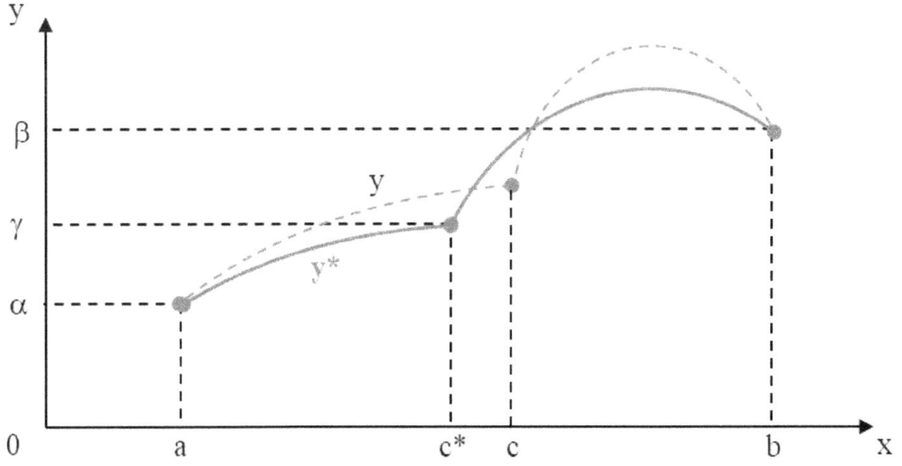

Figure 3-7: Functions C^1 by pieces.

The problem of calculus of variations is formulated as

$$\min_{y,c} J(y,c) = \int_a^b L(x,y,y')dx \quad \text{with} \begin{cases} y(a) = \alpha \\ y(b) = \beta \end{cases} \text{ and } y'(c^-) \neq y'(c^+) \qquad (3.51)$$

The unknowns in this problem are the function y and the position c of the corner. Theorem 3-5 completes theorem 3-1 by adding a condition associated with a corner.

Theorem 3-5: Corner necessary conditions

Assume that y* is a minimum of problem (3.51) with a corner in $x = c^*$.
Then y* satisfies the conditions of theorem 3-1.
Furthermore, the functions L_z and $L_z y' - L$ are continuous in $x = c^*$.

Demonstration (see [R14])

The demonstration consists in expressing the first variation of the functional J by applying perturbations on the function y and on the corner position c, and then in writing the conditions so that this first variation is zero.

Consider a function y of class C^1 by pieces on $[a;b]$.
This function is assumed to have a single corner in $x = c$ (figure 3-8).

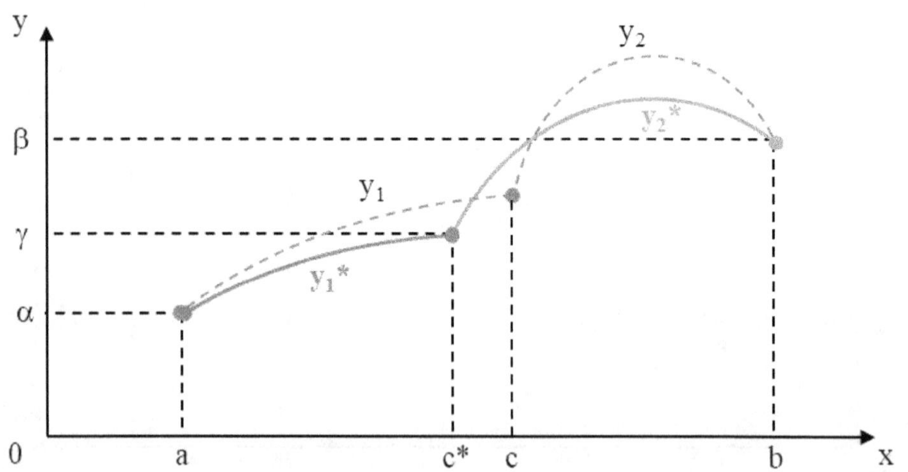

Figure 3-8: Functions y_1 and y_2 of class C^1 on each interval.

The functions y_1 and y_2 are the restrictions of y to the intervals $[a;c]$ and $[c;b]$.

$$y(x)=\begin{cases} y_1(x) & \text{if } x\in[a;c] \\ y_2(x) & \text{if } x\in[c;b] \end{cases}$$

The function y is continuous in $x=c$: $y_1(c)=y_2(c)=y(c)$.

Its derivative is discontinuous in $x=c$: $\begin{cases} y_1'(c)=y'(c^-) \\ y_2'(c)=y'(c^+) \end{cases}$.

Definition of perturbations

Assume that the function y* with a corner in c^* is a weak minimum of J.

Let us apply perturbations η_1 and η_2 of class C^1 to the functions $y_1{}^*$ and $y_2{}^*$, as well as a perturbation δc on the position of the corner c^*.

$$\begin{cases} y_1=y_1{}^*+\varepsilon\eta_1 & \text{on } [a;c] \quad \text{with } \eta_1(a)=0 \\ y_2=y_2{}^*+\varepsilon\eta_2 & \text{on } [c;b] \quad \text{with } \eta_2(b)=0 \end{cases} \quad \text{and } c=c^*+\varepsilon\delta c$$

In the case $c>c^*$, we extend the function $y_1{}^*$ by a segment on $[c^*;c]$.

$$y_{1p}{}^*(x)=y^*(c^*)+y^{*\prime}(c^{*-})(x-c^*) \quad \text{for } c^*\le x\le c$$

In the case $c<c^*$, we extend the function $y_2{}^*$ by a segment on $[c;c^*]$.

$$y_{2p}{}^*(x)=y^*(c^*)+y^{*\prime}(c^{*-})(x-c^*) \quad \text{for } c\le x\le c^*$$

These perturbations are shown in figure 3-9.

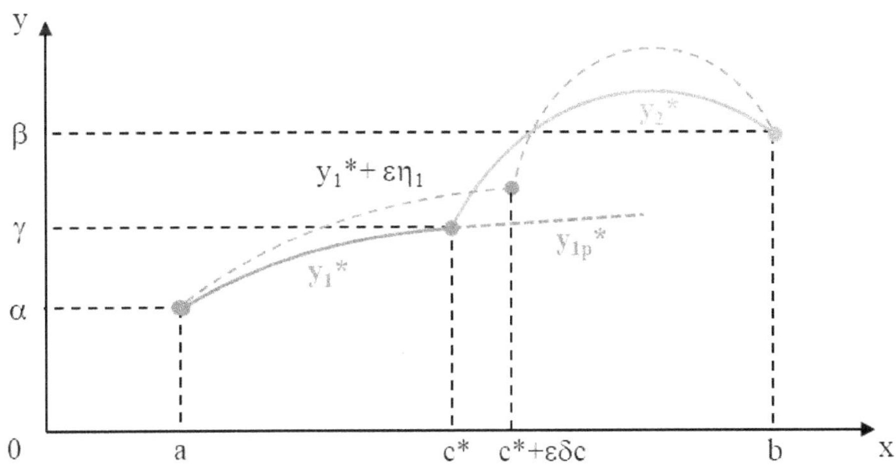

Figure 3-9: Perturbed function with linear extension.

The function y is of class C^1 on each interval and satisfies the endpoint conditions: $y(a)=\alpha$, $y(b)=\beta$. We must impose its continuity in $x=c$.

The function y* is continuous at c*: $y_1{}^*(c^*)=y_2{}^*(c^*)$.

Let us express the values of $y_1(c)$ and $y_2(c)$: $\begin{cases} y_1(c)=y_1{}^*(c^*)+\delta y_{1\eta}+\delta y_{1c} \\ y_2(c)=y_2{}^*(c^*)+\delta y_{2\eta}+\delta y_{2c} \end{cases}$

with the following notations shown in figure 3-10.

$\delta y_{1\eta}$ = variation in c* of $y_1{}^*$ due to $\varepsilon\eta_1$ $\rightarrow \delta y_{1\eta}=\varepsilon\eta_1(c^*)$

$\delta y_{2\eta}$ = variation in c* of $y_2{}^*$ due to $\varepsilon\eta_2$ $\rightarrow \delta y_{2\eta}=\varepsilon\eta_2(c^*)$

δy_{1c} = variation in c* of $y_1{}^*$ due to $\varepsilon\delta c$ $\rightarrow \delta y_{1c}=y^{*\prime}(c^{*-})\varepsilon\delta c$

δy_{2c} = variation in c* of $y_2{}^*$ due to $\varepsilon\delta c$ $\rightarrow \delta y_{2c}=y^{*\prime}(c^{*+})\varepsilon\delta c$

The variations δy_{1c} and δy_{2c} result from the linear extensions of $y_1{}^*$ and $y_2{}^*$.

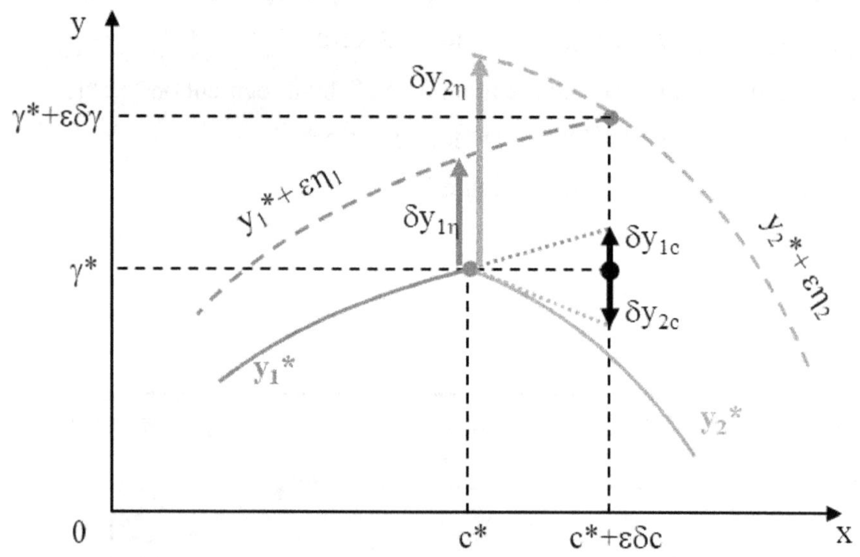

Figure 3-10: Continuity condition.

Putting these contributions together, the continuity of y in c gives the relation

$y_1(c)=y_2(c) \;\Rightarrow\; \delta y_{1\eta}+\delta y_{1c}=\delta y_{2\eta}+\delta y_{2c}$

$\Rightarrow\; \eta_1(c^*)+y^{*\prime}(c^{*-})\delta c=\eta_2(c^*)+y^{*\prime}(c^{*+})\delta c$ **(equation C)**

This continuity condition (C) will be used below.

Calculation of the first variation

Let us now calculate the first variation of J with these perturbations. Decomposing on the intervals $[a;c^*]$ and $[c^*;b]$, J is expressed as

$$J(y^*,c^*)=\int_a^b L(x,y^*,y^{*\prime})dx=\int_a^{c^*} L(x,y_1^*,y_1^{*\prime})dx+\int_{c^*}^b L(x,y_2^*,y_2^{*\prime})dx$$

$$=J_1(y_1^*,c^*)+J_2(y_2^*,c^*)$$

Let us expand to order 1 the functional J_1 on the interval $[a;c]$.

$$J_1(y_1^*+\varepsilon\eta_1,c^*+\varepsilon\delta c)=\int_a^{c^*+\varepsilon\delta c} L(x,y_1^*+\varepsilon\eta_1,(y_1^*+\varepsilon\eta_1)')dx$$

$$=\int_a^{c^*}(L+\varepsilon\eta_1 L_y+\varepsilon\eta_1' L_z)(x,y_1^*,y_1^{*\prime})dx+\int_{c^*}^{c^*+\varepsilon\delta c} L(x,y_1^*,y_1^{*\prime})dx$$

The second integral yields

$$\int_{c^*}^{c^*+\varepsilon\delta c} L(x,y_1^*,y_1^{*\prime})dx=L(c^*,y_1^*(c^*),y_1^{*\prime}(c^*))\varepsilon\delta c$$

$$=L(c^*,y^*(c^*),y^{*\prime}(c^{*-}))\varepsilon\delta c \underset{\text{noted}}{=} L(c^{*-})\varepsilon\delta c$$

Integrating by parts the term in $\eta_1' L_z$ in the first integral, with the condition: $\eta_1(a)=0$, we obtain for the first variation of J_1

$$\delta J_{1y^*}(\eta_1,\delta c)=\int_a^{c^*}\eta_1(L_y-L_z')dx+\eta_1(c^*)L_z(c^{*-})+L(c^{*-})\delta c$$

Similarly, the first variation of J_2 is given by

$$\delta J_{2y^*}(\eta_2,\delta c)=\int_{c^*}^{b}\eta_2(L_y-L_z')dx-\eta_2(c^*)L_z(c^{*+})-L(c^{*+})\delta c$$

The first variation of J is the sum of these two contributions.

$$\delta J_{y^*}(\eta_1,\eta_2,\delta c)=\delta J_{1y^*}(\eta_1,\delta c)+\delta J_{2y^*}(\eta_2,\delta c)$$

$$=\int_a^{c^*}\eta_1(L_y-L_z')dx+\int_{c^*}^{b}\eta_2(L_y-L_z')dx$$

$$+\eta_1(c^*)L_z(c^{*-})-\eta_2(c^*)L_z(c^{*+})+\left(L(c^{*-})-L(c^{*+})\right)\delta c$$

The values of $\eta_1(c^*)$, $\eta_2(c^*)$, δc are linked by the continuity condition (C). By noting $\delta\gamma$ the value of the two members of equation (C), we express the values of $\eta_1(c^*)$, $\eta_2(c^*)$ as a function of the two <u>independent</u> perturbations $\delta c, \delta\gamma$.

$$\begin{cases}\eta_1(c^*)=\delta\gamma-y^{*\prime}(c^{*-})\delta c\\ \eta_2(c^*)=\delta\gamma-y^{*\prime}(c^{*+})\delta c\end{cases}$$

This yields δJ_{y*} as a function of independent perturbations $\eta_1, \eta_2, \delta c, \delta\gamma$.

$$\delta J_{y*}(\eta_1,\eta_2,\delta c,\delta\gamma) = \int_a^{c^*} \eta_1\left(L_y - L_z'\right)dx + \int_{c^*}^{b} \eta_2\left(L_y - L_z'\right)dx - \left[L_z\right]_{c^*-}^{c^*+}\delta\gamma + \left[L_z y' - L\right]_{c^*-}^{c^*+}\delta c$$

The necessary condition of minimum in y* is that the first variation is zero.

$\delta J_{y*}(\eta_1,\eta_2,\delta c,\delta\gamma) = 0 \;,\; \forall \eta_1,\eta_2$ feasible $,\; \forall \delta c, \delta\gamma$

The coefficient for each perturbation must be zero:

- the terms in η_1, η_2 give the Euler-Lagrange equation on $[a;b]$: $L_y = L_z'$;
- the term in $\delta\gamma$ imposes the continuity of L_z in $x = c^*$;
- the term in δc imposes the continuity of $L_z y' - L$ in $x = c^*$.

The continuity conditions of L_z and $L_z y' - L$ at a corner are called necessary conditions of Weierstrass-Erdmann (established in 1865 and 1877 respectively). They apply at each discontinuity point of the derivative.
Example 3-6 shows the use of these corner conditions to obtain piecewise C^1 solutions.

Example 3-6: Surface of revolution of minimum area/Catenary

We are looking for the surface of revolution with the minimum area generated by rotating a curve (AB) around the axis (Ox), as shown in figure 3-11a.
The equation of the curve (AB) is: $y(x)$, $a \le x \le b$.
The endpoint conditions are: $y(a) = \alpha$, $y(b) = \beta$.

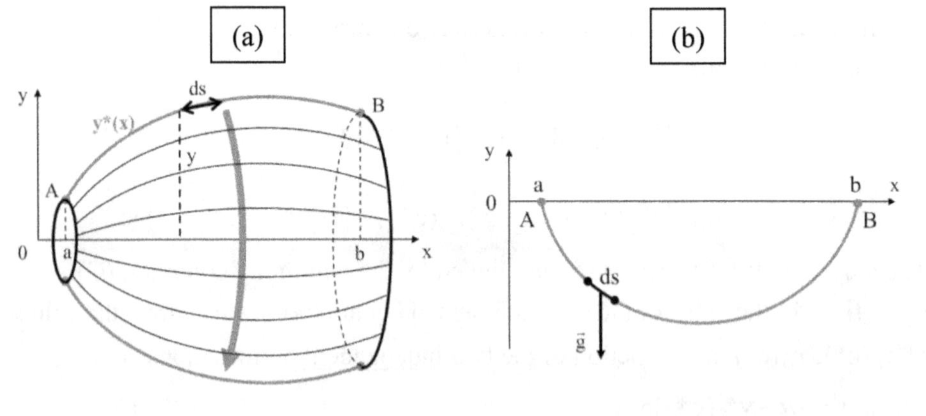

Figure 3-11: Minimum area of revolution/catenary.

Formulation of the problem

The curve element at abscissa x has a length: $ds = \sqrt{1+y'(x)^2}dx$.

By rotation around Ox, this element generates the surface: $dS = 2\pi y ds$.

The problem is formulated as: $\min\limits_{y} S = 2\pi\int_{A}^{B} y ds = 2\pi\int_{a}^{b} y(x)\sqrt{1+y'(x)^2}dx$.

This problem of surface of revolution is equivalent to the problem of the catenary.

Catenary problem

We are looking for the shape taken by a homogeneous weighted wire attached at A and B as shown in figure 3-11b. The density of the wire is noted ρ.

The length element ds at y has a potential energy: $dE_P = \rho g y ds$.

The wire shape minimizes the total potential energy: $E_P = \int_{a}^{b} \rho g y(x)\sqrt{1+y'(x)^2}dx$.

The problem is formulated as: $\min\limits_{y} E_P = \rho g\int_{a}^{b} y(x)\sqrt{1+y'(x)^2}dx$.

This formulation is analogous to the previous surface of revolution problem.

Note: The length of the curve (AB) is free.
In the case of a given length curve, the resolution must account for an integral constraint as in section 3.3.2.

Solutions of class C^1

Let us start by looking for solutions of class C^1 on $[a;b]$.

The Lagrangian $L(x,y,z) = y\sqrt{1+z^2}$ is not explicit in x. As shown in (3.30), the Euler-Lagrange equation admits as a first integral: $L_{y'}y' - L = H = C^{te}$,

which leads to a differential equation satisfied by y.

$$\frac{yy'^2}{\sqrt{1+y'^2}} - y\sqrt{1+y'^2} = H \quad \Rightarrow \quad y' = \pm\frac{\sqrt{y^2-H^2}}{H}$$

This equation is integrated by separating the variables x and y.

$$dx = \frac{\pm H dy}{\sqrt{y^2-H^2}} \quad \Rightarrow \quad x = x_0 \pm H.\ln\frac{y+\sqrt{y^2-H^2}}{H} = x_0 \pm H.\text{Ach}\left(\frac{y}{H}\right)$$

(Ach = inverse hyperbolic cosine)

The curve (AB) is a chain

equation: $y = H.ch\left(\frac{x - x_0}{H}\right)$.

The surface of revolution generated is a catenoid, shown opposite ("catena" = chain).

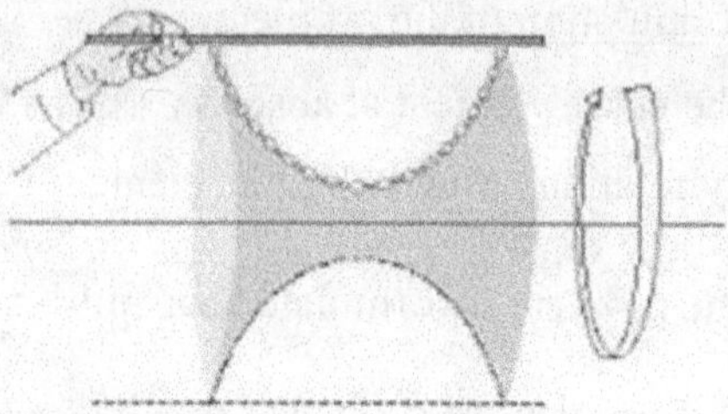

This solution considers only curves of class C^1.

<u>Solutions of class C^1 by pieces</u>

Let us now look for solutions of class C^1 piecewise.

The coordinates of the endpoints are $A(-a;\alpha)$ and $B(a;\alpha)$.

The solutions are sought in the form of parametric curves: $\begin{cases} x = x(t) \\ y = y(t) \end{cases}$.

The problem is formulated as: $\min\limits_{x,y} \int_a^b y(t)\sqrt{\dot{x}(t)^2 + \dot{y}(t)^2}\,dt$.

The Lagrangian depends on two functions x and y: $L(t, x, y, \dot{x}, \dot{y}) = y\sqrt{\dot{x}^2 + \dot{y}^2}$.

Each function $x(t)$ and $y(t)$ satisfies the Euler-Lagrange equation: $\begin{cases} L_x = \dot{L}_{\dot{x}} \\ L_y = \dot{L}_{\dot{y}} \end{cases}$.

This two-function problem $x(t)$ and $y(t)$ of one variable t is equivalent to the one-function problem $y(x)$ in variable x, as shown in (3.40).

The two Euler-Lagrange equations are therefore not independent.

We choose to solve the equation in L_x which is simpler.

$L_x = 0 \Rightarrow L_{\dot{x}} = K = C^{te} \Rightarrow y\dot{x} = K\sqrt{\dot{x}^2 + \dot{y}^2}$

Two forms of extremals are obtained depending on the value of the constant K.

- 1st case: $K \neq 0$

We retrieve class C^1 solutions of the chain type by posing: $x = t \Rightarrow \dot{x} = 1$.

The constants are determined by the conditions in $A(-a;\alpha)$ and $B(a;\alpha)$.

The chain solution is: $y = H.ch\left(\frac{x}{H}\right)$ with H determined by: $\alpha = H.ch\left(\frac{a}{H}\right)$.

The constant H exists only if $\alpha > 1.509a$ (numerically obtained minimum).

Physically, the ordinate α must be high enough to keep the low point above the axis (Ox), which is the axis of rotation for the surface problem or the ground level for the catenary problem.

- 2nd case: $K=0$

The Euler-Lagrange equation has two solutions: $y\dot{x}=0 \Rightarrow y=0$ or $x=C^{te}$.
A composite solution is formed connecting these two solutions at $C\,(x_C\,;y_C=0)$.
This composite solution must satisfy the corner conditions in C.

<u>Corner conditions</u>
The first condition is the continuity of $p=L_{\dot{x}}$ in C.
This condition is well satisfied, as $p=L_{\dot{x}}=K=0 \;\Rightarrow\; p^-=p^+=0$.
The second condition is the continuity of $H=L_{\dot{x}}\dot{x}-L$ in C.
This condition is well satisfied at point C.

$$\begin{cases} L_{\dot{x}}=K=0 \\ L=y\sqrt{\dot{x}^2+\dot{y}^2}=0 \quad \text{because } y_C=0 \end{cases} \Rightarrow H^-=H^+=0$$

The corner conditions are satisfied by connecting the extremals: $x=C^{te}$ and $y=0$.
The conditions in A and B impose $x=\pm a$ for solutions of the form $x=C^{te}$.
The solution curve consists of three segments [AA'] $(x=-a)$, [A'B'] $(y=0)$ and [B'B] $(x=+a)$.
These segments form three sides of a rectangle as shown in figure 3-12.

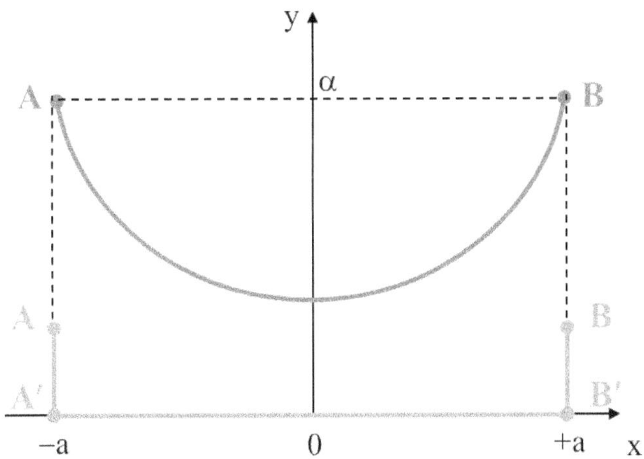

Figure 3-12: Chain and rectangle solutions.

Let us now compare the "chain" and "rectangle" solutions in terms of the value of the functional and the feasibility.

The area of the rectangle solution is obtained by integrating over each segment.

$$S = 2\pi\left(\int_A^{A'} + \int_{A'}^{B'} + \int_B^{B'} y\sqrt{\dot{x}^2+\dot{y}^2}\,dt\right) = 2\pi\left(\int_{t=\alpha}^{t=0} y|\dot{y}|dt + \int_{t=-a}^{t=+a} 0dt + \int_{t=0}^{t=\alpha} y|\dot{y}|dt\right)$$

$$= \pi\left(\left[-y^2\right]_\alpha^0 + \left[y^2\right]_0^\alpha\right) = 2\pi\alpha^2$$

This surface corresponds to two discs generated by the segments [AA'] and [BB']. If the ordinate α is sufficient ($\alpha > 1.509a$), the chain solution is better (lower surface). If the ordinate α is too small, the chain solution is not feasible, as the curve would pass below the axis (Ox). The "rectangle" solution is the only feasible one in this case.

We will see in section 3.4 that the Weierstrass-Erdmann corner conditions reflect the continuity of the canonical variables of problem (3.51).
The conditions established here consider neighboring functions in norm 1. The next section extends the conditions to neighboring functions in norm 0 to define a strong minimum.

3.2.4 Strong minimum necessary conditions

In this section, we consider functions of class C^1 by pieces on $[a;b]$ with the neighborhood in norm 0. The Weierstrass function $E:\mathbb{R}^4 \to \mathbb{R}$ is defined by

$$E(x,y,z,w) = L(x,y,w) - L(x,y,z) - (w-z)L_z(x,y,z) \tag{3.52}$$

Theorem 3-6 gives a necessary condition that an extremal y* must satisfy to be a strong local minimum of J.

> **Theorem 3-6:** Necessary condition for a strong minimum
>
> Assume that y* is a strong local minimum of problem (3.22).
>
> Then y* is an extremal satisfying the Weierstrass condition
>
> $$E(x,y^*,y^{*\prime},w) \geq 0 \;,\; \forall w \in \mathbb{R} \tag{3.53}$$
>
> in any point $x \in [a;b]$ that is not a corner.

Recall that an extremal satisfies the Euler-Lagrange equation: $L_y = L'_z$.

Demonstration (from [R14])

Assume that the function y^* is a strong local minimum of J and consider two points x_0 and x_1 in $[a;b]$, such that the interval $[x_0;x_1]$ does not contain a corner. We define functions y parametrized by a real $\varepsilon \ge 0$ and another real w.

$$y(x,\varepsilon)=\begin{cases} y^*(x) & \text{if } a \le x \le x_0 \\ y^*(x_0)+w(x-x_0) & \text{if } x_0 \le x \le x_0+\varepsilon \\ y^*(x)+\dfrac{x_1-x}{x_1-(x_0+\varepsilon)}\left(y^*(x_0)+w\varepsilon-y^*(x_0+\varepsilon)\right) & \text{if } x_0+\varepsilon \le x \le x_1 \\ y^*(x) & \text{if } x_1 \le x \le b \end{cases}$$

The shape of such a function $y(x,\varepsilon)$ is shown in figure 3-13.

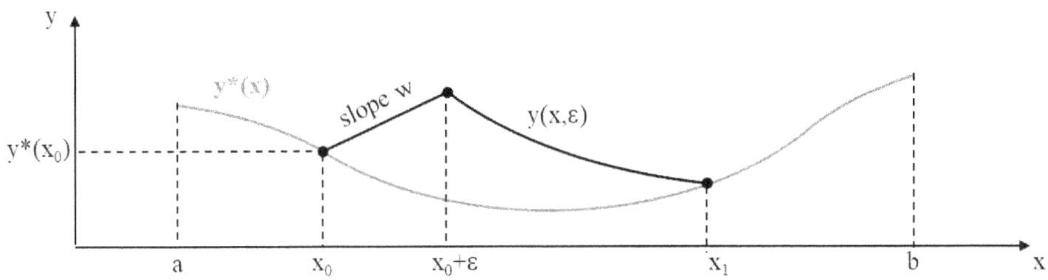

Figure 3-13: Perturbed function on $[x_0 ; x_1]$.

The functions $y(x,\varepsilon)$ are of class C^1 with discontinuous derivatives in the three points x_0 , $x_0+\varepsilon$ and x_1.
For $\varepsilon \ge 0$ small, the functions $y(.,\varepsilon)$ are close to y^* in norm 0.
For $\varepsilon = 0$, the function $y(.,0)$ coincides with y^*.

y^* being a strong minimum of J, we have: $J(y^*) = J\left(y(.,0)\right) \le J\left(y(.,\varepsilon)\right)$, $\forall \varepsilon \ge 0$.
The function $F:\varepsilon \mapsto F(\varepsilon) = J\left(y(.,\varepsilon)\right)$ defined for $\varepsilon \ge 0$ therefore admits a minimum in $\varepsilon = 0$. Its right derivative at 0 must be positive: $F'(0^+) \ge 0$.
This derivative $F'(0^+)$ is calculated below to show that
$F'(0^+) = E(x_0, y^*(x_0), y^{*\prime}(x_0), w)$
(Weierstrass function evaluated for the function y^* at point x_0).
The point x_0 and the real w being arbitrary, we will thus show the condition (3.53).

Calculation of the derivative of the function F

$$F(\varepsilon)=J(y(.,\varepsilon))=\int_a^b L\big(x,y(x,\varepsilon),y_x(x,\varepsilon)\big)dx=\int_a^{x_0}+\int_{x_0}^{x_0+\varepsilon}+\int_{x_0+\varepsilon}^{x_1}+\int_{x_1}^{b}$$

The function $y(x,\varepsilon)$ depends on ε only on the intervals $[x_0;x_0+\varepsilon]$ and $[x_0+\varepsilon;x_1]$. The derivatives of the integrals on these two intervals are calculated:

- derivative with respect to ε of the integral: $I_0=\int_{x_0}^{x_0+\varepsilon} L\big(x,y(x,\varepsilon),y_x(x,\varepsilon)\big)dx$

On $[x_0;x_0+\varepsilon]$: $y(x,\varepsilon)=y^*(x_0)+w(x-x_0) \Rightarrow y_x(x,\varepsilon)=w$

$$\rightarrow L\big(x,y(x,\varepsilon),y_x(x,\varepsilon)\big)=L\big(x,y^*(x_0)+w(x-x_0),w\big)$$

For ε small, we get

$$\frac{d}{d\varepsilon}(I_0)=\frac{d}{d\varepsilon}\left(\int_{x_0}^{x_0+\varepsilon} L\big(x,y^*(x_0)+w(x-x_0),w\big)dx\right)=L\big(x_0+\varepsilon,y^*(x_0)+w\varepsilon,w\big);$$

- derivative with respect to ε of the integral: $I_1=\int_{x_0+\varepsilon}^{x_1} L\big(x,y(x,\varepsilon),y_x(x,\varepsilon)\big)dx$

$$\frac{d}{d\varepsilon}(I_1)=\frac{d}{d\varepsilon}\left(\int_{x_0+\varepsilon}^{x_1} L\big(x,y(x,\varepsilon),y_x(x,\varepsilon)\big)dx\right)$$

$$=-L\big(x_0+\varepsilon,y(x_0+\varepsilon,\varepsilon),y_x(x_0+\varepsilon,\varepsilon)\big)+\int_{x_0+\varepsilon}^{x_1}\frac{d}{d\varepsilon}\Big(L\big(x,y(x,\varepsilon),y_x(x,\varepsilon)\big)\Big)dx$$

by separating the derivative with respect to the lower bound (1st term) and the derivative under the integral. The derivative under the integral is

$$\frac{d}{d\varepsilon}L\big(x,y(x,\varepsilon),y_x(x,\varepsilon)\big)=L_y\big(x,y(x,\varepsilon),y_x(x,\varepsilon)\big)y_\varepsilon(x,\varepsilon)$$
$$+L_z\big(x,y(x,\varepsilon),y_x(x,\varepsilon)\big)y_{x\varepsilon}(x,\varepsilon)$$

By replacing in the expression of $\frac{d}{d\varepsilon}(I_1)$, we have

$$\frac{d}{d\varepsilon}(I_1)=-L\big(x_0+\varepsilon,y(x_0+\varepsilon,\varepsilon),y_x(x_0+\varepsilon,\varepsilon)\big)$$
$$+\int_{x_0+\varepsilon}^{x_1} L_y\big(x,y(x,\varepsilon),y_x(x,\varepsilon)\big)y_\varepsilon(x,\varepsilon)dx$$
$$+\int_{x_0+\varepsilon}^{x_1} L_z\big(x,y(x,\varepsilon),y_x(x,\varepsilon)\big)y_{x\varepsilon}(x,\varepsilon)dx$$

Let us note I_2 the second integral above.

$$I_2 = \int_{x_0+\varepsilon}^{x_1} L_z\big(x, y(x,\varepsilon), y_x(x,\varepsilon)\big) y_{x\varepsilon}(x,\varepsilon)dx$$

By integrating by parts, we have

$$I_2 = \Big[L_z\big(x, y(x,\varepsilon), y_x(x,\varepsilon)\big) y_\varepsilon(x,\varepsilon)\Big]_{x_0+\varepsilon}^{x_1} - \int_{x_0+\varepsilon}^{x_1} \frac{d}{dx} L_z\big(x, y(x,\varepsilon), y_x(x,\varepsilon)\big) y_\varepsilon(x,\varepsilon)dx$$

Let us calculate the term in square brackets using the expression of $y(x,\varepsilon)$.

On $[x_0+\varepsilon; x_1]$: $y(x,\varepsilon) = y^*(x) + \dfrac{x_1 - x}{x_1 - (x_0+\varepsilon)}\big(y^*(x_0) + w\varepsilon - y^*(x_0+\varepsilon)\big)$

In x_1 : $\quad y(x_1,\varepsilon) = y^*(x_1) \quad \Rightarrow \quad y_\varepsilon(x_1,\varepsilon) = 0$

In $x_0+\varepsilon$: $y(x_0+\varepsilon,\varepsilon) = y^*(x_0) + w\varepsilon \quad \Rightarrow \quad y_x(x_0+\varepsilon,\varepsilon) + y_\varepsilon(x_0+\varepsilon,\varepsilon) = w$

$$\Rightarrow \quad y_\varepsilon(x_0+\varepsilon,\varepsilon) = w - y_x(x_0+\varepsilon,\varepsilon)$$

The term in square brackets is

$$\Big[L_z\big(x, y(x,\varepsilon), y_x(x,\varepsilon)\big) y_\varepsilon(x,\varepsilon)\Big]_{x_0+\varepsilon}^{x_1} =$$
$$-L_z\big(x_0+\varepsilon, y(x_0+\varepsilon,\varepsilon), y_x(x_0+\varepsilon,\varepsilon)\big)\big(w - y_x(x_0+\varepsilon,\varepsilon)\big)$$

Let us now group together all the terms of $F'(\varepsilon)$ coming from I_0 and I_1.

$$\begin{aligned} F'(\varepsilon) = \; & L\big(x_0+\varepsilon, y^*(x_0)+w\varepsilon, w\big) - L\big(x_0+\varepsilon, y(x_0+\varepsilon,\varepsilon), y_x(x_0+\varepsilon,\varepsilon)\big) \\ & - L_z\big(x_0+\varepsilon, y(x_0+\varepsilon,\varepsilon), y_x(x_0+\varepsilon,\varepsilon)\big)\big(w - y_x(x_0+\varepsilon,\varepsilon)\big) \\ & + \int_{x_0+\varepsilon}^{x_1} \left(L_y\big(x, y(x,\varepsilon), y_x(x,\varepsilon)\big) - \frac{d}{dx} L_z\big(x, y(x,\varepsilon), y_x(x,\varepsilon)\big)\right) y_\varepsilon(x,\varepsilon)dx \end{aligned}$$

We are interested in the value in $\varepsilon = 0$, for which: $y^* = y(.,0)$. By assumption:

- y^* is an extremal: $L_y\big(x, y^*, y^{*\prime}\big) - \dfrac{d}{dx} L_z\big(x, y^*, y^{*\prime}\big) = 0$;
- x_0 is not a corner, so y^* is derivable at x_0 : $y_x(x_0, 0) = y^{*\prime}(x_0)$.

With these assumptions, we obtain

$$\begin{aligned} F'(0^+) = \; & L\big(x_0, y^*(x_0), w\big) - L\big(x_0, y^*(x_0), y^{*\prime}(x_0)\big) \\ & - L_z\big(x_0, y^*(x_0), y^{*\prime}(x_0)\big)\big(w - y^{*\prime}(x_0)\big) \\ = \; & E\big(x_0, y^*(x_0), y^{*\prime}(x_0), w\big) \end{aligned}$$

which is the expected relation.

The Weierstrass condition (1879) reflects the local convexity of the Lagrangian with respect to the variable $z = y'$.

$$L(x, y^*, w) \geq L(x, y^*, y^{*\prime}) + (w - y^{*\prime})L_z(x, y, y^{*\prime}) \quad , \quad \forall w \in \mathbb{R} \tag{3.54}$$

The Weierstrass excess function E is the difference between the function L(x,y,z) and its tangent at the point (x; y).
Using the canonical variables p and H (which will be studied in section 3.4)

$$\begin{cases} p = L_z \\ H = L_z y' - L \end{cases} \tag{3.55}$$

the Weierstrass function can be written as

$$\begin{aligned} E(x, y^*, y^{*\prime}, w) &= L(x, y^*, w) - L(x, y^*, y^{*\prime}) - (w - y^{*\prime})L_z(x, y^*, y^{*\prime}) \\ &= L(x, y^*, w) - L(x, y^*, y^{*\prime}) - (w - y^{*\prime})p \\ &= \left(py^{*\prime} - L(x, y^*, y^{*\prime})\right) - \left(pw - L(x, y^*, w)\right) \\ &= \quad H(x, y^*, y^{*\prime}, p) \quad - \quad H(x, y^*, w, p) \end{aligned} \tag{3.56}$$

which shows that the Hamiltonian H is maximum when $w = y^{*\prime}$.
We will see in the context of an optimal control problem (chapter 4) that the Weierstrass condition is directly linked to the maximum principle.

3.2.5 Summary

The different minimum requirements are summarized below:

- weak minimum: with functions C^1 and neighborhood in norm 1;
- strong minimum: with functions C^1 by pieces and neighborhood in norm 0.

<u>Necessary conditions for a weak local minimum</u>

$L_y = L_z'$	Euler-Lagrange (1740, 1755)
$L_{zz} \geq 0$	Legendre (1786)

<u>Sufficient conditions for a weak local minimum</u>

$L_y = L_z'$	Euler-Lagrange (extremal)
$L_{zz} > 0$	Legendre (strong condition)
No conjugate point	Jacobi (1837)

Necessary conditions for a strong local minimum

$L_y = L'_z$	Euler-Lagrange (extremal)
$L_{zz} \geq 0$	Legendre
L_z and $L_z y' - L$ continuous	Weierstrass-Erdmann (1865, 1877) in every corner
$E(x, y^*, y^{*\prime}, w) \geq 0$, $\forall w$	Weierstrass (1879) everywhere but corners

3.3 Constraints

The previous section established optimality conditions for the standard problem

$$\min_y J(y) = \int_a^b L(x, y, y')dx \quad \text{with} \quad \begin{cases} y(a) = \alpha \\ y(b) = \beta \end{cases} \tag{3.57}$$

The only constraints on the solution $y^*(x)$ were on the endpoint values.
This section extends the optimality conditions to different types of constraints.

3.3.1 Final constraint

It is assumed in this section that the endpoint coordinates can vary.

The changes in abscissa and ordinate are noted: $\begin{cases} \delta x(a) = \delta a \\ \delta x(b) = \delta b \end{cases}$ and $\begin{cases} \delta y(a) = \delta\alpha \\ \delta y(b) = \delta\beta \end{cases}$.

The first variation of the functional J is then given by

$$\delta J = \int_a^b \eta\left(L_y - L'_{y'}\right)dx + \left[L_{y'}\delta y\right]_a^b + \left[(L - L_{y'}y')\delta x\right]_a^b \tag{3.58}$$

Demonstration

The functional J depends on the function y and on the abscissas a and b.

$$J(y, a, b) = \int_a^b L(x, y, y')dx$$

Let us apply a perturbation η on the y-function and perturbations δa and δb on the abscissas a and b as shown in figure 3-14.

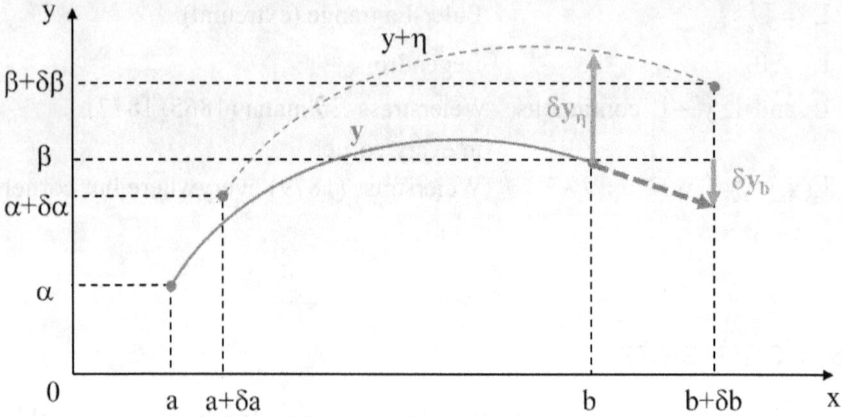

Figure 3-14: Function with free endpoints.

Let us calculate $J(y+\eta, a+\delta a, b+\delta b)$ by expanding the Lagrangian to order 1.

$$J(y+\eta, a+\delta a, b+\delta b) = \int_{a+\delta a}^{b+\delta b} L\big(x, y+\eta, (y+\eta)'\big)dx = \int_{a+\delta a}^{b+\delta b} \big(L+\eta L_y + \eta' L_{y'}\big)dx$$

$$= \int_{a+\delta a}^{b+\delta b} Ldx + \int_{a+\delta a}^{b+\delta b} \big(\eta L_y + \eta' L_{y'}\big)dx$$

The first integral gives at order 1:

$$\int_{a+\delta a}^{b+\delta b} Ldx = \int_{a+\delta a}^{a} Ldx + \int_a^b Ldx + \int_b^{b+\delta b} Ldx = \int_a^b Ldx + L(b)\delta b - L(a)\delta a = \int_a^b Ldx + [L\delta x]_a^b$$

The second integral is integrated by parts, neglecting the terms in $\eta\delta a, \eta\delta b$, which are of order 2.

$$\int_{a+\delta a}^{b+\delta b} \big(\eta L_y + \eta' L_{y'}\big)dx = \int_a^b \big(\eta L_y + \eta' L_{y'}\big)dx = \int_a^b \eta\big(L_y - L'_{y'}\big)dx + \big[\eta L_{y'}\big]_a^b$$

The first variation δJ is obtained by grouping these terms.

$$\delta J = J(y+\eta, a+\delta a, b+\delta b) - J(y,a,b) = \int_a^b \eta\big(L_y - L'_{y'}\big)dx + \big[\eta L_{y'}\big]_a^b + [L\delta x]_a^b$$

We now seek to make the ordinate variations $\delta y(a), \delta y(b)$ appear.

The variation δy at the endpoints results on the one hand from the function η, on the other hand from the abscissa variation δa or δb as shown in figure 3-14.

We have the relations

$$\begin{cases}\delta y(a)=\eta(a)+y'(a)\delta a\\ \delta y(b)=\eta(b)+y'(b)\delta b\end{cases} \Rightarrow \begin{cases}\eta(a)=\delta y(a)-y'(a)\delta a\\ \eta(b)=\delta y(b)-y'(b)\delta b\end{cases} \text{ with } \begin{cases}\delta a=\delta x(a)\\ \delta b=\delta x(b)\end{cases}$$

By replacing in the expression of δJ, we obtain the formula (3.58).

$$\delta J=\int_a^b \eta\left(L_y-L'_{y'}\right)dx+\left[L_{y'}\delta y\right]_a^b+\left[(L-L_{y'}y')\delta x\right]_a^b$$

The formula (3.58) is used below for a problem with a free final ordinate and then for a problem with a constrained final position on a curve.

Problem with free final ordinate

Assume that the coordinates a,α,b are fixed and the ordinate β is free.
The problem is formulated as

$$\min_y J(y)=\int_a^b L(x,y,y')dx \quad \text{with} \quad y(a)=\alpha \tag{3.59}$$

Figure 3-15 shows three functions y_1,y_2,y^* of values β_1,β_2,β^* in $x=b$.

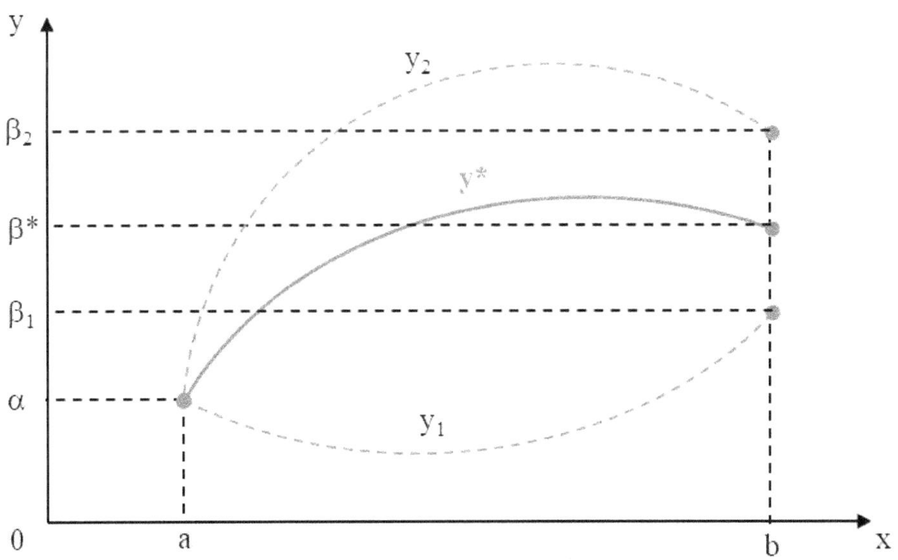

Figure 3-15: Function with free final ordinate.

Theorem 3-7 gives necessary conditions that a function y* must satisfy in order to be a weak local minimum of J with free ordinate in $x = b$.

Theorem 3-7: Necessary conditions of minimum with free final ordinate

If y* is a weak local minimum of problem (3.59), then y* satisfies

- the Euler-Lagrange equation

$$L_y - L_z' = 0 \tag{3.60}$$

- the condition of transversality

$$L_z(b) = 0 \tag{3.61}$$

Demonstration

In the case where a, α, b are fixed, we have: $\delta x(a) = \delta y(a) = \delta x(b) = 0$.

The first variation (3.58) reduces to: $\delta J = \int_a^b \eta\left(L_y - L_{y'}'\right)dx + L_{y'}(b)\delta y(b)$.

The condition of minimum: $\delta J = 0$, $\forall \eta, \forall \delta y(b)$ requires: $\begin{cases} L_y - L_{y'}' = 0 \\ L_{y'}(b) = 0 \end{cases}$.

The Euler-Lagrange equation (3.60) is a second order differential equation with two integration constants.
These constants are determined by the initial condition $y(a) = \alpha$ and the transversality condition (3.61), which replaces the final condition $y(b) = \beta$.
Example 3-7 shows the use of this transversality condition.

Example 3-7: Minimum distance in the plane with free end

Let us retrieve example 3-1 assuming now the ordinate is free at $x = b$.
We are looking for the curve y(x) minimizing the distance between the fixed point A and a free point B of abscissa b.
The extremals (example 3-1) are still straight lines: $y' = C^{te}$.

The initial condition requires that we pass through point A $(a;\alpha)$.

The problem is formulated as

$$\min_y \int_a^b \sqrt{1+y'(x)^2}\,dx$$

with the Lagrangian:

$$L(x,y,z)=\sqrt{1+z^2}$$

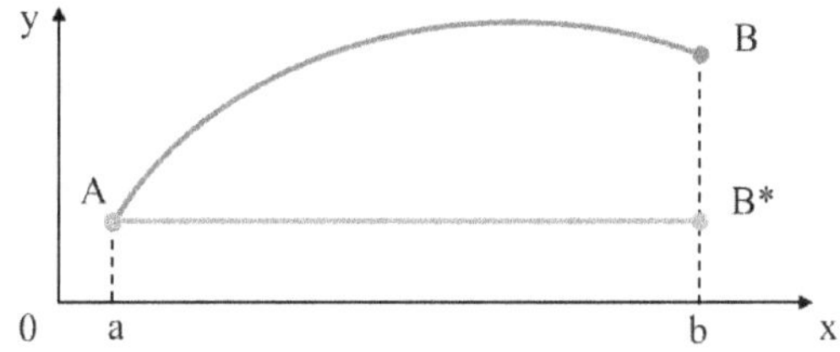

The transversality condition gives $\frac{\partial L}{\partial y'}(b)=0 \Rightarrow \frac{y'}{\sqrt{1+y'^2}}(b)=0 \Rightarrow y'(b)=0$.

The solution is the horizontal segment [AB*] passing through A, which gives indeed the minimum distance to the imposed abscissa $x=b$.

Problem with constrained final position on a curve

Assume that $(a;\alpha)$ are fixed and that the endpoint must lie on the curve with equation $y=\psi(x)$. The abscissa b is free. The problem is formulated as

$$\min_{y,b} J(y)=\int_a^b L(x,y,y')dx \quad \text{with} \quad \begin{cases} y(a)=\alpha \\ y(b)=\psi(b) \end{cases} \tag{3.62}$$

Figure 3-16 shows three functions y_1, y_2, y^* arriving on the curve $y=\psi(x)$.

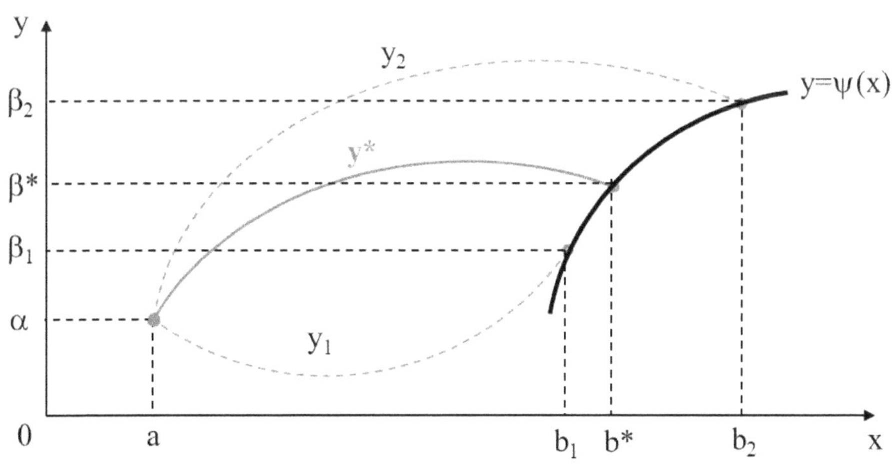

Figure 3-16: Function with end position on a curve.

Theorem 3-8 gives necessary conditions satisfied by a function y* to be a weak minimum of J with the final position such that $y(b) = \psi(b)$.

Theorem 3-8: Necessary conditions of minimum with final constraint

If y* is a weak local minimum of problem (3.62), then y* satisfies

- the Euler-Lagrange equation

$$L_y - L'_z = 0 \tag{3.63}$$

- the condition of transversality

$$\left[L + L_z(\psi' - y^{*\prime})\right](b) = 0 \tag{3.64}$$

Demonstration

$(a;\alpha)$ being fixed: $\delta x(a) = \delta y(a) = 0$, the first variation (3.58) reduces to

$$\delta J = \int_a^b \eta\left(L_y - L'_{y'}\right)dx + \left[L_{y'}\delta y\right](b) + \left[(L - L_{y'}y')\delta x\right](b)$$

The variation $\delta y(b)$ depends on function η and variation δb. These variations are linked, as the endpoint must remain on the curve $y = \psi(x)$ as in figure 3-17.

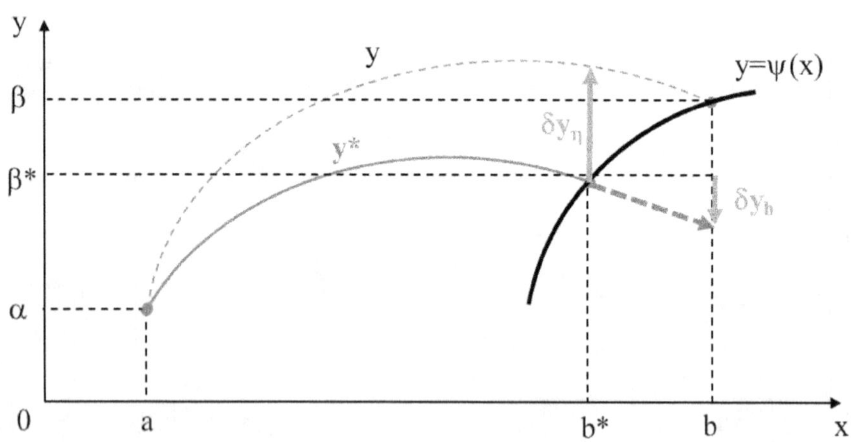

Figure 3-17: Constrained variation at the endpoint.

The final constraint imposes: $\delta y(b)=\psi'(b)\delta b$. Replacing in δJ, we obtain

$$\delta J=\int_a^b \eta\left(L_y-L'_{y'}\right)dx+\left[L+L_{y'}(\psi'-y')\delta x\right](b)$$

The condition of minimum: $\delta J=0$, $\forall\eta$, $\forall\delta x(b)$ requires

$$\begin{cases} L_y-L'_{y'}=0 \\ \left[L+L_{y'}(\psi'-y')\right](b)=0 \end{cases}$$

The Euler-Lagrange equation (3.63) yields two integration constants and the free final abscissa b is also to be determined. These three unknowns must satisfy the three equations concerning respectively the initial condition $y(a)=\alpha$, the final constraint $y(b)=\psi(b)$ and the transversality condition (3.64).
Example 3-8 shows the use of this transversality condition.

Example 3-8: Minimum distance to a curve

We are looking for the curve minimizing the distance between the fixed point A and the curve of equation $y=\psi(x)$ as shown in figure 3-18.

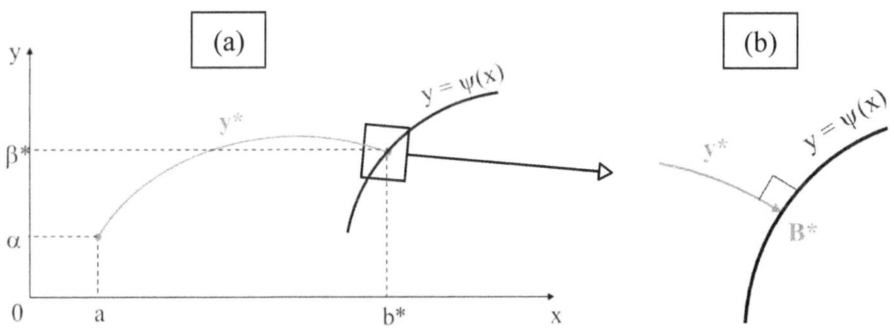

Figure 3-18: Minimum distance to a curve.

The endpoint on the curve is free and the distance is weighted by a function $w(x,y)$. In particular, the function $w(x,y)$ can represent the inverse of the velocity with respect to the position. The solution in this case is the fastest path to reach the target curve $y=\psi(x)$.

The problem is of the form: $\min_{y,b} \int_a^b w(x,y)\sqrt{1+y'(x)^2}\,dx \quad \text{s.t.} \quad y(b)=\psi(b)$.

Let us write the transversality condition (3.64) in b with the Lagrangian given by $L(x,y,z)=w(x,y)\sqrt{1+z^2}$

$$L+L_z(\psi'-y^{*\prime})=0 \;\Rightarrow\; w\sqrt{1+y'^2}+\frac{wy'}{\sqrt{1+y'^2}}(\psi'-y')=0 \;\Rightarrow\; 1+y'\psi'=0$$

The transversality condition imposes in b: $y'(b)\psi'(b)=-1$.
It reflects the orthogonality of the solution curve and the constraint curve at the end point, as shown in figure 3-18b.

3.3.2 Integral constraint

It is assumed in this section that the endpoints are fixed, but that the solution is subject to an integral form constraint. The problem is formulated as

$$\min_y J(y)=\int_a^b L(x,y,y')dx \;\text{ s.t. }\; C(y)=\int_a^b M(x,y,y')dx=0\,,\; \begin{cases} y(a)=\alpha \\ y(b)=\beta \end{cases} \tag{3.65}$$

To solve this type of problem, we define the augmented cost as

$$K(y)=\lambda_0 J(y)+\lambda C(y)=\int_a^b (\lambda_0 L+\lambda M)(x,y,y')dx \tag{3.66}$$

The real numbers λ_0 and λ are the Lagrange multipliers of the cost and the constraint. The function $N=\lambda_0 L+\lambda M$ is called the augmented Lagrangian.

Theorem 3-9 gives necessary conditions that a function y* must satisfy to be a weak minimum of J with the integral constraint $C(y)=0$.

Theorem 3-9: Necessary conditions of minimum with integral constraint

Assume that y* is a weak local minimum of problem (3.65).

Then there exist $(\lambda_0\,;\lambda)\neq(0\,;0)$ such that y* satisfies

$$N_y-N_z'=0 \tag{3.67}$$

We recognize the Euler-Lagrange equation applied to the augmented Lagrangian.

Demonstration (from [R14])

Assume that y* is a minimum of J under the constraint $C(y)=0$.

Apply perturbations of the form $y=y^*+\varepsilon_1\eta_1+\varepsilon_2\eta_2$ with any functions η_1,η_2 of class C^1.

The function $F:(\varepsilon_1,\varepsilon_2)\in\mathbb{R}^2 \mapsto \big(J(y),C(y)\big)\in\mathbb{R}^2$ is of class C^1 of $\mathbb{R}^2$ in $\mathbb{R}^2$.

Let us study this function in the vicinity of $(\varepsilon_1;\varepsilon_2)=(0;0)$.

Its value in $(\varepsilon_1;\varepsilon_2)=(0;0)$ is $F(0,0)=\big(J(y^*),C(y^*)=0\big)$

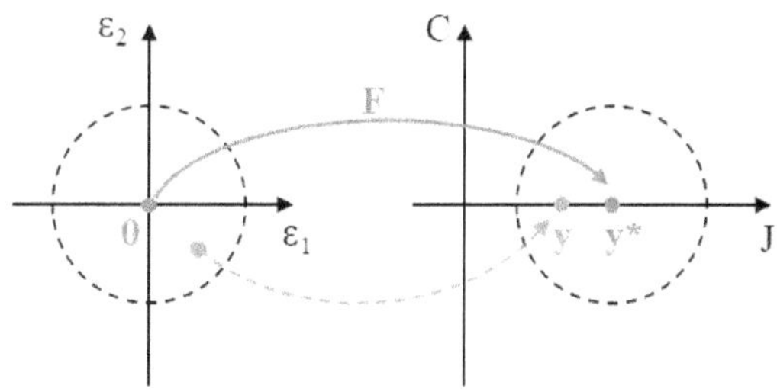

Its gradient matrix in $(0,0)$ is of the form: $\nabla F(0,0)=\begin{pmatrix}\dfrac{\partial J}{\partial\varepsilon_1} & \dfrac{\partial C}{\partial\varepsilon_1}\\ \dfrac{\partial J}{\partial\varepsilon_2} & \dfrac{\partial C}{\partial\varepsilon_2}\end{pmatrix}$.

The implicit function theorem states that if the gradient matrix $\nabla F(0,0)$ is invertible, then the function F is a bijection from a neighborhood of $(0,0)$ into a neighborhood of $\big(J(y^*),C(y^*)\big)$.

A pair $(\varepsilon_1;\varepsilon_2)\neq(0;0)$ can then be found such that $C(y)=0$ (feasible) and $J(y)<J(y^*)$. This situation is illustrated in the figure above with a solution y on the x-axis: $C(y)=0$ and a value $J(y)$ lower than $J(y^*)$.

This contradicts the assumption that y* is a minimum.

Therefore, the matrix $\nabla F(0,0)$ is necessarily singular, which means that its columns are linearly dependent.

There exist $(\lambda_0;\lambda)\neq(0;0)$ such that: $\lambda_0\dfrac{\partial J}{\partial\varepsilon}+\lambda\dfrac{\partial C}{\partial\varepsilon}=0$, noting $\varepsilon=(\varepsilon_1,\varepsilon_2)\in\mathbb{R}^2$.

(equation E)

Let us calculate the derivatives of J and C by applying (3.11) and (3.21) for any perturbation η.

$$J(y+\varepsilon\eta) = \int_a^b L(x,y+\varepsilon\eta,(y+\varepsilon\eta)')dx = \varepsilon\int_a^b \eta\left(L_y - L_z'\right)dx$$

$$\Rightarrow \frac{\partial J}{\partial \varepsilon} = \int_a^b \eta\left(L_y - L_z'\right)dx$$

$$C(y+\varepsilon\eta) = \int_a^b M(x,y+\varepsilon\eta,(y+\varepsilon\eta)')dx = \varepsilon\int_a^b \eta\left(M_y - M_z'\right)dx$$

$$\Rightarrow \frac{\partial C}{\partial \varepsilon} = \int_a^b \eta\left(M_y - M_z'\right)dx$$

Replacing these derivatives in (E) with N defined by: $N=\lambda_0 L+\lambda M$, we obtain

$$\int_a^b \eta\left(N_y - N_z'\right)dx = 0 \,,\ \forall\eta$$

Since the perturbation η is arbitrary, its coefficient in the integral must be zero.
We conclude that there exist $(\lambda_0\,;\lambda) \neq (0\,;0)$ such as $N_y - N_z' = 0$.

The values of $(\lambda_0\,;\lambda)$ are determined to within one multiplicative factor.
The multiplier λ_0 is called the **abnormal** multiplier.
If $\lambda_0 = 0$ (abnormal case), there is only one function y* satisfying the constraint.
If $\lambda_0 \neq 0$ (normal case), the constraint $C(y^*) = 0$ determines the ratio λ/λ_0.
The general choice in the normal case is to set $\lambda_0 = 1$.

An integral constraint is also called an isoperimetric constraint in reference to the problem of Dido (Queen of Carthage in 814 BC) presented below.

Example 3-9: Problem of Dido

The maximum area enclosed by a curve of given length is sought.

We will solve this problem in Cartesian coordinates, then in polar coordinates, which will allow an extension to the maximum volume problem in dimension 3.

<u>Formulation in Cartesian coordinates</u>

The points A and B are given and define the x axis.
The curve with equation y(x) joins points A and B.
The area between the curve (AB) and the x axis is: $S = \int_a^b y dx$.

The length of the curve (AB) is fixed: $L = \int_a^b \sqrt{1+y'(x)^2} dx = L_0$.

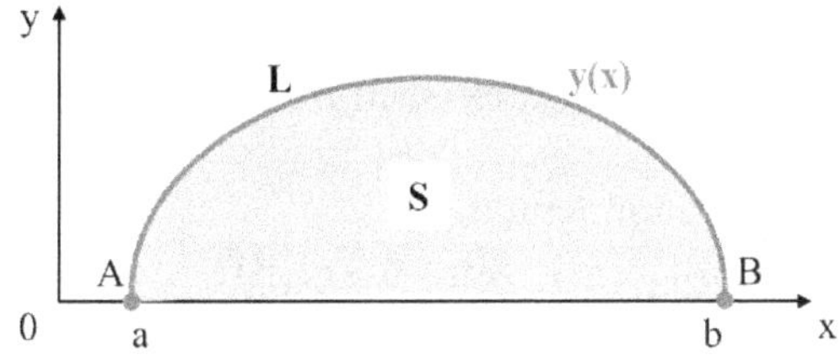

The problem is formulated as

$$\min_y J(y) = -\int_a^b y dx \quad \text{s.t.} \quad C(y) = \int_a^b \sqrt{1+y'(x)^2} dx - L_0 = 0$$

Let us introduce the multiplier λ of the integral constraint by setting $\lambda_0 = 1$.

The augmented Lagrangian is: $N(x,y,y') = -y + \lambda\sqrt{1+y'^2}$.

Equation (3.67) yields: $\frac{\partial N}{\partial y} = \frac{d}{dx}\left(\frac{\partial N}{\partial y'}\right) \Rightarrow \frac{d}{dx}\left(\frac{\lambda y'}{\sqrt{1+y'^2}}\right) = -1$.

This second-order differential equation can be solved analytically.

$$\frac{\lambda y'}{\sqrt{1+y'^2}} = -(x-x_0) \Rightarrow y' = \frac{x-x_0}{\sqrt{\lambda^2-(x-x_0)^2}}$$

$$\Rightarrow y = y_0 - \sqrt{\lambda^2-(x-x_0)^2} \Rightarrow (x-x_0)^2 + (y-y_0)^2 = \lambda^2$$

The curve around the maximum area is a circular arc. The coordinates of the center $(x_0; y_0)$ and the radius λ are determined by a, b and the length L_0.

<u>Remark on the formulation</u>

- Let us consider on the one hand the problem:
  "Maximize the area enclosed by a curve of given length"
  $\max_y S(y)$ s.t. $L(y) - L_0 = 0 \rightarrow$ Lagrangian $N_1(x,y,y') = -y + \lambda_L\sqrt{1+y'^2}$
- Let us consider on the other hand the problem:
  "Minimize the length of the curve enclosing a given area"
  $\min_y L(y)$ s.t. $S(y) - S_0 = 0 \rightarrow$ Lagrangian $N_2(x,y,y') = \sqrt{1+y'^2} + \lambda_S y$

The Lagrangians are identical by posing: $\lambda_S = -\frac{1}{\lambda_L}$.

These two problems have the same extremals which are arcs of circle.

Formulation in polar coordinates

Let us take the problem in polar coordinates (r,φ).

The equation of the curve (AB) is $r(\varphi)$.

An element of length dL and normal $\vec{n}$ at the position $\vec{r}$ generates (seen from the origin O):

- an apparent length $dL_r = dL\cos\theta$ with $\theta = (\vec{r},\vec{n})$;
- an angle at the center $d\varphi = \dfrac{dL_r}{r}$;
- a triangle of area $dS = \dfrac{1}{2}h\,dL$ with $h = \vec{r}\cdot\vec{n} = r\cos\theta$ (triangle with base dL, vertex O and height h).

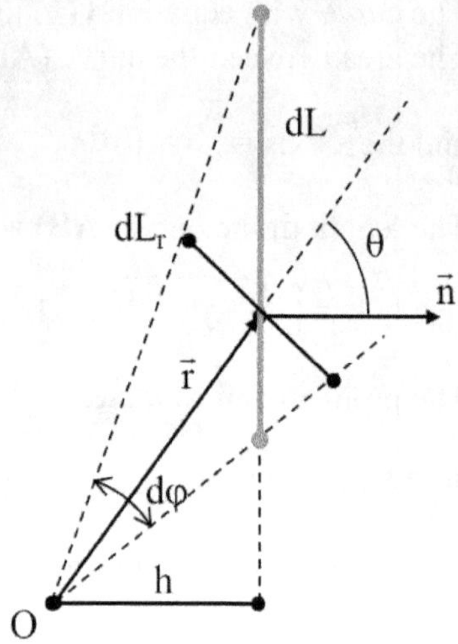

dL and dS are expressed in terms of the angle $d\varphi$: $dL = \dfrac{r}{\cos\theta}d\varphi$, $dS = \dfrac{1}{2}r^2 d\varphi$.

The problem is formulated with the two unknown functions $r(\varphi)$, $\theta(\varphi)$.

$$\max_{r,\theta} S(r,\theta) = \int_{\varphi} \frac{1}{2}r^2 d\varphi \quad \text{s.t.} \quad L(r,\theta) = \int_{\varphi} \frac{r}{\cos\theta}d\varphi = L_0$$

This is a parametric curve problem as in (3.36). Each function satisfies the Euler-Lagrange equation (3.40), here concerning the augmented Lagrangian.

By introducing the multiplier λ of the integral constraint, with $\lambda_0 = 1$, we have

$$N = \frac{1}{2}r^2 + \lambda\frac{r}{\cos\theta} \rightarrow \begin{cases} N_\theta = (N_{\theta'})' \rightarrow \lambda r\dfrac{\sin\theta}{\cos^2\theta} = 0 \rightarrow \theta = 0 \ (\vec{n} \text{ parallel to } \vec{r}) \\ N_r = (N_{r'})' \rightarrow r + \lambda\dfrac{1}{\cos\theta} = 0 \rightarrow r = -\dfrac{\lambda}{\cos\theta} = -\lambda = C^{te} \end{cases}$$

The curve enclosing the maximum area is thus a circle of radius $r_0 = -\lambda$.

The value of the radius is determined by the constrained length, equal to L_0.

$$L = \int_0^{2\pi} \frac{r}{\cos\theta}d\varphi = L_0 \quad \text{with} \quad \begin{cases} r = r_0 \\ \theta = 0 \end{cases} \Rightarrow r_0 = \frac{L_0}{2\pi}$$

As with a continuous optimization problem, the sensitivity of the cost (area) to the constraint (length) is given by the opposite of the multiplier.

$$S = \pi r_0^2 = \pi\left(\frac{L_0}{2\pi}\right)^2 \Rightarrow \frac{\partial S}{\partial L_0} = \left(\frac{L_0}{2\pi}\right) = r_0 = -\lambda$$

Maximum volume problem

The formulation in polars allows a generalization to dimension 3.
The "spatial" Dido problem consists in finding the maximum volume enclosed by a surface of given area.
Let us repeat the previous approach, now in spherical coordinates (r,λ,φ) .
The surface has the equation $r(\lambda,\varphi)$.

An element of surface dS and normal $\vec{n}$ at the position $\vec{r}$ generates (seen from the origin O):

- an apparent surface $dS_r = dS\cos\theta$ with $\theta = (\vec{r},\vec{n})$;
- a solid angle $d\Omega = \dfrac{dS_r}{r}$;
- a cone of volume $dV = \dfrac{1}{3} h dS$ with $h = \vec{r}\cdot\vec{n} = r\cos\theta$ (cone with base dS , vertex O and height h).

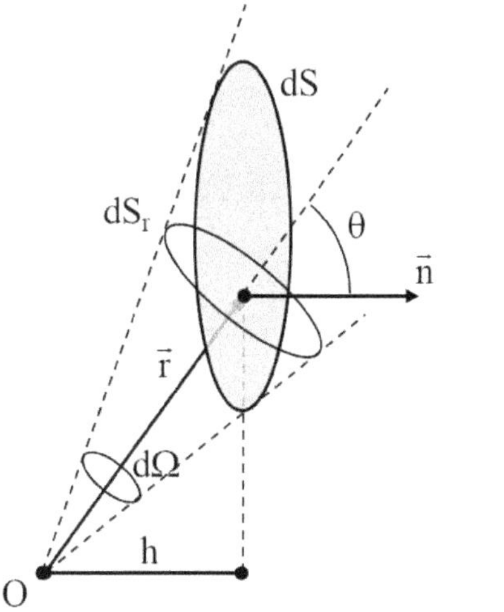

dS and dV are expressed in terms of the solid angle $d\Omega$.

$$dS = \frac{r^2}{\cos\theta} d\Omega \ , \ dV = \frac{1}{3} r^3 d\Omega$$

The total volume and surface are obtained by integrating on the solid angle $d\Omega$.
The entire space corresponds to a solid angle $\Omega = 4\pi$.
The problem is formulated with the two unknown functions $r(\lambda,\varphi)$, $\theta(\lambda,\varphi)$.

$$\max_{r,\theta} V(r,\theta) = \int_\Omega \frac{1}{3} r^3 d\Omega \quad \text{s.t.} \quad S(r,\theta) = \int_\Omega \frac{r^2}{\cos\theta} d\Omega = S_0$$

As in dimension 2, we obtain a parametric curve problem.
Let us solve the Euler-Lagrange equation with the augmented Lagrangian.

$$N = \frac{1}{3} r^3 + \lambda \frac{r^2}{\cos\theta} \rightarrow \begin{cases} N_\theta = (N_{\theta'})' \rightarrow \lambda r^2 \dfrac{\sin\theta}{\cos^2\theta} = 0 & \rightarrow \theta = 0 \ (\vec{n} \text{ parallel to } \vec{r}) \\ N_r = (N_{r'})' \rightarrow r^2 + 2\lambda \dfrac{r}{\cos\theta} = 0 & \rightarrow r = -\dfrac{2\lambda}{\cos\theta} = -2\lambda = C^{te} \end{cases}$$

The surface enclosing the maximum volume is thus a sphere of radius $r_0 = -2\lambda$.

The value of the radius is determined by the constrained surface, equal to S_0 .

$$S = \int_0^{4\pi} \frac{r^2}{\cos\theta} d\Omega = S_0 \quad \text{with} \quad \begin{cases} r = r_0 \\ \theta = 0 \end{cases} \Rightarrow r_0 = \sqrt{\frac{S_0}{4\pi}}$$

As with a continuous optimization problem, the sensitivity of the cost (volume) to the constraint (area) is given by the opposite of the multiplier.

$$V=\frac{4}{3}\pi r_0^3=\frac{4}{3}\pi\left(\frac{S_0}{4\pi}\right)^{\frac{3}{2}} \Rightarrow \frac{\partial V}{\partial S_0}=\frac{1}{2}\left(\frac{S_0}{4\pi}\right)^{\frac{1}{2}}=\frac{1}{2}r_0=-\lambda$$

3.3.3 Path constraint

It is assumed in this section that the endpoints are fixed, but that the solution is constrained at each point. The problem is formulated as

$$\min_y J(y)=\int_a^b L(x,y,y')dx \quad \text{s.t.} \quad M(x,y,y')=0\,,\ \forall x\in[a;b]\,,\ \begin{cases} y(a)=\alpha \\ y(b)=\beta \end{cases} \tag{3.68}$$

To solve this type of problem, we define the augmented cost.

$$K(y)=J(y)+\int_a^b \lambda(x)M(x,y,y')dx=\int_a^b \big(L+\lambda(x)M\big)(x,y,y')dx \tag{3.69}$$

The function $\lambda(x)$ acts as a multiplier of the constraint at each point x.
The function $N=L+\lambda(x)M$ is called the augmented Lagrangian.
Theorem 3-10 gives necessary conditions that a function y* must satisfy to be a weak minimum of J with the path constraint $M(x,y,y')=0$.

Theorem 3-10: Necessary conditions of minimum with path constraint

Assume that y* is a weak minimum of the problem (3.68).

Then there is a function $\lambda(x)$ such that y* satisfies

$$N_y-N_z'=0 \tag{3.70}$$

The demonstration will be given in the context of optimal control (chapter 4).
The value of $\lambda(x)$ at each point is determined by the constraint $M(x,y,y')=0$.
A path constraint of the form $M(x,y)=0$ is called a holonomic constraint.
This type of constraint allows in some cases to reduce the dimension of the problem by expressing y as a function of x.

Example 3-10: Pendulum problem

Consider a pendulum of length d in a uniform gravity field g.

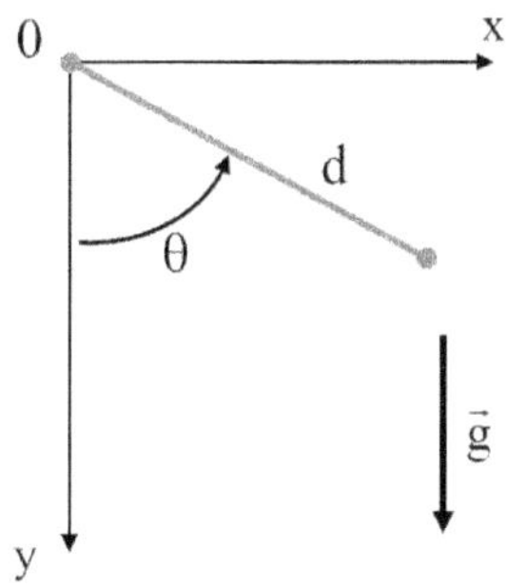

The potential energy per unit mass is: $U = -gy$.
The kinetic energy per unit mass is
$T = \frac{1}{2}(\dot{x}^2 + \dot{y}^2)$.
The motion must satisfy the holonomic constraint
$M = x^2 + y^2 - d^2 = 0$.
We will see in section 3.4 that the motion minimizes the action integral defined as

$$S(x,y) = \int_{t_0}^{t_f} L(t, x, \dot{x}, y, \dot{y})dt \quad \text{with} \quad L = T - U .$$

The problem is formulated as

$$\min_{x,y} \int_{t_0}^{t_f} \left(\frac{1}{2}(\dot{x}^2 + \dot{y}^2) + gy \right) dx \quad \text{s.t.} \quad M(x,y) = x^2 + y^2 - d^2 = 0 .$$

Let us introduce the multiplier $\lambda(t)$ associated with the current constraint.
The Euler-Lagrange equations deal with the augmented Lagrangian which depends on the two functions $x(t)$ and $y(t)$.

$$N = \frac{1}{2}(\dot{x}^2 + \dot{y}^2) + gy + \lambda(x^2 + y^2 - d^2) \quad \to \quad \begin{cases} N_x = \dot{N}_{\dot{x}} & \to \; 2\lambda x = \ddot{x} \\ N_y = \dot{N}_{\dot{y}} & \to \; 2\lambda y + g = \ddot{y} \end{cases}$$

The equations of motion are thus: $\begin{cases} \ddot{x} = 2\lambda x \\ \ddot{y} = 2\lambda y + g \\ x^2 + y^2 = d^2 \end{cases}$

with 3 unknown functions $x(t), y(t), \lambda(t)$.

By posing: $\begin{cases} x = d\sin\theta \\ y = d\cos\theta \end{cases}$, we eliminate $\lambda(t)$ and find the equation of the pendulum

$$\ddot{\theta} = -\frac{g}{d}\sin\theta .$$

3.4 Canonical form

This section presents the canonical (or Hamiltonian) formulation of a calculus of variations problem and its applications in mechanics.

3.4.1 Change of variables

A calculus of variations problem can be formulated in various ways depending on the choice of the unknowns and the functional. Here we establish some properties concerning the changes of variables.

Equivalent functionals

Consider two problems with respective functionals J and K and their Euler-Lagrange equations.

$$\begin{aligned} &\min_y J(y) = \int_a^b L(x,y,y')dx \quad \rightarrow L_y = L'_{y'} \\ &\min_y K(y) = \int_a^b M(x,y,y')dx \rightarrow M_y = M'_{y'} \end{aligned} \qquad (3.71)$$

The functionals J and K are said to be equivalent if they have the same extremals (which are the solutions of the Euler-Lagrange equations).

Property 3-11: Sufficient condition for equivalence of two functionals

If the Lagrangians L and M differ by one total derivative, then the functionals J and K are equivalent and differ by one constant.

Demonstration

Suppose that $M(x,y,y') = L(x,y,y') + \dfrac{d\Phi(x,y)}{dx}$ where $\Phi(x,y)$ is any function.

Let $F(x,y,y')$ be the function equal to the total derivative of $\Phi(x,y)$.

$$F(x,y,y') = \frac{d\Phi(x,y)}{dx} = \Phi_x(x,y) + \Phi_y(x,y)y'(x)$$

$$\Rightarrow \begin{cases} F_y = \Phi_{xy} + \Phi_{yy}y' \\ F_{y'} = \Phi_y \qquad \Rightarrow F'_{y'} = \Phi'_y = \Phi_{yx} + \Phi_{yy}y' \end{cases}$$

The derivatives M_y and $M'_{y'}$ are given by

$$\begin{cases} M_y = L_y + F_y = L_y + \Phi_{xy} + \Phi_{yy}y' \\ M'_{y'} = L'_{y'} + F'_{y'} = L'_{y'} + \Phi_{yx} + \Phi_{yy}y' \end{cases} \Rightarrow M_y - M'_{y'} = L_y - L'_{y'}$$

The Euler-Lagrange equations in L and M therefore have the same solutions.
Let us calculate the value of $K(y)$.

$$K(y) = \int_a^b M dx = \int_a^b (L+F)dx = \int_a^b \left(L + \frac{d\Phi}{dx} \right) dx = J(y) + \Phi(b) - \Phi(a) = J(y) + C^{te}$$

The two functionalities do differ by one constant.

Change of variable

Consider the change of variable defined by: $y(x) = f\left(x, Y(x)\right)$.

Assume that $y(x)$ is an extremal of: $J(y) = \int_a^b L(x, y, y')dx$.

Then $Y(x)$ is an extremal of: $K(Y) = \int_a^b M\left(x, Y, Y'\right)dx$ with M defined by

$$M(x, Y, Y') \underset{def}{=} L\left(x, f(x, Y), f_x + f_Y Y'\right) \tag{3.72}$$

Demonstration

Let us calculate the derivatives M_Y and $M'_{Y'}$ with M defined by (3.72).

$$\begin{cases} M_Y = L_y f_Y + L_{y'}(f_{xY} + f_{YY}Y') \\ M_{Y'} = L_{y'} f_Y \end{cases}$$

Deriving the 2nd equation with respect to x and using the 1st equation, we have

$$M'_{Y'} = \frac{d}{dx} M_{Y'} = \left(\frac{d}{dx} L_{y'} \right) f_Y + L_{y'} \left(\frac{d}{dx} f_Y \right) = L'_{y'} f_Y + L_{y'}(f_{Yx} + f_{YY}Y')$$
$$= L'_{y'} f_Y + (M_Y - L_y f_Y)$$

from which we deduce: $M'_{Y'} - M_Y = (L'_{y'} - L_y) f_Y$.

If $y(x)$ is an extremal of the functional J, then $L'_{y'} - L_y = 0$ and $Y(x)$ is an extremal of the functional K because $M'_{Y'} - M_Y = 0$.

Change of variable with parameter

Consider the functional J in standard form:

$$J(y) = \int_a^b L(x,y,y')dx \tag{3.73}$$

and apply a change of variable depending on a parameter ε :

$$(x,y) \mapsto (x^*,y^*) \quad \text{with} \quad \begin{cases} x^* = X(x,y,y',\varepsilon) \\ y^* = Y(x,y,y',\varepsilon) \end{cases} \quad \text{and} \quad \begin{cases} x = X(x,y,y',0) \\ y = Y(x,y,y',0) \end{cases} \tag{3.74}$$

For $\varepsilon = 0$, the transformation $(x,y) \mapsto (x^*,y^*)$ corresponds to the identity.

The functional J is **invariant** by the transformation if it keeps its value.

$$J(y^*) = J(y) \quad \Leftrightarrow \quad \int_{x^*=a^*}^{x^*=b^*} L(x^*,y^*,y^{*\prime})dx^* = \int_{x=a}^{x=b} L(x,y,y')dx \tag{3.75}$$

Noether's theorem gives a first integral in the case of an invariant functional. This theorem is used to establish conservation laws in mechanics (section 3.4.4).

Property 3-12: Noether's theorem

If the functional J (3.73) is invariant by the transformation (3.74) whatever the integral bounds a and b,

then the function $L_{y'} \left.\dfrac{\partial Y}{\partial \varepsilon}\right|_{\varepsilon=0} + \left(L - L_{y'}y'\right) \left.\dfrac{\partial X}{\partial \varepsilon}\right|_{\varepsilon=0}$ is a first integral.

Demonstration

Formula (3.58) gives the variation of J as a function of perturbations $\eta, \delta x(a), \delta x(b), \delta y(a), \delta y(b)$.

$$\delta J = \int_a^b \eta\left(L_y - L'_{y'}\right)dx + [L_{y'}\delta y]_a^b + [(L - L_{y'}y')\delta x]_a^b$$

Let us calculate the variations δx, δy by expanding (3.74) to order 1 in ε.

$$\begin{cases} x^* = X(x,y,y',\varepsilon) = X(x,y,y',0) + \varepsilon X_\varepsilon\big|_{\varepsilon=0} = x + \varepsilon X_\varepsilon\big|_{\varepsilon=0} \Rightarrow \delta x = \varepsilon X_\varepsilon\big|_{\varepsilon=0} \\ y^* = Y(x,y,y',\varepsilon) = Y(x,y,y',0) + \varepsilon\, Y_\varepsilon\big|_{\varepsilon=0} = y + \varepsilon\, Y_\varepsilon\big|_{\varepsilon=0} \Rightarrow \delta y = \varepsilon\, Y_\varepsilon\big|_{\varepsilon=0} \end{cases}$$

The first variation is in the form

$$\delta J = \int_a^b \eta\left(L_y - L'_{y'}\right)dx + \varepsilon\left[L_{y'} Y_\varepsilon\big|_{\varepsilon=0} + \left(L - L_{y'}y'\right)X_\varepsilon\big|_{\varepsilon=0}\right]_a^b$$

Along an extremal, we have

$$\begin{cases} L_y - L'_{y'} = 0 \\ \delta J = 0 \end{cases} \Rightarrow \left[L_{y'} Y_\varepsilon\big|_{\varepsilon=0} + \left(L - L_{y'}y'\right)X_\varepsilon\big|_{\varepsilon=0}\right]_a^b = 0$$

By assumption, the functional J must be invariant whatever the bounds a and b. The term in square brackets must therefore be constant.

$$L_{y'} Y_\varepsilon\big|_{\varepsilon=0} + \left(L - L_{y'}y'\right)X_\varepsilon\big|_{\varepsilon=0} = C^{te}$$

This constant quantity along an extremal is indeed a first integral.

Noether's theorem remains valid for a functional depending on several unknown functions $(y_j)_{j=1 \text{ to } n}$ as in (3.33). Indeed, the above proof applies to each function which satisfies its own Euler-Lagrange equation (3.34).

3.4.2 Canonical variables

Consider a problem that depends on n functions $y = (y_1, \cdots, y_n)$.

$$\min_y J(y) = \int_a^b L(x, y, y')dx \tag{3.76}$$

We define n functions noted $p = (p_1, \cdots, p_n)$ and the function $H(x, y, p)$ by

$$\begin{cases} p = L_{y'} \\ H = py' - L \end{cases} \tag{3.77}$$

The n functions $(y_j)_{j=1 \text{ to } n}$ of $\mathbb{R}$ in $\mathbb{R}$ are called **coordinate**.

The n functions $(p_j)_{j=1 \text{ to } n}$ of $\mathbb{R}$ in $\mathbb{R}$ are called **momentum**.

The function H of $\mathbb{R}^{2n+1}$ in $\mathbb{R}$ is called the **Hamiltonian** and is expressed as

$$H(x, y, p) = \sum_{i=1}^n p_i y'_i - L(x, y_1, \ldots, y_n, y'_1, \ldots, y'_n) \tag{3.78}$$

(y, p, H) are called canonical variables associated with the J functional.

Note on Legendre's transformation

The canonical variables can be defined by the transformation of Legendre .

The Legendre transform of the function $f:\mathbb{R}^n \to \mathbb{R}$ is the function $h:\mathbb{R}^n \to \mathbb{R}$ defined by: $h(p) \underset{def}{=} \max\limits_{z\in\mathbb{R}^n}\left(pz - f(z)\right)$.

By denoting z_p the value of z that maximizes $pz - f(z)$ for fixed p, we have the following properties:

- $\frac{d}{dz}(pz - f(z)) = 0$ in $z_p \Rightarrow p = f'(z_p) \Rightarrow h'(p) = z_p$ and $h''(p) = \left(f''(z_p)\right)^{-1}$;
- the function h is convex;
- if f is convex, the Legendre transform of h gives back f (involution).

In order to define the canonical variables associated with the functional J, we apply the Legendre transformation to the Lagrangian $L(x,y,z)$ considered as a function of z: $\begin{cases} H(x,y,p) \underset{def}{=} \max\limits_{z}\left(pz_p - L(x,y,z_p)\right) = pz_p - L(x,y,z_p) \\ p \underset{def}{=} L_z(x,y,z_p) \end{cases}$.

The canonical variables are used below to transform the second-order Euler-Lagrange equation into a system of first-order differential equations.

<u>Canonical equations</u>

The derivatives of $H = py' - L$ are given by: $\begin{cases} H_y = -L_y \\ H_p = y' \end{cases}$.

The Euler-Lagrange equation of problem (3.76) gives

$$L_y = L'_{y'} \Rightarrow -H_y = p' \tag{3.79}$$

We call canonical (or Hamiltonian) equations the following differential system.

$$\begin{cases} y' = H_p \\ p' = -H_y \end{cases} \tag{3.80}$$

This differential system in variables (y,p) is of order 1 and of dimension 2n.

It is equivalent to the Euler-Lagrange equation in y variables (differential system of order 2 and dimension n).

First integral

A first integral of the functional J is a function that keeps a constant value along any extremal y^*.

Let us calculate the total derivative of the Hamiltonian $H(x,y,p)$ along an extremal. Using the canonical equations (3.80) of the extremal, we obtain

$$\frac{dH}{dx} = \frac{\partial H}{\partial x} + \frac{\partial H}{\partial y}y' + \frac{\partial H}{\partial p}p' = \frac{\partial H}{\partial x} \tag{3.81}$$

If H does not depend explicitly on x, then H is a first integral.

Note on Poisson's brackets

Let us calculate the total derivative of a function $\Phi(x,y,p)$ along an extremal. Using the canonical equations (3.80) of the extremal, we obtain

$$\frac{d\Phi}{dx} = \frac{\partial \Phi}{\partial x} + \frac{\partial \Phi}{\partial y}y' + \frac{\partial \Phi}{\partial p}p' = \frac{\partial \Phi}{\partial x} + \frac{\partial \Phi}{\partial y}\frac{\partial H}{\partial p} - \frac{\partial \Phi}{\partial p}\frac{\partial H}{\partial y}$$

We define the Poisson bracket of Φ and H by: $[\Phi, H] \underset{\text{def}}{=} \frac{\partial \Phi}{\partial y}\frac{\partial H}{\partial p} - \frac{\partial \Phi}{\partial p}\frac{\partial H}{\partial y}$.

A function $\Phi(y,p)$ not explicit in x is a first integral if $[\Phi, H] = 0$.

Poisson's brackets are used widely in celestial mechanics.

Canonical transformation

Consider two functionals $J(y)$ and $K(Y)$.

The functional J of Lagrangian $L(x,y,y')$ has the canonical variables (y,p) and the Hamiltonian $H(x,y,p)$.

$$J = \int_a^b L(x,y,y')dx = \int_a^b \left(py' - H(x,y,p)\right)dx \quad \text{with} \begin{cases} p = L_{y'} \\ H = py' - L \end{cases} \tag{3.82}$$

The functional K of Lagrangian $M(x,Y,Y')$ has the canonical variables (Y,P) and the Hamiltonian $H^*(x,Y,P)$.

$$K = \int_a^b M(x,Y,Y')dx = \int_a^b \left(PY' - H^*(x,Y,P)\right)dx \quad \text{with} \begin{cases} P = M_{y'} \\ H^* = pY' - M \end{cases} \tag{3.83}$$

Suppose that the two functionals J and J* are equivalent, which means that they have the same extremals. A sufficient condition for this is that their Lagrangians differ by one total derivative (property 3-11).

$$py' - H(x,y,p) = PY' - H^*(x,Y,P) + \frac{d\Phi}{dx} \tag{3.84}$$

$\Phi(x,y,Y)$ is any function of the y and Y coordinates.
Expressing from (3.84) the differential $d\Phi$ as a function of dx, dy, dY:

$$d\Phi = pdy - PdY + \big(H^*(x,Y,P) - H(x,y,p)\big)dx \tag{3.85}$$

we obtain the variables p, P, H^* as a function of the partial derivatives of Φ.

$$p = \frac{\partial\Phi}{\partial y} \;,\; P = -\frac{\partial\Phi}{\partial Y} \;,\; H^* = H + \frac{\partial\Phi}{\partial x} \tag{3.86}$$

The function $\Phi(x,y,Y)$ is called the **generating function** of the canonical transformation $(y,p) \mapsto (Y,P)$. The problems in variables (y,p,H) or (Y,P,H^*) are equivalent. Their respective canonical systems are

$$\begin{cases} y' = H_p \\ p' = -H_y \end{cases} \text{ with } H(x,y,p) \quad \text{and} \quad \begin{cases} Y' = H^*_P \\ P' = -H^*_Y \end{cases} \text{ with } H^*(x,Y,P) \tag{3.87}$$

Note on generating functions

The generating function can be expressed in variables (y,Y), (y,P), (p,Y) or (p,P). We note $\Phi_1(x,y,Y)$ the function in variables (y,Y) defined above (3.84). The transformation from one form to the other is done as follows.

$\Phi_1(x,y,Y) \quad \Rightarrow d\Phi_1 = pdy - PdY + (H^* - H)dx$

$\quad \rightarrow p = \dfrac{\partial\Phi_1}{\partial y}, P = -\dfrac{\partial\Phi_1}{\partial Y}$

$\Phi_2(x,y,P) = \Phi_1(x,y,Y) + PY \quad \Rightarrow d\Phi_2 = pdy + YdP + (H^* - H)dx$

$\quad \rightarrow p = \dfrac{\partial\Phi_2}{\partial y}, Y = \dfrac{\partial\Phi_2}{\partial P}$

$\Phi_3(x,p,Y) = \Phi_1(x,y,Y) - py \quad \Rightarrow d\Phi_3 = -ydp - PdY + (H^* - H)dx$

$\quad \rightarrow y = -\dfrac{\partial\Phi_3}{\partial p}, P = -\dfrac{\partial\Phi_3}{\partial Y}$

$$\Phi_4(x,p,P) = \Phi_1(x,y,Y) - py + PY \;\Rightarrow\; d\Phi_4 = -ydp + YdP + (H^* - H)dx$$
$$\rightarrow\; y = -\frac{\partial \Phi_4}{\partial p}\,,\; Y = \frac{\partial \Phi_4}{\partial P}$$

3.4.3 Hamilton-Jacobi-Bellman equation

Consider a problem with fixed endpoint $A(a;\alpha)$ and variable endpoint $B(b;\beta)$.

$$\min_y J(y) = \int_a^b L(x,y,y')dx \quad \text{with} \quad \begin{cases} y(a) = \alpha \\ y(b) = \beta \end{cases} \tag{3.88}$$

The **value function** $V(b,\beta)$ is the value of J for the extremal y* from A to B.

$$V(b,\beta) = \int_a^b L\left(t, y^*, y^{*\prime}\right)dt \tag{3.89}$$

Theorem 3-13: Hamilton-Jacobi-Bellman equation

The value function satisfies the equation of Hamilton-Jacobi-Bellman (HJB).

$$\frac{\partial V}{\partial x} + H(x,y,\frac{\partial V}{\partial y}) = 0 \tag{3.90}$$

Demonstration

We calculate the partial derivatives of the value function with respect to b and β. To do this, we start from the first variation of the functional J given by (3.58).

$$\delta J = \int_a^b \eta\left(L_y - L'_{y'}\right)dx + [L_{y'}\delta y]_a^b + [(L - L_{y'}y')\delta x]_a^b$$

If y* is an extremal $(L_y = L'_{y'})$ and A is fixed $(\delta x(a) = \delta y(a) = 0)$, then

$\delta J = L_{y'}\delta y(b) + (L - L_{y'}y')\delta x(b) = p\delta y(b) - H\delta x(b)$

with the canonical variables p and H defined in (3.77).

This variation δJ represents the differential dV of the value function.

By noting $(x;y)$ instead of $(b;\beta)$ the coordinates of point B, we have

$$dV = pdy - Hdx \Rightarrow \frac{\partial V}{\partial x} = -H \ , \ \frac{\partial V}{\partial y} = p$$

The Hamiltonian H is a function of (x,y,p) with $p = \frac{\partial V}{\partial y}$, which leads to

$$\frac{\partial V}{\partial x} = -H(x,y,p) = -H\left(x,y,\frac{\partial V}{\partial y}\right)$$

Solving the equation HJB yields the value function $V(x,y)$ at any point M $(x;y)$ starting for a fixed point $A(a;\alpha)$. Once the value function has been determined, the extremals can be found using property 3-14.

Property 3-14: First integral deduced from the value function

Suppose that the value function $V(x,y,\gamma)$ depends on constants noted γ.

Then the functions $\frac{\partial V}{\partial \gamma}$ are first integrals of problem (3.88).

Demonstration

It must be shown that $\frac{d}{dx}\left(\frac{\partial V}{\partial \gamma}\right) = \frac{\partial^2 V}{\partial x \partial \gamma} + \frac{\partial^2 V}{\partial y \partial \gamma}\frac{dy}{dx}$ is zero along an extremal.

$V(x,y,\gamma)$ satisfies the equation HJB: $\frac{\partial V}{\partial x} + H\left(x,y,\frac{\partial V}{\partial y}\right) = 0$.

Deriving w.r.t γ, we have: $\frac{\partial}{\partial \gamma}\left(\frac{\partial V}{\partial x} + H(x,y,\frac{\partial V}{\partial y})\right) = \frac{\partial^2 V}{\partial \gamma \partial x} + \frac{\partial H}{\partial p}\frac{\partial^2 V}{\partial \gamma \partial y} = 0$.

Along an extremal $y(x)$, we have: $y' = \frac{\partial H}{\partial p} \Rightarrow \frac{\partial^2 V}{\partial \gamma \partial x} + \frac{dy}{dx}\frac{\partial^2 V}{\partial \gamma \partial y} = 0$.

Analytical solution of the HJB equation is generally not possible. In simple cases such as example 3-11, special form solutions can be obtained by separation of variables.

Example 3-11: Hamilton-Jacobi-Bellman equation

Consider a functional of the form: $J(y)=\int_a^b f(y)\sqrt{1+y'^2}dx$.

Let us express the canonical variables from: $L(x,y,y')=f(y)\sqrt{1+y'^2}$.

$$\begin{cases} H=py'-L=py'-f(y)\sqrt{1+y'^2} \\ p=L_{y'}=f(y)\dfrac{y'}{\sqrt{1+y'^2}} \end{cases}$$

The 2nd equation is used to express y' as a function of (x,y,p), then replace it in the 1st equation to express H as a function of (x,y,p).

$$p=\frac{f(y)y'}{\sqrt{1+y'^2}} \Rightarrow y'=\frac{p}{\sqrt{f(y)^2-p^2}} \rightarrow H(x,y,p)=-\sqrt{f(y)^2-p^2}$$

The equation HBJ is then: $V_x+H(x,y,V_y)=0 \Rightarrow V_x^2+V_y^2=f(y)^2$.

Solution by separation of variables

Let us look for a solution of the form: $V(x,y)=v(x)+w(y)$.

$$v'(x)^2+w'(y)^2=f(y)^2 \Rightarrow v'(x)^2=f(y)^2-w'(y)^2=\gamma^2=C^{te}$$

because the equality must be true for all (x,y).

Integrating each member with the initial condition: $V(a,\alpha)=0$, we obtain

$$\begin{cases} v(x)=\gamma(x-a) \\ w(y)=\int_\alpha^y \sqrt{f(u)^2-\gamma^2}du \end{cases} \Rightarrow V(x,y)=\gamma(x-a)+\int_\alpha^y \sqrt{f(u)^2-\gamma^2}du$$

To determine the extremals starting from A $(a;\alpha)$, we use property 3-14 indicating that $\dfrac{\partial V}{\partial \gamma}$ is a first integral.

$$\frac{\partial V}{\partial \gamma}=x-a-\int_\alpha^y \frac{\gamma du}{\sqrt{f(u)^2-\gamma^2}}=C^{te}=0 \quad (\text{for } x=a\ ,\ y=\alpha)$$

The extremals therefore have the equation: $x-\int_\alpha^y \dfrac{\gamma du}{\sqrt{f(u)^2-\gamma^2}}=a$.

Distance functional

In the case $f(y)=1$, the functional J is the distance in the plane.
The value function and the equations of the extremals are simplified.

Value function: $V(x,y)=\gamma(x-a)+\sqrt{1-\gamma^2}(y-\alpha)$

Extremal: $x-\dfrac{\gamma}{\sqrt{1-\gamma^2}}(y-\alpha)=a$

We find the equation of a line through A.

For a given point $M(x;y)$, the constant γ is: $\gamma=\dfrac{(x-a)}{\sqrt{(x-a)^2+(y-\alpha)^2}}$.

Let us replace γ in the value function.

$V(x,y)=\gamma(x-a)+\sqrt{1-\gamma^2}(y-\alpha)=\sqrt{(x-a)^2+(y-\alpha)^2}$

We find indeed that the value function represents the minimal AM distance.

3.4.4 Application to mechanics

The laws of mechanics can be established from the calculus of variations.
Consider a system of n material points (particles) in a given potential field:

- particle masses: $m=(m_1,\cdots,m_n)$;
- particle positions: $\vec{r}(t)=(\vec{r}_1,\cdots,\vec{r}_n) \underset{\text{noted}}{=} q$;
- particle velocities: $\vec{v}(t)=\dfrac{d\vec{r}}{dt}=(\vec{v}_1,\cdots,\vec{v}_n) \underset{\text{noted}}{=} \dot{q}$;
- particle accelerations: $\vec{a}(t)=\dfrac{d\vec{v}}{dt}=(\vec{a}_1,\cdots,\vec{a}_n) \underset{\text{noted}}{=} \ddot{q}$;
- total kinetic energy: $T=\dfrac{1}{2}m\dot{q}^2=\sum_{i=1}^{n}\dfrac{1}{2}m_i\dot{q}_i^{\,2}$;
- total potential energy: $U(t,q)$;
- force applied to particle i: $\vec{F}_i=-\dfrac{\partial U}{\partial q_i}$ (derived from the potential).

The **action** of the system is defined by

$$S=\int_{t_0}^{t_f} L(t,q,\dot{q})dt \quad \text{with} \quad L=T-U=\frac{1}{2}m\dot{q}^2-U(t,q) \tag{3.91}$$

The Lagrangian L of the particle system is the difference between the kinetic energy and the potential energy. The laws of mechanics are based on the following postulate.

<u>Least action postulate</u>: the trajectories of the particles minimize the S action.

We can therefore find the trajectories of the particles by looking for the extremals of the functional S. The Euler-Lagrange equation of the functional (3.91) yields

$$L_q = \dot{L}_{\dot{q}} \Rightarrow -\frac{\partial U}{\partial q} = \frac{d}{dt}(m\dot{q}) \Rightarrow \vec{F} = m\ddot{q} \tag{3.92}$$

We find Newton's second law (fundamental principle of dynamics). The postulate of least action "contains" Newton's second law. It also allows us to find the conservation laws. Let us express the canonical variables associated with the functional S (3.91).

$$\begin{cases} p = L_{\dot{q}} = m\dot{q} \\ H = p\dot{q} - L = m\dot{q}^2 - L = 2T - (T - U) = T + U \end{cases} \tag{3.93}$$

The variables p are the quantities of motion (or momentum) of the particles. The Hamiltonian H is the total energy of the system.

Let us look at some specific systems.

<u>Conservative system</u>

For a conservative system, the potential U is not explicitly time dependent. The Hamiltonian H (3.93) is then a first integral from (3.81).

$$\frac{dH}{dt} = \frac{\partial H}{\partial t} = \frac{\partial}{\partial t}(T + U) = 0 \Rightarrow H = T + U = C^{te} \tag{3.94}$$

We find the law of conservation of energy.

<u>Translation invariant system</u>

Consider the translation of parameter ε defined by

$$\begin{cases} t^* = X(t, q, \dot{q}, \varepsilon) = t \\ q^* = Y(t, q, \dot{q}, \varepsilon) = q + \varepsilon \end{cases} \tag{3.95}$$

As the system is invariant by this transformation, so is the functional (3.91).

Noether's theorem (property 3-12) then gives the first integral

$$L_{y'}\left.\frac{\partial Y}{\partial \varepsilon}\right|_{\varepsilon=0}+\left(L-L_{y'}y'\right)\left.\frac{\partial X}{\partial \varepsilon}\right|_{\varepsilon=0}=C^{te} \tag{3.96}$$

or by using the canonical variables $p=L_{y'}$, $H=L-L_{y'}y'$

$$p\left.\frac{\partial Y}{\partial \varepsilon}\right|_{\varepsilon=0}+H\left.\frac{\partial X}{\partial \varepsilon}\right|_{\varepsilon=0}=C^{te} \tag{3.97}$$

For the transformation (3.95), we obtain: $p=m\dot{q}=C^{te}$.
We find the law of conservation of momentum for a translationally invariant system.

Rotation invariant system

An infinitesimal rotation of an angle ε about an axis $\vec{n}$ transforms the vector $\vec{q}$ into a vector $\vec{q}^*=\vec{q}+\delta\vec{q}$ defined by

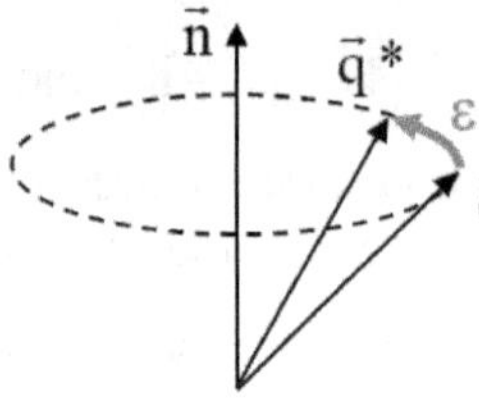

$$\delta\vec{q}=\varepsilon\vec{n}\wedge\vec{q} \quad \rightarrow \quad \vec{q}^*=\vec{q}+\varepsilon\vec{n}\wedge\vec{q}$$

Consider the rotation of parameter ε defined by

$$\begin{cases} t^* = X(t,q,\dot{q},\varepsilon)=t \\ \vec{q}^* = Y(t,q,\dot{q},\varepsilon)=\vec{q}+\varepsilon\vec{n}\wedge\vec{q} \end{cases} \tag{3.98}$$

Since the system is invariant by this transformation, the functional (3.91) is also invariant. Noether's theorem (property 3-12) then gives the first integral

$$p\left.\frac{\partial Y}{\partial \varepsilon}\right|_{\varepsilon=0}+H\left.\frac{\partial X}{\partial \varepsilon}\right|_{\varepsilon=0}=C^{te} \tag{3.99}$$

For the transformation (3.98), we obtain: $\vec{p}\cdot(\vec{n}\wedge\vec{q})=\vec{n}\cdot(\vec{q}\wedge\vec{p})=C^{te}$.

Since the relations holds for any axis of rotation $\vec{n}$, we can deduce: $\vec{q}\wedge\vec{p}=C^{te}$.

This quantity is the angular momentum (moment of momentum): $\vec{h}\underset{def}{=}\vec{q}\wedge m\dot{\vec{q}}$.

We find the law of conservation of angular momentum for a rotation invariant system.

Kepler problem

Kepler's problem is the motion of a particle in a 1/r potential.
Let us use the previous results to find the main properties of the Kepler problem.

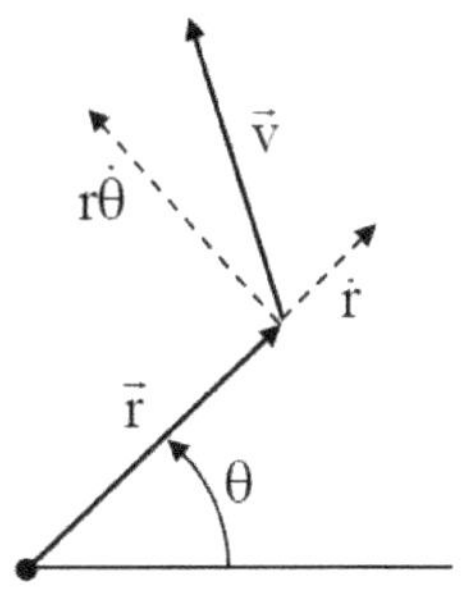

The movement is planar because the acceleration is central.
We place ourselves in polar coordinates (r,θ) as shown opposite.

The velocity components are: $\vec{v} = \begin{pmatrix} \dot{r} \\ r\dot{\theta} \end{pmatrix}$.

Let us express the Lagrangian of the system:

- potential energy: $U = -\frac{\mu}{r}$;
- kinetic energy: $T = \frac{1}{2}v^2 = \frac{1}{2}(\dot{r}^2 + r^2\dot{\theta}^2)$;
- Lagrangian: $L(t,r,\theta,\dot{r},\dot{\theta}) = T - U = \frac{1}{2}(\dot{r}^2 + r^2\dot{\theta}^2) + \frac{\mu}{r}$.

The Euler-Lagrange equations for the functions (r,θ) yield

$$\begin{cases} L_r = \dot{L}_{\dot{r}} \Rightarrow r\dot{\theta}^2 - \frac{\mu}{r^2} = \ddot{r} \\ L_\theta = \dot{L}_{\dot{\theta}} \Rightarrow 0 = \frac{d}{dt}(r^2\dot{\theta}) \Rightarrow r^2\dot{\theta} = C^{te} \end{cases}$$

The second equation involves the modulus of angular momentum $\vec{h}$.

$$\vec{h} = \vec{r} \wedge \vec{v} = \begin{pmatrix} r \\ 0 \\ 0 \end{pmatrix} \wedge \begin{pmatrix} \dot{r} \\ r\dot{\theta} \\ 0 \end{pmatrix} = \begin{pmatrix} 0 \\ 0 \\ r^2\dot{\theta} \end{pmatrix}$$

The Lagrangian L does not depend explicitly on the time t.
The system is conservative and its energy (Hamiltonian) is then constant.
$H = T + U = C^{te}$

The Lagrangian L does not depend explicitly on the polar angle θ.
The system is rotationally invariant and its angular momentum is then constant.
$\vec{h} = \vec{r} \wedge \vec{v} = C^{te}$

3.5 Dynamic system

This section introduces the state formulation of a system which allows us to move from the calculus of variations to more general problems of optimal control.

3.5.1 State formulation

The calculus of variations considers problems of the form

$$\min_y J(y) = \int_a^b L(x, y, y')dx \tag{3.100}$$

where the unknown is a derivable function $y(x)$.

A more general approach is to formulate a control problem by introducing an unknown function $u(x)$ and expressing the derivative as a differential equation: $y' = f(x, u, t)$.

Control problem

The variables are changed as follows:

- the variable x is replaced by the time t;
- the function $y(x)$ is replaced by the state $x(t)$;
- the unknown function is the command or control $u(t)$;
- the state follows a differential equation of the form: $\dot{x} = f(x, u, t)$.

This differential equation is called state equation or dynamic equation or evolution equation. The control problem is then formulated as

$$\min_u J(u) = \int_a^b L(x, u, t)dt \quad \text{with} \quad \dot{x} = f(x, u, t) \tag{3.101}$$

This formulation is much more general than (3.100):

- the unknown function $u(t)$ is arbitrary and the curve $x(t)$ results;
- the dimensions of $u(t)$ and $x(t)$ are arbitrary;
- it is easier to impose all kinds of constraints on $u(t)$ and $x(t)$.

Example 3-12 illustrates the reformulation of the brachistochrone problem into a control problem.

Example 3-12: Brachistochrone as a control problem

The problem of the brachistochrone was addressed by the calculus of variations in example 3-2. Recall that the aim is to find the curve (AB) that minimizes the travel time of a particle under the effect of gravity.

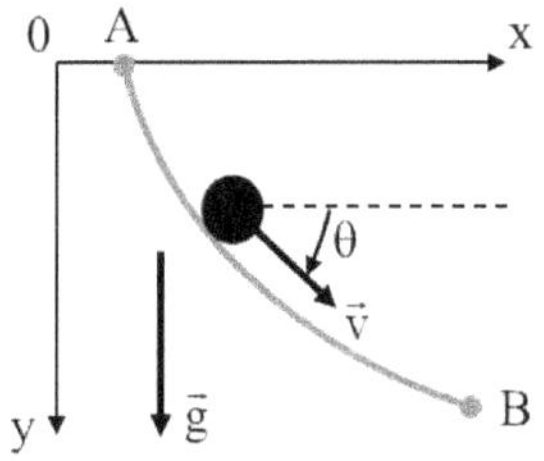

The formulation in calculus of variations is

$$\min_y \int_a^b \sqrt{\frac{1+y'(x)^2}{y(x)}}dx \quad \text{with} \quad \begin{cases} y(a)=0 \\ y(b)=\beta \end{cases}$$

Let us formulate an equivalent control problem:
- the variable is the time t;
- the two-component state vector is the position of the particle $x(t), y(t)$;
- the control $\theta(t)$ is the angle of the curve with the axis (Ox).

The conservation of energy gives: $\frac{1}{2}v^2 - gy = 0 \Rightarrow v = \sqrt{2gy}$.

The equation of state is: $\begin{cases} \dot{x} = v\cos\theta = \sqrt{2gy}\cos\theta \\ \dot{y} = v\sin\theta = \sqrt{2gy}\sin\theta \end{cases}$ with $\begin{cases} x(t_0)=x_A \,,\, y(t_0)=y_A \\ x(t_f)=x_B \,,\, y(t_f)=y_B \end{cases}$

The unknowns here are the control $\theta(t)$ and the final time t_f.
The control problem is formulated as

$$\min_{\theta,t_f} J = t_f - t_0 = \int_{t_0}^{t_f} dt \quad \text{with} \quad \begin{cases} \dot{x} = v\cos\theta = \sqrt{2gy}\cos\theta \\ \dot{y} = v\sin\theta = \sqrt{2gy}\sin\theta \end{cases} , \quad \begin{cases} x(t_0)=x_A \,,\, y(t_0)=y_A \\ x(t_f)=x_B \,,\, y(t_f)=y_B \end{cases}$$

State equation

For a given system, the choice of the state vector depends on the characteristics one wishes to study. The state vector must completely define the system at a given time, independently of its previous evolution.

The evolution of the system from a time t_0 depends only on the initial state $x(t_0)$ and is obtained by integrating the state equation $\dot{x} = f(x,u,t)$ from the initial condition. This differential equation has a solution as soon as the functions f and u are sufficiently regular (for example if f is of class C^1 in x, f and f_x are continuous with respect to u and t, and u is continuous by pieces).

The set of successive states $x(t)$ forms the **trajectory** of the system.

The system can be autonomous, controlled or linear.

The system is **autonomous** if f does not depend explicitly on time: $\dot{x} = f(x,u)$.
The solution then depends of the duration $t - t_0$, independently of the initial time.

The system is uncontrolled if f does not depend on the control: $\dot{x} = f(x,t)$.
The solution then depends only on the initial state $x(t_0)$.

The system is **linear** if f is linear in x and u: $\dot{x}(t) = F(t)x(t) + G(t)u(t)$.
The solution is then a linear function of the initial state and the control.
The properties of linear systems are presented in section 3.5.3.

<u>Variational equation</u>
For a nonlinear system, we are interested in the behavior in the vicinity of a nominal (or reference) trajectory. The nominal trajectory noted $x^*(t)$ is obtained with the control $u^*(t)$ between the initial time t_0 and the final time t_f.
Perturbations of the control $\delta u(t)$ produce state perturbations $\delta x(t)$. The nominal and perturbed trajectories have the respective state equations

$$\begin{cases} \dot{x}^* = f(x^*, u^*, t) \\ \dot{x}^* + \delta\dot{x} = f(x^* + \delta x, u^* + \delta u, t) \end{cases} \tag{3.102}$$

Expanding the second equation to order 1 and eliminating the terms of order 0, we obtain the equation for the evolution of the deviation $\delta x(t)$.

$$\delta\dot{x} = F(t)\delta x + G(t)\delta u \quad \text{with} \quad \begin{cases} F(t) = \dfrac{\partial f}{\partial x}(x^*, u^*, t)^T \\ G(t) = \dfrac{\partial f}{\partial u}(x^*, u^*, t)^T \end{cases} \tag{3.103}$$

The matrices $F(t), G(t)$ are calculated on the nominal path (x^*, u^*).
The deviation $\delta x(t)$ follows a linear state equation that can be solved with the methods presented in section 3.5.3. This equation allows the study of stability in the vicinity of the nominal trajectory as presented below.

3.5.2 Stability

This section presents some elements on the stability of dynamical systems. For simplicity, we limit ourselves to an autonomous system of dimension 2 of the form

$$\begin{cases} \dot{x}_1 = f_1(x_1, x_2) \\ \dot{x}_2 = f_2(x_1, x_2) \end{cases} \tag{3.104}$$

Combining the equations in (3.104), we obtain an equation linking x_1 and x_2.

$$\frac{dx_2}{dx_1} = \frac{f_2(x_1,x_2)}{f_1(x_1,x_2)} \tag{3.105}$$

The trajectory of the system can be represented in the phase plane (x_1,x_2).
The set of possible trajectories forms the phase portrait of the system.
To draw the phase portrait, one can either:

- integrate the differential system (3.104) in $x_1(t)$ and $x_2(t)$, then eliminate t;
- or integrate the differential equation (3.105) linking x_1 and x_2;
- or use the tangent field (3.105) (method of isoclines).

Example 3-13: Phase portrait

Example of uncontrolled system
Consider the uncontrolled system defined by: $\ddot{\theta} + a\theta = 0$.

Defining the state $(x_1,x_2) = (\theta,\dot{\theta})$, we obtain the system: $\begin{cases} \dot{x}_1 = x_2 \\ \dot{x}_2 = -ax_1 \end{cases}$.

Eliminating time and integrating gives the relation between x_1 and x_2.

$$\frac{dx_2}{dx_1} = -\frac{ax_1}{x_2} \Rightarrow ax_1dx_1 + x_2dx_2 = 0 \Rightarrow ax_1^2 + x_2^2 = c$$

The phase portrait is the set of these curves as the constant c varies.
These are conics of center O (ellipses if $a > 0$, hyperbolas if $a < 0$).

Example of controlled system
Consider the controlled system defined by

$$\ddot{\theta} = u \quad \text{with} \begin{cases} u = -1 \text{ if } \theta > 0 \\ u = +1 \text{ if } \theta < 0 \end{cases}$$

By defining the state $(x_1,x_2) = (\theta,\dot{\theta})$,

we obtain the system: $\begin{cases} \dot{x}_1 = x_2 \\ \dot{x}_2 = u \end{cases}$.

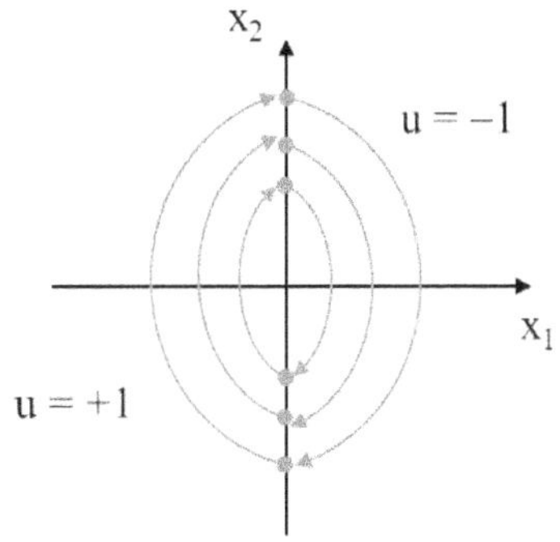

Eliminating time and integrating gives the relation between x_1 and x_2.

$$\frac{dx_2}{dx_1} = \frac{u}{x_2} \Rightarrow x_2 dx_2 = u dx_1 \Rightarrow x_2^2 = 2ux_1 + c$$

The phase portrait is formed by 2 families of parabolic arcs.

The control u is given by: $\begin{cases} u = -1 & \text{if } x_1 > 0 \\ u = +1 & \text{if } x_1 < 0 \end{cases}$.

Equilibrium points

An equilibrium point or singular point of the system (3.104) is a point such that

$$\begin{cases} \dot{x}_1 = f_1(x_1, x_2) = 0 \\ \dot{x}_2 = f_2(x_1, x_2) = 0 \end{cases} \Rightarrow \begin{cases} x_1 = C^{te} \\ x_2 = C^{te} \end{cases} \quad (3.106)$$

The stability depends on the behavior in the vicinity of the equilibrium points. The case of a linear system is analyzed first, followed by a nonlinear system.

Linear system

A linear system of order 1 and dimension 2 has a state equation of the form

$$\begin{cases} \dot{x}_1 = c_{11}x_1 + c_{12}x_2 \\ \dot{x}_2 = c_{21}x_1 + c_{22}x_2 \end{cases} \quad (3.107)$$

The only equilibrium point is the origin: $x_1 = x_2 = 0$.

To study the stability of the system in the vicinity of the origin, we form the equivalent second-order differential equation.

$$\begin{cases} x_1 = x \\ x_2 = \dot{x} \end{cases} \rightarrow \ddot{x} + a\dot{x} + bx = 0 \quad (3.108)$$

The general solution is of the form

$$\begin{cases} x(t) = k_1 e^{s_1 t} + k_2 e^{s_2 t} & \text{if } s_1 \neq s_2 \\ x(t) = k_1 e^{s_1 t} + k_2 t e^{s_1 t} & \text{if } s_1 = s_2 \end{cases} \quad (3.109)$$

where s_1 and s_2 are the real or complex roots of the characteristic equation

$$s^2 + as + b = 0 \quad (3.110)$$

Figure 3-19 shows the possible behaviors depending on the position of the roots s_1 and s_2 in the complex plane. Depending on the case, the equilibrium point is called stable/unstable node, inflection point, stable/unstable focus or center.

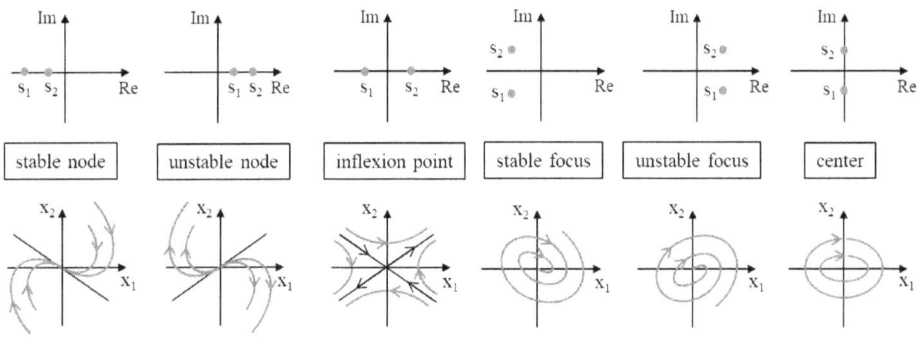

Figure 3-19: Equilibrium point of a linear system.

Nonlinear system

A nonlinear system may have several singular points, limit cycles and/or chaotic behavior.

A limit cycle is a solution forming a closed curve. The motion on this curve is periodic. Neighboring initial conditions can converge to this curve or diverge from it. The limit cycle is said:

- stable if all neighboring trajectories converge on the limit cycle;
- unstable if all neighboring trajectories diverge from the limit cycle;
- semi-stable if some trajectories converge and others diverge.

Example 3-14: Limit cycle

Consider the nonlinear system: $\begin{cases} \dot{x}_1 = x_2 + a x_1 (x_1^2 + x_2^2 - 1)^n \\ \dot{x}_2 = -x_1 + a x_2 (x_1^2 + x_2^2 - 1)^n \end{cases}$.

Let us switch to polar coordinates.

$$\begin{cases} x_1 = r\cos\theta \\ x_2 = r\sin\theta \end{cases} \rightarrow \begin{cases} \dot{r} = \dot{x}_1\cos\theta + \dot{x}_2\sin\theta \\ r\dot{\theta} = -\dot{x}_1\sin\theta + \dot{x}_2\cos\theta \end{cases} \rightarrow \begin{cases} \dot{r} = ar(r^2-1)^n \\ \dot{\theta} = -1 \end{cases}$$

The circle with center O and radius 1 is a limit cycle.

$r(t_0) = 1 \Rightarrow \dot{r} = 0 \Rightarrow r(t) = 1$

The stability of the limit cycle depends on n and a. Figure 3-20 shows the three possible evolutions from conditions close to the limit cycle.

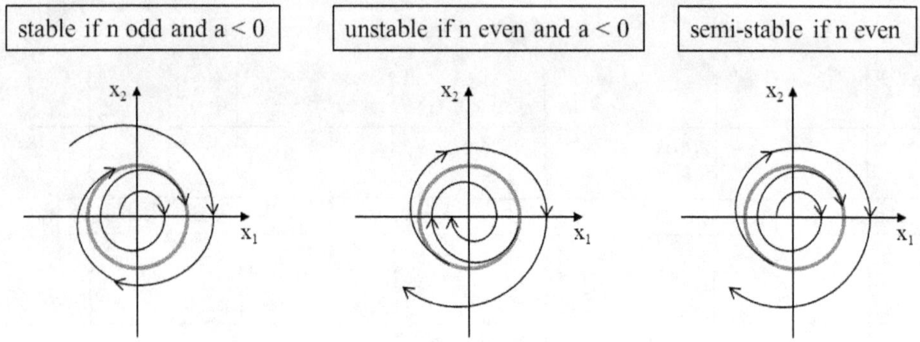

Figure 3-20: Limit cycle.

Chaotic behavior means high sensitivity to initial conditions.
The state of the system becomes unpredictable after some time as shown below.

Example 3-15: Chaotic behavior

Consider the nonlinear system of order 2: $\ddot{x}+0.1\dot{x}+x^5=6\sin t$.
Figure 3-21 shows the evolution of the system from two neighboring conditions.

Dotted lines
$$\begin{cases} x_0=2 \\ \dot{x}_0=3 \end{cases}$$

Solid lines
$$\begin{cases} x_0=2.01 \\ \dot{x}_0=3.01 \end{cases}$$

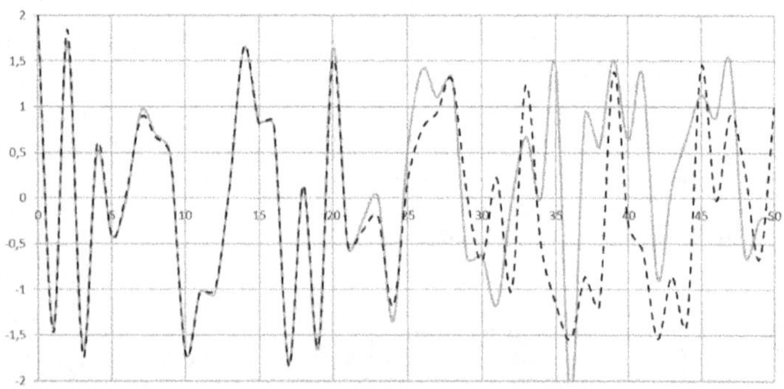

Figure 3-21: Chaotic behavior.

The evolution of the system becomes completely different from t = 30.
The perturbed path cannot be predicted from the dotted reference path.

Local and asymptotic stability

Note x_e an equilibrium point and $\delta x = x - x_e$ the deviation from it.

Different forms of stability are defined in the vicinity of the equilibrium point x_e:

- the point x_e is locally stable if an arbitrarily small deviation $\delta x(t)$ can be obtained at $t > t_0$ by choosing the initial condition $\delta x(t_0)$;
- the point x_e is asymptotically stable if $\lim\limits_{t \to \infty} \delta x(t) = 0$;
- the point x_e is exponentially stable if $\|\delta x(t)\| < \alpha \|\delta x(t_0)\| e^{-\lambda t}$, $\alpha, \lambda > 0$.

A nonlinear system can remain bounded without being stable, for example if it admits a stable limit cycle.
To analyze the stability of a system in the vicinity of an equilibrium point, one can proceed by linearization or by Lyapunov's method. It is assumed here that the system is autonomous with a state equation of the form $\dot{x} = f(x)$.

Study by linearization

The equation for the deviations in the vicinity of the equilibrium point x_e is

$$\delta \dot{x} = F \delta x \quad \text{with } F = \frac{\partial f}{\partial x}(x_e)^T \tag{3.111}$$

If all eigenvalues of F have strictly negative real parts, the system is exponentially stable. If one eigenvalue of F has a strictly positive real part, the system is unstable (= not stable). Otherwise, it is not possible to say anything about the stability of the nonlinear system.

Study by Lyapunov's method

A Lyapunov function of the system $\dot{x} = f(x)$ is a function $V(\delta x)$ of class C^1 defined positive: $V(\delta x) > 0$ if $\delta x \neq 0$ and decreasing: $\frac{dV}{dt}[\delta x(t)] < 0$.

The following sufficient stability conditions are established:

- if the system admits a Lyapunov function V, the point x_e is locally stable;
- if V is strictly decreasing, the point x_e is asymptotically stable;
- if $\lim\limits_{\|\delta x\| \to \infty} V(\delta x) = \infty$, the point x_e is globally asymptotically stable.

In most cases, one can define a Lyapunov function from the energy of the system and demonstrate stability by studying this function.

3.5.3 Linear system

This section deals with linear systems of the form

$$\dot{x}(t) = F(t)x(t) \tag{3.112}$$

The results also apply to nonlinear systems $\dot{x} = f(x,t)$ when considering their deviation $\delta x(t)$ from a nominal trajectory $x^*(t)$.
The deviation $\delta x(t)$ follows the linear variational equation

$$\delta\dot{x}(t) = F(t)\delta x(t) \quad \text{with} \quad F(t) = \left(\frac{\partial f}{\partial x}(x^*(t),t)\right)^T \tag{3.113}$$

where the Jacobian matrix $F(t)$ is evaluated along the nominal path.
Some important notions about linear systems are introduced below: transition and covariance matrix, adjoint state, observability, controllability.

Transition matrix
The transition matrix $\Phi(t,t_0)$ of the linear system $\dot{x}(t) = F(t)x(t)$ allows to go from the state at time t_0 to the state at time t.

$$x(t) = \Phi(t,t_0)x(t_0) \tag{3.114}$$

If the state is of dimension n, the transition matrix is of dimension $n \times n$.
The term $\Phi_{ij}(t,t_0)$ represents the value taken by the component $x_i(t)$ when all components of $x(t_0)$ are 0 except the component $x_j(t_0)$ which is 1 (pulse).

This correspondence is illustrated in figure 3-22. The transition matrix is also called the impulse response matrix.

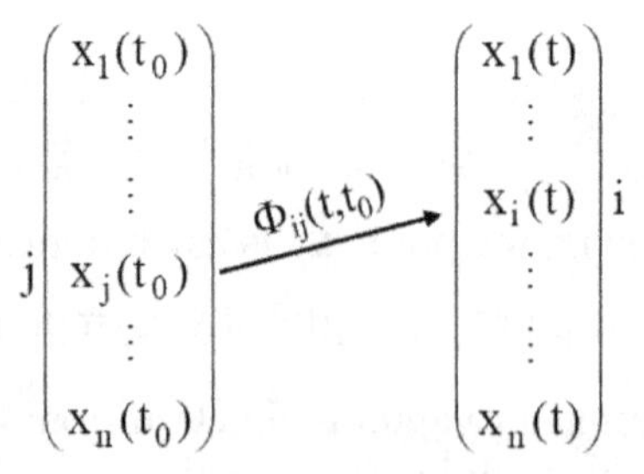

Figure 3-22: Transition matrix.

The transition matrix has the following properties:

- inverse and product

$$\begin{cases}\Phi(t,t)=I \\ \Phi(t_0,t_1)\Phi(t_1,t_0)=I \\ \Phi(t_2,t_1)\Phi(t_1,t_0)=\Phi(t_2,t_0)\end{cases} \tag{3.115}$$

- differential equation followed by $\Phi(t,t_0)$ for a fixed initial time t_0

$$\begin{cases}x(t) &= \Phi(t,t_0)x(t_0) \\ \dot{x}(t) &= F(t)x(t)\end{cases} \Rightarrow \frac{d}{dt}\Phi(t,t_0) = F(t)\Phi(t,t_0) \tag{3.116}$$

- differential equation followed by $\Phi(t_f,t)$ for a fixed final time t_f

$$\begin{cases}x(t_f) = \Phi(t_f,t)x(t) \\ \dot{x}(t) = F(t)x(t)\end{cases} \Rightarrow \frac{d}{dt}\Phi(t_f,t) = -\Phi(t_f,t)F(t) \tag{3.117}$$

Equation (3.117) is called the adjoint equation of (3.116).

<u>Symplectic matrix</u>

For some linear systems, the transition matrix is symplectic, which means that it satisfies the relation

$$\Phi^T J \Phi = J \quad \text{with } J = \begin{pmatrix} 0 & I \\ -I & 0 \end{pmatrix} \tag{3.118}$$

The interest of a symplectic matrix is that its inverse is obtained by

$$\Phi^{-1} = -J\Phi^T J \tag{3.119}$$

Let us calculate the derivative of $\Phi^T J\Phi$ using equation (3.116): $\dot{\Phi} = F\Phi$.

$$\frac{d}{dt}(\Phi^T J\Phi) = \dot{\Phi}^T J\Phi + \Phi^T J\dot{\Phi} = \Phi^T(F^T J + JF)\Phi \tag{3.120}$$

The matrix Φ is symplectic if: $F^T J + JF = 0$. Indeed, $\Phi^T J\Phi$ is then constant equal to its initial value J. For a partitioned matrix, we have

$$F = \begin{pmatrix} F_{11} & F_{12} \\ F_{21} & F_{22} \end{pmatrix} \Rightarrow F^T J + JF = \begin{pmatrix} F_{21} - F_{21}^T & F_{11}^T + F_{22} \\ -F_{11} - F_{22}^T & F_{12}^T - F_{12} \end{pmatrix} \tag{3.121}$$

Such a matrix is symplectic if F_{12} and F_{21} are symmetric and $F_{11}^T = -F_{22}$.

Example 3-16: Motion in a gravity field

Consider the motion of a material point in a central gravity field.
The position, velocity and gravity are noted respectively $\vec{r}, \vec{v}, \vec{g}$.

The equations of motion are: $\begin{cases} \dot{\vec{r}} = \vec{v} \\ \dot{\vec{v}} = \vec{g}(\vec{r}) \end{cases}$ with $\vec{g}(\vec{r}) = -\mu \dfrac{\vec{r}}{r^3}$.

The linearized equations in the vicinity of a reference trajectory are

$$\begin{cases} \delta\dot{\vec{r}} = \delta\vec{v} \\ \delta\dot{\vec{v}} = G\delta\vec{r} \end{cases} \quad \text{with } G = \frac{\partial \vec{g}}{\partial \vec{r}} = \frac{\mu}{r^5}(3\vec{r}\,\vec{r}^T - r^2 I) = \frac{\mu}{r^5}\begin{pmatrix} 3x^2 - r^2 & 3xy & 3xz \\ 3yx & 3y^2 - r^2 & 3yz \\ 3zx & 3zy & 3z^2 - r^2 \end{pmatrix}$$

The matrix G is the gravity gradient matrix.
The variational system is linear of the form

$$\delta\dot{x} = F\delta x \quad \text{with } \delta x = \begin{pmatrix} \delta\vec{r} \\ \delta\vec{v} \end{pmatrix} \text{ and } F = \begin{pmatrix} 0 & I \\ G & 0 \end{pmatrix}$$

The matrix F is partitioned (3.121) and so that: $\begin{cases} F_{12} = I \text{ and } F_{21} = G \text{ symmetric} \\ F_{11}^T = -F_{22} = 0 \end{cases}$

The transition matrix is therefore symplectic and it satisfies
$\Phi(t_0,t_1) = \Phi(t_1,t_0)^{-1} = -J\Phi(t_1,t_0)^T J$

The deviations are propagated by the relation: $\begin{pmatrix} \delta\vec{r} \\ \delta\vec{v} \end{pmatrix}_{t_1} = \Phi(t_1,t_0)\begin{pmatrix} \delta\vec{r} \\ \delta\vec{v} \end{pmatrix}_{t_0}$.

The symplectic matrix property simplifies the propagation calculations.

Covariance matrix

Consider a linear system $\dot{x}(t) = F(t)x(t)$ whose state is a random variable with mean $\bar{x}(t)$ and covariance $P(t)$.

$$\begin{cases} \bar{x}(t) = E\left[x(t)\right] \\ P(t) = E\left[(x(t) - \bar{x}(t))(x(t) - \bar{x}(t))^T\right] \end{cases} \tag{3.122}$$

To simplify the formulas, let us assume that the initial mean is zero.
The state propagates with the transition matrix: $x(t) = \Phi(t,t_0)x(t_0)$.

The mean and covariance then propagate according to the equations

$$\begin{cases} \bar{x}(t) = \Phi(t,t_0)\bar{x}(t_0) \\ P(t) = \Phi(t,t_0)P(t_0)\Phi(t,t_0)^T \\ \dot{P}(t) = F(t)P(t) + P(t)F(t)^T \end{cases} \tag{3.123}$$

Demonstration

For the mean

$\bar{x}(t) = E[\Phi(t,t_0)x(t_0)] = \Phi(t,t_0)E[x(t_0)] = \Phi(t,t_0)\bar{x}(t_0)$

For the covariance, assuming the initial mean is zero: $\bar{x}(t_0) = 0 \Rightarrow \bar{x}(t) = 0$

$P(t) = E\left[x(t)x(t)^T\right] = \Phi(t,t_0)E\left[x(t_0)x(t_0)^T\right]\Phi(t,t_0)^T = \Phi(t,t_0)P(t_0)\Phi(t,t_0)^T$

The covariance differential equation is obtained by deriving directly and using: $\dot{x}(t) = F(t)x(t)$.

$$\dot{P}(t) = \frac{d}{dt}E\left[x(t)x(t)^T\right] = E\left[\dot{x}(t)x(t)^T + x(t)\dot{x}(t)^T\right]$$
$$= E\left[F(t)x(t)x(t)^T + x(t)x(t)^T F(t)^T\right] = F(t)P(t) + P(t)F(t)^T$$

Adjoint state

Consider a linear system propagated from t to t_f with its transition matrix

$$x(t_f) = \Phi(t_f,t)x(t) \tag{3.124}$$

and a linear function of the final state

$$\varphi[x(t_f)] = c^T x(t_f) = c^T \Phi(t_f,t)x(t) \tag{3.125}$$

The adjoint state associated with the function φ is defined as the derivative of $\varphi[x(t_f)]$ with respect to the state $x(t)$ at time t.

$$p(t) \underset{def}{=} \frac{\partial \varphi[x(t_f)]}{\partial x(t)} = \Phi(t_f,t)^T c \tag{3.126}$$

The adjoint state $p(t)$, also called costate or influence function, represents the sensitivity of the function φ at the final time with respect to state variations along the trajectory. Its value at the final time is: $p(t_f) = \Phi(t_f, t_f)c = c$.

The propagation equations of the adjoint state form the adjoint equations.

They are obtained by deriving (3.126) from (3.117): $\frac{d}{dt}\Phi(t_f, t) = -\Phi(t_f, t)F(t)$.

$$\begin{cases} \dot{p}(t) = -F(t)^T p(t) \\ p(t_f) = c \end{cases} \tag{3.127}$$

The adjoint state, known at the final time, is calculated by backward integration.

A linear change of state leads to a change of adjoint given by

$$y(t) = G(t)x(t) \quad \rightarrow \quad q(t) = G(t)^{-T} p(t) \tag{3.128}$$

Demonstration

The state $x(t)$ has the transition matrix Φ: $x(t) = \Phi(t, t_0)x(t_0)$ and the adjoint: $p(t) = \Phi(t_f, t)^T c$.

We calculate the transition matrix $\Psi(t, t_0)$ for a state $y(t)$ defined by $y(t) = G(t)x(t)$

$$\begin{cases} x(t) = \Phi(t, t_0)x(t_0) & \Rightarrow \; G(t)^{-1}y(t) = \Phi(t, t_0)G(t_0)^{-1}y(t_0) \\ y(t) = \Psi(t, t_0)y(t_0) & \Rightarrow \; \Psi(t, t_0) = G(t)\Phi(t, t_0)G(t_0)^{-1} \end{cases}$$

We then define the linear function $\psi[y(t_f)]$ associated with the state $y(t)$.

$$\psi[y(t_f)] \underset{def}{=} \varphi[x(t_f)] = c^T x(t_f) = c^T G(t_f)^{-1} y(t_f) = d^T y(t_f) \quad \text{with } d = G(t_f)^{-T} c$$

The adjoint $q(t)$ associated with the function ψ is defined by (3.126).

$$q(t) \underset{def}{=} \frac{\partial \psi[y(t_f)]}{\partial y(t)} = \Psi(t_f, t)^T d$$

Replacing $\Psi(t, t_0)$ and d by their expressions in terms of $\Phi(t, t_0)$ and $G(t)$, we obtain (3.128).

$$\begin{aligned} q(t) &= \Psi(t_f, t)^T d = \left[G(t)^{-T} \Phi(t_f, t)^T G(t_f)^T\right] G(t_f)^{-T} c \\ &= G(t)^{-T} \Phi(t_f, t)^T c = G(t)^{-T} p(t) \end{aligned}$$

Observability

Consider a linear system of equation: $\dot{x}(t) = F(t)x(t)$ for which we have linear measurements of the form

$$y(t) = H(t)x(t) \quad \text{with} \quad x \in \mathbb{R}^n \, , \, y \in \mathbb{R}^m \, , \, H \in \mathbb{R}^{m \times n} \tag{3.129}$$

The system is observable if it is possible to assess the initial state vector $x(t_0)$ from a series of measurements $y(t)$ between t_0 and t_f.

We define the of observability matrix $M(t_0, t_f)$ by

$$M(t_0, t_f) = \int_{t_0}^{t_f} \Phi(t, t_0)^T H(t)^T H(t) \Phi(t, t_0) dt \tag{3.130}$$

A condition for observability is that the matrix $M(t_0, t_f)$ is invertible.

Justification

Integrate the measurement equation with the state given by: $x(t) = \Phi(t, t_0) x(t_0)$.

$$y(t) = H(t)x(t) = H(t)\Phi(t, t_0)x(t_0) \quad \Rightarrow \quad \int_{t_0}^{t_f} \Phi(t, t_0)^T H(t)^T y(t) dt = M(t_0, t_f) x(t_0)$$

If $M(t_0, t_f)$ is invertible, we can express $x(t_0)$ in terms of the integral. It is shown that this condition is also necessary. An equivalent observability condition is that the symmetric matrix $M(t_0, t_f)$ is positive definite.

Controllability

A controlled linear system has an equation of the form

$$\dot{x}(t) = F(t)x(t) + G(t)u(t) \tag{3.131}$$

where $u(t)$ is the control vector.

The general solution of this system is expressed with the transition matrix $\Phi(t, t_0)$ associated with the uncontrolled system: $\dot{x}(t) = F(t)x(t)$.

$$x(t_f) = \Phi(t_f, t_0) x(t_0) + \int_{t_0}^{t_f} \Phi(t_f, t) G(t) u(t) dt \tag{3.132}$$

Demonstration

Let us start with the equation of state: $\dot{x}(t) = F(t)x(t) + G(t)u(t)$.

Pre-multiplying by $\Phi(t_f,t)$: $\Phi(t_f,t)\dot{x}(t) = \Phi(t_f,t)F(t)x(t) + \Phi(t_f,t)G(t)x(t)$

and with $\frac{d}{dt}\Phi(t_f,t) = -\Phi(t_f,t)F(t)$, we get

$$\Phi(t_f,t)\dot{x}(t) = -\frac{d}{dt}\Phi(t_f,t)x(t) + \Phi(t_f,t)G(t)x(t)$$

This equation can be written as: $\frac{d}{dt}\left[\Phi(t_f,t)x(t)\right] = \Phi(t_f,t)G(t)x(t)$.

By integrating from t_0 to t_f: $\left[\Phi(t_f,t)x(t)\right]_{t_0}^{t_f} = \int_{t_0}^{t_f} \Phi(t_f,t)G(t)x(t)dt$

and with $\Phi(t_f,t_f) = I$, we get: $x(t_f) = \Phi(t_f,t_0)x(t_0) + \int_{t_0}^{t_f} \Phi(t_f,t)G(t)x(t)dt$.

Equation (3.132) is interpreted as the superposition of the contributions of the initial state $x(t_0)$ and the control $u(t)$.

The initial state generates the term $\Phi(t_f,t_0)x(t_0)$ as for an uncontrolled system.

The control $u(t)$ applied at t for a duration dt generates a state change $dx(t)$ given by: $\dot{x}(t) = F(t)x(t) + G(t)u(t) \rightarrow dx(t) = G(t)u(t)dt$.

This variation $dx(t)$ itself generates a final change: $dx(t_f) = \Phi(t_f,t)dx(t)$.

The final state is obtained by summing all these contributions from t_0 to t_f, which leads to the integral term in (3.132).

The controllable system is if any final state $x(t_f)$ is accessible from any initial state $x(t_0)$. We define the controllability matrix $C(t_0,t_f)$ by

$$C(t_0,t_f) = \int_{t_0}^{t_f} \Phi(t_f,t)G(t)G(t)^T\Phi(t_f,t)^T dt \tag{3.133}$$

A necessary and sufficient condition for controllability is that the matrix $C(t_0,t_f)$ is invertible.

Demonstration

The general solution of the equation of state is given by (3.132).

$$x(t_f)=\Phi(t_f,t_0)x(t_0)+\int_{t_0}^{t_f}\Phi(t_f,t)G(t)u(t)dt$$

Consider the control defined by: $u(t)=G(t)^T\Phi(t_f,t)^Tc$ with $c\in\mathbb{R}^n$.

This control yields the final state: $x(t_f)=\Phi(t_f,t_0)x(t_0)+C(t_0,t_f)c$:

- if $C(t_0,t_f)$ is invertible, we can go from $x(t_0)$ to $x(t_f)$ by taking $c=C(t_0,t_f)^{-1}\left[x(t_f)-\Phi(t_f,t_0)x(t_0)\right]$;
- if $C(t_0,t_f)$ is not invertible, there exists a non-zero $c\in\mathbb{R}^n$ such that $c^TC(t_0,t_f)c=0$.

We then have

$$c^TC(t_0,t_f)c=\int_{t_0}^{t_f}c^T\Phi(t_f,t)G(t)G(t)^T\Phi(t_f,t)^Tc\,dt=\int_{t_0}^{t_f}\left\|c^T\Phi(t_f,t)G(t)\right\|^2dt=0$$

The positive term under the integral must be constantly zero.

$c^T\Phi(t_f,t)G(t)=0$, $\forall t$

Let us then consider any control $u(t)$.

Using the general solution (3.132)

$$x(t_f)-\Phi(t_f,t_0)x(t_0)=\int_{t_0}^{t_f}\Phi(t_f,t)G(t)u(t)dt$$

and pre-multiplying by c^T, we obtain

$$c^T\left[x(t_f)-\Phi(t_f,t_0)x(t_0)\right]=\int_{t_0}^{t_f}c^T\Phi(t_f,t)G(t)u(t)dt=0$$

because the term $c^T\Phi(t_f,t)G(t)$ is zero.

This relation, which is true for any control $u(t)$, shows that the final state cannot be arbitrary, because the vector $x(t_f)-\Phi(t_f,t_0)x(t_0)$ belongs to the space orthogonal to the vector c. Not all possible final states can be reached.

3.5.4 Two-point boundary value problem

We have seen that the adjoint state in (3.127) is calculated by backward integration from the final condition, while the state is calculated by forward integration from the initial condition. This two-point boundary value problem occurs in optimal control (chapter 4).

We study here a solution method for a linear problem with two endpoint conditions of the form

$$\dot{x}(t) = F(t)x(t) + g(t) \quad \text{with} \begin{cases} Ax(t_0) = a \\ Bx(t_f) = b \end{cases} \tag{3.134}$$

The state vector $x(t)$ is of dimension n. The initial and final conditions are defined by linear systems.
The vector $a \in \mathbb{R}^m$ defines m initial conditions with the matrix $A \in \mathbb{R}^{m \times n}$.
The vector $b \in \mathbb{R}^{n-m}$ defines n–m final conditions with the matrix $B \in \mathbb{R}^{(n-m) \times n}$.
The objective is to find the initial state vector $x(t_0)$ that allows both the m initial conditions and the n–m final conditions to be met.

The transport (or sweep) method solves the problem in the following three stages:

- four matrices $C_1(t), C_2(t), C_3(t), C_4(t)$ are defined from $A, B, F(t)$:

$$\begin{pmatrix} A \\ B \end{pmatrix} F(t) \begin{pmatrix} A \\ B \end{pmatrix}^{-1} \underset{\text{def}}{=} \begin{pmatrix} C_1(t) & C_2(t) \\ C_3(t) & C_4(t) \end{pmatrix} \tag{3.135}$$

- the matrix $S(t) \in \mathbb{R}^{m \times (n-m)}$ and the vector $s(t) \in \mathbb{R}^m$ are calculated by integrating the following differential system from t_0 to t_f:

$$\begin{cases} \dot{S} = -SC_3S - SC_4 + C_1S + C_2 \\ \dot{s} = (C_1 - SC_3)s + (A - SB)g \end{cases} \text{with} \begin{cases} S(t_0) = 0 \\ s(t_0) = a \end{cases} \tag{3.136}$$

- the state $x(t)$ is obtained by backward integration of the state equation: $\dot{x}(t) = F(t)x(t) + g(t)$ starting from the final state $x(t_f)$ defined by

$$x(t_f) = \begin{pmatrix} A \\ B \end{pmatrix}^{-1} \begin{pmatrix} S(t_f)b + s(t_f) \\ b \end{pmatrix} \tag{3.137}$$

Justification

The principle of this method is to "transport" the initial condition $Ax(t_0)=a$ to the final time. To do this, the state vector $x(t)\in\mathbb{R}^n$ is partitioned into two sub-vectors $y(t)\in\mathbb{R}^m$ and $z(t)\in\mathbb{R}^{n-m}$.

These vectors are defined by: $\begin{cases} y(t)=Ax(t) \\ z(t)=Bx(t) \end{cases}$ with the matrices A and B of (3.134).

The initial and final conditions are: $\begin{cases} y(t_0)=a \\ z(t_f)=b \end{cases}$.

Let us express x as a function of y and z: $\begin{pmatrix} y \\ z \end{pmatrix}=\begin{pmatrix} A \\ B \end{pmatrix}x \;\Rightarrow\; x=\begin{pmatrix} A \\ B \end{pmatrix}^{-1}\begin{pmatrix} y \\ z \end{pmatrix}$

and replace in the state equation.

$$\dot{x}=Fx+g \;\Rightarrow\; \begin{pmatrix} \dot{y} \\ \dot{z} \end{pmatrix}=\begin{pmatrix} A \\ B \end{pmatrix}F\begin{pmatrix} A \\ B \end{pmatrix}^{-1}\begin{pmatrix} y \\ z \end{pmatrix}+\begin{pmatrix} A \\ B \end{pmatrix}g$$

Using the matrices C_1,C_2,C_3,C_4 defined in (3.135), this differential system is written as: $\begin{cases} \dot{y}=C_1y+C_2z+Ag \\ \dot{z}=C_3y+C_4z+Bg \end{cases}$.

Let us introduce a matrix $S(t)\in\mathbb{R}^{m\times(n-m)}$ and a vector $s(t)\in\mathbb{R}^m$ functions of time such that: $y(t)=S(t)z(t)+s(t)$.

The existence of $S(t)$ and $s(t)$ is ensured for example with: $S(t)=0$, $s(t)=y(t)$.

Let us replace $y=Sz+s$ in the differential system

$$\begin{cases} \dot{y}=C_1y+C_2z+Ag \\ \dot{z}=C_3y+C_4z+Bg \end{cases} \xrightarrow{y=Sz+s} \begin{cases} \dot{S}z+S\dot{z}+\dot{s}=C_1Sz+C_1s+C_2z+Ag \\ \dot{z}=C_3Sz+C_3s+C_4z+Bg \end{cases}$$

then substitute $\dot{z}$ in the first equation

$$(\dot{S}+SC_3S+SC_4-C_1S-C_2)z+\dot{s}+(SC_3-C_1)s+(SB-A)g=0$$

Now assume that $S(t)$ is chosen to cancel the z-term. The remaining term is then also zero.

This leads to the following conditions on $S(t)$ and $s(t)$.

$\begin{cases} \dot{S}=-SC_3S-SC_4+C_1S+C_2 \\ \dot{s}=(C_1-SC_3)s+(A-SB)g \end{cases}$ which gives the differential system (3.136).

To satisfy the initial condition: $y(t_0)=a$, we also impose: $\begin{cases} S(t_0)=0 \\ s(t_0)=a \end{cases}$.

Integrating the system from t_0 to t_f, we obtain $S(t_f)$ and $s(t_f)$, from which we deduce: $y(t_f)=S(t_f)z(t_f)+s(t_f)$.

Since the final value of z is known: $z(t_f)=b$, we obtain the full final state vector $x(t_f)$, which allows us to return to the initial state $x(t_0)$ by backward integration of the state equation.

Example 3-17 illustrates the solution of a two-point boundary value problem using the transport method.

Example 3-17: Double integrator with conditions at both ends

Consider the double integrator problem in dimension 1.

$$\begin{cases} \dot{x}_1 = x_2 \\ \dot{x}_2 = u \end{cases} \text{ with } \begin{cases} x_1(t_0)=a \\ x_2(t_f)=b \end{cases}$$

The initial condition is on the position x_1, the final condition on the velocity x_2.

The matrices associated with this problem are

$$F=\begin{pmatrix} 0 & 1 \\ 0 & 0 \end{pmatrix},\ g=\begin{pmatrix} 0 \\ u \end{pmatrix},\ \begin{matrix} A=\begin{pmatrix} 1 & 0 \end{pmatrix} \\ B=\begin{pmatrix} 0 & 1 \end{pmatrix} \end{matrix} \rightarrow \begin{pmatrix} C_1 & C_2 \\ C_3 & C_4 \end{pmatrix}=\begin{pmatrix} 0 & 1 \\ 0 & 0 \end{pmatrix}$$

The differential system in $S(t)$ and $s(t)$ is written as: $\begin{cases} \dot{S}=1 \\ \dot{s}=-Su \end{cases}$ with $\begin{cases} S(t_0)=0 \\ s(t_0)=a \end{cases}$

The solution is given by: $S(t)=t-t_0$ and $s(t)=a-\int_{t_0}^{t}(\tau-t_0)u(\tau)d\tau$.

We then have the final state vector

$$x(t_f)=\begin{pmatrix} A \\ B \end{pmatrix}^{-1}\begin{pmatrix} S_f b+s_f \\ b \end{pmatrix}=\begin{pmatrix} a+b(t_f-t_0)-\int_{t_0}^{t_f}(\tau-t_0)u(\tau)d\tau \\ b \end{pmatrix}$$

which is integrated backwards to obtain the complete solution from t_0 to t_f.

3.6 Conclusion

3.6.1 The key points

- The extremals of a calculus of variations problem are the solutions of the Euler-Lagrange equation. These extremals may or may not be minima;
- the sufficient conditions for optimality are based on the absence of conjugate points, determined by solving the Jacobi equation;
- the final constraints (free endpoint or endpoint constrained on a curve) lead to transversality conditions;
- the extremals are solutions of the canonical system formed from the Hamiltonian of the functional to be minimized;
- the value function (= value of the functional at any point) is a solution of the Hamilton-Jacobi-Bellman partial differential equation.

3.6.2 To go further

- *Calculus of variations and optimal control theory (D. Liberzon, Princeton University Press 2012)*

  This book presents the calculus of variations in chapters 1 to 3. The concepts are introduced progressively in a rigorous manner, while avoiding too heavy formalism. Chapter 3 makes the transition between the calculus of variations and optimal control. This book is highly educational and is to be recommended as an introduction to these subjects.

- *Calculus of variations (I.M. Gelfand, S.V. Fomin, Dover 1963)*

  This book is entirely devoted to the calculus of variations. The different conditions of optimality (order 1 and 2, weak or strong minimum,) are demonstrated in chapters 1 to 6. Chapter 7 presents the generalization to more complex problems formulated by multiple integrals.

- *Calculus of variations with applications (G.M. Ewing, Dover 1985)*

  Like the previous one, this book is entirely devoted to the calculus of variations, with a more formal presentation. The optimality conditions are proved with additions on parametric problems.

- *Geometric optimal control (H. Schättler, U. Ledzewicz, Springer 2012)*

  This comprehensive book includes a first chapter on the calculus of variations.

- *The variational principles of mechanics (C. Lanczos-Urruty, Dover 1949)*

This book presents the Hamiltonian approach to mechanics. The theory of calculus of variations is recalled in chapter 3, and then applied to derive the canonical equations and the Hamilton-Jacobi-Bellman equation.

4. Optimal control

An optimal control problem consists in finding the control law of a dynamic system in order to achieve a trajectory at the best cost. This formulation generalizes the calculus of variations formulation presented in chapter 3 and allows a wide variety of problems in engineering or economics to be addressed.

Section 1 presents the formulation of a control problem and the conditions for optimality. The Pontryaguin minimum principle (historically called the maximum principle by a choice of opposite sign) states that the optimal control is a global minimum of the Hamiltonian of the problem. Its determination requires the introduction of an adjoint state and the solution of a two-point boundary value problem. The variational approach establishes only the weak version (local minimum of the Hamiltonian), but it interest is to generalize easily to problems with constraints.

Section 2 introduces the consideration of constraints. The terminal constraints induce so-called transversality conditions on the adjoint and the Hamiltonian. The interior constraints apply at an intermediate time and allow to account for discontinuities of dynamics or state of the system. They induce jump conditions on the adjoint and the Hamiltonian. The current and state constraints must be respected on the whole trajectory. The inequality constraints can induce junction conditions with control discontinuities.

Section 3 studies the extremals which are the trajectories satisfying the first-order optimality conditions. For the same problem, there may exist normal or abnormal (cost-independent), regular or singular (not determined by the first-order conditions) extremals. Linearization in the vicinity of a nominal solution allows the construction of a state feedback control, useful for real-time applications. The value function that gives the optimal cost from any initial condition is a solution of the Hamilton-Jacobi-Bellman equation.

Section 4 presents the second-order optimality conditions. These conditions are established from the second variation of the augmented cost and lead to a local linear quadratic problem that is solved by the transport method. The sufficient minimum conditions are based on the absence of a conjugate point. Singular arcs may exist in the frequent case of a linear control. Higher order conditions are needed to establish the optimality of these arcs and their possible junctions to regular arcs.

4.1 Optimality conditions

4.1.1 Control problem

The standard formulation of a control problem is as follows.

$$\min_{u,t_f} J(u) = \int_{t_0}^{t_f} L[x(t),u(t),t]dt + \varphi[x(t_f),t_f] \quad \text{with} \begin{cases} \dot{x}(t) = f[x(t),u(t),t] \\ x(t_0) = x_0 \end{cases} \tag{4.1}$$

The state $x(t)$ of dimension n follows the **state equation**: $\dot{x}(t) = f[x(t),u(t),t]$.
The initial conditions (x_0, t_0) are fixed, the final conditions (x_f, t_f) are free.
The control $u(t)$ of dimension m is any function of time.
The trajectory $\{x(t)\, ,\, t_0 \le t \le t_f\}$ associated with the control $\{u(t)\, ,\, t_0 \le t \le t_f\}$ is obtained by integrating the equation of state from the initial state x_0.
When there is no ambiguity, we write the problem (4.1) in a more compact form by omitting the time variable.

$$\min_{u,t_f} J(u) = \int_{t_0}^{t_f} L(x,u,t)dt + \varphi(x_f,t_f) \quad \text{with} \begin{cases} \dot{x} = f(x,u,t) \\ x(t_0) = x_0 \end{cases} \tag{4.2}$$

To simplify the formulas, the partial derivatives will be noted $f_x = \dfrac{\partial f}{\partial x}$, $f_u = \dfrac{\partial f}{\partial u}$.
We are looking for the control law minimizing the functional J which is composed of an **integral cost** term: $\int_{t_0}^{t_f} L(x,u,t)dt$ and a **final cost** term: $\varphi(x_f,t_f)$.

The problem is stated in the form:
- of Lagrange if the cost is purely integral $(\varphi = 0)$;
- of Mayer if the cost is purely final $(L = 0)$;
- of Bolza in the general case.

These three forms are equivalent. The transition from one to the other is made by the following transformations.

Transition from Mayer to Lagrange form

Assume that the cost is purely final

$$J(u) = \varphi(x_f, t_f) \quad \text{with} \quad \dot{x} = f(x,u,t)\ ,\ x(t_0) = x_0 \tag{4.3}$$

and define the function $L(x,u,t)$ by

$$L(x,u,t) \underset{\text{def}}{=} \frac{\partial \varphi}{\partial x}(x,t)^T f(x,u,t) + \frac{\partial \varphi}{\partial t}(x,t) \tag{4.4}$$

The final cost can then be expressed as

$$\begin{aligned}\varphi(x_f, t_f) &= \varphi(x_0, t_0) + \int_{t_0}^{t_f} \frac{d\varphi}{dt}(x,t)dt \\ &= \varphi(x_0, t_0) + \int_{t_0}^{t_f} \left(\frac{\partial \varphi}{\partial x}(x,t)^T \dot{x} + \frac{\partial \varphi}{\partial t}(x,t) \right) dt \\ &= \varphi(x_0, t_0) + \int_{t_0}^{t_f} L(x,u,t)dt\end{aligned} \tag{4.5}$$

The integral cost $\int_{t_0}^{t_f} L(x,u,t)dt$ equates the final cost (4.3) to within one constant.

Transition from Lagrange to Mayer form

Assume that the cost is purely integral

$$J(u) = \int_{t_0}^{t_f} L(x,u,t)dt \quad \text{with} \quad \dot{x} = f(x,u,t)\ ,\ x(t_0) = x_0 \tag{4.6}$$

and define the additional state variable ξ by

$$\dot{\xi} = L(x,u,t) \quad \text{with} \quad \xi(t_0) = 0 \tag{4.7}$$

The new state $y \underset{\text{def}}{=} \begin{pmatrix} x \\ \xi \end{pmatrix}$ of dimension n+1 has the equation: $\dot{y} = \begin{pmatrix} \dot{x} \\ \dot{\xi} \end{pmatrix} = \begin{pmatrix} f(x,u,t) \\ L(x,u,t) \end{pmatrix}$

The integral cost can then be expressed as

$$\int_{t_0}^{t_f} L(x,u,t)dt = \int_{t_0}^{t_f} \frac{d\xi}{dt} dt = \xi(t_f) - \xi(t_0) = \xi(t_f) \tag{4.8}$$

The final cost defined by $\varphi(y_f, t_f) = \xi(t_f)$ equates the integral cost (4.6).

Constraints

Problem (4.2) has fixed initial conditions and free final conditions.
The formulation can be extended to different types of constraints:

- fixed or free initial time and state, fixed or free final time and state
- constraints on the initial state: $\psi_0\left[x(t_0),t_0\right]=0$;
- constraints on the final state: $\psi_f\left[x(t_f),t_f\right]=0$;
- point constraints on the state: $\psi_1\left[x(t_1),t_1\right]=0$ at $t_1\in[t_0;t_f]$;
- discontinuities in dynamics: $\begin{cases}\dot{x}=f_1(x,u,t) & \text{if } t<t_1\\ \dot{x}=f_2(x,u,t) & \text{if } t>t_1\end{cases}$;
- discontinuities in state: $x(t_1^+)=x(t_1^-)+\delta x_1$ at $t_1\in[t_0;t_f]$;
- path equality constraints: $\psi_c(x,u,t)=0$, $\forall t$;
- path inequality constraints: $\psi_c(x,u,t)\le 0$, $\forall t$.

The optimality conditions in the presence of these different types of constraints will be presented in section 4.2.

4.1.2 Principle of the minimum

Consider a standard control problem with a fixed final time.

$$\min_u J(u)=\int_{t_0}^{t_f}L(x,u,t)dt+\varphi(x_f) \quad \text{with} \quad \begin{cases}\dot{x}=f(x,u,t)\\ x(t_0)=x_0\end{cases} \tag{4.9}$$

The state $x(t)$ is of dimension n, the control $u(t)$ is of dimension m.
The final cost $\varphi(x_f)$ depends only on the state x_f, as the final time t_f is fixed.
We define the **adjoint state** $p(t)$, also called costate, and the **Hamiltonian** $H(x,u,p,t)$ by the following equations.

$$\begin{cases}H(x,u,p,t)=p_0L(x,u,t)+p^Tf(x,u,t)\\ \dot{p}=-\dfrac{\partial H}{\partial x} \quad \text{with} \quad p(t_f)=p_0\dfrac{d\varphi}{dx}(x_f)\end{cases} \tag{4.10}$$

The adjoint $p(t)$ has the same dimension as the state $x(t)$ and it is calculated by backward integration from its final value $p(t_f)$. The real p_0 is called the abnormal multiplier. The Hamiltonian is a function of $\mathbb{R}^{n+m+n+1}$ in $\mathbb{R}$.
We then have the following necessary conditions of optimality.

Theorem 4-1: Pontryaguin minimum principle (PMP)

Let $u^*(t)$ be a solution of problem (4.9) and $x^*(t)$ the associated trajectory. Then there exist $p_0 \geq 0$ and $p^*(t)$, non-simultaneously zero, satisfying the equations (4.10) and such that the control $u^*(t)$ minimizes the Hamiltonian at each time.

$$\min_u H(x^*,u,p^*,t) \quad \rightarrow \ u^*(t) \tag{4.11}$$

The proof of the existence of $(p_0\,;\,p^*) \neq (0\,;\,0)$ is difficult. We limit ourselves here to verifying the minimum property of the Hamiltonian when $p_0 \neq 0$.

Verification of the minimization condition (see [R14])

Consider a control law u*(t) from t_0 to t_f. The associated trajectory x*(t) is obtained by integration of the state equation: $\dot{x}^* = f(x^*,u^*,t)$.
Let us apply a control perturbation defined as follows.

$$u(t)=\begin{cases} u^*(t) & \text{if } t_0 \leq t < t_1-\varepsilon \\ v & \text{if } t_1-\varepsilon \leq t \leq t_1 \\ u^*(t) & \text{if } t_1 < t \leq t_f \end{cases}$$

The time $t_1 \in]t_0;t_f]$ is arbitrary.
The value $\varepsilon > 0$ is assumed to be small and the constant v is arbitrary.

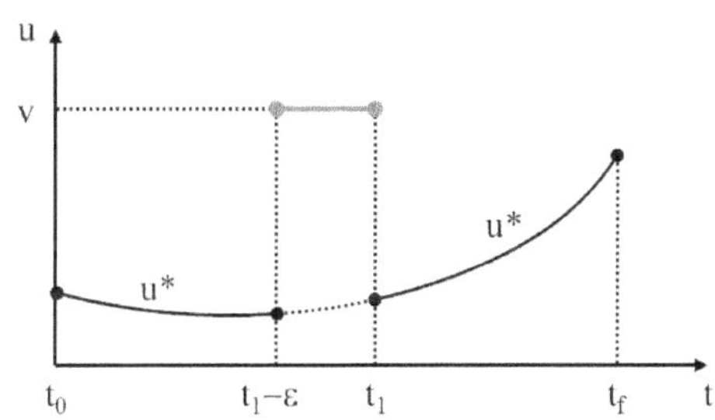

Figure 4-1: Needle variation.

This perturbation, called **needle variation** is shown in figure 4-1.
It only changes the control u* on $[t_1-\varepsilon\,;\,t_1]$. The interval is assumed to be small, but the perturbation v on this interval can be arbitrarily large.
The needle variation on $[t_1-\varepsilon\,;\,t_1]$ changes the trajectory from $t_1-\varepsilon$.
The perturbed trajectory is of the form

$$x(t)=\begin{cases} x^*(t) & \text{if} \quad t_0 \leq t \leq t_1-\varepsilon \\ x^*(t)+\delta x & \text{if } \ t_1-\varepsilon < t \leq t_f \end{cases}$$

The proof of (4.11) proceeds in four stages: calculation of the state change at t_1, propagation to t_f, resulting final cost and transition to an integral cost.

First stage: variation of the state at time t_1

We calculate the variation $\delta x(t_1)$ resulting from the variation of u on $[t_1-\varepsilon\,;t_1]$.

The state at $t_1-\varepsilon$ is nominal: $x(t_1-\varepsilon)=x^*(t_1-\varepsilon)$.

This nominal state at t_1 is obtained with the nominal control u* on $[t_1-\varepsilon\,;t_1]$.

$$\dot{x}^*=f(x^*,u^*,t) \;\text{ on }\; [t_1-\varepsilon\,;t_1] \;\Rightarrow\; x^*(t_1)=x^*(t_1-\varepsilon)+\int_{t_1-\varepsilon}^{t_1} f(x^*,u^*,t)dt$$

The perturbed state at t_1 is obtained with the perturbed control v on $[t_1-\varepsilon\,;t_1]$.

$$\dot{x}=f(x,v,t) \quad \text{on }\; [t_1-\varepsilon\,;t_1] \;\Rightarrow\; x(t_1)=x^*(t_1-\varepsilon)+\int_{t_1-\varepsilon}^{t_1} f(x^*+\delta x,v,t)dt$$

By difference, and for ε small, we have at the first order

$$\delta x(t_1)=x(t_1)-x^*(t_1)=\varepsilon\left[f(x^*+\delta x,v,t)-f(x^*,u^*,t)\right]_{t_1}+o(\varepsilon)$$

As the interval $[t_1-\varepsilon\,;t_1]$ is small, so is the perturbation $\delta x(t_1)$.

We expand to order 1 in δx : $f(x^*+\delta x,v,t)=f(x^*,v,t)+f_x^T\delta x+o(\delta x)$

and replace in $\delta x(t_1)$: $\delta x(t_1)=\varepsilon\left[f(x^*,v,t)-f(x^*,u^*,t)\right]_{t_1}$ ($\varepsilon\delta x$ is of order 2).

Second stage: propagation of the state perturbation to the final time t_f

The control on $[t_1;t_f]$ is identical to the nominal control u*.

Let us write the propagation equation of the deviation $\delta x=x-x^*$ on $[t_1;t_f]$.

$$\begin{cases}\dot{x}=f(x,u^*,t)=f(x^*+\delta x,u^*,t)=f(x^*,u^*,t)+f_x^T\delta x+o(\delta x)\\ \dot{x}^*=f(x^*,u^*,t)\end{cases} \rightarrow \delta\dot{x}=f_x^T\delta x$$

The solution of the equation $\delta\dot{x}=f_x^T\delta x$ is expressed with its transition matrix Φ (section 3.5.3).

$$\delta x(t_f)=\Phi(t_f,t)\delta x(t_1) \;\text{ with }\; \begin{cases}\dot{\Phi}(t_f,t_1)=-\Phi(t_f,t_1)f_x^T\\ \Phi(t_f,t_f)=I\end{cases} \;\text{ as established in (3.117)}$$

Replacing $\delta x(t_1)$ (found at stage 1), we have

$$\delta x(t_f)=\Phi(t_f,t_1)\delta x(t_1)=\varepsilon\Phi(t_f,t_1)\left[f(x^*,v,t)-f(x^*,u^*,t)\right]_{t_1}$$

Third stage: variation of the final cost
Assume the control problem in Mayer form: $J(u) = \varphi(x_f)$.
A variation δx_f produces at order 1 a cost variation δJ : $\delta J = \varphi_x(x_f{}^*)\delta x_f$.
By replacing $\delta x(t_f)$ (found at stage 2), we have

$$\delta J = \varepsilon\varphi_x(x_f{}^*)\Phi(t_f,t_1)\left[f(x^*,v,t)-f(x^*,u^*,t)\right]_{t_1}$$

Let us define the adjoint $p(t)$ and the Hamiltonian $H(x,u,p,t)$ by

$$\begin{cases} p(t) = p_0\Phi(t_f,t)^T\varphi_x(x_f) \\ H(x,u,p,t) = p^T f(x,u,t) \end{cases} \quad \text{with } p_0 > 0$$

The adjoint satisfies: $\begin{cases} \dot{p} = -f_x p & \text{because } \dot{\Phi}(t_f,t) = -\Phi(t_f,t)f_x^T \\ p(t_f) = p_0\varphi_x(x_f) & \text{because } \Phi(t_f,t_f) = I \end{cases}$,

or by using H: $\begin{cases} \dot{p} = -H_x \\ p(t_f) = p_0\varphi_x(x_f) \end{cases}$.

This definition is the one in (4.10) in the case of a purely final cost $(L = 0)$.
Let p^* be the adjoint associated with (u^*, x^*) and let δJ be expressed in terms of p and H.

$$\delta J = \frac{\varepsilon}{p_0}p^*(t_1)^T\left[f(x^*,v,t)-f(x^*,u^*,t)\right]_{t_1} = \frac{\varepsilon}{p_0}\left[H(x^*,v,p^*,t)-H(x^*,u^*,p^*,t)\right]_{t_1}$$

A necessary condition for optimality of u* is: $\delta J \geq 0$, $\forall t_1$, $\forall\varepsilon > 0$, $\forall v$.
The control u* thus achieves a minimum of $H(x^*,u,p^*,t)$ (since $\varepsilon>0$, $p_0>0$).

Fourth stage: extension to an integral cost
Let us now consider the case of a problem in Bolza form.

$$J(u) = \int_{t_0}^{t_f} L(x,u,t)dt + \varphi(x_f).$$

We return to the Mayer form by adding a state variable ξ as in (4.7).

With this variable defined by: $\begin{cases} \dot{\xi} = L(x,u,t) \\ \xi(t_0) = 0 \end{cases}$, the augmented state vector is

$$y = \begin{pmatrix} x \\ \xi \end{pmatrix} \rightarrow \dot{y} = \begin{pmatrix} f \\ L \end{pmatrix}$$

The final cost is expressed as: $J = \xi_f + \varphi(x_f) \underset{\text{def}}{=} \psi(y_f)$.

The augmented adjoint q associated with the augmented state y and the final cost $\psi(y_f)$ is defined as in stage 3.

$$q=\begin{pmatrix}p\\ \pi\end{pmatrix} \rightarrow \dot{q}=\begin{pmatrix}\dot{p}\\ \dot{\pi}\end{pmatrix}=\begin{pmatrix}-f_x p\\ -L_\xi \pi=0\end{pmatrix} \text{ with } q(t_f)=\begin{pmatrix}p_f\\ \pi_f\end{pmatrix}=p_0\begin{pmatrix}\psi_x=\varphi_x\\ \psi_\xi=1\end{pmatrix}$$

We observe that the additional adjoint component noted π is constant.

$\dot{\pi}=0 \;\Rightarrow\; \pi(t)=\pi(t_f)=p_0$

The Hamiltonian of this problem is: $H=\begin{pmatrix}p\\ \pi\end{pmatrix}^T\begin{pmatrix}f\\ L\end{pmatrix}=p_0L+p^Tf$.

This definition corresponds to the one given in (4.10) in the case of a Bolza problem. The adjoint p still satisfies the equations (4.10) as in stage 3.
The result obtained in stage 3 thus extends to the case of an integral and final cost with the Hamiltonian defined by: $H=p_0L+p^Tf$.

The statement of theorem 4-1 calls for some remarks.

<u>Note 1</u>: *strong/weak minimum principle*

The above demonstration considers arbitrarily large control perturbations on a small duration. It shows that the optimal control u achieves a <u>global minimum of the Hamiltonian</u> at each time. For this reason, the minimum principle is called strong.*

The variational approach (section 4.1.3) will consider small control perturbations over the whole trajectory. This approach will show that the optimal control u only achieves a <u>local minimum of the Hamiltonian</u> at each time. This result will constitute the weak minimum principle.*

<u>Note 2</u>: *abnormal multiplier*

The multiplier p_0 allows solutions with $p_0=0$ to be taken into account.
These so-called abnormal extremals are presented in section 4.3.2. In the normal case $(p_0\neq 0)$, the value of p_0 can be freely chosen (it is usually set to 1).

<u>Note 3</u>: *sign convention*

Theorem 4-1 *is stated assuming a positive abnormal multiplier: $p_0\geq 0$, which leads to a minimum of the Hamiltonian (in stage 3 of the proof). The opposite choice $p_0\leq 0$ would lead to a maximum of the Hamiltonian.*

The original theorem (1962) adopts the convention $p_0 \le 0$ and is therefore called "Pontryaguin maximum principle". This name remains in use even when $p_0 \ge 0$. The use of the acronym PMP hides this wording ambiguity. The less natural convention $p_0 \le 0$ comes from the Weierstrass condition established in calculus of variations (section 3.2.4), as explained below.

Link with the calculus of variations

Consider a problem of calculus of variations: $\min\limits_y J(y) = \int_a^b L(x,y,z=y')dx$.

The canonical variables (3.55) are defined by: $\begin{cases} p = L_z \\ H = L_z y' - L \end{cases}$ **(equation E)**

The Euler-Lagrange necessary condition is: $L_y = \dot{L}_z \Rightarrow \dot{p} = L_y$.

The Weierstrass strong minimum condition is: $\max\limits_w H(x,y,w,p) \rightarrow y'$.

<u>Equivalent control problem</u> (with the convention $p_0 = -1$)

By renaming the variables: $\begin{cases} x \rightarrow t \\ y(x) \rightarrow x(t) \\ y'(x) \rightarrow \dot{x}(t) \end{cases}$ and with the state equation: $\dot{x} = u$,

an equivalent control problem is formulated with a control $u \in \mathbb{R}$.

$$\min_u J(u) = \int_a^b L(t,x,u)dx \quad \text{with} \quad \dot{x} = u$$

The PMP with $p_0 = -1$ yields: $\begin{cases} H = -L + p^T u \\ \dot{p} = L_x \end{cases}$

and $\max\limits_u H \Rightarrow H_u = -L_u + p = 0 \Rightarrow p = L_u$

Replacing $p = L_u$, we have: $\begin{cases} H = -L + L_u u \\ \dot{L}_u = L_x \end{cases}$ (with u of dimension 1).

We retrieve the canonical variables (equation E) and the Euler-Lagrange equation with $u \rightarrow y'$. The Weierstrass condition is thus equivalent to the PMP with the convention $p_0 = -1$, which leads to the maximization of the Hamiltonian.

Example 4-1 illustrates the stages involved in solving a control problem using the PMP.

Example 4-1: Ballistic range

Consider the trajectory of a ballistic missile with a flat Earth and a constant gravity model. The trajectory in the (Oxz) plane is shown in figure 4-2.

<u>Modelling</u>

The assumptions on propulsion and control are as follows:

- the missile takes off at $t_0 = 0$ and stops propelling at the fixed time t_f;
- the thrust acceleration is constant with modulus a;
- the thrust direction is defined by the pitch angle θ (angle with the horizontal);
- the evolution law $\theta(t)$ is free and represents the control of the missile.

The components of position, velocity and acceleration are given below.

Position: $\vec{r} = \begin{pmatrix} x \\ z \end{pmatrix}$

Velocity: $\vec{v} = \begin{pmatrix} v_x \\ v_z \end{pmatrix}$

Gravity: $\vec{g} = \begin{pmatrix} 0 \\ -g \end{pmatrix}$

Acceleration: $\vec{a} = a\begin{pmatrix} \cos\theta \\ \sin\theta \end{pmatrix}$

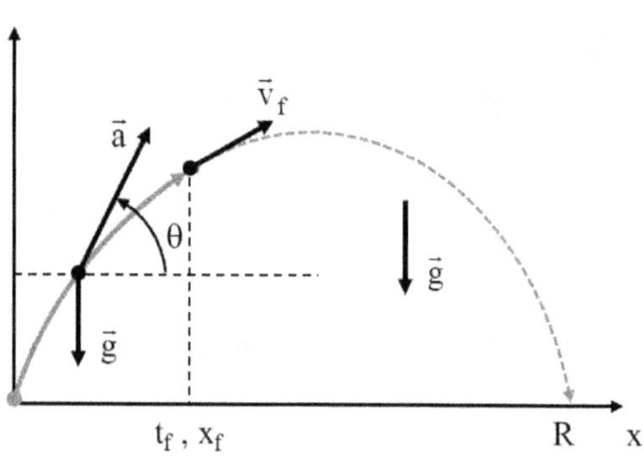

Figure 4-2: Ballistic missile trajectory.

The position-velocity state: $\begin{pmatrix} x \\ z \\ v_x \\ v_z \end{pmatrix}$ evolves according to: $\begin{cases} \dot{x} = v_x \\ \dot{z} = v_z \\ \dot{v}_x = a\cos\theta \\ \dot{v}_z = a\sin\theta - g \end{cases}$.

The range R depends on the conditions at the time of injection t_f.

$$R = x_f + \frac{v_{xf}}{g}\left[v_{zf} + \sqrt{v_{zf}^2 + 2gz_f}\right]$$

Control problem

The control problem is in Mayer form (purely final cost).

$$\min_{\theta} \varphi(x_f, z_f, v_{xf}, v_{zf}) = -R(x_f, z_f, v_{xf}, v_{zf}) \quad \text{with} \quad \begin{cases} \dot{x} = v_x \\ \dot{z} = v_z \\ \dot{v}_x = a\cos\theta \\ \dot{v}_z = a\sin\theta - g \end{cases} \quad \text{and} \quad \begin{cases} x_0 = 0 \\ z_0 = 0 \\ v_{x0} = 0 \\ v_{z0} = 0 \end{cases}$$

Application of the PMP

The Hamiltonian is: $H = p_x v_x + p_z v_z + p_{vx} a\cos\theta + p_{vz}(a\sin\theta - g)$

with the adjoint $(p_x, p_z, p_{vx}, p_{vz})$

The application of the PMP gives: $\min_{\theta} H \quad \rightarrow \quad \tan\theta = \dfrac{p_{vz}}{p_{vx}}$ **(equation 1)**

To determine $\theta(t)$, we solve the adjoint equations with their final conditions.

$$\text{Adjoint equations:} \begin{cases} \dot{p}_x = -H_x = 0 & \rightarrow p_x(t) = A \\ \dot{p}_z = -H_z = 0 & \rightarrow p_z(t) = B \\ \dot{p}_{vx} = -H_{vx} = -p_x & \rightarrow p_{vx}(t) = C + A(t_f - t) \\ \dot{p}_{vz} = -H_{vz} = -p_z & \rightarrow p_{vz}(t) = D + B(t_f - t) \end{cases}$$

$$\text{Final conditions:} \begin{cases} p_x(t_f) = \varphi_x = \dfrac{\partial R}{\partial x_f} & \rightarrow A = -1 \\ p_z(t_f) = \varphi_z = \dfrac{\partial R}{\partial z_f} & \rightarrow B = -\dfrac{v_{xf}}{\sqrt{v_{zf}^2 + 2gz_f}} \\ p_{vx}(t_f) = \varphi_{vx} = \dfrac{\partial R}{\partial v_{xf}} & \rightarrow C = -\dfrac{v_{zf} + \sqrt{v_{zf}^2 + 2gz_f}}{g} \\ p_{vz}(t_f) = \varphi_{vz} = \dfrac{\partial R}{\partial v_{zf}} & \rightarrow D = -\dfrac{v_{xf}}{g}\left(1 + \dfrac{v_{zf}}{\sqrt{v_{zf}^2 + 2gz_f}}\right) \end{cases}$$

Observing that: $D = -BC$ and with $A = -1$, equation 1 yields

$$\tan\theta(t) = \frac{p_{vz}(t)}{p_{vx}(t)} = \frac{D + B(t_f - t)}{C + A(t_f - t)} = -B\frac{C - (t_f - t)}{C - (t_f - t)} = -B = \frac{v_{xf}}{\sqrt{v_{zf}^2 + 2gz_f}} \quad \textbf{(equation 2)}$$

The maximum range is therefore obtained with a constant direction of thrust.
It can be noted that this result obtained with a flat Earth and constant acceleration remains true if the acceleration varies according to a predetermined law $a(t)$.

Optimum thrust angle

To find the value of θ, we integrate the trajectory (with a and θ constant).

$$\begin{cases} \dot{x} = v_x \\ \dot{z} = v_z \\ \dot{v}_x = a\cos\theta \\ \dot{v}_z = a\sin\theta - g \end{cases} \rightarrow \begin{cases} x_f = (a\cos\theta)t_f^2/2 \\ z_f = (a\sin\theta - g)t_f^2/2 \\ v_{xf} = (a\cos\theta)t_f \\ v_{zf} = (a\sin\theta - g)t_f \end{cases}$$

Replacing z_f, v_{xf}, v_{zf} into equation 2, we obtain

$$\tan\theta\sqrt{(a\sin\theta - g)^2 + g(a\sin\theta - g)} = a\cos\theta,$$

which can be written as: $\sin^3\theta - 2n\sin^2\theta + n = 0$ with $n = \dfrac{a}{g}$ **(equation 3)**

The variable n is the acceleration of the missile expressed in g.

Search for solutions

We search the solutions of equation 3 as a function of the acceleration level n.
The angle θ is between 0 deg (vertical) and 90 deg (horizontal).

Figure 4-3 shows the function $f(\theta) = \sin^3\theta - 2n\sin^2\theta + n$ plotted for four different values of n.

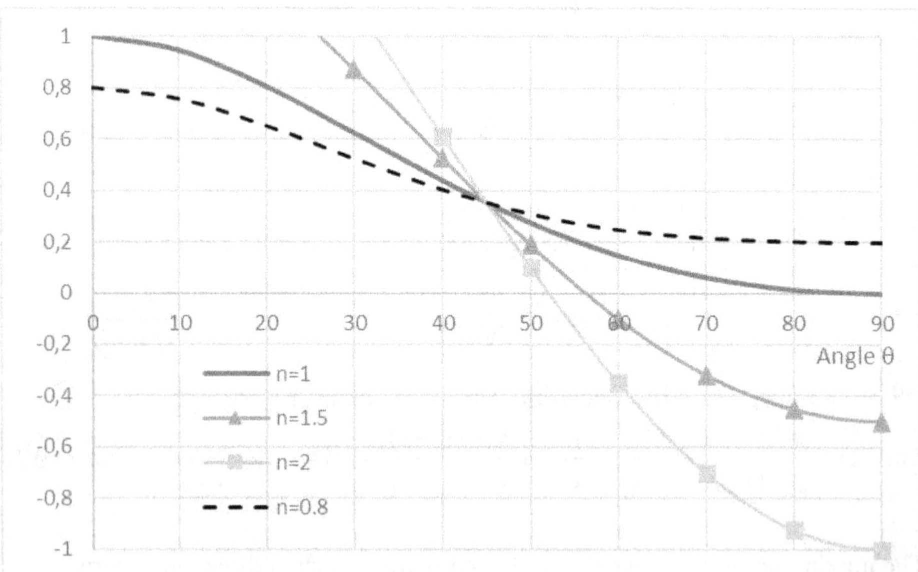

Figure 4-3: Solution as a function of acceleration level.

We observe that the equation does not have a solution if $n<1$. Indeed, take-off is impossible with a thrust that is less than the weight $(a<g)$.
When $n\geq 1$, the equation has a unique solution with a thrust angle comprised between 45 and 90 deg.
If $n=1$, the optimum angle is 90 deg. It is indeed necessary to thrust vertically to take off $(a=g)$.
If $n\to\infty$, the optimum angle is 45 deg. This corresponds to an instantaneous velocity impulse. We retrieve a known result: the launch direction maximizing the range from the ground is indeed 45 deg.

4.1.3 Variational method

The variational method consists in expressing the variation of the cost J as a function of small variations δu applied along the trajectory. This approach based on an augmented cost leads to a weaker result than the PMP, but it has the advantage of being easily generalized when the control problem is subject to various constraints.

Augmented cost
Consider a control problem with free initial and final conditions.

$$\min_{u,x_0,t_0,x_f,t_f} J=\int_{t_0}^{t_f} L(x,u,t)dt+\varphi(x_0,t_0,x_f,t_f) \quad \text{with} \quad \dot{x}=f(x,u,t) \tag{4.12}$$

The state $x(t)$ depends on the control $u(t)$ by the state equation: $\dot{x}=f(x,u,t)$.
Control variations $\delta u(t)$ produce state variations $\delta x(t)$ which must be taken into account in the calculation of δJ. The Lagrange method is used:
- the equation $\dot{x}=f(x,u,t)$ represents a constraint applied at each time t;
- the constraint with multipliers $p(t)$ called adjoints is added to the cost J.

The **augmented cost** J_a is thus defined by

$$J_a=J+\int_{t_0}^{t_f} p^T\left[f(x,u,t)-\dot{x}\right]dt \tag{4.13}$$

By introducing the Hamiltonian defined by

$$H(x,u,p,t)=L(x,u,t)+p^T f(x,u,t) \tag{4.14}$$

the augmented cost is expressed as

$$J_a = \int_{t_0}^{t_f} \left[H(x,u,p,t) - p^T \dot{x} \right] dt + \varphi(x_0, t_0, x_f, t_f) \tag{4.15}$$

The integration by parts of the term $p^T \dot{x}$ leads to

$$J_a = \int_{t_0}^{t_f} \left[H(x,u,p,t) + \dot{p}^T x \right] dt - \left[p^T x \right]_{t_0}^{t_f} + \varphi(x_0, t_0, x_f, t_f) \tag{4.16}$$

<u>First variation</u>

Let us apply variations on the control $\delta u(t)$, on the state $\delta x(t)$, on the initial time δt_0 and on the final time δt_f. By expanding formula (4.16) to the first order (small variations), we obtain the following first variation of the augmented cost.

$$\begin{aligned} \delta J_a = & \int_{t_0}^{t_f} H_u^T \delta u dt + \int_{t_0}^{t_f} (H_x + \dot{p})^T \delta x dt \\ & + \left[\varphi_{x_0} + p(t_0) \right]^T \delta x_0 + \left[\varphi_{t_0} - H(t_0) \right] \delta t_0 \\ & + \left[\varphi_{x_f} - p(t_f) \right]^T \delta x_f + \left[\varphi_{t_f} + H(t_f) \right] \delta t_f \end{aligned} \tag{4.17}$$

Demonstration

Let us apply variations $\delta u(t), \delta x(t), \delta t_0$, δt_f to the augmented cost (4.16).

$$J_a = \int_{t_0}^{t_f} \left[H(x,u,p,t) + \dot{p}^T x \right] dt + p(t_0)^T x(t_0) - p(t_f)^T x(t_f) + \varphi(x_0, t_0, x_f, t_f)$$

By expanding each term to order 1, we have

$$\begin{aligned} \delta J_a = & \int_{t_0}^{t_f} (H_x^T \delta x + H_u^T \delta u + \dot{p}^T \delta x) dt && \rightarrow \text{variation in the integral} \\ & + \left[H + \dot{p}^T x \right](t_f) \delta t_f - \left[H + \dot{p}^T x \right](t_0) \delta t_0 && \rightarrow \text{variation of bounds} \\ & + \delta p(t_0)^T x(t_0) + p(t_0)^T \delta x(t_0) && \rightarrow \text{variation of } p(t_0)^T x(t_0) \\ & - \delta p(t_f)^T x(t_f) - p(t_f)^T \delta x(t_f) && \rightarrow \text{variation of } p(t_f)^T x(t_f) \\ & + \varphi_{x_0}^T \delta x(t_0) + \varphi_{t_0} \delta t_0 + \varphi_{x_f}^T \delta x(t_f) + \varphi_{t_f} \delta t_f && \rightarrow \text{variation of } \varphi(x_0, t_0, x_f, t_f) \end{aligned}$$

The variations of the adjoint at t_0 and t_f are expressed as: $\begin{cases} \delta p(t_0) = \dot{p}(t_0)\delta t_0 \\ \delta p(t_f) = \dot{p}(t_f)\delta t_f \end{cases}$.

The terms in $\dot{p}(t_0)$ and $\dot{p}(t_f)$ are eliminated and we obtain

$$\delta J_a = \int_{t_0}^{t_f} H_u^T \delta u dt + \int_{t_0}^{t_f} (H_x + \dot{p})^T \delta x dt + \left[\varphi_{x_0} + p(t_0)\right]^T \delta x_0 + \left[\varphi_{t_0} - H(t_0)\right]\delta t_0$$
$$+ \left[\varphi_{x_f} - p(t_f)\right]^T \delta x_f + \left[\varphi_{t_f} + H(t_f)\right]\delta t_f$$

In formula (4.17), the variations of the state $\delta x(t)$ and of the control $\delta u(t)$ are not independent. They are indeed linked by the linearized state equation

$$\dot{x} = f(x,u,t) \quad \Rightarrow \quad \delta\dot{x} = f_x \delta x + f_u \delta u \tag{4.18}$$

The explicit expression of $\delta x(t)$ as a function of $\delta u(t)$ is not known.
To get rid of this dependence, we choose the adjoint in order to cancel the integral in δx in formula (4.17). We thus find the adjoint equations (4.10).

$$\dot{p} = -H_x \tag{4.19}$$

With this choice, the first variation (4.17) no longer depends on $\delta x(t)$. It depends only on $\delta u(t), \delta x_0, \delta t_0, \delta x_f, \delta t_f$.

$$\delta J_a = \int_{t_0}^{t_f} H_u^T \delta u dt + \left[\varphi_{x_0} + p(t_0)\right]^T \delta x_0 + \left[\varphi_{t_0} - H(t_0)\right]\delta t_0$$
$$+ \left[\varphi_{x_f} - p(t_f)\right]^T \delta x_f + \left[\varphi_{t_f} + H(t_f)\right]\delta t_f \tag{4.20}$$

This formula makes it possible to express the necessary conditions for optimality and to calculate the sensitivities of the cost functional.

<u>Necessary conditions for optimality</u>
A necessary condition of first order for optimality is

$\delta J_a = 0 \,,\ \forall \delta u, \delta x_0, \delta t_0, \delta x_f, \delta t_f$

Since the variations $\delta u(t), \delta x_0, \delta t_0, \delta x_f, \delta t_f$ are independent, their coefficients must be zero in (4.20).

The condition $(\delta J_a = 0 \,,\, \forall \delta u)$ requires

$$H_u = 0 \quad , \quad \forall t \tag{4.21}$$

The optimal control is a stationary point of the Hamiltonian at each time.

If the initial conditions are free, the condition $(\delta J_a \geq 0 \,,\, \forall \delta x_0 , \forall \delta t_0)$ requires

$$\begin{cases} p(t_0) = -\varphi_{x_0} & \to \text{ for } \delta x_0 \\ H(t_0) = \varphi_{t_0} & \to \text{ for } \delta t_0 \end{cases} \tag{4.22}$$

If the final conditions are free, the condition $(\delta J_a \geq 0 \,,\, \forall \delta x_f , \forall \delta t_f)$ requires

$$\begin{cases} p(t_f) = \varphi_{x_f} & \to \text{ for } \delta x_f \\ H(t_f) = -\varphi_{t_f} & \to \text{ for } \delta t_f \end{cases} \tag{4.23}$$

These relations at the initial or final point are called **transversality conditions**. Each transversality condition provides a number of equations equal to the number of unknowns: n equations for $\delta x_{0/f}$, 1 equation for $\delta t_{0/f}$. These conditions lead to a two-point boundary value problem discussed in section 4.1.4.

Influence functions

Assume that the problem is in Lagrangian form $(\varphi = 0)$. We can always reduce to this formulation by the transformation (4.4).
On the optimal trajectory: $H_u = 0$, $J_a = J^*$ and the formula (4.20) reduces to

$$\delta J^* = p(t_0)^T \delta x_0 - H(t_0)\delta t_0 - p(t_f)^T \delta x_f + H(t_f)\delta t_f \tag{4.24}$$

This formula shows the partial derivatives of the optimal cost J* with respect to the initial and final conditions. These derivatives are called **influence functions**.

$$\begin{cases} \dfrac{\partial J^*}{\partial x_0} = p(t_0) \\ \dfrac{\partial J^*}{\partial t_0} = -H(t_0) \end{cases} \quad \text{and} \quad \begin{cases} \dfrac{\partial J^*}{\partial x_f} = -p(t_f) \\ \dfrac{\partial J^*}{\partial t_f} = H(t_f) \end{cases} \tag{4.25}$$

If the initial/final conditions are free (hence cost-optimal), the corresponding derivatives of J* are zero. This is consistent with the optimality conditions (4.22) and (4.23) in the case of a Lagrangian form $(\varphi = 0)$.
When the initial/final conditions are fixed, the formulas (4.25) allow us to know the effect of initial or final perturbations on the optimal cost.

Let us assume that the optimal trajectory from $(x_0;t_0)$ to $(x_f;t_f)$ is known and place ourselves at any point $(x;t)$ on this trajectory. According to Bellman principle of optimality (developed in section 4.3.6), the part of the trajectory from $(x;t)$ to $(x_f;t_f)$ is optimal and we can apply the formulas (4.25) to it.
In particular, at each point of the optimal trajectory we have:

$$\frac{\partial J^*}{\partial x(t)} = p(t) \tag{4.26}$$

Change of state vector
Assume that we make a change of state vector: $y = g(x)$. Let p and q be the adjoints associated with x and y respectively. According to (4.26), we have:

$$\begin{cases} p = \dfrac{\partial J^*}{\partial x} = \left(\dfrac{dy}{dx}\right)\dfrac{\partial J^*}{\partial y} \\ q = \dfrac{\partial J^*}{\partial y} \end{cases} \quad \text{with } \frac{dy}{dx} = \frac{dg}{dx} \underset{\text{noted}}{=} g_x = \begin{pmatrix} \dfrac{\partial g_1}{\partial x_1} & \cdots & \dfrac{\partial g_n}{\partial x_1} \\ \vdots & \ddots & \vdots \\ \dfrac{\partial g_1}{\partial x_n} & \cdots & \dfrac{\partial g_n}{\partial x_n} \end{pmatrix} \tag{4.27}$$

where g_x is the gradient matrix (transpose of the Jacobian) of $y = g(x)$.
We deduce thus the adjoint change formula associated with the change of state.

$$y = g(x) \;\Rightarrow\; q = g_x^{-1} p \tag{4.28}$$

Example 4-2: Switching from polar to Cartesian coordinates

The polar-Cartesian transformation is defined as: $\begin{pmatrix} x \\ y \end{pmatrix} = g\begin{pmatrix} r \\ \theta \end{pmatrix} = \begin{pmatrix} r\cos\theta \\ r\sin\theta \end{pmatrix}$.

The gradient matrix G of the transformation is: $G = \begin{pmatrix} \cos\theta & \sin\theta \\ -r\sin\theta & r\cos\theta \end{pmatrix}$.

The adjoints are transformed as follows.

$$\begin{pmatrix} p_x \\ p_y \end{pmatrix} = G^{-1}\begin{pmatrix} p_r \\ p_\theta \end{pmatrix} = \frac{1}{r}\begin{pmatrix} r\cos\theta & -\sin\theta \\ r\sin\theta & \cos\theta \end{pmatrix}\begin{pmatrix} p_r \\ p_\theta \end{pmatrix} \quad \rightarrow \quad \begin{cases} p_x = p_r\cos\theta - \dfrac{1}{r}p_\theta\sin\theta \\ p_y = p_r\sin\theta + \dfrac{1}{r}p_\theta\cos\theta \end{cases}$$

Weak minimum principle

Let us expand formula (4.20) to order 2 with respect to δu in order to express the second variation of the augmented cost.

$$\delta^2 J_a = \frac{1}{2}\int_{t_0}^{t_f} \begin{pmatrix} \delta x \\ \delta u \end{pmatrix}^T \begin{pmatrix} H_{xx} & H_{ux} \\ H_{xu} & H_{uu} \end{pmatrix} \begin{pmatrix} \delta x \\ \delta u \end{pmatrix} dt \tag{4.29}$$

The state variations $\delta x(t)$ follow the linearized state equation: $\delta\dot{x} = f_x \delta x + f_u \delta u$.

A necessary condition of order 2 for optimality is: $\delta^2 J_a \geq 0$, $\forall \delta u$.

We will see in section 4.4.2 that this requires

$$H_{uu} \geq 0 \ , \ \forall t \tag{4.30}$$

This condition associated with (4.21) shows that the optimal control u* achieves a local minimum of the Hamiltonian at each time. This result is weaker than the PMP (global minimum) and constitutes the weak minimum principle.

The variational approach is applied in mechanics to derive the Lagrange equations for a non-conservative system.

Application to mechanics: Lagrange equations (from [R3] and [R13])

The Lagrange equations in mechanics are established using the above variational techniques.

For a conservative system, the principle of least action is used.

For a non-conservative system, the principle of virtual work is used.

Conservative system

Consider a system defined by a state vector q (generalized coordinates).

For a conservative system, the forces are derived from a potential.

The potential energy noted $V(q)$ is a function of the coordinates.

The kinetic energy noted $T(q,\dot{q})$ is a function of the coordinates and their derivatives (or velocities).

The following quantities are defined from $V(q)$ and $T(q,\dot{q})$:

- the Lagrangian of the system: $L(q,\dot{q}) = T(q,\dot{q}) - V(q)$;
- the Hamiltonian of the system: $H(q,\dot{q}) = T(q,\dot{q}) + V(q)$;
  ("mechanical Hamiltonian" or energy)
- the action between t_0 and t_f : $S = \int_{t_0}^{t_f} L(q,\dot{q}) dt$.

The <u>principle of least action</u> postulates that the trajectory minimizes the action.

$$\min_q S = \int_{t_0}^{t_f} L(q,\dot{q})dt \qquad \rightarrow q(t)$$

By posing $u = \dot{q}$, we formulate an equivalent control problem.

$$\min_u S = \int_{t_0}^{t_f} L(q,u)dt \ \text{ s.t. } \ \dot{q} = u$$

The "control Hamiltonian" associated with this problem is denoted H_c.

$$H_c(q,u,p) = L(q,u) + p^T u$$

Applying the PMP results in: $\begin{cases} \min_u H_c \Rightarrow \dfrac{\partial L}{\partial u} + p = 0 \\ \dot{p} = -\dfrac{\partial H_c}{\partial q} = -\dfrac{\partial L}{\partial q} \end{cases} \Rightarrow \dfrac{d}{dt}\left(\dfrac{\partial L}{\partial u}\right) = \dfrac{\partial L}{\partial q}$.

Returning to $\dot{q} = u$, we obtain the equations: $\dfrac{d}{dt}\left(\dfrac{\partial L}{\partial \dot{q}}\right) = \dfrac{\partial L}{\partial q}$.

Let us now express the control Hamiltonian as a function of V and T with

$$p = -\frac{\partial L}{\partial u} = -\frac{\partial L}{\partial \dot{q}}$$

$$H_c = L + p^T u = L - u^T \frac{\partial L}{\partial u} = (T - V) - u^T \frac{\partial T}{\partial u} \quad \text{because} \quad L(q,u) = T(q,u) - V(q)$$

Since the kinetic energy is quadratic with respect to the velocity, we have

$$\dot{q}^T \frac{\partial T}{\partial \dot{q}} = 2T \iff u^T \frac{\partial T}{\partial u} = 2T$$

The result is: $H_c = (T - V) - 2T = -(T + V) = -H$.

The control Hamiltonian is the opposite of the mechanical Hamiltonian which represents the energy.

The sign change comes from the choice $p_0 = +1$ (note 3 on theorem 4-1).

<u>Non-conservative system</u>

For a non-conservative system, the forces noted $Q(q)$ do not derive from a potential and the principle of least action does not apply.

The work of the forces is expressed as: $W = \int_{q_0}^{q_f} Q(q)^T dq$ and it depends on the path followed between q_0 and q_f.

The principle of virtual works postulates that the trajectory $q(t)$ is such that

$$\delta\left(\int_{t_0}^{t_f} T(q,\dot{q})dt\right)+\int_{t_0}^{t_f} Q(q)^T \delta q dt=0 \quad , \quad \forall \delta q \, , \forall \delta \dot{q}$$

The first term is the change in kinetic energy due to changes in δq and $\delta \dot{q}$.
The second term is the work associated with the displacement δq.
The displacement δq is arbitrary (virtual) regardless of the characteristics of the system.

By expanding to order 1: $\int_{t_0}^{t_f}\left(T_q^T \delta q+T_{\dot{q}}^T \delta \dot{q}+Q^T \delta q\right)dt=0$ and posing $\delta \dot{q}=\delta u$,

it comes down to: $\int_{t_0}^{t_f}\left(T_q^T \delta q+T_u^T \delta u+Q^T \delta q\right)dt=0 \quad \text{s.t.} \quad \delta \dot{q}=\delta u$.

Let us apply the variational method by introducing an adjoint p associated with the constraint $\delta \dot{q}=\delta u$. We add the constraint with its adjoint in the integral.

$$\int_{t_0}^{t_f}\left[T_q^T \delta q+T_u^T \delta u+Q^T \delta q+p^T(\delta u-\delta \dot{q})\right]dt=0$$

Then we integrate by parts.

$$\int_{t_0}^{t_f}\left[\left(T_u+p\right)^T \delta u+\left(T_q+Q+\dot{p}\right)^T \delta q\right]dt+(p\delta q)_{t_0}-(p\delta q)_{t_f}=0$$

To cancel the term in δq, the adjoint is required to satisfy: $T_q+Q+\dot{p}=0$.

The integral must cancel for any variation δu : $T_u+p=0$.

Combining these two equations with $\dot{q}=u$, we obtain: $\dfrac{d}{dt}\left(\dfrac{\partial T}{\partial \dot{q}}\right)-\dfrac{\partial T}{\partial q}=Q$.

These Lagrange equations deduced from the principle of virtual work generalize those obtained for a conservative system from the principle of least action.

In the case of a conservative system: $Q=-V_q$, we retrieve indeed

$$\frac{d}{dt}\left(\frac{\partial T}{\partial \dot{q}}\right)=\frac{\partial (T-V)}{\partial q}=\frac{\partial L}{\partial q} \quad \text{with} \quad L(q,\dot{q})=T(q,\dot{q})-V(q) \quad \Rightarrow \quad \frac{\partial L}{\partial \dot{q}}=\frac{\partial T}{\partial \dot{q}}$$

Parameters

Assume that the system depends on the control u(t) and real parameters α.
The hybrid control problem is formulated as

$$\min_{u,\alpha} J(u) = \int_{t_0}^{t_f} L(x,u,t,\alpha)dt + \varphi(x_f,\alpha) \quad \text{with} \quad \dot{x} = f(x,u,t,\alpha) \tag{4.31}$$

The aim here is to write optimality conditions associated with the parameters α. For this reason, we do not specify whether the initial and final conditions are free or fixed. The treatment of these conditions remains the same as in (4.17).
The Hamiltonian and the adjoint are defined as in (4.14) and (4.19).

$$\begin{cases} H(x,u,p,t,\alpha) = L(x,u,t,\alpha) + p^T f(x,u,t,\alpha) \\ \dot{p} = -H_x \end{cases} \tag{4.32}$$

Let us apply variations of control $\delta u(t)$ and parameters $\delta\alpha$.
The first variation of the augmented cost J_a (4.16) is then given by

$$\delta J_a = \int_{t_0}^{t_f} H_u^T \delta u dt + \left(\int_{t_0}^{t_f} H_\alpha dt + \varphi_\alpha \right)^T \delta\alpha \tag{4.33}$$

Demonstration

Let us apply variations $\delta u(t), \delta x(t), \delta\alpha$ to the augmented cost defined by (4.16).

$$J_a = \int_{t_0}^{t_f} \left[H(x,u,p,t,\alpha) + \dot{p}^T x \right] dt + p(t_0)^T x(t_0) - p(t_f)^T x(t_f) + \varphi(x_f,\alpha)$$

Each term is expanded to the first order (ignoring variations in the initial and final conditions).

$$\delta J_a = \int_{t_0}^{t_f} (H_x^T \delta x + H_u^T \delta u + H_\alpha^T \delta\alpha + \dot{p}^T \delta x) dt + \varphi_\alpha^T \delta\alpha$$

With the choice $\dot{p} = -H_x$, there remains: $\delta J_a = \int_{t_0}^{t_f} H_u^T \delta u dt + \left(\int_{t_0}^{t_f} H_\alpha^T dt + \varphi_\alpha^T \right) \delta\alpha$.

Formula (4.33) allows us to express necessary conditions of optimality for a control problem with parameters.

If the control $u(t)$ is optimal, then $\delta J_a = 0$, $\forall \delta u$, which requires

$$H_u = 0 \ , \ \forall t \tag{4.34}$$

If the parameters α are optimal, then $\delta J_a = 0$, $\forall \delta\alpha$, which requires

$$\int_{t_0}^{t_f} H_\alpha dt + \varphi_\alpha = 0 \tag{4.35}$$

Each parameter gives an equation (4.35), which completes equations (4.22) and (4.23) associated with the initial and final conditions (if they are free).

Formula (4.33) is also of interest when the reals α represent fixed parameters of the dynamic system. The optimized trajectory for the nominal values α gives a control u^* and an optimal cost J^*.
The optimal cost derivative with respect to these parameters is then given by

$$\frac{\partial J^*}{\partial \alpha} = \int_{t_0}^{t_f} H_\alpha dt + \varphi_\alpha \tag{4.36}$$

where the integral is calculated on the optimal trajectory with the control u^*.
This formula allows to evaluate the first-order effect of model uncertainties, without re-solving the control problem with parameters $\alpha + \delta\alpha$.

Model parameters

An alternative method for proving condition (4.35) is considered here, as well as its use in dealing with a free final time problem.

<u>Additional state component</u>
An alternative method of proving (4.35) is to add a state component x_{n+1} associated with the parameter α. The equation of state for this component is $\dot{x}_{n+1} = 0$, $x_{n+1}(t_0) = \alpha$.
By replacing α by x_{n+1}, the problem (4.31) takes the form

$$\min_{u, x_{n+1}(t_0)} J(u) = \int_{t_0}^{t_f} L(x_{1:n+1}, u, t)dt + \varphi[x_{1:n}(t_f), x_{n+1}(t_0)] \quad \text{with} \quad \begin{cases} \dot{x}_{1:n} = f(x_{1:n}, u, t) \\ \dot{x}_{n+1} = 0 \end{cases}$$

The adjoint p_{n+1} associated with x_{n+1} follows the equation: $\dot{p}_{n+1} = -H_{x_{n+1}} = -H_\alpha$.

The conditions of transversality on the adjoint p_{n+1} are:

- $p_{n+1}(t_0) = -\varphi_{x_{n+1}}(t_0) = -\varphi_\alpha$, because $x_{n+1}(t_0)$ is free;
- $p_{n+1}(t_f) = \varphi_{x_{n+1}}(t_f) = 0$, because φ does not depend on $x_{n+1}(t_f)$.

Integrating on $[t_0;t_f]$, we get: $p_{n+1}(t_f) = p_{n+1}(t_0) + \int_{t_0}^{t_f} \dot{p}_{n+1}dt \Rightarrow \varphi_\alpha + \int_{t_0}^{t_f} H_\alpha dt = 0$.

The optimality condition (4.35) on α is retrieved.

Free final time

The free final time can be optimized by the following change of variable.

$$\tau = \frac{t-t_0}{t_f-t_0} = \frac{t-t_0}{\alpha} \rightarrow dt = \alpha d\tau$$

The normalized time τ varies from 0 to 1. The parameter α is to be optimized. The control problem is then with fixed final time $\tau_f = 1$ and it is formulated as

$$\min_{u,\alpha} J = \int_{\tau_0}^{\tau_f} L(x,u,t_0+\alpha\tau)\alpha d\tau + \varphi(x_f, t_0+\alpha\tau_f) \quad \text{with} \quad \frac{dx}{d\tau} = \alpha f(x,u,t_0+\alpha\tau)$$

The Hamiltonian is: $H(x,u,p,t,\alpha) = \alpha L(x,u,t_0+\alpha\tau) + p^T \alpha f(x,u,t_0+\alpha\tau)$.

The optimality condition (4.35) on the parameter α is: $\int_{\tau_0}^{\tau_f} \frac{\partial H}{\partial \alpha} d\tau + \frac{\partial \varphi}{\partial \alpha} = 0$.

Let us calculate the partial derivatives of the Hamiltonian and the final cost:

- derivative of the Hamiltonian using $\frac{dH}{dt} = \frac{\partial H}{\partial t}$ shown below (4.44):

$$\frac{\partial H}{\partial \alpha} = L + p^T f + \alpha\tau\frac{\partial L}{\partial t} + p^T \alpha\tau\frac{\partial f}{\partial t} = H + \alpha\tau\frac{\partial H}{\partial t} = H + (t-t_0)\frac{dH}{dt} = \frac{d}{dt}\left[(t-t_0)H\right],$$

which yields: $\int_{\tau_0}^{\tau_f} \frac{\partial H}{\partial \alpha} d\tau = \int_{t_0}^{t_f} \frac{\partial H}{\partial \alpha}\frac{dt}{\alpha} = \frac{1}{\alpha}\left[(t-t_0)H\right]_{t_0}^{t_f} = \frac{t_f-t_0}{\alpha}H(t_f) = H(t_f)$;

- derivative of final cost:

$\frac{\partial \varphi}{\partial \alpha} = \frac{\partial \varphi}{\partial t}(t_f)$ because $\alpha = t_f - t_0$.

With these derivatives, the optimality condition on α gives: $H(t_f) + \frac{\partial \varphi}{\partial t}(t_f) = 0$.

We retrieve the transversality condition (4.23) associated with the free final time.

4.1.4 Two-point boundary value problem

Consider a free final time control problem.

$$\min_{u,t_f} J(u) = \int_{t_0}^{t_f} L(x,u,t)dt + \varphi(x_f,t_f) \quad \text{with} \quad \begin{cases} \dot{x} = f(x,u,t) \\ x(t_0) = x_0 \end{cases} \tag{4.37}$$

The state and adjoint equations form a Hamiltonian system.

$$\begin{cases} \dot{x} = H_p \\ \dot{p} = -H_x \end{cases} \quad \text{with} \quad H = L + p^T f \tag{4.38}$$

The optimal control is determined by minimizing the Hamiltonian.

$$\min_u H(x,u,p,t) \quad \Rightarrow \begin{cases} H_u = 0 \\ H_{uu} \geq 0 \end{cases} \quad \rightarrow \; u^*(x,p,t) \tag{4.39}$$

The initial conditions relate to the state and the final conditions to the adjoint.

$$\begin{cases} x(t_0) = x_0 \\ p(t_f) = \varphi_x(x_f,t_f) \end{cases} \tag{4.40}$$

If the final time is free, we have the additional final condition

$$H(t_f) = -\varphi_t(x_f,t_f) \tag{4.41}$$

The initial and final conditions form the transversality conditions.

The optimal trajectory is obtained by integrating (4.38) with the control determined at each time by (4.39). The difficulty comes from the endpoint conditions (4.40): the initial state $x(t_0)$ is known, but not the adjoint $p(t_0)$.
This initial adjoint must be found in order to satisfy the final condition on $p(t_f)$.
The unknowns of the problem are:
- the initial adjoint $p(t_0)$, with n components;
- the final time t_f (if free).

The conditions to be satisfied are:
- the transversality condition on the final adjoint $p(t_f)$, with n conditions;
- the transversality condition on the final Hamiltonian $H(t_f)$ (if t_f is free).

This two-point boundary value problem (TPBVP) with n + 1 unknowns and n + 1 equations is depicted in figure 4-4.

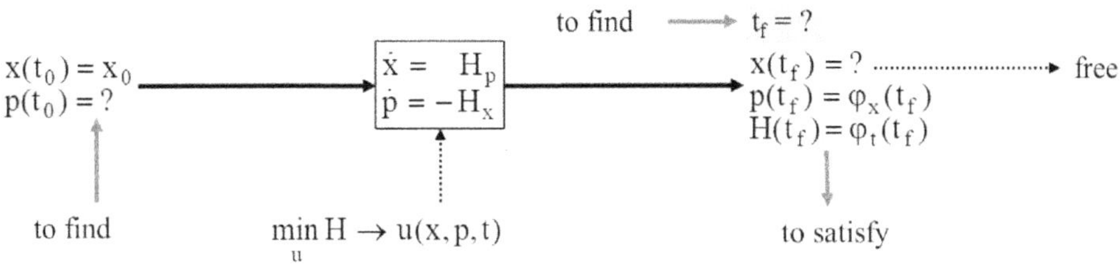

Figure 4-4: Two-point boundary value problem.

The optimality conditions of a control problem systematically lead to a two-point boundary value problem, with one part of the conditions on (x,p) expressed at the initial time and the other part expressed at the final time.
Methods for solving this type of problem are presented in chapter 5.

In some cases, the existence of a first integral simplifies the solution.

First integral

Let us calculate the total derivative of the Hamiltonian on the optimal trajectory.

$$\frac{dH}{dt}(x,u,p,t) = \frac{\partial H}{\partial x}\frac{dx}{dt} + \frac{\partial H}{\partial u}\frac{du}{dt} + \frac{\partial H}{\partial p}\frac{dp}{dt} + \frac{\partial H}{\partial t} \tag{4.42}$$

Using equations (4.38) and (4.39)

$$\frac{\partial H}{\partial x} = -\frac{dp}{dt} \quad , \quad \frac{\partial H}{\partial p} = \frac{dx}{dt} \quad , \quad \frac{\partial H}{\partial u} = 0 \tag{4.43}$$

the total derivative reduces to

$$\frac{dH}{dt} = \frac{\partial H}{\partial t} \tag{4.44}$$

In the frequent case of an autonomous system the dynamics (function f) and the cost (function L) do not depend explicitly on time. We then have

$$H(x,u,p,t) = L(x,u) + p^T f(x,u) \quad \Rightarrow \quad \frac{\partial H}{\partial t} = 0 \tag{4.45}$$

In this case, the Hamiltonian is constant along the optimal trajectory and it provides a first integral (constant quantity along the trajectory). If the final time is free, its value is given by the transversality condition (4.41).

$$H(t) = H(t_f) = -\varphi_t(x_f, t_f) \tag{4.46}$$

Minimum time problem

The problem of reaching a target in the minimum time is formulated as

$$\min_{u,t_f} J = \int_{t_0}^{t_f} 1 dt \quad \text{with} \quad \begin{cases} \dot{x} = f(x,u,t) \\ x(t_0) = x_0 \end{cases} \tag{4.47}$$

The transversality condition (4.41) gives: $H(t_f) = 0$. If the system is autonomous, the Hamiltonian is constant along the whole trajectory.

$$H = 1 + p^T f(x,u) = 0 \;\; , \;\; \forall t \tag{4.48}$$

This relation between x and p can replace a final condition, which simplifies the solution of the two-point boundary value problem.

The following two examples illustrate the solution of a minimum time problem in dimension 2.

Example 4-3 concerns a problem of navigation in a fluid stream.

Example 4-3: Navigation in a fluid stream (from [R3])

We consider the plane trajectory of a vehicle in a moving fluid, for example an aircraft in the presence of wind or a boat in the presence of current.

Modelling

The assumptions about the motion are as follows:

- the velocity $\vec{e}$ of the fluid is a function of the position (drive velocity);
- the velocity $\vec{w}$ of the vehicle relative to the fluid has a constant modulus w;
- the control is the orientation θ of the relative velocity.

Figure 4-5 shows the relative $\vec{w}$, absolute $\vec{v}_a$ and drive $\vec{e}$ velocities.

Fluid drive velocity: $\vec{e} = \begin{pmatrix} u(x,y) \\ v(x,y) \end{pmatrix}$

Vehicle relative velocity: $\vec{w} = w \begin{pmatrix} \cos\theta \\ \sin\theta \end{pmatrix}$

State equations: $\begin{cases} \dot{x} = w\cos\theta + u(x,y) \\ \dot{y} = w\sin\theta + v(x,y) \end{cases}$

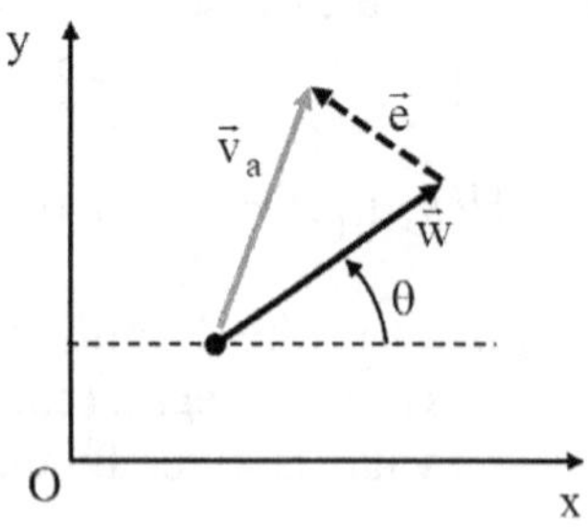

Figure 4-5: Relative and absolute velocity.

Control problem

The objective is to return to the origin O in the shortest possible time from a given position $(x_0;y_0)$. The final velocity is free. The problem is formulated as

$$\min_{\theta,t_f} J = \int_{t_0}^{t_f} 1\,dt \quad \text{with} \quad \begin{cases} \dot{x} = w\cos\theta + u \\ \dot{y} = w\sin\theta + v \end{cases} \quad \text{and} \quad \begin{cases} x(t_f) = 0 \\ y(t_f) = 0 \end{cases}$$

Application of the PMP

The Hamiltonian is: $H = 1 + p(w\cos\theta + u) + q(w\sin\theta + v)$ with the adjoint (p,q)

The application of the PMP results in: $\min_\theta H \;\rightarrow\; H_\theta = -p\sin\theta + q\cos\theta = 0$

The system is autonomous: $H(t) = H(t_f) = 0$ (minimum time)

The adjoint equations are: $\begin{cases} \dot{p} = -H_x = -pu_x - qv_x \\ \dot{q} = -H_y = -pu_y - qv_y \end{cases}$

Control and trajectory determination

We want to express the control as a function of the position. Let us use the two equations on H to express the adjoint (p,q) as a function of the control θ.

$$\begin{cases} H = 0 \\ H_\theta = 0 \end{cases} \Rightarrow \begin{cases} p(w\cos\theta + u) + q(w\sin\theta + v) = -1 \\ p\sin\theta - q\cos\theta = 0 \end{cases}$$

$$\Rightarrow \begin{cases} pW = -\cos\theta \\ qW = -\sin\theta \end{cases} \quad \text{with} \quad W = w + u\cos\theta + v\sin\theta$$

Deriving these equations: $\begin{cases} pW = -\cos\theta \\ qW = -\sin\theta \end{cases} \rightarrow \begin{cases} \dot{p}W + p\dot{W} = \dot{\theta}\sin\theta \\ \dot{q}W + q\dot{W} = -\dot{\theta}\cos\theta \end{cases}$

with $(\dot{p},\dot{q})$ given by: $\begin{cases} \dot{p} = -H_x = -pu_x - qv_x \\ \dot{q} = -H_y = -pu_y - qv_y \end{cases}$ (adjoint equations)

we obtain the system: $\begin{cases} Wu_x\cos\theta + Wv_x\sin\theta - \dot{W}\cos\theta = W\dot{\theta}\sin\theta \\ Wu_y\cos\theta + Wv_y\sin\theta - \dot{W}\sin\theta = -W\dot{\theta}\cos\theta \end{cases}$

Then we eliminate $\dot{W}$: $\quad \dot{\theta} = v_x\sin^2\theta + (u_x - v_y)\sin\theta\cos\theta - u_y\cos^2\theta$

The optimal trajectory is found by integrating the differential system in (x,y,θ).

$$\begin{cases} \dot{x} = w\cos\theta + u(x,y) \\ \dot{y} = w\sin\theta + v(x,y) \\ \dot{\theta} = v_x\sin^2\theta + (u_x - v_y)\sin\theta\cos\theta - u_y\cos^2\theta \end{cases}$$

The initial conditions $(x_0;y_0)$ are given. The initial control θ_0 is sought such that the trajectory passes through the origin O.
Let us look at two simple special cases that allow for an analytical solution.

Case of a constant stream
If u and v are constant, the derivatives u_x, u_y, v_x, v_y are zero and the equation for θ reduces to $\dot{\theta}=0$. The solution: $\theta=\theta_0=C^{te}$ corresponds to a straight-line trajectory towards the origin.

Case of a stream parallel to (Ox) and linear as a function of y: $u=cy$, $v=0$
The equation in θ integrates.

$$\dot{\theta}=-u_y\cos^2\theta=-c\cos^2\theta \;\Rightarrow\; \frac{d\theta}{\cos^2\theta}=-cdt \;\Rightarrow\; \tan\theta=\tan\theta_f+c(t_f-t)$$

The control $\theta(t)$ follows a linear tangent law as a function of time.

We rewrite the equations of motion by changing the variable to θ instead of t.

Change of variable: $\dfrac{d\theta}{dt}=-c\cos^2\theta \;\Rightarrow\; dt=-\dfrac{d\theta}{c\cos^2\theta}$

Equation in y

$$\frac{dy}{dt}=w\sin\theta \;\Rightarrow\; \frac{dy}{d\theta}=-\frac{w\sin\theta}{c\cos^2\theta} \;\Rightarrow\; y=\frac{w}{c}\left(\frac{1}{\cos\theta_f}-\frac{1}{\cos\theta}\right)$$

Equation in x

$$\frac{dx}{dt}=w\cos\theta+cy \;\Rightarrow\; \frac{dx}{d\theta}=-\frac{w}{c\cos\theta}-\frac{w}{c\cos^2\theta}\left(\frac{1}{\cos\theta_f}-\frac{1}{\cos\theta}\right)$$

These differential equations can be integrated analytically.
After (long) calculations, we obtain

$$\begin{cases} x=\dfrac{w}{2c}\left[\dfrac{\tan\theta-\tan\theta_f}{\cos\theta_f}+\tan\theta\left(\dfrac{1}{\cos\theta_f}-\dfrac{1}{\cos\theta}\right)+\ln\left(\tan\theta+\dfrac{1}{\cos\theta}\right)-\ln\left(\tan\theta_f+\dfrac{1}{\cos\theta_f}\right)\right] \\ y=\dfrac{w}{c}\left[\dfrac{1}{\cos\theta_f}-\dfrac{1}{\cos\theta}\right] \end{cases}$$

The initial conditions $(x_0;y_0)$ allow us to determine the unknowns $(\theta_0;\theta_f)$ by solving a nonlinear system of dimension 2.

Numerical application

Figure 4-6 shows the application for a stream defined by: $u=-y$, $v=0$.

The initial conditions are: $x_0 = y_0 = -4.5\,m$.

The solution: $\theta_0 = 98.8$ deg , $\theta_f = 240.0$ deg gives a final time: $t_f = 8.2\,s$.

The arrows indicate the direction of the relative velocity every second.

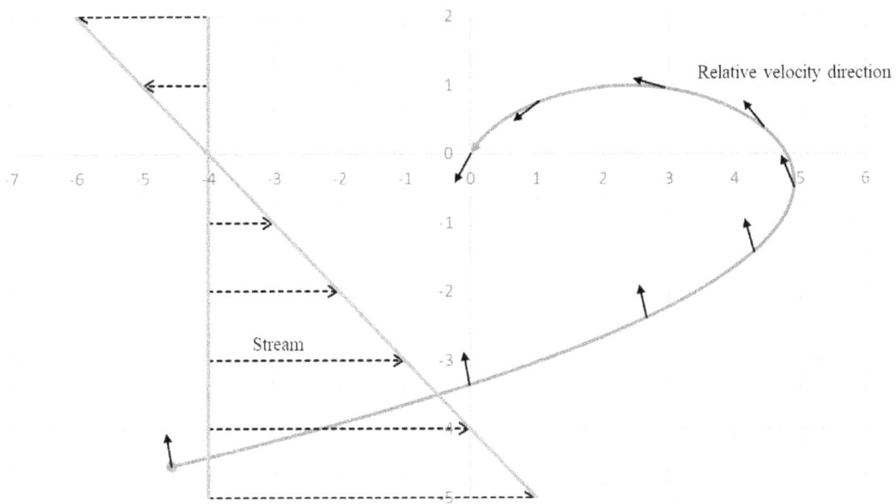

Figure 4-6: Optimal trajectory in a linear stream.

Example 4-4 is a double integrator problem in dimension 2.

Example 4-4: Double integrator in dimension 2 (from [R3])

We consider a plane trajectory controlled in acceleration.

Modelling

The assumptions about the motion are as follows:

- the acceleration $\vec{a}$ is of constant modulus;
- the control is the direction $\vec{u}$ of the acceleration oriented by an angle θ .

Position: $\vec{r} = \begin{pmatrix} x \\ y \end{pmatrix} \rightarrow \dot{\vec{r}} = \vec{v}$

Velocity: $\vec{v} = \begin{pmatrix} v_x \\ v_y \end{pmatrix} \rightarrow \dot{\vec{v}} = \vec{a}$

Control: $\vec{u} = \begin{pmatrix} \cos\theta \\ \sin\theta \end{pmatrix}$

Acceleration: $\vec{a} = a\vec{u}$

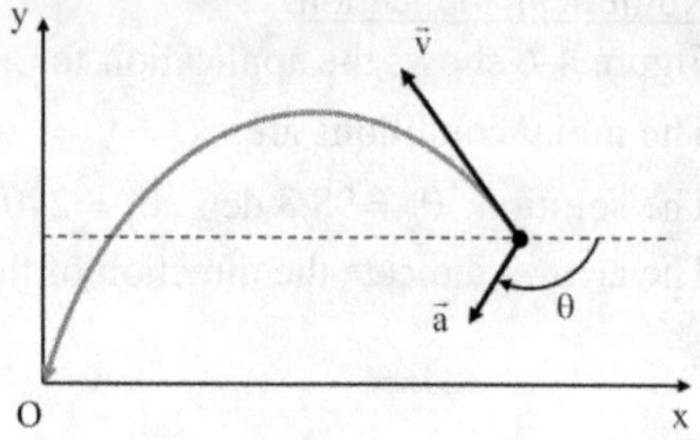

Figure 4-7: Double integrator model.

Control problem

The objective is to return and stop in O in the minimum time from given initial conditions $(\vec{r}_0 ; \vec{v}_0)$. The problem is formulated as

$$\min_{\vec{u},t_f} J = \int_{t_0}^{t_f} 1\, dt \quad \text{with} \quad \begin{cases} \dot{\vec{r}} = \vec{v} \\ \dot{\vec{v}} = a\vec{u} \end{cases}, \quad \begin{cases} \vec{r}(t_0) = \vec{r}_0 \\ \vec{v}(t_0) = \vec{v}_0 \end{cases}, \quad \begin{cases} \vec{r}(t_f) = 0 \\ \vec{v}(t_f) = 0 \end{cases}$$

Application of the PMP

The Hamiltonian is: $H = 1 + \vec{p}_r.\vec{v} + a\vec{p}_v.\vec{u}$ with the adjoint components $(\vec{p}_r ; \vec{p}_v)$.

The application of the PMP gives: $\min_{\vec{u}} H \quad \rightarrow \quad \vec{u} = -\dfrac{\vec{p}_v}{\|\vec{p}_v\|}$ (unit vector).

The adjoint equations are: $\begin{cases} \dot{\vec{p}}_r = -H_{\vec{r}} = \vec{0} & \rightarrow \vec{p}_r = \vec{B} \\ \dot{\vec{p}}_v = -H_{\vec{v}} = -\vec{p}_r & \rightarrow \vec{p}_v = \vec{B}(t_f - t) + \vec{A} \end{cases}$

with constant vectors $\vec{A}$ and $\vec{B}$

Control and trajectory determination

We replace the adjoints in the expression of $\vec{u}$ and derive the angle θ.

$$\vec{u} = -\frac{\vec{p}_v}{\|\vec{p}_v\|} = -\frac{\vec{A} + \vec{B}(t_f - t)}{\|\vec{A} + \vec{B}(t_f - t)\|} \quad \rightarrow \quad \tan\theta = \frac{u_y}{u_x} = \frac{A_y + B_y(t_f - t)}{A_x + B_x(t_f - t)}$$

The control $\theta(t)$ follows a bilinear tangent law as a function of time. In order to integrate the trajectory, we reduce it to an equivalent linear tangent law.

$$\tan\theta = \frac{A_y + B_y(t_f - t)}{A_x + B_x(t_f - t)} \quad \Leftrightarrow \quad \tan(\theta - \alpha) = \tan(\theta_f - \alpha) + e(t_f - t)$$

The relations between the constants (A_x, A_y, B_x, B_y) and $(\alpha, e, \theta_0, \theta_f)$ are found by expanding $\tan(\theta-\alpha)$ and identifying the terms.

$$\tan\theta = \frac{\tan(\theta_f-\alpha)+\tan\alpha+e(t_f-t)}{1-\tan(\theta_f-\alpha)\tan\alpha-e(t_f-t)\tan\alpha} \rightarrow \begin{cases} A_x = 1-\tan(\theta_f-\alpha)\tan\alpha \\ A_y = \tan(\theta_f-\alpha)+\tan\alpha \\ B_x = -e\tan\alpha \\ B_y = e \end{cases}$$

We write the equations of motion by changing the variable to θ instead of t.
The change of variable is defined by

$$\tan(\theta-\alpha) = \tan(\theta_f-\alpha)+e(t_f-t) \quad \Rightarrow \quad dt = -\frac{d\theta}{e\cos^2(\theta-\alpha)}$$

For the following calculations, we denote:
- (x, y) the Cartesian components of the position;
- (u, v) the Cartesian components of the velocity;
- $(\dot{x}, \dot{y}, \dot{u}, \dot{v})$ the derivatives with respect to t;
- (x', y', u', v') the derivatives with respect to θ.

From the equations in t: $\begin{cases} \dot{x} = u \;,\; \dot{u} = a\cos\theta \\ \dot{y} = v \;,\; \dot{v} = a\sin\theta \end{cases}$ and with $dt = -\dfrac{d\theta}{e\cos^2(\theta-\alpha)}$

we get the equations in θ: $\begin{cases} x' = \dfrac{-u}{e\cos^2(\theta-\alpha)} \;,\; u' = \dfrac{-a\cos\theta}{e\cos^2(\theta-\alpha)} \\ y' = \dfrac{-v}{e\cos^2(\theta-\alpha)} \;,\; v' = \dfrac{-a\sin\theta}{e\cos^2(\theta-\alpha)} \end{cases}$

These equations are integrated analytically. After (long) calculations, we obtain

$$\begin{cases} x - x_f = \dfrac{a}{e^2}(f_x\cos\alpha - f_y\sin\alpha) - u_f(t_f-t) \;,\; u - u_f = \dfrac{a}{e}(f_u\cos\alpha - f_v\sin\alpha) \\ y - y_f = \dfrac{a}{e^2}(f_x\sin\alpha + f_y\cos\alpha) - v_f(t_f-t) \;,\; v - v_f = \dfrac{a}{e}(f_u\sin\alpha + f_v\cos\alpha) \end{cases}$$

$$\text{with} \begin{cases} f_x = -f_u\tan(\theta-\alpha)+f_v \\ f_y = -\dfrac{1}{2}f_u - \dfrac{1}{2}f_v\tan(\theta-\alpha) + \dfrac{1}{2}\dfrac{\tan(\theta_f-\alpha)-\tan(\theta-\alpha)}{\cos(\theta_f-\alpha)} \\ f_u = \ln\left|\tan(\theta_f-\alpha)+\dfrac{1}{\cos(\theta_f-\alpha)}\right| - \ln\left|\tan(\theta-\alpha)+\dfrac{1}{\cos(\theta-\alpha)}\right| \\ f_v = \dfrac{1}{\cos(\theta_f-\alpha)} - \dfrac{1}{\cos(\theta-\alpha)} \end{cases}$$

The current values (x,y,u,v,θ) along the trajectory are thus expressed as a function of the final values $(x_f,y_f,u_f,v_f,\theta_f)$, of the final time t_f and of the constants α and e.

Final conditions

The aim is to return to the origin $(x_f = y_f = 0)$ and to stop there $(u_f = v_f = 0)$. With these final conditions, the equations of the trajectory simplify to

$$\begin{cases} x = \dfrac{a}{e^2}(f_x \cos\alpha - f_y \sin\alpha) \quad , \quad u = \dfrac{a}{e}(f_u \cos\alpha - f_v \sin\alpha) \\ y = \dfrac{a}{e^2}(f_x \sin\alpha + f_y \cos\alpha) \quad , \quad v = \dfrac{a}{e}(f_u \sin\alpha + f_v \cos\alpha) \end{cases}$$

with (f_x, f_y, f_u, f_v) expressed as a function of $(\theta - \alpha, \theta_f - \alpha)$.

Let us place ourselves at any point (x,y,u,v). The above equations form a system of dimension 4 whose unknowns are $(\alpha, e, \theta_f, \theta)$.

System reduction

To solve this system, we switch to polar coordinates.

The position $\vec{r}$ is defined by $(r;\sigma)$:
- r is the polar radius;
- σ is called the sight angle.

The velocity $\vec{v}$ is defined by $(w;\gamma)$:
- w is the velocity modulus;
- γ is the angle of the velocity with the line of sight.

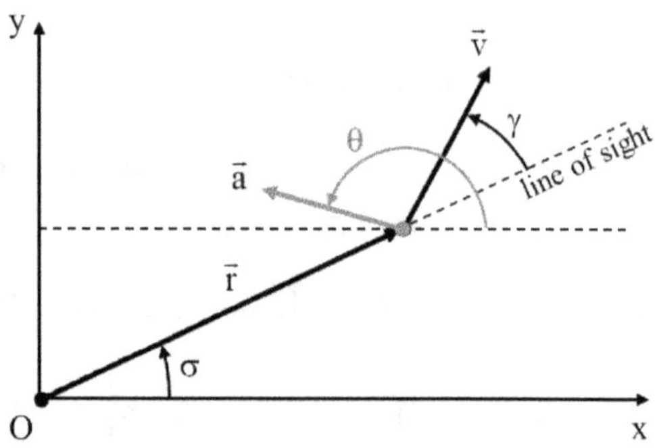

Figure 4-8: Polar coordinates.

The transformation to polar coordinates is defined by

$$\begin{cases} x = r\cos\sigma \quad , \quad u = w\cos(\sigma+\gamma) \\ y = r\sin\sigma \quad , \quad v = w\sin(\sigma+\gamma) \end{cases}$$

Let us express from (x,y,u,v) the two quantities $\dfrac{w^2}{ar}$ and γ.

$$\begin{cases} \dfrac{w^2}{ar} = \dfrac{u^2+v^2}{a\sqrt{x^2+y^2}} \\ \cos\gamma = \dfrac{xu+yv}{rw} \\ \sin\gamma = \dfrac{xv-yu}{rw} \end{cases} \text{ with } \begin{cases} x = \dfrac{a}{e^2}(f_x\cos\alpha - f_y\sin\alpha) \\ y = \dfrac{a}{e^2}(f_x\sin\alpha + f_y\cos\alpha) \\ u = \dfrac{a}{e}(f_u\cos\alpha - f_v\sin\alpha) \\ v = \dfrac{a}{e}(f_u\sin\alpha + f_v\cos\alpha) \end{cases} \rightarrow \begin{cases} \dfrac{w^2}{ar} = \dfrac{f_u^2+f_v^2}{\sqrt{f_x^2+f_y^2}} \\ \cos\gamma = \dfrac{a^2}{e^3}\dfrac{f_xf_u+f_yf_v}{rw} \\ \sin\gamma = \dfrac{a^2}{e^3}\dfrac{f_xf_v-f_yf_u}{rw} \end{cases}$$

The functions (f_x, f_y, f_u, f_v) depend only on the 2 unknowns $(\theta-\alpha, \theta_f-\alpha)$. We are thus reduced to a nonlinear system of dimension 2.

$$\begin{cases} \dfrac{w^2}{ar} = \dfrac{f_u^2+f_v^2}{\sqrt{f_x^2+f_y^2}} \\ \gamma = \text{Atan2}(f_xf_u+f_yf_v\ ,\ f_xf_v-f_yf_u) \end{cases} \quad \text{where Atan2 denotes Arctan(x,y)}$$

The values of (w, r, γ) depend on (x, y, u, v) which are known (current point). The numerical resolution of the system gives $(\theta-\alpha, \theta_f-\alpha)$.

We then deduce: $e = \dfrac{a\sqrt{f_u^2+f_v^2}}{w}$ and $t_f - t = \dfrac{\tan(\theta-\alpha)-\tan(\theta_f-\alpha)}{e}$.

The value of α is obtained from the transversality condition: $H(t_f)=0$.

$$H(t_f) = \left(1+\vec{p}_r.\vec{v}+a\vec{p}_v.\vec{u}\right)_{t_f} = 1-a\left\|\vec{A}\right\|$$

$$\text{using } \vec{v}_f = \vec{0}\ ,\ \vec{u} = -\frac{\vec{p}_v}{\|\vec{p}_v\|}\ ,\ \vec{p}_v = \vec{B}(t_f-t)+\vec{A}$$

The vector $\vec{A}$ has the components: $\begin{cases} A_x = 1-\tan(\theta_f-\alpha)\tan\alpha \\ A_y = \tan(\theta_f-\alpha)+\tan\alpha \end{cases}$.

The equation: $H(t_f) = 1-a\left\|\vec{A}\right\| = 0$ gives: $\cos\alpha = \dfrac{\sqrt{a}}{\cos(\theta_f-\alpha)}$,

which gives α from $(\theta_f-\alpha)$ calculated above.

<u>Numerical application</u>

The initial conditions are: $\begin{cases} x = 10 \text{ m} \\ y = 4 \text{ m} \end{cases}$ and $\begin{cases} u = 0 \text{ m}\cdot\text{s}^{-1} \\ v = 0.5 \text{ m}\cdot\text{s}^{-1} \end{cases}$.

The acceleration is: $a = 0.1 \text{ m}\cdot\text{s}^{-2}$. The solution gives a final time: $t_f = 24.13$ s.

Figure 4-9 shows the evolution of the angle θ. The dashed curve is the angle $\theta_v = 180° + \sigma - \theta$ between the line of sight (towards O) and the direction of the acceleration. There is a phase of acceleration towards the origin ($\theta_v < 90°$) up to $t \approx 14$ s, then a phase of deceleration ($\theta_v \approx 180°$) to reduce the velocity and stop at the origin. The value of θ during the deceleration indicates the direction of approach to the target (here $\theta \approx 50°$).

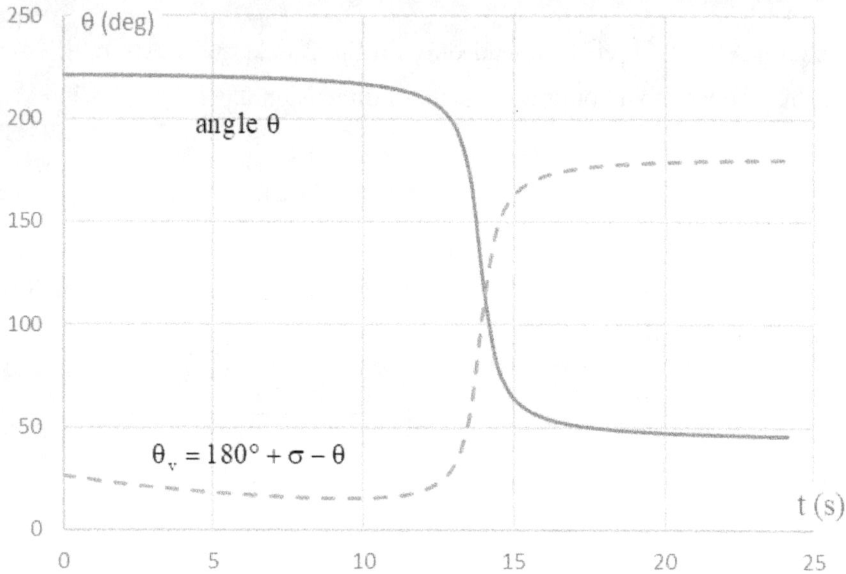

Figure 4-9: Double integrator control in dimension 2.

Figure 4-10 shows the position $(x;y)$ and the velocity $(u;v)$. The final approach is almost straight along the angle $\theta \approx 50°$ seen in figure 4-9.

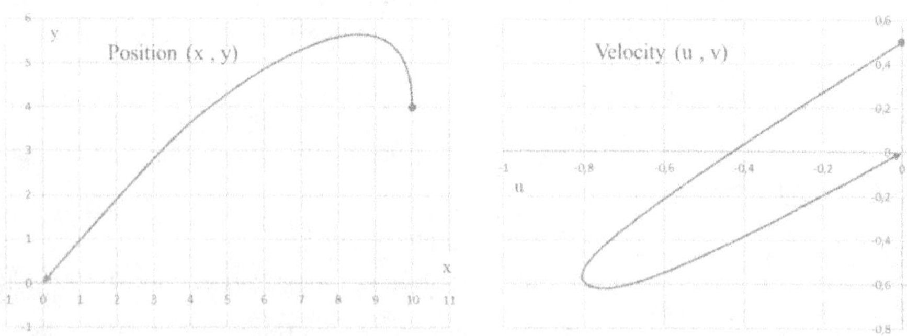

Figure 4-10: Double integrator trajectory in dimension 2.

Figure 4-11 summarizes the set of solutions as a function of γ and $\frac{w^2}{2ar}$.

These two variables depend on the position and velocity coordinates (x, y, u, v). The angle $\gamma_v = 180° - \gamma$ (abscissa) between the direction of the target and the velocity varies from 0° to 180°. An angle $\gamma_v < 90°$ (resp. $\gamma_v > 90°$) indicates approaching (resp. moving away from) the target. The ratio $\frac{w^2}{2ar}$ (ordinate) or its inverse varies from 0 to 1.

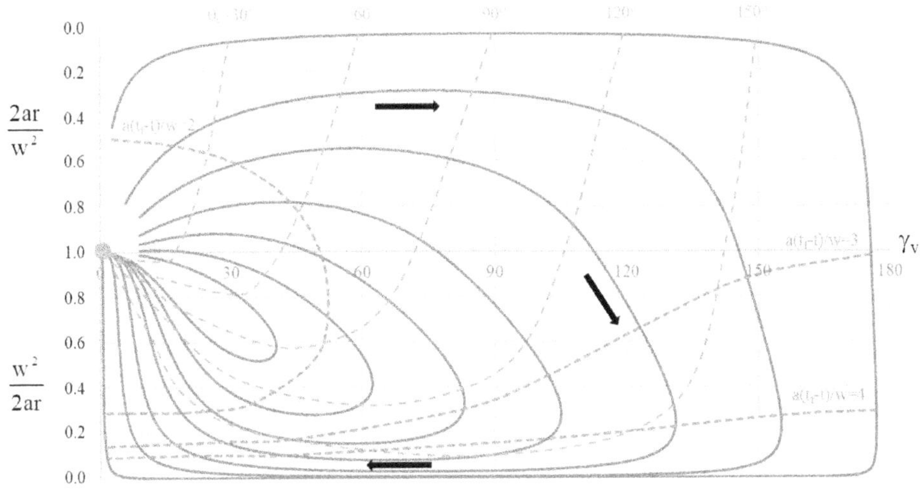

Figure 4-11: Summary of double integrator solutions.

The solid lines are trajectories following the arrows (clockwise). The dashed lines are the level lines of $\theta_v = 180° + \sigma - \theta$ and $\frac{a(t_f - t)}{w}$. An angle $\theta_v < 90°$ (resp. $\theta_v > 90°$) indicates accelerating towards (respectively away from) the target.

The trajectories end in the lower left quadrant and converge on the point of coordinates $(0\,;1)$. On the arrival phase near the target:

- the velocity is low and the distance decreases: the ratio $\frac{w^2}{2ar} < 1$ increases;
- the velocity is directed towards the target: the angle $\gamma_v < 90°$ decreases to 0°, (corresponding to a straight path);
- the acceleration is always opposite to the target: $\theta_v > 90°$ (to brake).

Depending on the initial conditions, the trajectory may start with an acceleration phase towards the target: $\theta_v < 90°$, before the arrival braking phase described above. This occurs especially when the initial point is in the upper left quadrant, with a high initial velocity $\left(\frac{w^2}{2ar} > 1\right)$ and directed towards the target $(\gamma_v < 90°)$.

4.2 Constraints

This section presents the optimality conditions when the control problem is subject to various types of constraints or discontinuities.

4.2.1 Terminal constraints

Consider a control problem with initial and final constraints.

$$\min_{u,x_0,t_0,x_f,t_f} J = \int_{t_0}^{t_f} L(x,u,t)dt + \varphi(x_0,t_0,x_f,t_f) \quad \text{with} \quad \dot{x} = f(x,u,t) \\ \text{and} \quad \psi(x_0,t_0,x_f,t_f) = 0 \tag{4.49}$$

The initial $(x_0;t_0)$ and final $(x_f;t_f)$ conditions may vary, but the solution must satisfy q constraints of the form: $\psi(x_0,t_0,x_f,t_f) = 0$.
As in (4.10), we introduce the adjoint $p(t)$ and the Hamiltonian $H(x,u,p,t)$.

$$\begin{cases} H(x,u,p,t) = L(x,u,t) + p^T f(x,u,t) \\ \dot{p} = -\dfrac{\partial H}{\partial x} \end{cases} \tag{4.50}$$

The control $u(t)$ is obtained by minimizing the Hamiltonian at each time.

$$\min_u H(x,u,p,t) \quad \rightarrow \quad u(t) \tag{4.51}$$

The Hamiltonian considered here is normal (with the convention: $p_0 = 1$), which excludes abnormal extremals. We restrict ourselves here to normal extremals.

We then have the following necessary conditions of optimality.

Property 4-2: Necessary conditions for optimality with terminal constraints

Assume that the control $u(t)$ defined by (4.51) is a solution of (4.49).
Then there are q multipliers λ associated with the q constraints $\psi=0$ such that:

- the terminal constraints are satisfied

$$\psi(x_0,t_0,x_f,t_f)=0 \tag{4.52}$$

- the initial adjoint (if x_0 free) and the final adjoint (if x_f free) satisfy

$$\begin{cases} p(t_0) &= -\varphi_x(t_0) - \psi_x(t_0)\lambda \\ p(t_f) &= \varphi_x(t_f) + \psi_x(t_f)\lambda \end{cases} \tag{4.53}$$

- the initial Hamiltonian (if t_0 free) and the final Hamiltonian (if t_f free) satisfy

$$\begin{cases} H(t_0) = \varphi_t(t_0) + \psi_t(t_0)\lambda \\ H(t_f) = -\varphi_t(t_f) - \psi_t(t_f)\lambda \end{cases} \tag{4.54}$$

These conditions form a two-point boundary value problem:

- the q equations (4.52) are linked to the unknowns λ;
- the n conditions of transversality on $p(t_0)$ are linked to the unknowns $x(t_0)$;
- the n transversality conditions on $p(t_f)$ are linked to the unknowns $x(t_f)$;
- the transversality condition on $H(t_0)$ is linked to the unknown t_0;
- the transversality condition on $H(t_f)$ is linked to the unknown t_f.

Demonstration

We use the variational method by adding the q constraints ψ with multipliers λ to the augmented cost (4.13). Formula (4.16) becomes

$$J_a = \int_{t_0}^{t_f} \left[H(x,u,p,t) + \dot{p}^T x \right] dt + p(t_0)^T x(t_0) - p(t_f)^T x(t_f)$$
$$+ \varphi(x_0,t_0,x_f,t_f) + \lambda^T \psi(x_0,t_0,x_f,t_f)$$

Applying the variations $\delta u(t), \delta x(t), \delta t_0$, δt_f to the augmented cost gives formula (4.17) with additional terms from $\lambda^T \psi(x_0,t_0,x_f,t_f)$.

$$\delta J_a = \int_{t_0}^{t_f} H_u^T \delta u dt \quad + \left[\varphi_{x_0} + \psi_{x_0}\lambda + p(t_0)\right]^T \delta x_0 + \left[\varphi_{t_0} + \psi_{t_0}\lambda - H(t_0)\right]\delta t_0$$
$$+ \int_{t_0}^{t_f} (H_x + \dot{p})^T \delta x dt + \left[\varphi_{x_f} + \psi_{x_f}\lambda - p(t_f)\right]^T \delta x_f + \left[\varphi_{t_f} + \psi_{t_f}\lambda + H(t_f)\right]\delta t_f$$

The first-order optimality condition is: $\delta J_a = 0$, $\forall \delta u, \delta x_0, \delta t_0, \delta x_f, \delta t_f$.
As the variations $\delta u, \delta x_0, \delta t_0, \delta x_f, \delta t_f$ are independent, their coefficients must be zero. This leads to the choice (4.50) to eliminate δx and to the condition $H_u = 0$ for the term in δu . The transversality conditions (4.53) and (4.54) eliminate the terms in $(\delta x_0 ; \delta x_f)$ and in $(\delta t_0 ; \delta t_f)$.

Example 4-5 shows the application to a launch into orbit problem.

Example 4-5: Launch into orbit (from [R3])
We consider a trajectory controlled in acceleration in the vertical plane.

Modelling
The assumptions about the motion are as follows:
- the Earth is flat and the gravity $\vec{g}$ is constant and vertical;
- the target orbit is defined by an altitude z_c and a horizontal velocity v_c ;
- the acceleration $\vec{a}$ of constant modulus is along $\vec{u}$ defined by the angle θ .

Figure 4-12 shows the trajectory in the vertical plane (Oxz).

Position: $\vec{r} = \begin{pmatrix} x \\ y \end{pmatrix} \quad \rightarrow \quad \dot{\vec{r}} = \vec{v}$

Velocity: $\vec{v} = \begin{pmatrix} v_x \\ v_y \end{pmatrix} \quad \rightarrow \quad \dot{\vec{v}} = \vec{g} + a\vec{u}$

Gravity: $\vec{g} = \begin{pmatrix} 0 \\ -g \end{pmatrix}$

Control: $\vec{u} = \begin{pmatrix} \cos\theta \\ \sin\theta \end{pmatrix}$, $\vec{a} = a\vec{u}$

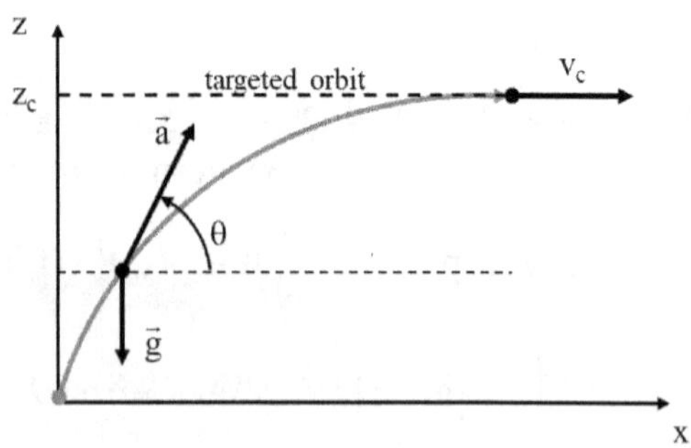

Figure 4-12: Launch into orbit.

Control problem

The aim is to reach the orbit in the minimum time by taking off at zero velocity. The injection abscissa $x(t_f)$ is free. The problem is formulated as

$$\min_{\vec{u},t_f} J = \int_{t_0}^{t_f} 1\,dt \quad \text{with} \quad \begin{cases} \dot{\vec{r}} = \vec{v} \\ \dot{\vec{v}} = \vec{g} + a\vec{u} \end{cases} \quad \text{and} \quad \begin{cases} \vec{r}(t_0) = \vec{0} \\ \vec{v}(t_0) = \vec{0} \end{cases}, \quad \begin{cases} z(t_f) = z_c \\ v_x(t_f) = v_c \\ v_z(t_f) = 0 \end{cases}$$

Application of the PMP

The Hamiltonian is: $H = 1 + \vec{p}_r.\vec{v} + \vec{p}_v.(\vec{g} + a\vec{u})$ with the adjoint $(\vec{p}_r, \vec{p}_v)$.

The application of the PMP gives: $\min_{\vec{u}} H \;\rightarrow\; \vec{u} = -\dfrac{\vec{p}_v}{\|\vec{p}_v\|}$ (unit vector).

The adjoint equations are: $\begin{cases} \dot{\vec{p}}_r = -H_{\vec{r}} = \vec{0} & \rightarrow \vec{p}_r = \vec{B} \\ \dot{\vec{p}}_v = -H_{\vec{v}} = -\vec{p}_r & \rightarrow \vec{p}_v = \vec{B}(t_f - t) + \vec{A} \end{cases}$.

with constant $\vec{A}$ and $\vec{B}$ vectors

The injection abscissa $x(t_f)$ being free, we have the transversality condition

$p_x(t_f) = 0 \;\Rightarrow\; B_x = 0$

Control and trajectory determination

We replace the adjoints in the expression of $\vec{u}$ and derive the angle θ.

$$\vec{u} = -\frac{\vec{p}_v}{\|\vec{p}_v\|} = -\frac{\vec{A} + \vec{B}(t_f - t)}{\|\vec{A} + \vec{B}(t_f - t)\|} \;\rightarrow\; \tan\theta = \frac{u_z}{u_x} = \frac{A_z + B_z(t_f - t)}{A_x}$$

The control $\theta(t)$ follows a linear tangent law of the form: $\tan\theta = \tan\theta_0 - ct$.

We rewrite the equations of motion by changing the variable to θ instead of t.

$$\tan\theta = \tan\theta_0 - ct \;\Rightarrow\; dt = -\frac{d\theta}{c\cos^2\theta} \;\rightarrow\; \begin{cases} \dfrac{d\vec{v}}{dt} = -c\cos^2\theta \dfrac{d\vec{v}}{d\theta} = \vec{g} + a\vec{u} \\ \dfrac{d\vec{r}}{dt} = -c\cos^2\theta \dfrac{d\vec{r}}{d\theta} = \vec{v} \end{cases}$$

First, the velocity components are integrated into the variable θ.

$$\begin{cases} \dfrac{dv_x}{d\theta} = -\dfrac{a}{c\cos\theta} \\ \dfrac{dv_z}{d\theta} = -\dfrac{a\sin\theta}{c\cos^2\theta} - g\dfrac{dt}{d\theta} \end{cases} \;\rightarrow\; \begin{cases} v_x = \dfrac{a}{c}\ln\left(\tan\theta_0 + \dfrac{1}{\cos\theta_0}\right) - \dfrac{a}{c}\ln\left(\tan\theta + \dfrac{1}{\cos\theta}\right) \\ v_z = \dfrac{a}{c}\left(\dfrac{1}{\cos\theta_0} - \dfrac{1}{\cos\theta}\right) - gt \end{cases}$$

The components of the position are then integrated into θ (long calculations).

$$\begin{cases} x = \dfrac{a}{c^2}\left[\dfrac{1}{\cos\theta_0} - \dfrac{1}{\cos\theta} + \tan\theta \ln\left(\tan\theta + \dfrac{1}{\cos\theta}\right) - \tan\theta \ln\left(\tan\theta_0 + \dfrac{1}{\cos\theta_0}\right)\right] \\ z = \dfrac{a}{2c^2}\left[\dfrac{\tan\theta_0}{\cos\theta_0} + \dfrac{\tan\theta}{\cos\theta} - \dfrac{2\tan\theta}{\cos\theta_0} + \ln\left(\tan\theta + \dfrac{1}{\cos\theta}\right) - \ln\left(\tan\theta_0 + \dfrac{1}{\cos\theta_0}\right)\right] - \dfrac{gt^2}{2} \end{cases}$$

Final conditions

The final conditions are on v_x, v_z, z with the control $\tan\theta_f = \tan\theta_0 - ct$.

This forms a system of dimension 4 whose unknowns are $(\theta_0, \theta_f, t_f, c)$.

$$\begin{cases} \tan\theta_f = \tan\theta_0 - ct_f \\ v_x(t_f) = \dfrac{a}{c}\ln\left(\tan\theta_0 + \dfrac{1}{\cos\theta_0}\right) - \dfrac{a}{c}\ln\left(\tan\theta_f + \dfrac{1}{\cos\theta_f}\right) = v_c \\ v_z(t_f) = \dfrac{a}{c}\left(\dfrac{1}{\cos\theta_0} - \dfrac{1}{\cos\theta_f}\right) - gt_f = 0 \\ z(t_f) = \dfrac{a}{2c^2}\left[\dfrac{\tan\theta_0}{\cos\theta_0} + \dfrac{\tan\theta_f}{\cos\theta_f} - \dfrac{2\tan\theta_f}{\cos\theta_0} + \ln\left(\tan\theta_f + \dfrac{1}{\cos\theta_f}\right) - \ln\left(\tan\theta_0 + \dfrac{1}{\cos\theta_0}\right)\right] - \dfrac{gt_f^2}{2} = z_c \end{cases}$$

System reduction

We eliminate (c, t_f) with the first 2 equations. This leaves a system in (θ_0, θ_f).

$$\begin{cases} \dfrac{a}{g}\left(\dfrac{1}{\cos\theta_0} - \dfrac{1}{\cos\theta_f}\right) = \tan\theta_0 - \tan\theta_f \\ \dfrac{\tan\theta_0}{\cos\theta_f} - \dfrac{\tan\theta_f}{\cos\theta_0} - \ln\left[\dfrac{\tan\theta_0 + \dfrac{1}{\cos\theta_0}}{\tan\theta_f + \dfrac{1}{\cos\theta_f}}\right] - \dfrac{2az_c}{v_c^2}\ln^2\left[\dfrac{\tan\theta_0 + \dfrac{1}{\cos\theta_0}}{\tan\theta_f + \dfrac{1}{\cos\theta_f}}\right] = 0 \end{cases}$$

The first equation gives θ_0 as a function of θ_f (equation of degree 2 in $\cos\theta_0$). There remains an equation in θ_f solved numerically. We deduce θ_0, then t_f by

$$t_f \ln\left[\dfrac{\tan\theta_0 + \dfrac{1}{\cos\theta_0}}{\tan\theta_f + \dfrac{1}{\cos\theta_f}}\right] = \dfrac{v_c}{a}(\tan\theta_0 - \tan\theta_f)$$

Numerical application

The final target conditions are: $\begin{cases} z_c = 200\,\text{km} \\ v_c = 7\,784\ \text{m}\cdot\text{s}^{-1} \end{cases}$ (circular orbit at 200 km)

The constant gravity is $g = 9\ \text{m}\cdot\text{s}^{-2}$.

The total impulse delivered during the trajectory is given by $\Delta V = at_f$.

The table opposite shows the solutions obtained for acceleration levels of 2g, 3g and 5g respectively.

Higher acceleration reduces the final time and decreases the total impulse required.

Acceleration	a = 2g	a = 3g	a = 5g
θ_0 (°)	45.64	43.90	49.24
θ_f (°)	11.46	−9.20	−31.75
t_f (s)	509.5	319.1	195.4
ΔV (m/s)	9 171	8 616	8 792

The corresponding trajectories are plotted in figure 4-13. The arrows indicate the thrust direction on the 3g trajectory.

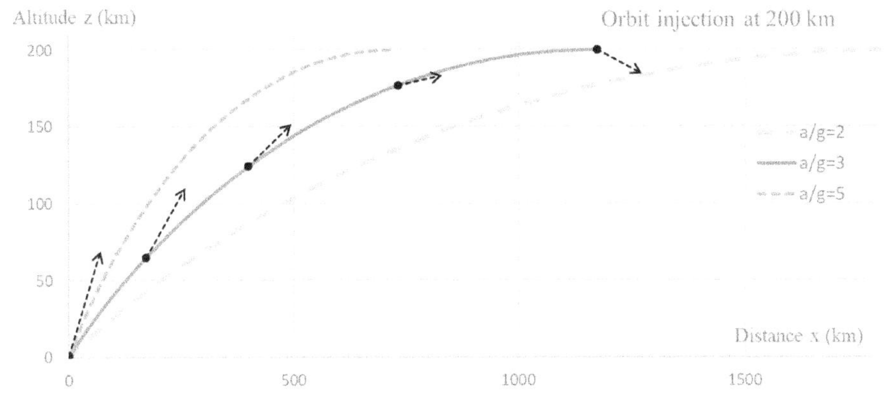

Figure 4-13: Optimal launch trajectories.

We then consider the determination of the acceleration level that minimizes the total impulse $\Delta V = at_f$. The problem is solved for acceleration levels between 1.5g and 6g and for orbits of altitude 200, 300 and 400 km.

The velocity in circular orbit at altitude z_c is given by: $v_c = \sqrt{g(R_T + z_c)}$, with the Earth radius $R_T = 6\,378$ km and the average gravity $g = 9\ \text{m}\cdot\text{s}^{-2}$.

The results plotted in figure 4-14 show that the optimum acceleration depends on the target altitude and that the impulse is quite sensitive to the level of acceleration, especially if it is low.

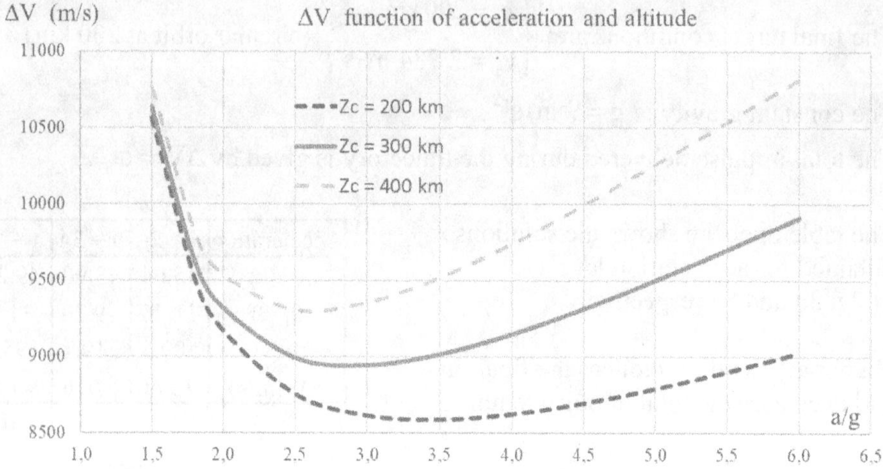

Figure 4-14: Required impulse as a function of altitude and acceleration.

Property 4-2 also applies for an integral constraint of the form

$$\int_{t_0}^{t_f} g(x,u,t)dt = c \tag{4.55}$$

For this purpose, we define an additional state variable x_{n+1} such that

$$\dot{x}_{n+1} = g(x,u,t) \quad \text{with} \quad \begin{cases} x_{n+1}(t_0) = 0 \\ x_{n+1}(t_f) = c \end{cases} \tag{4.56}$$

The Hamiltonian and the adjoint equations become

$$H(x_{1:n+1}, u, p_{1:n+1}, t) = L(x_{1:n}, u, t) + p_{1:n}^T f(x_{1:n}, u, t) + p_{n+1} g(x_{1:n}, u, t)$$

$$\dot{p}_{1:n+1} = -H_{x_{1:n+1}} \quad \Rightarrow \quad \begin{cases} \dot{p}_{1:n} = -H_{x_{1:n}} \\ \dot{p}_{n+1} = -H_{x_{n+1}} = 0 \quad \Rightarrow \quad p_{n+1} = C^{te} \end{cases} \tag{4.57}$$

The adjoint p_{n+1} constant is an unknown linked to the constraint $x_{n+1}(t_f) = c$. This unknown and the associated equation complete the two-point boundary value problem.

4.2.2 Interior constraints

Consider a system with a discontinuity at a time $t_1 \in [t_0 ; t_f]$.

A state discontinuity results in different states at t_1^- and t_1^+.

$$\begin{cases} x(t_1^-) = x_1^- \\ x(t_1^+) = x_1^+ \end{cases} \tag{4.58}$$

A discontinuity in dynamics results in a change in the state equation.

$$f(x,u,t) = \begin{cases} f^-(x,u,t) & \text{if} \quad t_0 \le t < t_1 \\ f^+(x,u,t) & \text{if} \quad t_1 \le t \le t_f \end{cases} \tag{4.59}$$

The integral cost functional may also differ on $[t_0 ; t_1[$ and $[t_1 ; t_f]$.

$$L(x,u,t) = \begin{cases} L^-(x,u,t) & \text{if} \quad t_0 \le t < t_1 \\ L^+(x,u,t) & \text{if} \quad t_1 \le t \le t_f \end{cases} \tag{4.60}$$

It is further assumed that q constraints are to be met at time t_1.

$$\psi(x_1^-, x_1^+, t_1) = 0 \tag{4.61}$$

The so-called **hybrid** control problem with these interior constraints (also called point or intermediate constraints) is formulated as

$$\min_{u, x_1^-, x_1^+, t_1} J = \int_{t_0}^{t_1^-} L^-(x,u,t)dt + \int_{t_1^+}^{t_f} L^+(x,u,t)dt + \varphi(x_1^-, x_1^+, t_1)$$
$$\text{with} \begin{cases} \dot{x} = f^-(x,u,t) & \text{if} \quad t_0 \le t < t_1 \\ \dot{x} = f^+(x,u,t) & \text{if} \quad t_1 \le t \le t_f \end{cases} \quad \text{and} \quad \psi(x_1^-, x_1^+, t_1) = 0 \tag{4.62}$$

The variables (x_1^-, x_1^+, t_1) are to be optimized in the same way as the control $u(t)$.

The derivatives of φ (or ψ) with respect to x_1^- or x_1^+ are noted: $\dfrac{\partial \varphi}{\partial x_1^\pm} \underset{\text{noted}}{=} \varphi_x(t_1^\pm)$.

The objective here is to write optimality conditions associated with (x_1^-, x_1^+, t_1).

For simplicity, it is assumed that there is only one interior point at time t_1 and it is not specified whether the terminal conditions are constrained or not. These terminal conditions do not appear in either the cost or the constraints of (4.62).

The generalization to several interior points and terminal conditions is discussed on page 314.

We define the Hamiltonians H^- on $[t_0;t_1[$ and H^+ on $[t_1;t_f]$.

$$H(x,u,p,t)=\begin{cases} H^-(x,u,p,t)=L^-(x,u,t)+p^Tf^-(x,u,t) & \text{if } t_0 \le t < t_1 \\ H^+(x,u,p,t)=L^+(x,u,t)+p^Tf^+(x,u,t) & \text{if } t_1 \le t \le t_f \end{cases} \quad (4.63)$$

The adjoint equations and the optimal control are defined on each interval with the corresponding Hamiltonian.

$$\begin{cases} \dot{p}=-\dfrac{\partial H^-}{\partial x} \;, \; \min\limits_u H^-(x,u,p,t) \;\to\; u(t) \;\; \text{if } t_0 \le t < t_1 \\ \dot{p}=-\dfrac{\partial H^+}{\partial x} \;, \; \min\limits_u H^+(x,u,p,t) \;\to\; u(t) \;\; \text{if } t_1 \le t \le t_f \end{cases} \quad (4.64)$$

We then have the following necessary conditions of optimality.

Property 4-3: Necessary conditions for optimality with interior constraints

Suppose that the control $u(t)$ defined by (4.64) is a solution of (4.62).
Then there are q multipliers λ associated with the q constraints $\psi=0$ such that

- the interior constraints are satisfied

$$\psi(x_1^-,x_1^+,t_1)=0 \quad (4.65)$$

- the adjoints at t_1^- and t_1^+ satisfy

$$\begin{cases} p(t_1^-)= \;\; \varphi_x(t_1^-)+\psi_x(t_1^-)\lambda \\ p(t_1^+)=-\varphi_x(t_1^+)-\psi_x(t_1^+)\lambda \end{cases} \quad (4.66)$$

- the Hamiltonians at t_1^- and t_1^+ satisfy

$$H^+(t_1^+)-H^-(t_1^-)=\varphi_t(t_1)+\psi_t(t_1)\lambda \quad (4.67)$$

These conditions form a two-point boundary value problem:
- the q equations (4.65) are linked to the unknowns λ;
- the n transversality conditions on $p(t_1^-)$ are linked to the unknowns $x(t_1^-)$;
- the n transversality conditions on $p(t_1^+)$ are linked to the unknowns $x(t_1^+)$;
- the transversality condition on $H(t_1^\pm)$ is linked to the unknown t_1.

Case of a continuous state

If the state is continuous: $x(t_1^+) = x(t_1^-) = x(t_1)$, condition (4.66) is replaced by

$$p(t_1^+) - p(t_1^-) = -\varphi_x(t_1) - \psi_x(t_1)\lambda \tag{4.68}$$

These n transversality conditions on $p(t_1^\pm)$ are linked to the n unknowns $x(t_1)$. The adjoint is discontinuous due to the constraint, although the state is continuous.

Demonstration

We use the variational method by adding to the augmented cost (4.13) the q constraints ψ with multipliers λ. Integrating by parts over each interval $[t_0; t_1^-]$ and $[t_1^+; t_f]$, the formula (4.16) becomes

$$J_a = \int_{t_0}^{t_1^-} \left[H^-(x,u,p,t) + \dot{p}^T x\right] dt + p(t_0)^T x(t_0) - p(t_1^-)^T x(t_1^-) + \lambda^T \psi(x_1^-, x_1^+, t_1)$$

$$+ \int_{t_1^+}^{t_f} \left[H^+(x,u,p,t) + \dot{p}^T x\right] dt + p(t_1^+)^T x(t_1^+) - p(t_f)^T x(t_f) + \varphi(x_1^-, x_1^+, t_1)$$

Applying variations $\delta u(t), \delta x(t), \delta x_1^-, \delta x_1^+, \delta t_1$ to the augmented cost, we obtain the equivalent of formula (4.17) on each interval $[t_0; t_1^-]$ and $[t_1^+; t_f]$.

On $[t_0; t_1^-]$, the changes in t_1^- act like final changes $(\delta x_f, \delta t_f)$.

On $[t_1^+; t_f]$, the changes in t_1^+ act like initial changes $(\delta x_0, \delta t_0)$.

We add the variations coming from $\lambda^T \psi(x_1^-, x_1^+, t_1)$ and $\varphi(x_1^-, x_1^+, t_1)$ to get

$$\delta J_a = \int_{t_0}^{t_1^-} H_u^{-T} \delta u dt + \int_{t_0}^{t_1^-} (H_x^- + \dot{p})^T \delta x dt + \int_{t_1^+}^{t_f} H_u^{+T} \delta u dt + \int_{t_1^+}^{t_f} (H_x^+ + \dot{p})^T \delta x dt$$

$$+ \left[\varphi_x(t_1^-) + \psi_x(t_1^-)\lambda - p(t_1^-)\right]^T \delta x_1^- + \left[\psi_{t_1}\lambda + H(t_1^-)\right] \delta t_1$$

$$+ \left[\varphi_x(t_1^+) + \psi_x(t_1^+)\lambda + p(t_1^+)\right]^T \delta x_1^+ + \left[\varphi_{t_1} - H(t_1^+)\right] \delta t_1$$

The first-order optimality condition is: $\delta J_a = 0$, $\forall \delta u(t), \delta x(t), \delta x_1^-, \delta x_1^+, \delta t_1$.

As the variations $\delta u(t), \delta x(t), \delta x_1^-, \delta x_1^+, \delta t_1$ are independent, their coefficients must be zero.

This leads to (4.64) for eliminating δx and to $H_u = 0$ for the term in δu.

Conditions (4.66) and (4.67) eliminate the terms in $(\delta x_1^-; \delta x_1^+)$ and δt_1.

Case of a continuous state

If the state is continuous in t_1, the state variations are equal: $\delta x_1^- = \delta x_1^+ = \delta x_1$.

The term in δx_1 comes from integration by parts, from $\lambda^T \psi(x_1, t_1)$ and $\varphi(x_1, t_1)$.

Its expression: $\left[p(t_1^+) - p(t_1^-) + \varphi_{x_1} + \psi_{x_1}\lambda\right]^T \delta x_1$ leads to the condition (4.68).

The conditions (4.66-67) are called **jump conditions**, because they generate discontinuities on the adjoint and Hamiltonian, and consequently on the control.
The generalization to several interior points in $t_1, t_2, \cdots, t_k$ is immediate:

- the constraints are of the form: $\psi(x_1^-, x_1^+, t_1, x_2^-, x_2^+, t_2, \cdots, x_k^-, x_k^+, t_k) = 0$;
- the final cost is of the form: $\varphi(x_1^-, x_1^+, t_1, x_2^-, x_2^+, t_2, \cdots, x_k^-, x_k^+, t_k)$;
- the integral cost is decomposed into k intervals with functions L_k and f_k.

At time t_k, the conditions (4.65-67) are associated with the unknowns (x_k^-, x_k^+, t_k).
Conditions (4.66) and (4.67) also allow us to find the transversality conditions (4.53) and (4.54) associated with the terminal constraints.

At the initial time, by posing: $\begin{cases} t_1^+ \to t_0 \\ x_1^+ \to x_0 \end{cases}$, we have: $\begin{cases} p(t_0) = -\varphi_x(t_0) - \psi_x(t_0)\lambda \\ H(t_0) = \varphi_t(t_0) + \psi_t(t_0)\lambda \end{cases}$.

At the final time, by posing: $\begin{cases} t_1^- \to t_f \\ x_1^- \to x_f \end{cases}$, we have: $\begin{cases} p(t_f) = \varphi_x(t_f) + \psi_x(t_f)\lambda \\ H(t_f) = -\varphi_t(t_f) - \psi_t(t_f)\lambda \end{cases}$.

Example 4-6 shows the effect of jump conditions in simple cases.

Example 4-6: Jump conditions

Assume that the final cost φ does not depend explicitly on the conditions at t_1.

1st case: assume that the state has a fixed discontinuity at a free time t_1.

The interior constraint is of the form: $\psi(x_1^-, x_1^+, t_1) = x_1^+ - x_1^- - c = 0$.

Conditions (4.66-67) yield: $\begin{cases} H(t_1^+) - H(t_1^-) = \psi_t(t_1)\lambda = 0 \\ p(t_1^+) = -\psi_x(t_1^+)\lambda = -\lambda \\ p(t_1^-) = \psi_x(t_1^-)\lambda = -\lambda \end{cases} \Rightarrow \begin{cases} H(t_1^+) = H(t_1^-) \\ p(t_1^+) = p(t_1^-) \end{cases}$

The Hamiltonian and the adjoint are continuous at t_1.
The constraint does not generate any jump.

2nd case: assume that the state has a fixed discontinuity at $t_1 = \tau$.

The two interior constraints are of the form: $\begin{cases} \psi_1(x_1^-, x_1^+, t_1) = x_1^+ - x_1^- - c = 0 \\ \psi_2(x_1^-, x_1^+, t_1) = t_1 - \tau = 0 \end{cases}$.

Conditions (4.66-67) yield

$$\begin{cases} H(t_1^+) - H(t_1^-) = \psi_{1,t}(t_1)\lambda_1 + \psi_{2,t}(t_1)\lambda_2 = \lambda_2 \\ p(t_1^+) = -\psi_{1,x}(t_1^+)\lambda_1 - \psi_{2,x}(t_1^+)\lambda_2 = -\lambda_1 \\ p(t_1^-) = \psi_{1,x}(t_1^-)\lambda_1 + \psi_{2,x}(t_1^-)\lambda_2 = -\lambda_1 \end{cases} \Rightarrow \begin{cases} H(t_1^+) = H(t_1^-) + \lambda_2 \\ p(t_1^+) = p(t_1^-) \end{cases}$$

The adjoint is continuous, but the Hamiltonian is discontinuous.

3rd case: assume that the state is continuous and that the constraint does not depend explicitly on t_1.

The interior constraint is of the form: $\psi(x_1) = 0$ and the time t_1 is free.

Conditions (4.66-67) yield

$$\begin{cases} H(t_1^+) - H(t_1^-) = \psi_t(t_1)\lambda = 0 \\ p(t_1^+) - p(t_1^-) = -\psi_x(t_1)\lambda \end{cases} \Rightarrow \begin{cases} H(t_1^+) = H(t_1^-) \\ p(t_1^+) = p(t_1^-) - \psi_x(t_1)\lambda \end{cases}$$

The Hamiltonian is continuous, but the adjoint is discontinuous.

4.2.3 Path constraints

Consider a control problem with equality and inequality constraints.

$$\min_u J = \int_{t_0}^{t_f} L(x,u,t)dt + \varphi(x_f) \quad \text{with} \quad \dot{x} = f(x,u,t)$$
$$\text{and} \begin{cases} \psi_E(x,u,t) = 0 \\ \psi_I(x,u,t) \le 0 \end{cases}, \quad \forall t \tag{4.69}$$

The q_E equality constraints ψ_E and the q_I inequality constraints ψ_I are called path constraints. They must be satisfied at all points along the path.

The principle of the minimum (theorem 4-1) then applies in the form

$$\min_u H(x,u,p,t) \quad \text{s.t.} \begin{cases} \psi_E(x,u,t) = 0 \\ \psi_I(x,u,t) \le 0 \end{cases} \rightarrow u^*(t) \tag{4.70}$$

The Hamiltonian is minimized under the constraints $\psi_E = 0$ and $\psi_I \le 0$.

This constrained optimization problem has to be solved at each time.
By introducing Lagrange multipliers λ_E multipliers and λ_I associated with the constraints ψ_E and ψ_I respectively, we form the Lagrangian of problem (4.70). This Lagrangian is called here the **augmented Hamiltonian** and it is noted H_a in order not to confuse it with the integral cost function $L(x,u,t)$.

$$H_a(x,u,p,t,\lambda_E,\lambda_I) = H(x,u,p,t) + \lambda_E^T \psi_E(x,u,t) + \lambda_I^T \psi_I(x,u,t) \tag{4.71}$$

The first-order KKT conditions of problem (4.70) are

$$\frac{\partial H_a}{\partial u} = 0 \quad \text{and} \quad \begin{cases} \psi_E(x,u,t) = 0 \\ \psi_I(x,u,t) \le 0 \\ \lambda_I \ge 0 \;\text{ if }\; \psi_I = 0 \\ \lambda_I = 0 \;\text{ if }\; \psi_I < 0 \end{cases} \tag{4.72}$$

An inactive inequality constraint at time t has a zero multiplier.
The system (4.72) is to be solved at each time to find the control $u^*(t)$:

- the m components of u^* are linked to the m equations $\dfrac{\partial H_a}{\partial u} = 0$;
- the q_E multipliers λ_E are linked to the q_E equations $\psi_E(x,u,t) = 0$;
- the q_I multipliers λ_I are zero or linked to the q_I equations $\psi_I(x,u,t) = 0$.

Example 4-7 is a double integrator with bounded control.

Example 4-7: Double integrator in dimension 1

We consider a material point moving on a straight line. The state consists of the position r and the velocity v. The control is the acceleration u between −1 and +1.

The state equations are: $\begin{cases} \dot{r} = v \\ \dot{v} = u \end{cases}$.

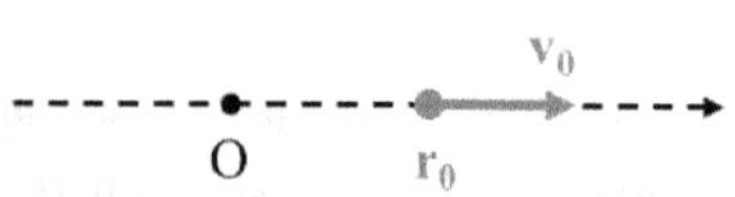

<u>Control problem</u>
The objective is to return and stop in O in the minimum time starting from given initial conditions (r_0, v_0). The minimum time control problem is formulated as

$$\min_{u,t_f} J = \int_{t_0}^{t_f} 1\,dt \quad \text{with} \quad \begin{cases} \dot{r} = v \\ \dot{v} = u \end{cases}, \quad \begin{cases} r(t_0) = r_0 \\ v(t_0) = v_0 \end{cases}, \quad \begin{cases} r(t_f) = 0 \\ v(t_f) = 0 \end{cases}, \quad -1 \le u \le +1$$

Application of the PMP

The Hamiltonian is: $H = 1 + p_r v + p_v u$ with the adjoint of components (p_r, p_v).

Applying the PMP gives: $\min_u H$ s.t. $-1 \le u \le +1 \;\rightarrow\; u = \begin{cases} -1 & \text{if } p_v < 0 \\ +1 & \text{if } p_v > 0 \end{cases}$

This control is of bang type $(u = \pm 1)$ with a switching (or commutation) function defined by the adjoint p_v.

The adjoint equations are: $\begin{cases} \dot{p}_r = -H_r = 0 & \rightarrow p_r = b \\ \dot{p}_v = -H_v = -p_r & \rightarrow p_v = b(t_f - t) + a \end{cases}$, a ,b constants

Control and trajectory determination

The adjoint p_v is a linear function of the time. Its sign changes at most once.
The path consists of at most two arcs controlled with $u = \pm 1$.
Let us integrate the equations for each arc with their respective controls:

- the first arc starts from the initial state (t_0, r_0, v_0) with the control $u_0 = \pm 1$.

$$\begin{cases} \dot{v} = u_0 \\ \dot{r} = v \end{cases} \Rightarrow \begin{cases} v = v_0 + u_0(t - t_0) \\ r = r_0 + v_0(t - t_0) + u_0(t - t_0)^2 / 2 \end{cases} \Rightarrow v^2 - 2u_0 r = v_0^2 - 2u_0 r_0$$

- the second arc arrives at the final state (t_f, r_f, v_f) with the control $u_f = \pm 1$.

$$\begin{cases} \dot{v} = u_f \\ \dot{r} = v \end{cases} \Rightarrow \begin{cases} v = v_f + u_f(t - t_f) \\ r = r_f + v_f(t - t_f) + u_f(t - t_f)^2 / 2 \end{cases} \Rightarrow v^2 - 2u_f r = v_f^2 - 2u_f r_f = 0$$

These arcs are represented by parabola arcs in the phase plane (r, v).
If $u_f = -1$, the final arc is on the parabola $\Gamma^- : v^2 = -2r$ (quadrant $r \le 0, v \ge 0$).
If $u_f = +1$, the final arc is on the parabola $\Gamma^+ : v^2 = +2r$ (quadrant $r \ge 0, v \le 0$).
The Γ curve formed by the half parabolas Γ^- and Γ^+ is the switching curve.

This curve Γ of equation: $v^2 \operatorname{sgn}(v) = -2r$ is drawn in solid lines in figure 4-15.
It separates the plane into two regions.

The control depends on the current position (r, v) in relation to the curve Γ:

- if the point is above Γ, the control is $u = -1$.
  This control decreases the velocity and the point joins the curve Γ^+;
- if the point is below Γ, the control is $u = +1$.
  This control increases the velocity and the point joins the curve Γ^-.

Two trajectories are drawn in dotted lines from each of the two regions.

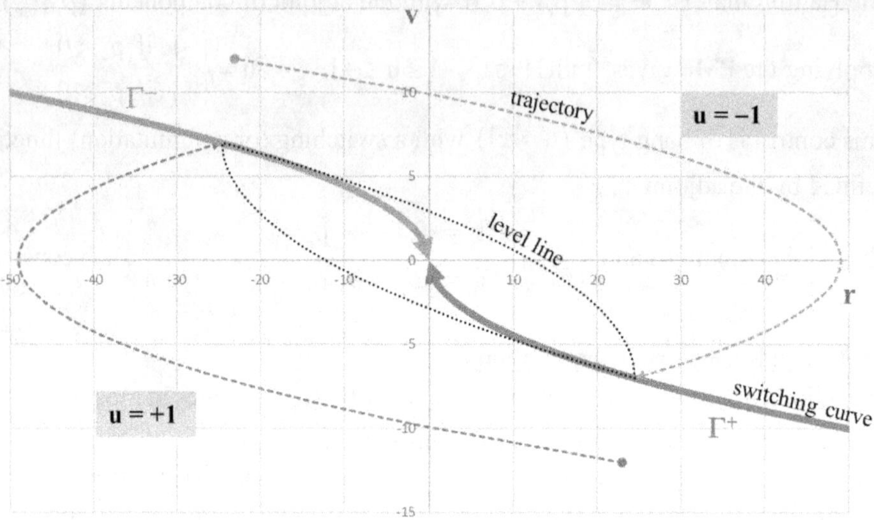

Figure 4-15: Double integrator in the phase plane.

Let us calculate the final time as a function of the initial conditions (r_0, v_0).
The initial and final arc intersect at (t_c, r_c, v_c). These coordinates are solution of

$$\begin{cases} v_c^2 - 2u_0 r_c = v_0^2 - 2u_0 r_0 & \text{(initial: } u = u_0) \\ v_c^2 + 2u_0 r_c = 0 & \text{(final: } u = u_f = -u_0 \text{ and } r_f = v_f = 0) \end{cases} \Rightarrow v_c^2 = \frac{1}{2}v_0^2 - u_0 r_0$$

The time spent on the initial arc is obtained by

$$v_c = v_0 + u_0(t_c - t_0) \Rightarrow t_c - t_0 = \frac{v_c - v_0}{u_0}$$

The time spent on the final arc is obtained by

$$v_f = v_c + u_f(t_f - t_c) \Rightarrow t_f - t_c = \frac{v_f - v_c}{u_f} = \frac{v_c}{u_0}$$

We deduce the total time: $\Delta t = t_f - t_0 = \dfrac{2v_c - v_0}{u_0} \Rightarrow 2v_c = v_0 + u_0 \Delta t$.

Replacing v_c and (r_0, v_0), we have: $2\Delta t^2 - 4u_0 r_0 + (v_0 - u_0 \Delta t)^2 = 0$.
This equation gives the total time as a function of (r_0, v_0). The initial control u_0 is determined by the sign of $v_0^2 \operatorname{sgn}(v_0) + 2r_0$ (position relative to Γ).
Figure 4-15 shows the level line of the final time (time to return to the origin) associated with the switching points of the two trajectories already drawn.

If a path constraint is not explicit in u, the approach (4.70) does not apply, as the minimization of the Hamiltonian would not take the constraints into account. These non-explicit constraints in u called state constraints are of the form

$$\begin{cases} \psi_E(x,t)=0 \\ \psi_I(x,t)\le 0 \end{cases} \tag{4.73}$$

Let us consider successively the case of an equality constraint, then an inequality.

Equality state constraint: $\psi(x,t)=0$

The constraint $\psi(x,t)=0$ has to be satisfied at each time of the trajectory.
Since the function $\psi(x,t)$ is constant, its derivative with respect to time is zero.
Using the state equation: $\dot{x}=f(x,u,t)$, this derivative is expressed as

$$\frac{d\psi}{dt}=\left(\frac{\partial\psi}{\partial x}\right)^T\frac{dx}{dt}+\frac{\partial\psi}{\partial t}=\psi_x^T(x,t)f(x,u,t)+\psi_t(x,t) \tag{4.74}$$

The function $\psi(x,t)$ is zero if and only if its derivative is zero and its initial or final value is zero. We can therefore replace the constraint $\psi(x,t)=0$ by

$$\begin{cases} \psi_x^T(x,t)f(x,u,t)+\psi_t(x,t)=0 \ , \quad \forall t \\ \psi(x_0,t_0)=0 \quad \text{or} \quad \psi(x_f,t_f)=0 \end{cases} \tag{4.75}$$

If the derivative is explicit in u, the approach (4.70) can be applied by replacing the constraint $\psi(x,t)=0$ by its derivative (4.74) and adding a constraint on the initial or final value of ψ. This initial or final constraint is taken into account in the transversality condition on $p(t_0)$ or $p(t_f)$ with a multiplier λ.

$$\begin{aligned} &\psi(x_0,t_0)=0 \quad \to \ p(t_0)=-\varphi_x(t_0)-\psi_x(t_0)\lambda \\ \text{or} \ &\psi(x_f,t_f)=0 \quad \to \ p(t_f)=\ \ \varphi_x(t_f)+\psi_x(t_f)\lambda \end{aligned} \tag{4.76}$$

If the derivative of ψ is not explicit in u, we continue to derive until we obtain an explicit function in u. The constraint is said of order k if it is necessary to derive k times the function ψ to obtain an explicit expression in u.
The constraint is then replaced by

$$\psi(x,t)=0 \ , \quad \forall t \quad \Leftrightarrow \quad \begin{cases} \psi^{(k)}(x,t)=0 \ , \quad \forall t \\ \psi^{(j)}(x_0,t_0)=0 \quad \text{for } 0\le j\le k-1 \end{cases} \tag{4.77}$$

with the notation $\psi^{(k)} \underset{\text{def}}{=} \dfrac{d^k\psi}{dt^k}$.

The initial constraints are then on all derivatives from order 0 to $k-1$.

These initial constraints are taken into account in the transversality conditions with their respective multipliers.

<u>Inequality state constraint</u>: $\psi(x,t) \le 0$
An inequality constraint can be active $(\psi = 0)$ or inactive $(\psi < 0)$ on successive arcs of the path. Suppose that the constraint becomes active at t_1.
On the part $t < t_1$, the constraint is not taken into account in (4.70).
On the part $t \ge t_1$, the constraint is treated as an equality.
Deriving k times to make the control u appear, we replace the constraint with the conditions (4.77) applied from t_1.

$$\psi(x,t) = 0 \;,\; \forall t \ge t_1 \quad \Leftrightarrow \quad \begin{cases} \psi^{(k)}(x,t) = 0 \;,\; \forall t \ge t_1 \\ \psi^{(j)}(x_1,t_1) = 0 \quad \text{for } 0 \le j < k \end{cases} \tag{4.78}$$

The k constraints at time t_1 are interior constraints.
They are taken into account in transversality conditions of the form (4.66) and (4.67) with their respective multipliers.

The change of a constraint from active to inactive is handled in the same way.
Connections between saturated $(\psi = 0)$ and unsaturated $(\psi < 0)$ arcs are called **junctions**. An important difficulty is that the optimal number of junctions is not known in advance (except in simple cases). An arbitrary assumption on this number is usually required to introduce the corresponding times $t_1, t_2, \cdots$ in the formulation of the two-point boundary value problem.
Example 4-8 shows a problem with an inequality state constraint leading to junctions between saturated and unsaturated arcs.

Example 4-8: Minimum distance with junction

We are looking for the shortest path between two given points in the plane without entering the circle of center O and radius a as shown in figure 4-16.
We pass to an equivalent minimum time problem by assuming a constant velocity of direction $\vec{u}$. The state is the position $\vec{r}$ and the state equation is: $\dot{\vec{r}} = \vec{u}$.
The control is the unit vector $\vec{u}$: $\|\vec{u}\| = 1$.

The state constraint (not to enter the circle) is formulated as: $\psi(\vec{r})=a^2-r^2\leq 0$.

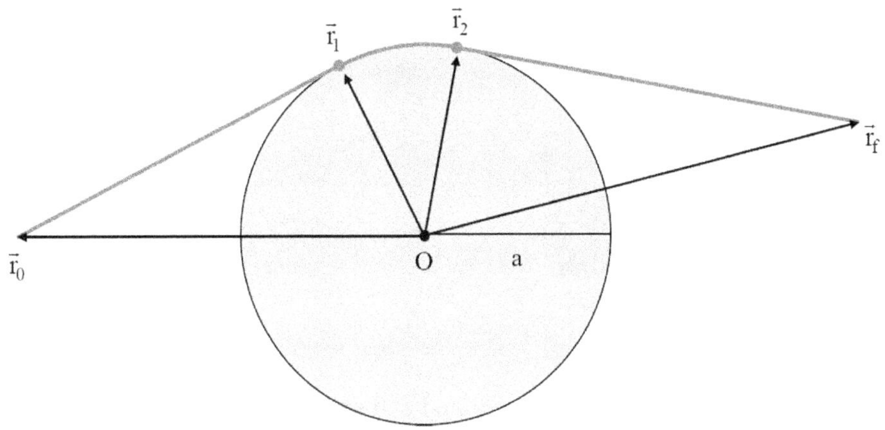

Figure 4-16: Minimum distance with junction.

Control problem

The minimum time control problem is formulated as

$$\min_{\vec{u},t_f} J=\int_{t_0}^{t_f} 1\,dt \quad \text{with} \quad \dot{\vec{r}}=\vec{u} \;,\; \begin{cases}\vec{r}(t_0)=\vec{r}_0\\ \vec{r}(t_f)=\vec{r}_f\end{cases} ,\; \begin{cases}\psi(\vec{r})=a^2-r^2\leq 0\\ \|\vec{u}\|=1\end{cases}$$

Application of the PMP

The Hamiltonian is: $H=1+\vec{p}\cdot\vec{u}$ with the adjoint $\vec{p}$.

The application of the PMP gives: $\min_{\vec{u}} H \;\text{ s.t. }\; \begin{cases}\|\vec{u}\|=1\\ \psi(\vec{r})=a^2-r^2\leq 0\end{cases}$.

We determine the order on the unsaturated $(\psi<0)$ and saturated arcs $(\psi=0)$, then we look for the junction conditions.

Unsaturated arc

The control is given by: $\min_{\vec{u}} H=1+\vec{p}\cdot\vec{u} \;\text{ s.t. }\; \|\vec{u}\|=1 \;\rightarrow\; \vec{u}=-\dfrac{\vec{p}}{\|\vec{p}\|}$

The adjoint equation is: $\dot{\vec{p}}=-H_{\vec{r}}=\vec{0} \;\Rightarrow\; \vec{p}$ constant.

The control $\vec{u}$ (direction of velocity) is constant and the trajectory is a straight line segment, which indeed minimizes the distance in the absence of constraints.

Saturated arc

The state constraint is derived until the control appears.

$\psi(\vec{r}) = a^2 - r^2 \;\Rightarrow\; \dot{\psi}(\vec{r}) = -2\vec{r}\cdot\dot{\vec{r}} = -2\vec{r}\cdot\vec{u}$

The constraint ψ is of order 1. It is replaced by its derivative.

$$\psi(\vec{r}) = 0 \;\Leftrightarrow\; \begin{cases} \dot{\psi}(\vec{r}) = 0 \\ \psi(\vec{r}_1) = 0 \end{cases}$$

We deduce the control on the saturated arc: $\dot{\psi}(\vec{r}) = -2\vec{r}\cdot\vec{u} = 0 \;\Rightarrow\; \vec{u} \perp \vec{r}$

The saturated arc follows the circle with the control $\vec{u}$ (velocity) constantly perpendicular to the circle radius.

Junction conditions

The junction conditions between unsaturated and saturated arc at a time t_1 are

$$\begin{cases} p(t_1^+) - p(t_1^-) = -\psi_{\vec{r}}(t_1)\lambda \\ H(t_1^+) - H(t_1^-) = \psi_t(t_1)\lambda \end{cases} \quad \text{with } \psi(\vec{r}) = a^2 - r^2 \;\Rightarrow\; \begin{cases} \psi_{\vec{r}} = -2\vec{r} \\ \psi_t = 0 \end{cases}$$

from which we can deduce: $\begin{cases} \vec{p}^+ = \vec{p}^- + 2\lambda\vec{r} \\ H^+ = H^- \end{cases}$.

Let us express the Hamiltonian before and after the junction.

$$\begin{cases} H^- = 1 + \vec{p}^-\cdot\vec{u}^- & \text{with } \vec{u}^- = -\dfrac{\vec{p}^-}{\|\vec{p}^-\|} \quad \text{(segment)} \\ H^+ = 1 + \vec{p}^+\cdot\vec{u}^+ & \text{with } \vec{u}^+ \perp \vec{r} \quad \text{(arc of circle)} \end{cases}$$

These expressions of H^- and H^+ are replaced in the junction conditions.

$$\begin{cases} \vec{p}^+ = \vec{p}^- + 2\lambda\vec{r} \\ \vec{p}^+\cdot\vec{u}^+ = \vec{p}^-\cdot\vec{u}^- \end{cases} \Rightarrow (\vec{p}^- + 2\lambda\vec{r})\cdot\vec{u}^+ = \vec{p}^-\cdot\vec{u}^- \Rightarrow \vec{u}^-\cdot\vec{u}^+ = 1 \quad \text{since} \begin{cases} \vec{p}^- = -p^-\vec{u}^- \\ \vec{u}^+\cdot\vec{r} = 0 \end{cases}$$

from which we deduce: $\vec{u}^+ = \vec{u}^-$ (unit vectors).

The junction is tangential to the circle.

This is the expected result for minimizing the distance around the circle.

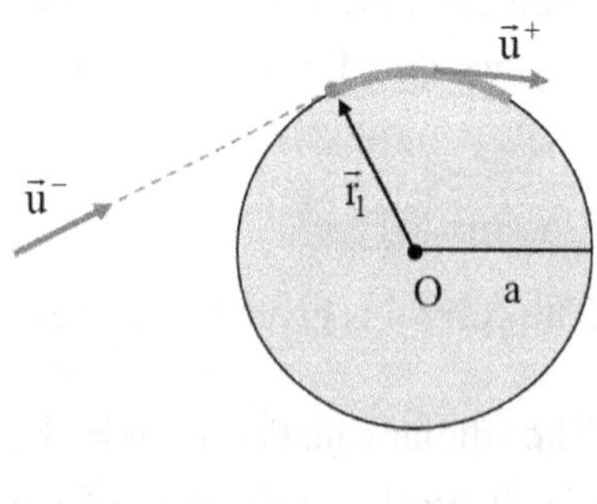

Note

The control is here continuous at the junction (which is not the case in general).

If we repeat the calculation assuming a different velocity v^- outside the circle and v^+ on the circle, we obtain the relation: $\vec{u}^- \cdot \vec{u}^+ = \frac{v^-}{v^+}$.

This is Descartes' law of refraction which reflects a change in direction when the medium changes index (different speed of propagation).

Control discontinuities

A discontinuity in the control is called a **corner** or edge.

Discontinuities are generally associated with:

- an interior point modifying the definition of the Hamiltonian as in (4.63);
- an interior constraint generating an adjoint jump as in (4.66) or (4.68);
- a state constraint passing from active to inactive or vice versa.

The control obtained by global minimization of the Hamiltonian (from the PMP) can be discontinuous even in the absence of interior or state constraints.

Figure 4-17 shows a Hamiltonian with two local minima as a function of u. The function H(u) is plotted at times t^-, t and t^+. The minima evolve with time. The global minimum changes from u^- to u^+ generating a control discontinuity at t.

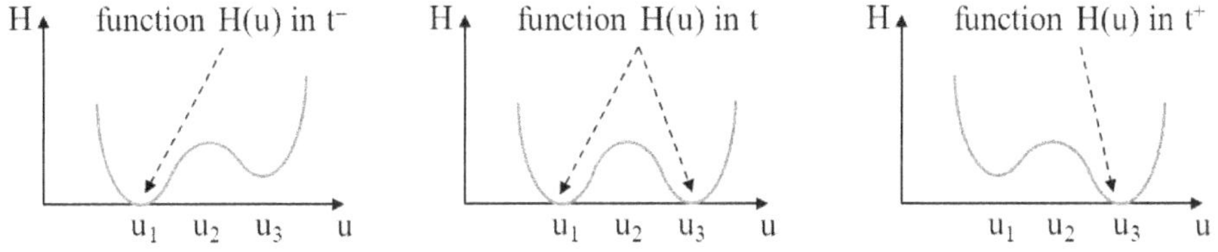

Figure 4-17: Jump of the global minimum.

4.2.4 Quadratic-linear problem

The standard formulation of a quadratic-linear (QL) control problem is

$$\min_u J = \frac{1}{2}(x^T S x)_{t_f} + \frac{1}{2}\int_{t_0}^{t_f}(x^T Q x + u^T R u)dt \quad \text{with} \quad \begin{cases} \dot{x} = Fx + Gu \\ x(t_0) = x_0 \end{cases} \tag{4.79}$$

The initial time t_0, the final time t_f and the initial state $x(t_0)$ are fixed.

The objective is to come close a zero final state $x(t_f)=0$, while limiting the state values $x(t)$ and control $u(t)$ along the trajectory. This justifies the choice of a quadratic cost composed of an integral term with weighting matrices Q and R on the state and control, and of a final term with a weighting matrix S_f on the final state. The assumptions on these matrices are as follows:

- the matrices F, G, Q and R are functions of time → F(t), G(t), Q(t), R(t);
- the matrices Q and S_f are positive semidefinite (they can be set to zero);
- the matrix R is positive definite.

The Q, R and S_f matrices are set according to the desired shape of the solution. Suppose we want to limit the state and control components by

$$\begin{cases} |x_i(t_f)| \le x_{i,max}(t_f) & , \ i=1 \text{ to } n \\ |x_i(t)| \le x_{i,max}(t) & , \ i=1 \text{ to } n \\ |u_j(t)| \le u_{j,max}(t) & , \ j=1 \text{ to } m \end{cases} \tag{4.80}$$

Each state and control component must remain below a threshold which can be variable along the trajectory. We can choose for example diagonal matrices: $S_f = \text{diag}(s_i)$, $Q = \text{diag}(q_i)$, $R = \text{diag}(r_i)$ with coefficients

$$s_i = \frac{1}{x_{imax}(t_f)^2} \ , \ q_i = \frac{1}{(t_f - t_0)x_{imax}(t)^2} \ , \ r_i = \frac{1}{(t_f - t_0)u_{imax}(t)^2} \tag{4.81}$$

These coefficients are to be set by trials and observing the effect on the solution.

Let us apply the minimum principle to the QL problem (4.79).
The Hamiltonian and the adjoint are given by

$$H = \frac{1}{2}x^T Q x + \frac{1}{2}u^T R u + p^T(Fx + Gu) \quad \text{with} \quad \begin{cases} \dot{p} = -Qx - F^T p \\ p(t_f) = S_f x(t_f) \end{cases} \tag{4.82}$$

The optimal control minimizes the Hamiltonian.

$$\min_u H \ \Rightarrow \ Ru + G^T p = 0 \ \Rightarrow \ u = -R^{-1}G^T p \tag{4.83}$$

The matrix R is assumed to be positive definite and therefore invertible.
By replacing u in the state equation, we form a two-point boundary value problem.

$$\begin{pmatrix} \dot{x} \\ \dot{p} \end{pmatrix} = \begin{pmatrix} F & -GR^{-1}G^T \\ -Q & -F^T \end{pmatrix} \begin{pmatrix} x \\ p \end{pmatrix} \quad \text{with} \quad \begin{cases} x(t_0) = x_0 \\ p(t_f) = S_f x(t_f) \end{cases} \tag{4.84}$$

Example 4-9 shows the application to an interception problem.

Example 4-9: Interception

Consider a mobile with position $\vec{r}$, velocity $\vec{v}$, acceleration $\vec{u}$: $\begin{cases} \dot{\vec{r}} = \vec{v} \\ \dot{\vec{v}} = \vec{u} \end{cases}$.

The initial time t_0, the final time t_f and the initial state $(\vec{r}_0, \vec{v}_0)$ are given.
The control is the acceleration. The objective is to approach a zero final position and velocity $(\vec{r}_f = \vec{0}, \vec{v}_f = \vec{0})$, while minimizing the energy consumed.

This energy is measured by the integral: $\int_{t_0}^{t_f} u^2 dt$ (L_2 norm of the acceleration).

QL control problem

The QL control problem is formulated as

$$\min_u J = \frac{1}{2} K_r r(t_f)^2 + \frac{1}{2} K_v v(t_f)^2 + \frac{1}{2} \int_{t_0}^{t_f} u^2 dt$$

The positive reals K_r and K_v are the weights on the final position and velocity.

Application of the PMP

The Hamiltonian is: $H = \frac{1}{2} u^2 + \vec{p}_r \cdot \vec{v} + \vec{p}_v \cdot \vec{u}$ with the adjoint $(\vec{p}_r, \vec{p}_v)$.

The adjoint equations are

$$\begin{cases} \dot{\vec{p}}_r = \vec{0} \\ \dot{\vec{p}}_v = -\vec{p}_r \end{cases} \text{with} \begin{cases} \vec{p}_r(t_f) = K_r \vec{r}(t_f) \\ \vec{p}_v(t_f) = K_v \vec{v}(t_f) \end{cases} \rightarrow \begin{cases} \vec{p}_r = K_r \vec{r}(t_f) = C^{te} \\ \vec{p}_v = (t_f - t) K_r \vec{r}(t_f) + K_v \vec{v}(t_f) \end{cases}$$

Applying the PMP gives: $\min_{\vec{u}} H \Rightarrow \vec{u} = -\vec{p}_v = (t - t_f) K_r \vec{r}(t_f) - K_v \vec{v}(t_f)$

The state equations are integrated with this control.

$$\begin{cases} \dot{\vec{v}} = \vec{u} \rightarrow \vec{v}(t) = \frac{1}{2}(t - t_f)^2 K_r \vec{r}(t_f) - (t - t_f) K_v \vec{v}(t_f) + \vec{v}(t_f) \\ \dot{\vec{r}} = \vec{v} \rightarrow \vec{r}(t) = \frac{1}{6}(t - t_f)^3 K_r \vec{r}(t_f) - \frac{1}{2}(t - t_f)^2 K_v \vec{v}(t_f) + (t - t_f) \vec{v}(t_f) + \vec{r}(t_f) \end{cases}$$

By expressing $\vec{r}(t_f), \vec{v}(t_f)$ in terms of $\vec{r}(t), \vec{v}(t)$ and $\tau = t_f - t$ (time to go)

$$\begin{cases} D\vec{r}(t_f) = (\tau + \frac{1}{2}\tau^2 K_v)\vec{v} + (1 + \tau K_v)\vec{r} \\ D\vec{v}(t_f) = (1 - \frac{1}{6}\tau^3 K_r)\vec{v} - \frac{1}{2}\tau^2 K_r \vec{r} \end{cases} \text{with } D = 1 + \tau K_v + \frac{1}{3}\tau^3 K_r + \frac{1}{12}\tau^4 K_r K_v$$

we obtain an expression for the control as a function of the state (feedback control)

$$\vec{u} = -\frac{\tau K_r(1+\frac{1}{2}\tau K_v)\vec{r} + (K_v + \tau^2 K_r + \frac{1}{3}\tau^3 K_r K_v)\vec{v}}{1+\tau K_v + \frac{1}{3}\tau^3 K_r + \frac{1}{12}\tau^4 K_r K_v} \quad \text{with} \quad \tau = t_f - t$$

Application to interception

Consider a mobile M aiming at a target C:

- M has a position $\vec{r}_M(t)$, a velocity $\vec{v}_M(t)$ and an acceleration $\vec{u}_M(t)$;
- C has a position $\vec{r}_C(t)$, a velocity $\vec{v}_C(t)$ and an acceleration $\vec{u}_C(t)$.

The position and velocity differences between C and M, noted $\vec{r}$ and $\vec{v}$, satisfy

$$\begin{cases} \vec{r} = \vec{r}_M - \vec{r}_C \\ \vec{v} = \vec{v}_M - \vec{v}_C \end{cases} \Rightarrow \begin{cases} \dot{\vec{r}} = \vec{v} \\ \dot{\vec{v}} = \vec{u}_M - \vec{u}_C = \vec{u} \end{cases}$$

We retrieve the previous equations with $\vec{u}$ representing here the difference of acceleration between C and M. The coefficients K_r and K_v are chosen according to the targeted final conditions.

- Velocity-to-gain

The aim is to reach a given velocity (for an injection into orbit or for a range).

$$\begin{cases} K_r = 0 \\ K_v = \infty \end{cases} \rightarrow \vec{u} = -\frac{1}{\tau}\vec{v}$$

This control is of the form **velocity-to-gain**.

The required acceleration is the velocity difference divided by the time-to-go.

- Perfect interception

The aim is to reach a given position. The final velocity is free.

$$\begin{cases} K_r = \infty \\ K_v = 0 \end{cases} \rightarrow \vec{u} = -\frac{3}{\tau^2}(\vec{r} + \tau\vec{v})$$

This control is of the form **proportional navigation** (discussed below).

- Perfect rendezvous

The aim is to reach a given position and velocity perfectly.

$$\begin{cases} K_r = \infty \\ K_v = \infty \end{cases} \rightarrow \vec{u} = -\frac{1}{\tau^2}(6\vec{r} + 4\tau\vec{v})$$

This control is also of the form proportional navigation.

<u>Geometric interpretation of proportional navigation</u>

The reference (or nominal) trajectory is the straight line D travelled at constant velocity v_s. The point I is the target intercept position at t_f.

The position and velocity are projected along D and the normal direction.

$$\begin{cases} \vec{r} = \vec{r}_s + \vec{r}_n \\ \vec{v} = \vec{v}_s + \vec{v}_n \end{cases}$$

The motion along D is not controlled.

$$\vec{v}_s = -\frac{\vec{r}_s}{\tau} = C^{te}$$

The control acts in the normal direction.

$$\begin{cases} \dot{r}_n = v_n \\ \dot{v}_n = u_n \end{cases}$$

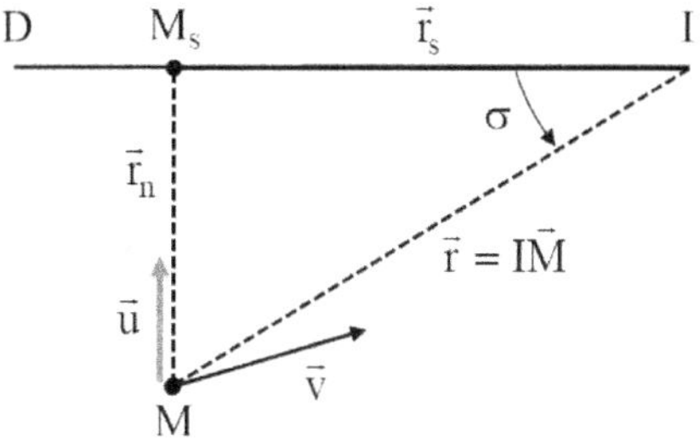

The sight angle σ between IM and D is given to order 1 by

$$\sigma = \frac{r_n}{r_s} = \frac{r_n}{\tau v_s} \Rightarrow \dot{\sigma} = \frac{r_n + \tau v_n}{\tau^2 v_s}$$

This gives us: $\begin{cases} r_n = \tau v_s \sigma \\ r_n + \tau v_n = \tau^2 v_s \dot{\sigma} \end{cases}$ which we replace in the control.

- Perfect interception

$$\vec{u} = -\frac{3}{\tau^2}(\vec{r} + \tau\vec{v}) \Rightarrow u_n = -\frac{3}{\tau^2}(r_n + \tau v_n) \Rightarrow u_n = -3v_s\dot{\sigma}$$

- Perfect rendezvous

$$\vec{u} = -\frac{1}{\tau^2}(6\vec{r} + 4\tau\vec{v}) \Rightarrow u_n = -\frac{1}{\tau^2}(6r_n + 4\tau v_n) \Rightarrow u_n = -v_s\left(\frac{2\sigma}{\tau} + 4\dot{\sigma}\right)$$

The acceleration u_n normal to the line of sight is thus expressed as a function of the sight angle σ.

This control manages the transverse deviation (normal to D). A correction u_s along D must be added to control the time of passage in I with respect to the target motion.

Transport method

To solve the linear system (4.84), we look for a solution similar to the final condition, that is of the form

$$p(t) = S(t)x(t) \tag{4.85}$$

Justification

The solution of the linear system (4.84) is expressed with its transition matrix.

$$\begin{pmatrix} x(t) \\ p(t) \end{pmatrix} = \begin{pmatrix} \Phi_{xx}(t,t_f) & \Phi_{xp}(t,t_f) \\ \Phi_{px}(t,t_f) & \Phi_{pp}(t,t_f) \end{pmatrix} \begin{pmatrix} x(t_f) \\ p(t_f) \end{pmatrix}$$

The transversality condition: $p(t_f) = S(t_f)x(t_f)$ is used to replace the final adjoint.

$$\begin{cases} x = (\Phi_{xx} + \Phi_{xp}S_f)x_f \\ p = (\Phi_{px} + \Phi_{pp}S_f)x_f \end{cases} \Rightarrow \begin{cases} x_f = (\Phi_{xx} + \Phi_{xp}S_f)^{-1}x \\ p = (\Phi_{px} + \Phi_{pp}S_f)(\Phi_{xx} + \Phi_{xp}S_f)^{-1}x \end{cases}$$

The adjoint is of the form: $p(t) = S(t)x(t)$.

with $S(t) \underset{def}{=} \left[\Phi_{px}(t,t_f) + \Phi_{pp}(t,t_f)S(t_f)\right]\left[\Phi_{xx}(t,t_f) + \Phi_{xp}(t,t_f)S(t_f)\right]^{-1}$

By replacing $p = Sx$ in equations (4.84)

$$\begin{cases} \dot{x} = Fx - GR^{-1}G^Tp \\ \dot{p} = -Qx - F^Tp \end{cases} \rightarrow \begin{cases} \dot{x} = (F - GR^{-1}G^TS)x \\ \dot{S}x + S\dot{x} = -(Q + F^TS)x \end{cases} \tag{4.86}$$

then eliminating $\dot{x}$, we obtain the differential equation satisfied by the matrix S.

$$\dot{S} = -SF - F^TS + SGR^{-1}G^TS - Q \tag{4.87}$$

This **Riccati equation** (quadratic term in S) is integrated backwards from the known final matrix $S(t_f) = S_f$, chosen for example by (4.81).

We deduce the initial adjoint: $p(t_0) = S(t_0)x(t_0)$ and then the trajectory by forward integration of the system (4.84). The matrix $S(t)$ allows to express the control (4.83) in state feedback: $u = -R^{-1}G^TSx$. The resolution process by the by the transport method is depicted in figure 4-18.

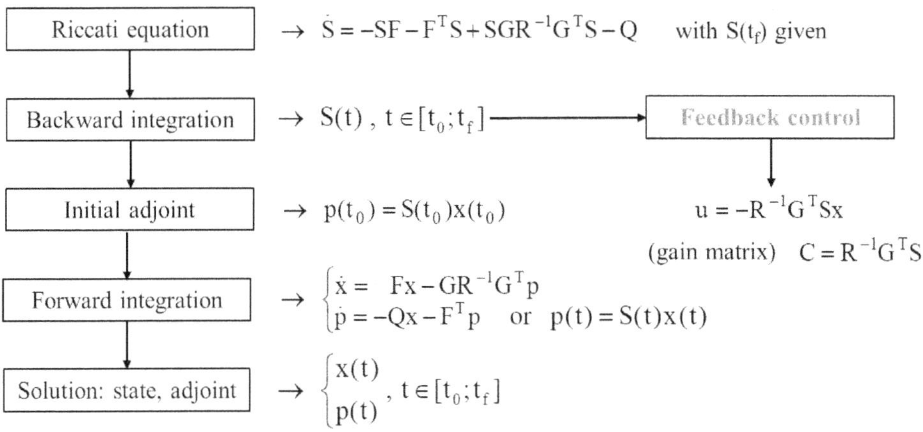

Figure 4-18: Solving the QL problem by transport method.

Regulator

For a stationary system (constant matrices F,G,Q,R), the Riccati equation (4.87) can admit a constant solution noted S_∞.

$$-S_\infty F - F^TS_\infty + S_\infty GR^{-1}G^TS_\infty - Q = 0 \tag{4.88}$$

This can happen especially when the final time t_f tends to infinity.

To calculate the matrix S_∞, one can solve the matrix equation (4.88) directly or integrate backwards the differential equation (4.87) starting from $S(t_f) = 0$ until a stabilization $(\dot{S} \approx 0)$ is obtained. This matrix must be positive.

The control is then expressed in state feedback with a constant gain matrix.

$$u = -C_\infty x \quad \text{with} \quad C_\infty = R^{-1}G^TS_\infty \tag{4.89}$$

This control is called a **regulator**. It limits the variations of $x(t)$ and $u(t)$ over an infinite time, as in example 4-10.

Example 4-10: Regulator

A regulator is sought for a first-order system: $\dot{x} = -\dfrac{1}{\tau}x + u$.

The objective is to keep x and u between the bounds: $\begin{cases} -x_m \le x \le +x_m \\ -u_m \le u \le +u_m \end{cases}$.

The quadratic problem is formulated as

$$\min_u J = \frac{1}{2}\int_{t_0}^{t_f}\left(\frac{x^2}{x_m^2}+\frac{u^2}{u_m^2}\right)dt \quad \text{with} \quad \dot{x} = -\frac{1}{\tau}x + u .$$

Equation (4.88) gives: $\frac{2}{\tau}S_\infty + u_m^2 S_\infty^2 - \frac{1}{x_m^2} = 0$

with the matrices: $F = -\frac{1}{\tau}$, $G = 1$, $Q = \frac{1}{x_m^2}$, $R = \frac{1}{u_m^2}$.

The only positive definite solution is: $S_\infty = \frac{1}{\tau u_m^2}\left(\sqrt{1+\tau^2\frac{u_m^2}{x_m^2}} - 1\right)$.

The regulator control is: $u(t) = -R^{-1}G^T S_\infty x(t) = -\frac{1}{\tau}\left(\sqrt{1+\tau^2\frac{u_m^2}{x_m^2}} - 1\right)x(t)$.

If x is not bounded $(x_m \to \infty)$, the optimal control is zero.

If u is not bounded $(u_m \to \infty)$, the optimal control is: $u(t) = -\frac{u_m}{x_m}x(t)$.

Final constraints

Assume that the final state is subject to linear constraints of the form

$$\psi(x_f) = M_f x_f = \psi_c \tag{4.90}$$

M_f is a given matrix, ψ_c is the target value.
The quadratic-linear problem is formulated as

$$\min_u J = \frac{1}{2}\int_{t_0}^{t_f}(x^T Q x + u^T R u)dt \quad \text{with} \quad \begin{cases} \dot{x} = Fx + Gu \ , \ x(t_0) = x_0 \\ \psi(x_f) = M_f x_f = \psi_c \end{cases} \tag{4.91}$$

Conditions (4.82 to 4.84) are unchanged. Only the transversality condition on $p(t_f)$ is modified and becomes, with the multipliers λ of the final constraints

$$p(t_f) = M(t_f)^T \lambda \tag{4.92}$$

The transport method consists in expressing the adjoint and the constraints as

$$\begin{cases} p(t) = S(t)x(t) + T(t)\lambda \\ \psi(x_f) = M(t)x(t) + N(t)\lambda = \psi_c \end{cases} \tag{4.93}$$

Justification

The solution of the linear system (4.84) is expressed with its transition matrix.

$$\begin{pmatrix} x(t) \\ p(t) \end{pmatrix} = \begin{pmatrix} \Phi_{xx}(t,t_f) & \Phi_{xp}(t,t_f) \\ \Phi_{px}(t,t_f) & \Phi_{pp}(t,t_f) \end{pmatrix} \begin{pmatrix} x(t_f) \\ p(t_f) \end{pmatrix}$$

Let us use the transversality condition: $p(t_f) = M(t_f)\lambda$ to replace the final adjoint.

$$\begin{cases} x = \Phi_{xx}x_f + \Phi_{xp}M_f^T\lambda \\ p = \Phi_{px}x_f + \Phi_{pp}M_f^T\lambda \end{cases} \Rightarrow \begin{cases} x_f = \Phi_{xx}^{-1}(x - \Phi_{xp}M_f^T\lambda) \\ p = \Phi_{px}\Phi_{xx}^{-1}x + (\Phi_{pp} - \Phi_{px}\Phi_{xx}^{-1}\Phi_{xp})M_f^T\lambda \end{cases}$$

The adjoint is of the form (4.93): $p(t) = S(t)x(t) + T(t)\lambda$

by posing $\begin{cases} S(t) \underset{def}{=} \Phi_{px}(t,t_f)\Phi_{xx}(t,t_f)^{-1} \\ T(t) \underset{def}{=} \left[\Phi_{pp}(t,t_f) - \Phi_{px}(t,t_f)\Phi_{xx}(t,t_f)^{-1}\Phi_{xp}(t,t_f)\right]M_f^T \end{cases}$

The final constraints are also of the form (4.93).

$\psi(x_f) = M(t)x(t) + N(t)\lambda$ by posing $\begin{cases} M(t) \underset{def}{=} \Phi_{xx}(t,t_f)^{-1} \\ N(t) \underset{def}{=} -\Phi_{xx}(t,t_f)^{-1}\Phi_{xp}(t,t_f)M_f^T \end{cases}$

By deriving the expressions (4.93), replacing in equations (4.84) and identifying the terms in x and λ, we obtain a differential system satisfied by the matrices S, T, M, N, with the final values given by the constraints (4.90).

$$\begin{cases} \dot{S} = -SF - F^TS - Q + SGR^{-1}G^TS \\ \dot{T} = -F^TT + SGR^{-1}G^TT \\ \dot{M} = -MF + MGR^{-1}G^TS \\ \dot{N} = MGR^{-1}G^TT \end{cases} \text{ with } \begin{cases} S(t_f) = 0 \\ T(t_f) = M(t_f)^T \\ M(t_f) = M_f \text{ given} \\ N(t_f) = 0 \end{cases} \tag{4.94}$$

We observe that: $T = M^T$, because these matrices satisfy the same equations.

The system (4.94) is integrated backwards from the known final conditions.

Suppose we solve problem (4.91) without the final constraints, which is equivalent to ignoring the multipliers λ in solution (4.93). The value taken by the final constraints would then be: $\psi(x_f) = M(t)x(t)$.
The function $\psi(x) = Mx$ thus yields the free (or natural) value of the constraints if we solve the problem (4.91) from the current state x without the constraints.
The multipliers λ are determined from (4.93) at the initial time.

$$\psi_c = M(t_0)x(t_0) + N(t_0)\lambda \;\Rightarrow\; \lambda = N(t_0)^{-1}(\psi_c - \psi_0) \tag{4.95}$$

The matrix N is negative.

$$\begin{cases} \dot{N} = MGR^{-1}G^TM^T \le 0 \\ N(t_f) = 0 \end{cases} \Rightarrow \; N(t) \le 0 \;\text{ for } t < t_f .$$

In the general case, $N(t_0)$ is invertible and the formula (4.95) gives λ. The initial adjoint is given by (4.93) and the trajectory is obtained by forward integration of the system (4.84). The matrices S, M, N give the control (4.83) in state feedback.

$$u = -R^{-1}G^TSx - R^{-1}G^TM^TN^{-1}(\psi_c - \psi) \quad \text{with} \quad \psi = Mx \tag{4.96}$$

The multipliers λ (4.95) and the control u (4.96) are functions of the deviation between ψ_c (target value) and ψ (free value without constraints). The solution of the unconstrained problem corresponds to zero multipliers: $\lambda = 0$.
The resolution process by the transport method is depicted in figure 4-19.

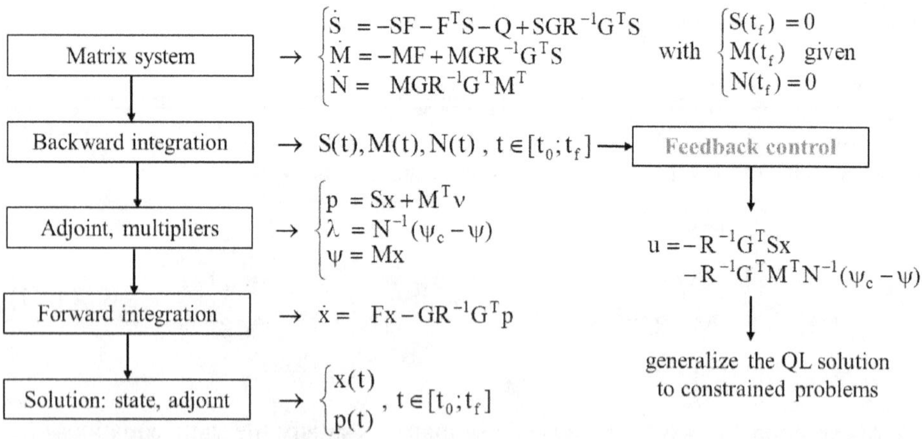

Figure 4-19: Constrained QL resolution by transport method.

4.2.5 Robust control

Consider a linear system of state x subjected to perturbations w. The observable outputs form a vector z which is a linear function of the state.

$$\begin{cases} \dot{x} = Ax + Dw \\ z = Cx \end{cases} \tag{4.97}$$

The matrices A, C and D are constant. The vectors x and z can represent the deviations from a nominal solution. The equations (4.97) are then the variational equations of a nonlinear system. We assume that the nominal output is $z = 0$ and look for worst perturbations in the sense of maximizing z.

For this, we consider the cost

$$\min_{w} J = \int_{t_0}^{\infty} (\frac{1}{\alpha} w^T w - \alpha z^T z) dt \quad \text{with} \quad \alpha > 0 \tag{4.98}$$

The objective is to maximize the norm of z (term $-\alpha z^T z$), while applying the smallest possible perturbations (term $+\frac{1}{\alpha} w^T w$). In the absence of this second term, the solution would not be bounded.

Property 4-4 gives a method for calculating the solution.

Property 4-4: Worst perturbations

Suppose there exists a symmetric positive definite matrix P such that:

- the matrix P satisfies the Riccati differential equation

$$PA + A^T P + \alpha C^T C + \alpha PDD^T P = 0 \tag{4.99}$$

- the matrix $A + \alpha DD^T P$ is a Hurwitz matrix, which means that all its eigenvalues have a negative real part.

Then the solution of problem (4.98) is given by

$$w^* = \alpha D^T P x^* \tag{4.100}$$

and the minimum cost is

$$J^* = -x_0^T P x_0 \tag{4.101}$$

Demonstration

The stated solution is checked by deriving the function: $V(x) = x^T P x$.

$$\dot{V} = \dot{x}^T P x + x^T P \dot{x} = (Ax + Dw)^T P x + x^T P(Ax + Dw)$$
$$= x^T(PA + A^T P)x + 2x^T PDw$$
$$= -\alpha x^T(C^T C + PDD^T P)x + 2x^T PDw \quad \text{since } PA + A^T P = -\alpha C^T C - \alpha PDD^T P$$
$$= -\alpha z^T z - \alpha x^T PDD^T Px + 2x^T PDw \quad \text{since } Z = Cx$$
$$= \frac{1}{\alpha} w^T w - \alpha z^T z - \frac{1}{\alpha}(w - \alpha D^T Px)^T (w - \alpha D^T Px) \quad \text{by factoring}$$

Let us integrate from t_0 to t_1.

$$V(x_1) - V(x_0) = \int_{t_0}^{t_1} (\frac{1}{\alpha}\|w\|^2 - \alpha\|z\|^2)dt - \int_{t_0}^{t_1} \frac{1}{\alpha}\|w - \alpha D^T Px\|^2 dt$$

Then take the limit $t_1 \to \infty$ to get the cost J.

$$J = \int_{t_0}^{\infty} (\frac{1}{\alpha}\|w\|^2 - \alpha\|z\|^2)dt = \lim_{t_1 \to \infty} \left[x_1^T P x_1 + \int_{t_0}^{t_1} \frac{1}{\alpha}\|w - \alpha D^T Px\|^2 dt \right] - x_0^T P x_0$$

The minimum of J is obtained for the perturbation: $w = \alpha D^T Px$.

With this solution, the state equation gives: $\dot{x} = Ax + Dw = (A + \alpha DD^T P)x$.

By assumption, the matrix $A + \alpha DD^T P$ has all its eigenvalues with negative real part. The system is therefore exponentially stable: $\lim_{t_1 \to \infty} x(t_1) = 0$.

The optimal cost finally reduces to: $J^* = -x_0^T P x_0$.

Interpretation

The solution can be interpreted with the L_2 norms of the perturbation w and of the output z. The cost (4.98) is expressed as

$$J = \frac{1}{\alpha}\|w\|_2^2 - \alpha\|z\|_2^2 \quad \text{with} \quad \|z\|_2 \underset{\text{def}}{=} \sqrt{\int_{t_0}^{\infty} z^T z\, dt} \tag{4.102}$$

Since the optimal cost is given by (4.101), we have for any perturbation w

$$J^* \le J \Rightarrow -x_0^T P x_0 \le \frac{1}{\alpha}\|w\|_2^2 - \alpha\|z\|_2^2 \tag{4.103}$$

For an initial state of zero, we obtain the inequality

$$\alpha\|z\|_2 \le \|w\|_2 \quad , \forall w \tag{4.104}$$

The **gain L_2** of the system (4.97) is defined as the maximum amplification factor between the perturbation w and the output z. This gain is denoted by k.

$$k \underset{def}{=} \sup_{w\neq 0} \frac{\|z\|_2}{\|w\|_2} \tag{4.105}$$

If the matrix P defined in property 4-4 exists, then the gain k is bounded by $\dfrac{1}{\alpha}$.

By a demonstration similar to above, we establish the following equivalence.

Property 4-5: Gain L_2

The gain L_2 is bounded by γ if and only if there is a symmetric positive definite matrix P such that

$$\begin{cases} PA + A^T P + \dfrac{1}{\gamma} C^T C + \dfrac{1}{\gamma} PDD^T P = 0 \\ A + \dfrac{1}{\gamma} DD^T P \text{ Hurwitz matrix} \end{cases} \tag{4.106}$$

A sufficient condition for the gain L_2 to be bounded by γ is obtained by relaxing equality (4.106) into a Riccati inequality.

$$\begin{cases} PA + A^T P + \dfrac{1}{\gamma} C^T C + \dfrac{1}{\gamma} PDD^T P \le 0 \\ A + \dfrac{1}{\gamma} DD^T P \text{ Hurwitz matrix} \end{cases} \tag{4.107}$$

To calculate the gain k defined by (4.105), we use the transfer matrix G of the system (4.97), which relates the Laplace transforms of w and z.

$$\begin{cases} \dot{x} = Ax + Dw \\ z = Cx \end{cases} \Rightarrow z = C(sI - A)^{-1} Dw \underset{def}{=} Gw \tag{4.108}$$

It is shown that the gain k is the largest singular value of the matrix $G(j\omega)$ for all possible frequencies $\omega \in \mathbb{R}$: $k = \|G\|_\infty = \sup_\omega \bar{\sigma}(G(j\omega))$.

This value is called the norm H_∞ of the transfer matrix.

Stabilization

Let us retrieve the linear system (4.97) and introduce a control u.

$$\begin{cases} \dot{x} = Ax + Bu + Dw \\ z = Cx \end{cases} \tag{4.109}$$

The aim is to stabilize the output z despite the perturbation w. For this purpose, let us look for a state feedback control: $u = Kx$. The system takes the form

$$\begin{cases} \dot{x} = A_f x + Dw \\ z = Cx \end{cases} \quad \text{with} \quad A_f = A + BK \tag{4.110}$$

with the matrix A_f of the closed loop system.

Stabilization consists in searching for a gain matrix K such that:

- the gain L_2 is less than a given bound γ;
- the closed-loop system is stable.

A sufficient condition is that there exists a symmetric positive definite P such that

$$PA_f + A_f^T P + \frac{1}{\gamma} C^T C + \frac{1}{\gamma} PDD^T P < 0 \tag{4.111}$$

This condition implies: $PA_f + A_f^T P < 0$ which is the Lyapunov condition for the matrix A_f to be stable (Hurwitz matrix).

We are thus led to search for a gain matrix K and a symmetric positive definite matrix P satisfying (4.111). This search is done by introducing symmetric positive definite matrices Q and R and a parameter $\varepsilon > 0$.

- inequality (4.111) is transformed into equality by adding εQ:

$$PA_f + A_f^T P + \frac{1}{\gamma} C^T C + \frac{1}{\gamma} PDD^T P + \varepsilon Q = 0 \tag{4.112}$$

- the gain K is sought in the form:

$$K = -\frac{1}{2\varepsilon} R^{-1} B^T P \tag{4.113}$$

Replacing in (4.112) the matrix $A_f = A + BK$ with K given by (4.113), we arrive at a Riccati equation.

$$PA + A^T P + \frac{1}{\gamma} C^T C + \frac{1}{\gamma} PDD^T P - \frac{1}{\varepsilon} PBR^{-1} B^T P + \varepsilon Q = 0 \tag{4.114}$$

It is then necessary to find setting parameters Q, R, ε such that this equation admits a solution P. This search can be done by a systematic procedure as described in reference [R14].

4.3 Extremals

This section presents the different types of solutions to a control problem.

4.3.1 Definitions

Consider a control problem in Lagrangian form without specifying the terminal conditions.

$$\min_u J = \int_{t_0}^{t_f} L(x,u,t)dt \quad \text{with} \quad \dot{x} = f(x,u,t) \tag{4.115}$$

The Hamiltonian depends on the adjoint $p(t)$ and the abnormal multiplier $p_0 \geq 0$.

$$H(x,u,p,t) = p_0 L(x,u,t) + p^T f(x,u,t) \tag{4.116}$$

An **extremal** is a quadruplet (x, p_0, p, u) satisfying the following Hamiltonian system where the control minimizes the Hamiltonian (by the convention $p_0 \geq 0$).

$$\begin{cases} \dot{x} = H_p \\ \dot{p} = -H_x \end{cases} \quad \text{with} \quad u \to \min_u H(x,u,p,t) \tag{4.117}$$

The extremal field (also called the Hamiltonian system flow) is the set of trajectories obtained by integration of (4.117) by varying the state and the initial adjoint. The solution of a control problem with specified terminal conditions is a particular extremal.

Extremals can be normal or abnormal, regular or singular:

- an extremal is said to be **normal** if $p_0 > 0$, **abnormal** if $p_0 = 0$.
  An abnormal extremal occurs, for example, when the solution is completely determined by the constraints. This solution is then indifferent to the cost function. For a normal extremal, we usually fix $p_0 = 1$;
- an extremal is said to be **regular** if H_u is explicit in u, **singular** otherwise.
  For a regular extremal, the optimal control is a solution of $H_u = 0$.
  A singular extremal requires the examination of higher order conditions.

An optimal trajectory can have regular arcs and singular arcs.

4.3.2 Abnormal extremal

Assume that there is locally only one control law u* satisfying the constraints of the control problem. The solution of the problem is then u*, independently of the cost function. According to the PMP, this control achieves the minimum of the Hamiltonian (4.116), for any function $L(x,u,t)$.

$$H_u = p_0 L_u + p^T f_u = 0 \ , \ \forall L \tag{4.118}$$

This is only possible if $p_0 = 0$, which yields an abnormal extremal.
Abnormal extremals may occur in other cases. Example 4-11 shows a control problem that admits normal or abnormal extremals depending on the initial conditions.

Example 4-11: Oscillator

The aim is to bring a harmonic oscillator to rest in the minimum time from given initial conditions. The control is the acceleration bounded between −1 and +1. The problem is formulated as

$$\min_{u,t_f} J = \int_{t_0}^{t_f} 1dt \quad \text{with} \quad \begin{cases} \dot{x}_1 = x_2 \\ \dot{x}_2 = -x_1 + u \end{cases} , \quad \begin{cases} x_1(t_0) = x_{10} \\ x_2(t_0) = x_{20} \end{cases} , \quad \begin{cases} x_1(t_f) = 0 \\ x_2(t_f) = 0 \end{cases} , \quad -1 \le u \le 1$$

Application of the PMP

The Hamiltonian is: $H = p_0 + p_1 x_2 + p_2(-x_1 + u)$ with the adjoint (p_1, p_2) and the abnormal multiplier p_0.

Applying the PMP gives: $\min_u H$ s.t. $-1 \le u \le +1 \quad \rightarrow u = \begin{cases} -1 & \text{if } p_2 < 0 \\ +1 & \text{if } p_2 > 0 \end{cases}$

The control $u = \pm 1$ is determined by the adjoint p_2 (switching function).

The adjoint equations are: $\begin{cases} \dot{p}_1 = p_2 \\ \dot{p}_2 = -p_1 \end{cases} \Rightarrow \ddot{p}_2 = -p_2 \Rightarrow p_2 = a\cos(t + \varphi)$

The switches (change of sign of p_2) are spaced by a duration of π.

Representation in the phase plane

We can represent in the phase plane the arcs of the trajectory for $u = \pm 1$.

$$\begin{cases} \dot{x}_1 = x_2 \\ \dot{x}_2 = -x_1 + u \end{cases} \Rightarrow \frac{dx_2}{dx_1} = \frac{-x_1 + u}{x_2} \Rightarrow (x_1 - u)dx_1 + x_2 dx_2 = 0 \Rightarrow (x_1 - u)^2 + x_2^2 = c^2$$

The arcs of the trajectory are semicircles with center $A^-(-1;0)$ for $u=-1$ and $A^+(+1;0)$ for $u=+1$. The extremals are constructed backwards from the origin.

The last arc to arrive at the origin is located on
- the semi-circle C^- (center A^-, radius 1) if $u=-1$;
- the semi-circle C^+ (center A^+, radius 1) if $u=+1$.

The previous arc is a semi-circle
- of center A^+ if we end up on C^-;
- of center A^- if we end up on C^+.

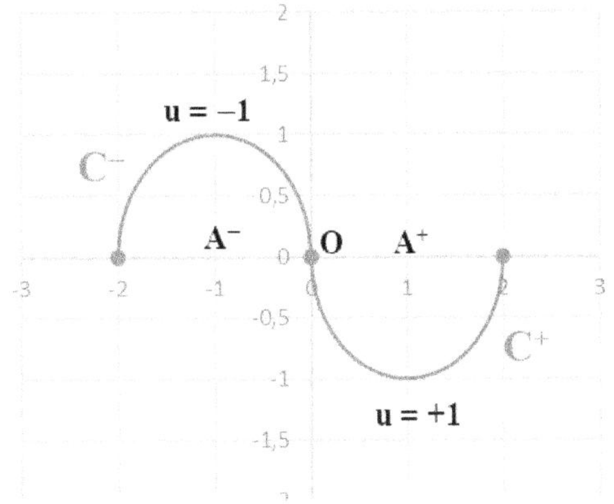

Figure 4-20 shows in the phase plane the set of initial conditions
- leading to an arrival on the semicircle C^+ (top);
- leading to an arrival on the semicircle C^- (bottom).

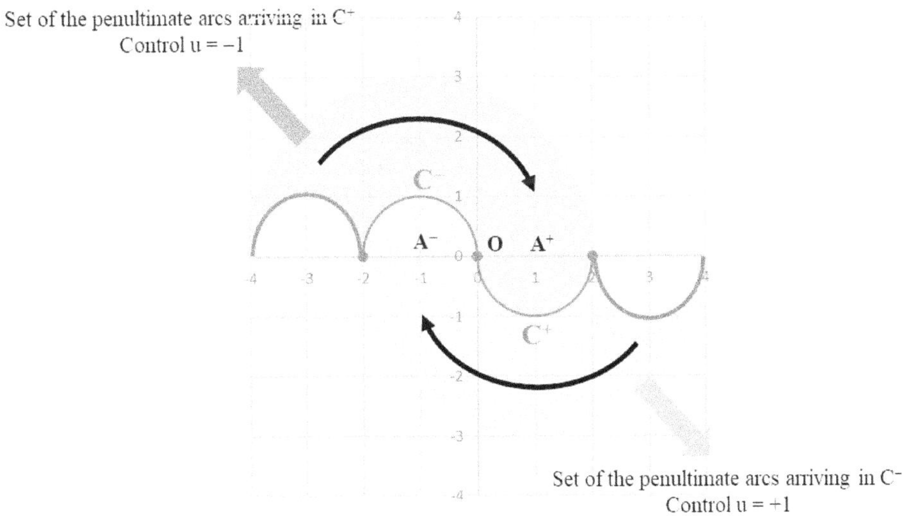

Figure 4-20: Penultimate arc.

Continuing the study backwards, we deduce that the previous arcs are semicircles of growing size with centers alternately in A^- and A^+. The commutations occur on a set of semicircles obtained by translation of C^- and C^+ along x_1. This set of semicircles plotted in figure 4-21 forms the switching curve C. The control is $u = -1$ above the curve C and $u = +1$ below it. By varying the last switch position on the semicircles C^- and C^+, we generate backwards the set of all possible extremals that constitute the phase portrait of the system.

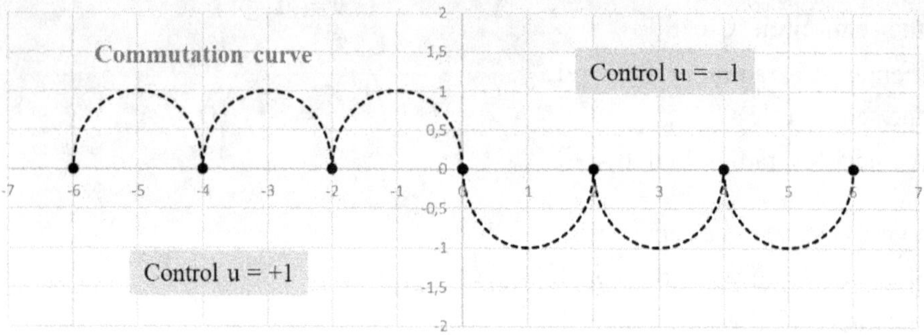

Figure 4-21: Switching curve.

Only one extremal passes through each point of the phase plane (associated with given initial conditions). Figure 4-22 shows the extremal starting from the initial conditions: $x_1(t_0) = -1.77$, $x_2(t_0) = -3.36$.

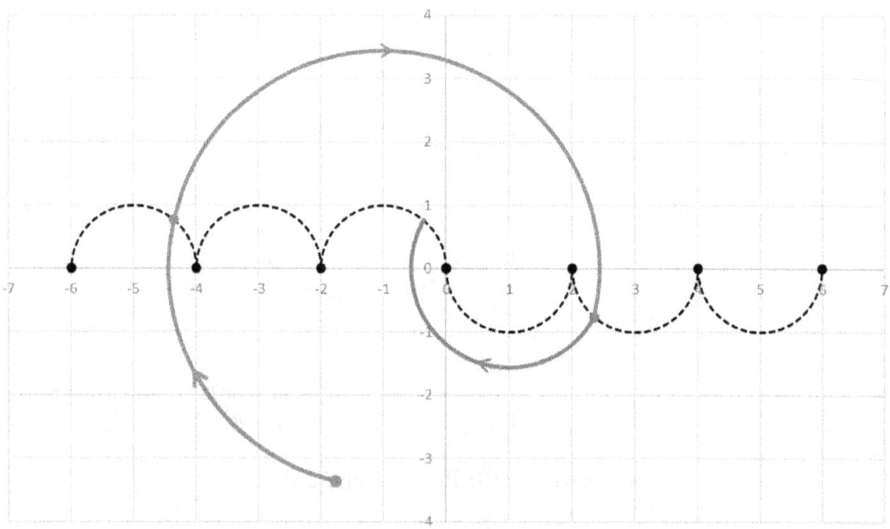

Figure 4-22: Extremal in the phase plane.

Abnormal extremals
Let us find out if the problem admits extremals.
The system is autonomous and has a free final time.
The Hamiltonian is therefore constant and zero: $H = p_0 + p_1x_2 + p_2(-x_1 + u) = 0$.

At the final time: $\begin{cases} x_1(t_f) = 0 \\ x_2(t_f) = 0 \end{cases} \Rightarrow H = p_0 + p_2u = 0$ with $u(t_f) = \pm 1$.

- if $p_2(t_f) \neq 0$, then $p_0 \neq 0$ and the extremal is normal;
- if $p_2(t_f) = 0$, then $p_0 = 0$ and the extremal is abnormal.

The switches are spaced by π and occur when p_2 cancels.

For an abnormal extremal, the last semicircle C^- or C^+ is completely traversed and the switches occur on the axis x_1.
The problem admits an infinite number of abnormal extremals, passing through the points with coordinates $(x_1 = 2k \,;\, x_2 = 0)$, $k \in \mathbb{Z}$. Depending on the initial condition, the extremal leading back to the origin can thus be normal or abnormal.

4.3.3 Singular extremal

The optimal control is usually obtained by solving the equation: $H_u = 0$.
A singular extremal occurs when the derivative H_u is not explicit in u.
This occurs in the frequent case of an affine controlled system of the form

$$\min_u J = \varphi(x_f, t_f) \quad \text{with} \quad \dot{x} = f(x,t) + g(x,t)u \tag{4.119}$$

In this case, higher order conditions are required to determine the optimal control. These conditions are presented in section 4.4.3.

The Goddard problem illustrates the determination of singular extremals.

Example 4-12: Goddard problem (from [R3])

The aim is to maximize the altitude reached by a rocket fired vertically from the ground.

Modelling
The assumptions about the motion are as follows:

- the state is composed of the rocket altitude h, velocity v and mass m;
- the rocket takes off from the ground $(h_0 = 0, v_0 = 0)$ and thrusts vertically;
- the initial total mass m_0 and the propellant mass m_e are fixed;
- the forces acting on the rocket are the weight W, the thrust T, the drag D;
- the control is the thrust level varying between 0 and T_{max}.
  When all the propellants are consumed, the thrust cancels.
  The mass flow rate is: $q = T / v_e$ where v_e is the gas ejection velocity.

The state equations are: $\begin{cases} \dot{h} = v \\ \dot{v} = (T - D - W) / m \\ \dot{m} = -T / v_e \end{cases}$.

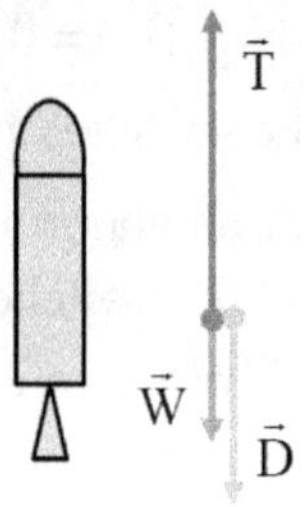

The weight is: $W = mg$.
The acceleration of gravity g is assumed to be constant.

The drag is: $D = \frac{1}{2}\rho v^2 S C_D$.

S is the reference surface, C_D is the drag coefficient, ρ is the atmospheric density depending on the altitude by an exponential model: $\rho = \rho_0 e^{-\frac{h}{H}}$.

ρ_0 is the density at the ground, H is the scale factor of the atmosphere.

Control problem

The objective is to find the thrust law that maximizes the altitude reached by the rocket. The control problem is formulated as

$$\min_{T, t_f} J = -h(t_f) \quad \text{with} \quad \begin{cases} \dot{h} = v \\ \dot{v} = (T - D - W) / m \\ \dot{m} = -T / v_e \end{cases} \quad \text{and} \quad \begin{cases} h(t_0) = 0 \\ v(t_0) = 0 \\ m(t_0) = m_0 \, , \, m(t_f) = m_0 - m_e \end{cases}$$

Application of the PMP

The Hamiltonian is: $H = p_h v + p_v \frac{T - D - W}{m} - p_m \frac{T}{v_e}$ with adjoint (p_h, p_v, p_m).

Noting $\Phi = v_e p_v - m p_m$, the Hamiltonian is: $H = p_h v - p_v \left(\frac{D}{m} + g \right) + \frac{\Phi}{m v_e} T$.

Applying the PMP gives: $\min_{0 \le T \le T_{max}} H \Rightarrow \begin{cases} T = 0 & \text{if } \Phi > 0 \\ T = T_{max} & \text{if } \Phi < 0 \end{cases}$ and $T = ?$ if $\Phi = 0$.

The optimal thrust depends on the switching function: $\Phi = v_e p_v - m p_m$.
A singular arc can occur if the function Φ cancels on an interval.

Search for singular arcs

Let us look for conditions on the adjoint (p_h, p_v, p_m) to obtain a singular arc.

If Φ cancels on an interval, its successive derivatives are zero. The Hamiltonian is moreover constantly zero, since the system is autonomous with free final time.

We have the system denoted (S): $$\begin{cases} H=0 & \rightarrow\ p_h v + p_v \dfrac{T-D-W}{m} - p_m \dfrac{T}{v_e} = 0 \\ \Phi = 0 & \rightarrow\ v_e p_v - m p_m = 0 \\ \dot{\Phi} = 0 & \rightarrow\ v_e \dot{p}_v - \dot{m} p_m - m \dot{p}_m = 0 \end{cases}$$

The adjoint equations are, denoting D_h and D_v the derivatives of the drag

$$\begin{cases} \dot{p}_h = -H_h = \dfrac{p_v D_h}{m} \\ \dot{p}_v = -H_v = -p_h + \dfrac{p_v D_v}{m} \\ \dot{p}_m = -H_m = p_v \dfrac{T-D}{m^2} \end{cases} \quad \text{with} \quad \begin{cases} D = \dfrac{1}{2}\rho v^2 S C_D = \dfrac{1}{2}\rho_0 e^{-\frac{h}{H}} v^2 S C_D \\ D_h = \dfrac{\partial D}{\partial h} = \dfrac{1}{2} v^2 S C_D \dfrac{\partial \rho}{\partial h} = -\dfrac{D}{H} \\ D_v = \dfrac{\partial D}{\partial v} = \rho v S C_D = \dfrac{2D}{v} \end{cases}.$$

We replace $\dot{m}, \dot{p}_v, \dot{p}_m$ in $\dot{\Phi}$, using also $\Phi = v_e p_v - m p_m = 0$ to replace p_m.

$$\dot{\Phi} = v_e \dot{p}_v - \dot{m} p_m - m \dot{p}_m = v_e\left(-p_h + \frac{p_v D_v}{m}\right) + \frac{T}{v_e} p_m - p_v \frac{T-D}{m} = \frac{p_v}{m}\left(1 + 2\frac{v_e}{v}\right) D - v_e p_h$$

The system (S) reduces to: $$\begin{cases} p_h v - p_v \dfrac{D+W}{m} = 0 \\ p_v v_e - p_m m = 0 \\ p_h m - p_v \left(\dfrac{2}{v} + \dfrac{1}{v_e}\right) D = 0 \end{cases}$$

According to the PMP the adjoint (p_h, p_v, p_m) cannot be identically zero. For a non-zero solution to exist, the matrix of the linear system (S) must be singular. Its determinant is

$$\begin{vmatrix} v & -\dfrac{D+W}{m} & 0 \\ 0 & v_e & -m \\ m & -\left(\dfrac{2}{v} + \dfrac{1}{v_e}\right) D & 0 \end{vmatrix} = mW - mD\left(1 + \frac{v}{v_e}\right) = 0$$

When the function $\Psi = W - D\left(1+\frac{v}{v_e}\right)$ cancels, a singular arc is possible.

The singular surface is the set of states (h, v, m) such that

$$\Psi = W - D\left(1+\frac{v}{v_e}\right) = 0 .$$

Let us study the evolution of the function Ψ along the flight.
The velocity is zero at the beginning of the flight (take-off) and at the end of the flight (apogee).

Since the drag is zero at these times, we have: $\begin{cases} \Psi(t_o) = m_0 g > 0 \\ \Psi(t_f) = m_f g > 0 \end{cases}$.

At the beginning of the flight, the weight W decreases (loss of mass), while the velocity and drag increase. The function Ψ is therefore decreasing and it may cancel or not.
According to the evolution of Φ and Ψ, the three possible control structures are

$$\begin{cases} T_{max} \to 0 \\ T_{max} \to T_{sing} \to 0 \\ T_{max} \to T_{sing} \to T_{max} \to 0 \end{cases}$$

<u>Expression of the singular thrust</u>
Let us find the expression for the thrust on a singular arc. To do this, we derive the switching function Φ until the control T appears.

The first derivative has already been calculated: $\dot{\Phi} = \frac{p_v}{m}\left(1+2\frac{v_e}{v}\right)D - v_e p_h$.

This derivative is constantly zero on the singular arc.

$$\dot{\Phi} = 0 \;\Rightarrow\; p_v G - m p_h = 0 \;\text{ with }\; G \underset{def}{=} \left(\frac{2}{v} + \frac{1}{v_e}\right) D$$

We derive a second time: $\dot{p}_v G + p_v \dot{G} - \dot{m} p_h - m \dot{p}_h = 0$ with $\dot{G} = G_h \dot{h} + G_v \dot{v}$.

Using the state and adjoint equations giving $(\dot{h}, \dot{v}, \dot{m}, \dot{p}_h, \dot{p}_v)$

$$T_{sing}\left(G_v + \frac{G}{v_e}\right) = \frac{D}{v_e} G + (D+W) G_v - mvG_h + mD_h$$

then replacing G and its derivatives, we get the thrust formula on a singular arc.

$$T_{sing}\left(1+4\alpha+2\alpha^2\right) = D(1+2\alpha) + \left(2D + 2W + \frac{mv^2}{H}\right)\alpha(1+\alpha) \quad \text{with } \alpha \underset{def}{=} \frac{v_e}{v}$$

The trajectory necessarily starts with a regular arc with maximum thrust T_{max} (because $\Psi(t_0) > 0$). A singular arc is possible if the functions Φ (switching) and Ψ (singular surface) cancel simultaneously. The thrust becomes then variable $T_{sing}(h,v,m)$ and the trajectory follows the singular surface. The singular arc is followed as long as $T_{sing} \le T_{max}$ or until the propellants are exhausted.

The following application shows a solution of the form $T_{max} \to T_{sing} \to T_{max} \to 0$.

<u>Numerical application</u>

The numerical data are summarized in the following table.

m_0	m_e	T_{max}	v_e	SC_D	ρ_0	H	g
100 kg	90 kg	2 000 N	2 000 m/s	0.1 m²	1.2 kg/m³	5 000 m	10 m/s²

The solution of the form $T_{max} \to T_{sing} \to T_{max} \to 0.$ is plotted in figure 4-23.

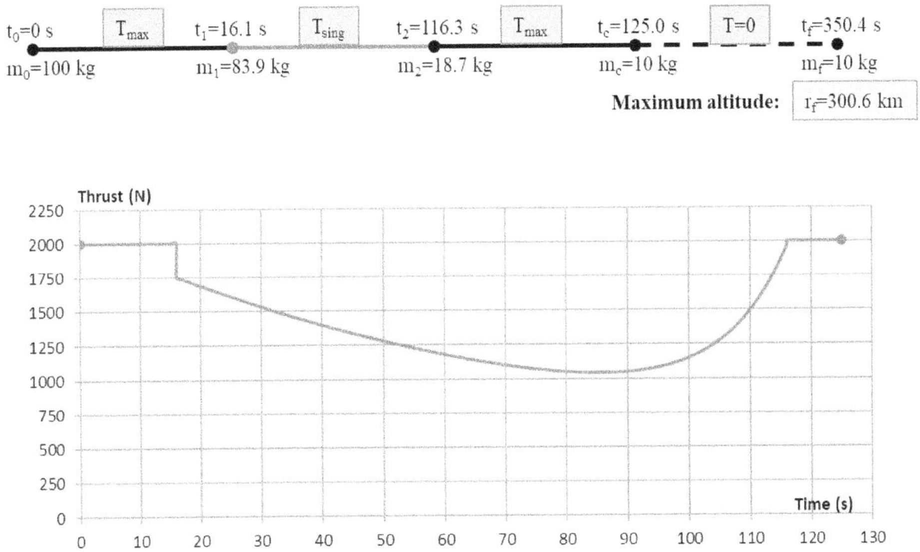

Figure 4-23: Thrust law with a singular arc (for $m_e = 90$ kg).

The structure of the solution depends on the available propellant mass:

- if $m_e > 81.3$ kg, the solution is of the form: $T_{max} \to T_{sing} \to T_{max} \to 0$;
- if $16.1 \text{ kg} < m_e < 81.3$ kg, the solution is of the form: $T_{max} \to T_{sing} \to 0$;
- if $m_e < 16.1$ kg, the solution is of the form: $T_{max} \to 0$.

In the case $m_e = 90\text{kg}$ shown in figure 4-23 a regular control $T_{max} \to 0$ would give a burn time of 90 s (instead of 125 s) and would terminate at a maximum altitude of 236.8 km (instead of 300.6 km).
The solution $T_{max} \to 0$ is in this case less good than $T_{max} \to T_{sing} \to T_{max} \to 0$.
These two solutions are compared in figure 4-24.

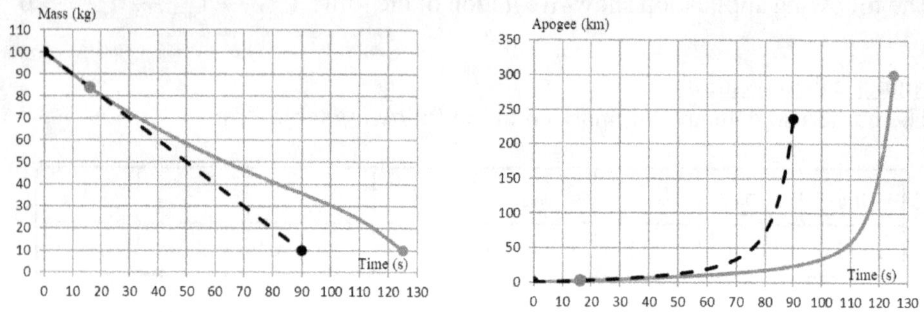

Figure 4-24: Comparison of singular solution and maximum thrust solution.

4.3.4 Neighboring extremals

Consider a free final time problem with final constraints.

$$\min_{u,t_f} J = \int_{t_0}^{t_f} L(x,u,t)dt + \varphi(x_f,t_f) \quad \text{with} \quad \begin{cases} \dot{x} = f(x,u,t) \\ x(t_0) = x_0 \\ \psi(x_f,t_f) = 0 \end{cases} \tag{4.120}$$

The solution is an extremal satisfying the equations

$$\begin{cases} \dot{x} = \ H_p \\ \dot{p} = -H_x \\ H_u = 0 \end{cases} \quad \text{with} \quad \begin{cases} \psi\left[x(t_f),t_f\right] = 0 \\ p(t_f) = \ \Omega_x[x(t_f),t_f,\lambda] \\ H(t_f) = -\Omega_t[x(t_f),t_f,\lambda] \end{cases} \tag{4.121}$$

$\lambda \in \mathbb{R}^q$ is the final constraint multiplier. The function Ω is defined by

$$\Omega[x(t_f),t_f,\lambda] \underset{def}{=} \varphi[x(t_f),t_f] + \lambda^T \psi[x(t_f),t_f] \tag{4.122}$$

This extremal is associated with the initial state x_0 and the constraints $\psi = 0$.
We will call it the nominal or reference extremal.

Assume that the initial state and the constraint threshold vary by δx_0 and $\delta\psi_f$.

$$\begin{cases} x(t_0)=x_0 \\ \psi(x_f,t_f)=0 \end{cases} \rightarrow \begin{cases} x(t_0)=x_0+\delta x_0 \\ \psi(x_f,t_f)=\delta\psi_f \end{cases} \tag{4.123}$$

The new extremal is close to the nominal extremal as shown in figure 4-25. We will call it a neighboring extremal.

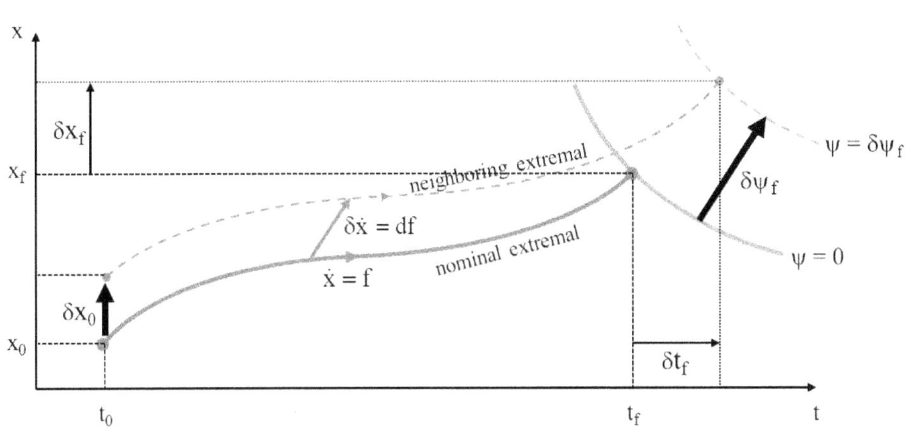

Figure 4-25: Neighboring extremal.

The neighboring extremal can be calculated from the nominal extremal as follows.

Property 4-6: Neighboring extremal

The deviations $(\delta x, \delta p, \delta\lambda)$ between the nominal extremal $(x_0, \psi=0)$ and the neighboring extremal $(x_0+\delta x_0, \psi=\delta\psi_f)$ follow the differential equations

$$\begin{cases} \delta\dot{x} = (f_x^T - f_u^T H_{uu}^{-1} H_{xu})\delta x - f_u^T H_{uu}^{-1} f_u \delta p \\ \delta\dot{p} = (H_{ux} H_{uu}^{-1} H_{xu} - H_{xx})\delta x + (H_{ux} H_{uu}^{-1} f_u - f_x)\delta p \end{cases} \tag{4.124}$$

and satisfy the final conditions

$$\begin{pmatrix} \Omega_{xx} & \psi_x & \Psi_x \\ \psi_x^T & 0 & \dot{\psi} \\ \Psi_x^T & \dot{\psi}^T & \dot{\Psi} \end{pmatrix} \begin{pmatrix} \delta x_f \\ \delta\lambda \\ \delta t_f \end{pmatrix} = \begin{pmatrix} \delta p_f \\ \delta\psi_f \\ 0 \end{pmatrix} \quad \text{with} \quad \begin{cases} \Omega \underset{def}{=} \varphi + \lambda^T \psi \\ \Psi \underset{def}{=} \Omega_t + H = \dot{\Omega} + L \end{cases} \tag{4.125}$$

Demonstration (see [R3])

We expand the equations of the extremal (4.121) to order 1.

Differential equations

For the differential equations and the derivative H_u, we have

$$\begin{cases} \dot{x} = H_p \\ \dot{p} = -H_x \\ H_u = 0 \end{cases} \rightarrow \begin{cases} \delta\dot{x} = H_{px}^T\delta x + H_{pu}^T\delta u + H_{pp}^T\delta p \\ \delta\dot{p} = -H_{xx}^T\delta x - H_{xu}^T\delta u - H_{xp}^T\delta p \\ 0 = H_{ux}^T\delta x + H_{uu}^T\delta u + H_{up}^T\delta p \end{cases} \rightarrow \begin{cases} \delta\dot{x} = f_x^T\delta x + f_u^T\delta u \\ \delta\dot{p} = -H_{xx}\delta x - H_{ux}\delta u - f_x\delta p \\ \delta u = -H_{uu}^{-1}(H_{xu}\delta x + f_u\delta p) \end{cases}$$

The last equation gives δu as a function of $\delta x, \delta p$ assuming H_{uu} is invertible. Replacing δu in the first two equations gives (4.124).

Final conditions

Variations δx_0 and $\delta\psi_f$ lead to variations $\delta t_f, \delta x_f, \delta p_f, \delta\lambda$.

The total changes in the state and the final adjoint are: $\begin{cases} dx_f = \dot{x}(t_f)\delta t_f + \delta x_f \\ dp_f = \dot{p}(t_f)\delta t_f + \delta p_f \end{cases}$.

The transversality condition on x_f is expressed with

$$\Omega = \varphi + \lambda^T\psi \rightarrow \Omega_x[x(t_f), t_f, \lambda] = p(t_f)$$

The transversality condition on t_f is expressed with

$$\Psi = \Omega_t + H \rightarrow \Psi[x(t_f), t_f, \lambda] = 0$$

We observe that Ψ can be written with the total derivative: $\dot{\Omega} = \Omega_x^T\dot{x} + \Omega_t$.

$$\Psi = \Omega_t + H = (\dot{\Omega} - \Omega_x^T\dot{x}) + (L + p^T\dot{x}) = \dot{\Omega} + L \quad \text{because } \Omega_x = p \text{ at } t_f$$

Each equation is expanded to order 1 using $\Omega_{x\lambda} = \psi_x$ and $\Psi_\lambda = \dot{\Omega}_\lambda = \dot{\psi}$.

$$\begin{cases} \psi[x(t_f), t_f] = \delta\psi_f \\ \Omega_x[x(t_f), t_f, \lambda] = p(t_f) \\ \Psi[x(t_f), t_f, \lambda] = 0 \end{cases} \rightarrow \begin{cases} \psi_x^T\delta x_f + \psi_x^T\dot{x}(t_f)\delta t_f + \psi_t^T\delta t_f = \delta\psi_f \\ \Omega_{xx}\delta x_f + \Omega_{xx}\dot{x}(t_f)\delta t_f + \Omega_{xt}\delta t_f + \psi_x\delta\lambda = \delta p_f + \dot{p}(t_f)\delta t_f \\ \Psi_x^T\delta x_f + \Psi_x^T\dot{x}(t_f)\delta t_f + \Psi_t^T\delta t_f + \dot{\psi}^T\delta\lambda = 0 \end{cases}$$

Grouping the terms into $\delta x_f, \delta t_f, \delta\lambda$, we obtain

$$\begin{cases} \psi_x^T\delta x_f + \dot{\psi}\delta t_f = \delta\psi_f \\ \Omega_{xx}\delta x_f + (\dot{\Omega}_x - \dot{p})\delta t_f + \psi_x\delta\lambda = \delta p \\ \Psi_x^T\delta x_f + \dot{\Psi}\delta t_f + \dot{\psi}^T\delta\lambda = 0 \end{cases}$$

The term $\dot{\Omega}_x - \dot{p}$ is simplified by using $\Omega_x = p$ at t_f.

$$\begin{cases} \dot{\Omega}_x = \Omega_{xt} + \Omega_{xx}\dot{x} = \Omega_{xt} + \Omega_{xx} f \\ -\dot{p} = H_x = L_x + f_x^T p = L_x + f_x^T \Omega_x \end{cases} \Rightarrow \dot{\Omega}_x - \dot{p} = \frac{\partial}{\partial x}(\Omega_t + L + \Omega_x f) = \frac{\partial}{\partial x}(\Omega_t + H) = \frac{\partial \Psi}{\partial x}$$

Replacing $\dot{\Omega}_x - \dot{p}$ in the second equation gives (4.125).

Equations (4.124) and (4.125) defining the neighboring extremal form a two-point boundary value problem with δx_0 given and δp_f constrained. This problem is solved by the transport method.

<u>Transport method</u>

The system (4.124) and the final constraints (4.125) are linear of the form

$$\begin{cases} \delta\dot{x} = A\delta x - B\delta p \\ \delta\dot{p} = -C\delta x - A^T\delta p \end{cases} \quad \text{with} \quad \begin{cases} A = f_x^T - f_u^T H_{uu}^{-1} H_{xu} \\ B = f_u^T H_{uu}^{-1} f_u \\ C = H_{xx} - H_{ux} H_{uu}^{-1} H_{xu} \end{cases} \tag{4.126}$$

$$\begin{pmatrix} \delta p_f \\ \delta\psi_f \\ 0 \end{pmatrix} = \begin{pmatrix} \Omega_{xx} & \psi_x & \Psi_x \\ \psi_x^T & 0 & \dot{\psi} \\ \Psi_x^T & \dot{\psi}^T & \dot{\Psi} \end{pmatrix} \begin{pmatrix} \delta x_f \\ \delta\lambda \\ \delta t_f \end{pmatrix} \quad \text{with} \quad \begin{cases} \Omega = \varphi + \lambda^T\psi \\ \Psi = \Omega_t + H = \dot{\Omega} + L \end{cases}$$

The variations $(\delta x_0, \delta\psi_f)$ are fixed, the unknowns are $(\delta p_0, \delta t_f, \delta\lambda)$.

The transport method (section 3.5.4) consists in searching for a linear solution in the same form as the final conditions (the solution is known to be of this form).

$$\begin{pmatrix} \delta p(t) \\ \delta\psi_f \\ 0 \end{pmatrix} = \begin{pmatrix} S & M & \alpha \\ M^T & N & \beta \\ \alpha^T & \beta^T & \gamma \end{pmatrix} \begin{pmatrix} \delta x(t) \\ \delta\lambda \\ \delta t_f \end{pmatrix} \tag{4.127}$$

The matrices $S, M, N, \alpha, \beta, \gamma$ satisfy the differential system

$$\begin{cases} \dot{S} = -SA - A^T S + SBS - C \\ \dot{M} = (SB - A^T)M \\ \dot{N} = M^T B M \\ \dot{\alpha} = (SB - A^T)\alpha \\ \dot{\beta} = M^T B \alpha \\ \dot{\gamma} = \alpha^T B \alpha \end{cases} \quad \text{with} \quad \begin{cases} S(t_f) = \Omega_{xx}(t_f) \\ M(t_f) = \psi_x(t_f) \\ N(t_f) = 0 \\ \alpha(t_f) = \Psi_x(t_f) \\ \beta(t_f) = \dot{\psi}(t_f) \\ \gamma(t_f) = \dot{\Psi}(t_f) \end{cases} \tag{4.128}$$

Demonstration (see [R3])

The final conditions can be deduced directly from (4.126) by identification.
The differential equations are obtained by deriving (4.127).

$$\begin{pmatrix}\delta p(t)\\ \delta\psi_f\\ 0\end{pmatrix}=\begin{pmatrix}S & M & \alpha\\ M^T & N & \beta\\ \alpha^T & \beta^T & \gamma\end{pmatrix}\begin{pmatrix}\delta x(t)\\ \delta\lambda\\ \delta t_f\end{pmatrix}$$

$$\Rightarrow\begin{pmatrix}\delta\dot{p}\\ 0\\ 0\end{pmatrix}=\begin{pmatrix}\dot{S} & \dot{M} & \dot{\alpha}\\ \dot{M}^T & \dot{N} & \dot{\beta}\\ \dot{\alpha}^T & \dot{\beta}^T & \dot{\gamma}\end{pmatrix}\begin{pmatrix}\delta x\\ \delta\lambda\\ \delta t_f\end{pmatrix}+\begin{pmatrix}S & M & \alpha\\ M^T & N & \beta\\ \alpha^T & \beta^T & \gamma\end{pmatrix}\begin{pmatrix}\delta\dot{x}\\ 0\\ 0\end{pmatrix}$$

and replacing $\delta\dot{x},\delta\dot{p}$: $\begin{cases}\delta\dot{x}= & A\delta x- & B\delta p\\ \delta\dot{p}= & -C\delta x-A^T\delta p\end{cases}$ with $\delta p(t)=S\delta x(t)+M\delta\lambda+\alpha\delta t_f$.

The nominal extremal has been previously calculated and the nominal values of (x_f,p_f,t_f,λ) are known. The matrices $S,M,N,\alpha,\beta,\gamma$ at the final time t_f are calculated by (4.128). Backward integration yields the matrices at the initial time. The unknowns $(\delta p_0,\delta t_f,\delta\lambda)$ are then solutions of the system (4.127) at t_0.

$$\begin{pmatrix}\delta p_0\\ \delta\psi_f\\ 0\end{pmatrix}=\begin{pmatrix}S & M & \alpha\\ M^T & N & \beta\\ \alpha^T & \beta^T & \gamma\end{pmatrix}(t_0)\begin{pmatrix}\delta x_0\\ \delta\lambda\\ \delta t_f\end{pmatrix} \tag{4.129}$$

The solution is expressed in terms of the given variations $(\delta x_0,\delta\psi_f)$.

$$\begin{cases}\delta\lambda & =\bar{N}^{-1}\left[\delta\psi_f-\bar{M}^T\delta x_0\right]\\ \delta t_f & =-\dfrac{\alpha^T\delta x_0+\beta^T\delta\lambda}{\gamma}\\ \delta p_0 & =(\bar{S}-\bar{M}\bar{N}^{-1}\bar{M}^T)\delta x_0+\bar{M}\bar{N}^{-1}\delta\psi_f\end{cases}\quad\text{with}\quad\begin{cases}\bar{M}=M-\dfrac{\alpha\beta^T}{\gamma}\\ \bar{N}=N-\dfrac{\beta\beta^T}{\gamma}\\ \bar{S}=S-\dfrac{\alpha\alpha^T}{\gamma}\end{cases} \tag{4.130}$$

Reduced transport method

A possible simplification is to eliminate δt_f from the third final condition (4.127).

$\delta t_f = -\dot{\Psi}^{-1}\left[\Psi_x^T \delta x_f + \dot{\psi}^T \delta\lambda\right]$.

The final conditions are reduced to

$$\begin{pmatrix} \delta p_f \\ \delta\psi \end{pmatrix} = \begin{pmatrix} \Omega_{xx} - \Psi_x \dot{\Psi}^{-1} \Psi_x^T & \psi_x - \Psi_x \dot{\Psi}^{-1} \dot{\psi}_x^T \\ \psi_x^T - \dot{\psi}\, \dot{\Psi}^{-1} \Psi_x^T & -\dot{\psi}\, \dot{\Psi}^{-1} \dot{\psi}^T \end{pmatrix} \begin{pmatrix} \delta x_f \\ \delta\lambda \end{pmatrix} \underset{\text{def}}{=} \begin{pmatrix} \bar{S}(t_f) & \bar{M}(t_f)^T \\ \bar{M}(t_f) & \bar{N}(t_f) \end{pmatrix} \begin{pmatrix} \delta x_f \\ \delta\lambda \end{pmatrix}$$

The transport method is then applied to the reduced form.

$$\begin{pmatrix} \delta p(t) \\ \delta\psi \end{pmatrix} = \begin{pmatrix} \bar{S} & \bar{M} \\ \bar{M}^T & \bar{N} \end{pmatrix} \begin{pmatrix} \delta x(t) \\ \delta\lambda \end{pmatrix} \quad \text{with} \quad \begin{cases} \dot{\bar{S}} = -\bar{S}A - A^T\bar{S} + \bar{S}B\bar{S} - C \\ \dot{\bar{M}} = (\bar{S}B - A^T)\bar{M} \\ \dot{\bar{N}} = \bar{M}^T B \bar{M} \end{cases}$$

The matrices $\bar{S}, \bar{M}, \bar{N}$ follow the same differential equations as S, M, N.

If the final time is fixed, we find: $\alpha = \beta = \gamma = 0$ and $\bar{S} = S$, $\bar{M} = M$, $\bar{N} = N$.

Figure 4-26 depicts the calculation of the neighboring extremal from the nominal extremal and from the variations $(\delta x_0, \delta\psi_f)$.

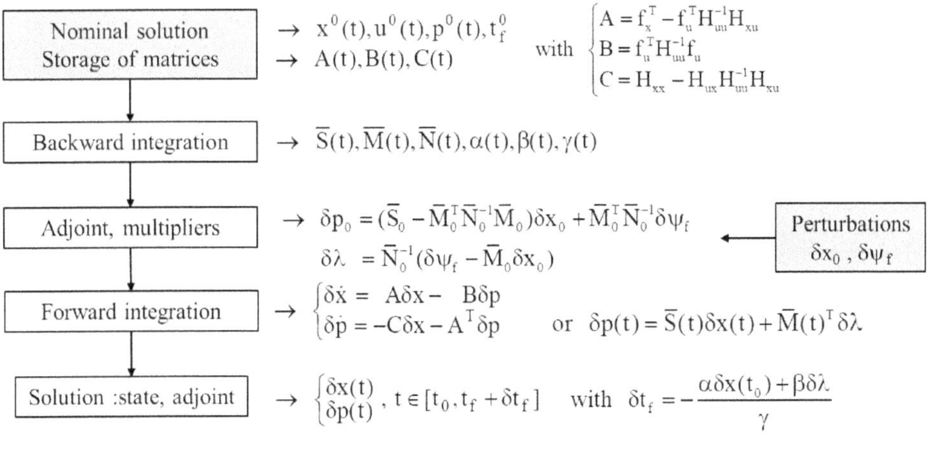

Figure 4-26: Calculation of the neighboring extremal.

4.3.5 State feedback control

The backward integration of (4.128) gives the matrices $\bar{S},\bar{M},\bar{N},\alpha,\beta,\gamma$ (4.130) as a function of time. These matrices allow to express the corrections on the control $\delta u(t)$ and on the final time δt_f to be applied at time t as a function of the state deviations $\delta x(t)$ and constraints deviations $\delta\psi$ observed at this same time.

$$\begin{cases} \delta u = -H_{uu}^{-1}\left[\left(H_{xu} + f_u(\bar{S} - \bar{M}\bar{N}^{-1}\bar{M}^T)\right)\delta x + f_u\bar{M}\bar{N}^{-1}\delta\psi\right] \\ \delta t_f = -\gamma^{-1}\left[(\alpha^T - \beta^T\bar{N}^{-1}\bar{M}^T)\delta x - \beta^T\bar{N}^{-1}\delta\psi\right] \end{cases} \quad (4.131)$$

Demonstration

The final time variation δt_f is given by (4.130) by replacing $\delta\lambda$.

The variation δu is given by $H_u = 0$ to order 1: $\delta u = -H_{uu}^{-1}(H_{xu}\delta x + f_u\delta p)$.

Equations (4.127) allow to express $\delta p,\delta\lambda,\delta t_f$ as a function of $\delta x,\delta\psi_f$ at time t. The calculation is similar to the one performed at the initial time, giving the formulas (4.130). Substituting $\delta p,\delta\lambda,\delta t_f$, we obtain the formula (4.131).

The formulas (4.131) can be put into the more compact form

$$\begin{cases} \delta u(t) = -K_x(t)\delta x(t) - K_\psi(t)\delta\psi \\ \delta t_f \;\; = -k_x(t)\delta x(t) - k_\psi(t)\,\delta\psi \end{cases} \quad (4.132)$$

The matrices K_x, K_ψ, k_x, k_ψ are the gains calculated along the nominal trajectory and stored as a function of the elapsed time $t - t_0$ or of the remaining time $t_f - t$ (also called time-to-go).

The state deviations $\delta x(t)$ between the actual and nominal states are estimated from measurements made by sensors. The constraint deviations $\delta\psi$ represent an optional change of target that may occur during the trajectory.

The feedback control in the form of (4.132) allows the trajectory to be corrected in real time according to the perturbations encountered.

Figure 4-27 illustrates the state feedback correction process.

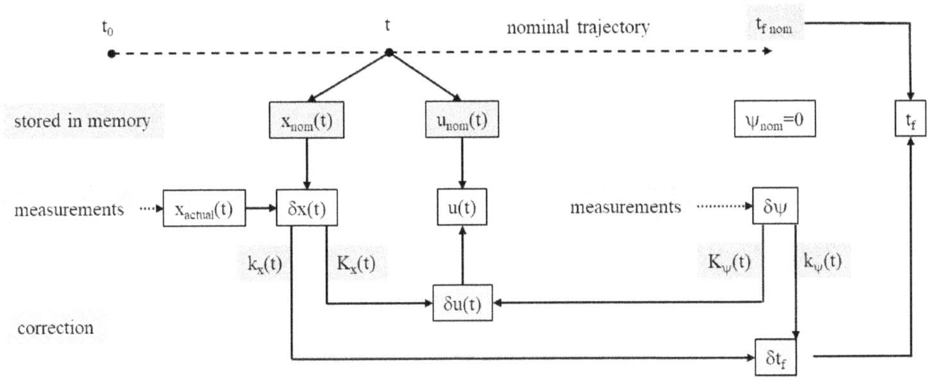

Figure 4-27: Control correction in state feedback.

Example 4-13: Launch into orbit (from [R3])

Let us retrieve the launch into orbit problem of example 4-5.
The aim here is to reach the orbit altitude z_c with a zero vertical velocity, while maximizing the final horizontal velocity at a fixed final time. The acceleration of constant modulus a is oriented by the pitch angle θ.
The problem is formulated as

$$\min_{\theta} J = -v_x(t_f) = -a\int_{t_0}^{t_f} \cos\theta \, dt \quad \text{with} \quad \begin{cases} \dot{z} = v_z \\ \dot{v}_z = a\sin\theta - g \end{cases} \quad \text{and} \quad \begin{cases} z(t_f) = z_c \\ v_z(t_f) = 0 \end{cases}$$

Nominal extremal
This problem is solved in the same way as the minimum time problem solved in example 4-5. The control $\theta(t)$ follows a linear tangent law of the form
$\tan\theta_f - \tan\theta = -c(t_f - t)$.
Switching to the variable θ, we obtain the trajectory equations of example 4-5.

Neighboring extremals
The matrices S, M, N involved in the transport method (4.128) must be computed with $\alpha = \beta = \gamma = 0$, as the problem is a fixed time problem. The useful matrices for this calculation are the following.

Derivatives of f (dynamics)

$$f = \begin{pmatrix} v \\ a\sin\theta - g \end{pmatrix} \Rightarrow f_x = \begin{pmatrix} 0 & 0 \\ 1 & 0 \end{pmatrix} , \quad f_u = (0 , a\cos\theta)$$

Derivatives of H (Hamiltonian)

$$H_{xx} = \begin{pmatrix} 0 & 0 \\ 0 & 0 \end{pmatrix} , \quad H_{ux} = \begin{pmatrix} 0 \\ 0 \end{pmatrix} , \quad H_{uu} = a\cos\theta - p_v a\sin\theta = \frac{a}{\cos\theta}$$

Derivatives of ψ (constraint)

$$\psi = \begin{pmatrix} z(t_f) - z_c \\ v(t_f) \end{pmatrix} \Rightarrow \psi_x = \begin{pmatrix} 1 & 0 \\ 0 & 1 \end{pmatrix}$$

Matrices A, B, C

$$\begin{cases} A = f_x^T - f_u^T H_{uu}^{-1} H_{xu} \\ B = f_u^T H_{uu}^{-1} f_u \\ C = H_{xx} - H_{ux} H_{uu}^{-1} H_{xu} \end{cases} \Rightarrow A = \begin{pmatrix} 0 & 1 \\ 0 & 0 \end{pmatrix} , \quad B = a\cos^3\theta \begin{pmatrix} 0 & 0 \\ 0 & 1 \end{pmatrix} , \quad C = \begin{pmatrix} 0 & 0 \\ 0 & 0 \end{pmatrix}$$

The matrices S, M, N are calculated by integrating backwards the system (4.128).

$$\begin{cases} \dot{S} = -SA - A^T S + SBS - C \\ S(t_f) = \Omega_{xx}(t_f) \end{cases} \rightarrow S(t) = \begin{pmatrix} 0 & 0 \\ 0 & 0 \end{pmatrix}$$

$$\begin{cases} \dot{M} = (SB - A^T)M \\ M(t_f) = \psi_x(t_f) \end{cases} \rightarrow \begin{cases} \dot{M}_{11} = 0 , & M_{11}(t_f) = 1 \\ \dot{M}_{21} = 0 , & M_{21}(t_f) = 0 \\ \dot{M}_{12} = -M_{11} , & M_{12}(t_f) = 0 \\ \dot{M}_{22} = -M_{21} , & M_{22}(t_f) = 1 \end{cases} \rightarrow \begin{cases} M_{11} = 1 \\ M_{21} = 0 \\ M_{12} = t_f - t \\ M_{22} = 1 \end{cases}$$

$$\rightarrow M(t) = \begin{pmatrix} 1 & t_f - t \\ 0 & 1 \end{pmatrix}$$

$$\begin{cases} \dot{N} = M^T B M \\ N(t_f) = 0 \end{cases} \rightarrow \begin{pmatrix} \dot{N}_{11} & \dot{N}_{12} \\ \dot{N}_{21} & \dot{N}_{22} \end{pmatrix} = a\cos^3\theta \begin{pmatrix} (t_f - t)^2 & t_f - t \\ t_f - t & 1 \end{pmatrix}$$

The terms of the matrix N are integrated by changing to the variable θ.

$$t_f - t = \frac{\tan\theta - \tan\theta_f}{c} \Rightarrow dt = -\frac{d\theta}{c\cos^2\theta}$$

$$\text{Term } N_{11}: \quad \frac{dN_{11}}{dt} = a(t_f - t)^2 \cos^3\theta \quad \Rightarrow \quad \frac{dN_{11}}{d\theta} = \frac{a}{c^3}(\tan\theta - \tan\theta_f)^2 \cos\theta$$

$$\rightarrow N_{11} = \frac{a}{c^3}\left[\frac{\sin\theta_f - \sin\theta}{\cos^2\theta_f} + \ln\left(\tan\theta_f + \frac{1}{\cos\theta_f}\right) - \ln\left(\tan\theta + \frac{1}{\cos\theta}\right)\right]$$

with $N_{11}(t_f) = 0$

$$\text{Term } N_{12}: \quad \frac{dN_{12}}{dt} = a(t_f - t)\cos^3\theta \quad \Rightarrow \quad \frac{dN_{12}}{d\theta} = \frac{a}{c^2}(\tan\theta - \tan\theta_f)\cos\theta$$

$$\rightarrow N_{12} = \frac{a}{c^2}\frac{\cos(\theta_f - \theta) - 1}{\cos\theta_f} \quad \text{with} \quad N_{12}(t_f) = 0$$

$$\text{Term } N_{22}: \quad \frac{dN_{22}}{dt} = a\cos^3\theta \quad \Rightarrow \quad \frac{dN_{22}}{d\theta} = -\frac{a}{c}\cos\theta$$

$$\rightarrow N_{22} = \frac{a}{c}(\sin\theta_f - \sin\theta) \quad \text{with} \quad N_{22}(t_f) = 0$$

Control correction

The control correction is given by (4.131), with $(\bar{S}, \bar{M}, \bar{N}) = (S, M, N)$.

$$\delta u = -H_{uu}^{-1}\left[\left(H_{xu} + f_u(S - MN^{-1}M^T)\right)\delta x(t) + f_u MN^{-1}\delta\psi\right]$$

This yields the correction $\delta\theta$ as a function of the state and constraint deviation.

Noting $v = v_z$, the state deviation is $\begin{pmatrix}\delta z \\ \delta v\end{pmatrix}$ and the constraint deviation is $\begin{pmatrix}\delta z_f \\ \delta v_f\end{pmatrix}$.

The correction is: $\delta\theta = K_z\left[\delta z + (t_f - t)\delta v - \delta z_f\right] + K_v\left[\delta v - \delta v_f\right]$

$$\text{with the gains: } \begin{cases} K_z = \dfrac{\cos^2\theta}{\det N}(N_{22}(t_f - t) - N_{12}) \\ K_v = \dfrac{\cos^2\theta}{\det N}(N_{11} - N_{12}(t_f - t)) \end{cases}$$

The time-dependent gains K_z and K_v are calculated on the nominal trajectory. They are associated with predicted deviations in final altitude and final velocity. If the targeted conditions do not change during the trajectory $(\delta z_f = 0, \delta v_f = 0)$, the correction takes the simpler form

$$\delta\theta = K_1(\delta v + K_2 \delta z) \quad \text{with} \quad \begin{cases} K_2 = K_z / K_1 \\ K_1 = K_v + K_z(t_f - t) \end{cases}.$$

Figure 4-28 shows the evolution of the gains K_1 and K_2 along a trajectory aiming at an altitude of 200 km in 300 s with the acceleration $a=3g$ ($g=9\ m\cdot s^{-2}$). The nominal solution gives (example 4-5): $\theta_0 = 28.08\text{deg}$ and $\beta=-0.0035$. The gains are plotted in logarithmic scale as $\log(-K)$.

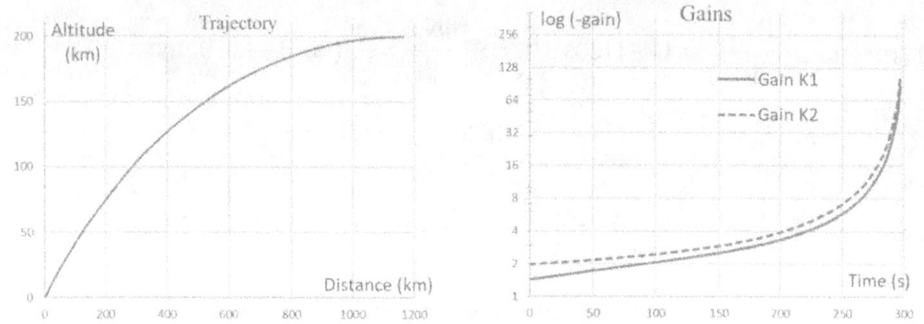

Figure 4-28: Gain evolution along the nominal trajectory.

Implicit guidance consists in applying a control correction of the form (4.132) with precalculated gains. This correction based on a local linearization in the neighborhood of the nominal solution does not give the true extremal from the current state. In contrast, explicit guidance consists in recalculating the true extremal from the current state. This strategy is more computationally expensive, but it can be more robust in the presence of strong perturbations.

4.3.6 Hamilton-Jacobi-Bellman equation

Consider a free final time problem with final constraints.

$$\min_{u,t_f} J = \int_{t_0}^{t_f} L(x,u,t)dt + \varphi(x_f,t_f) \quad \text{with} \quad \begin{cases} \dot{x} = f(x,u,t) \\ x(t_0) = x_0 \\ \psi(x_f,t_f) = 0 \end{cases} \tag{4.133}$$

The solution and the cost J depend on the initial point $(x_0;t_0)$. The value function $V(x,t)$ is defined as the minimum cost starting from $(x;t)$.

Figure 4-29 shows four different extremals obtained by varying the initial time and initial state.

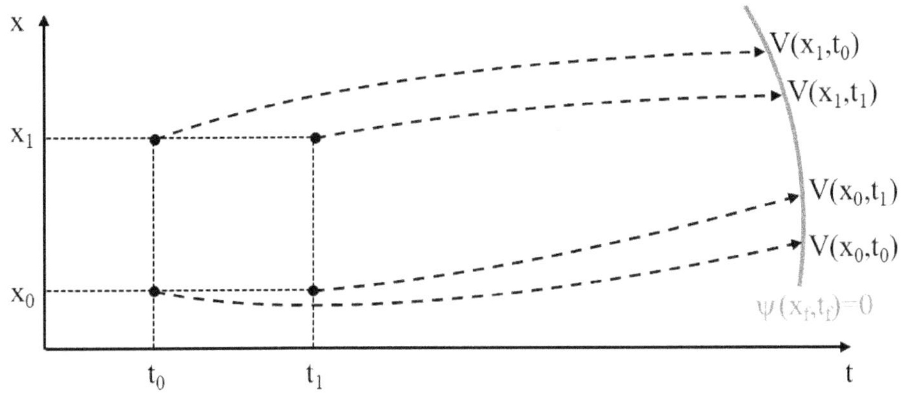

Figure 4-29: Endpoints and value function.

Property 4-7: Hamilton-Jacobi-Bellman equation

The function value $V(x,t)$ satisfies the equation (HJB).

$$\min_u H(x,u,V_x,t) + V_t = 0 \tag{4.134}$$

where $H(x,u,p,t) = L(x,u,t) + p^T f(x,u,t)$ is the Hamiltonian of problem (4.133) and $V_x = \frac{\partial V}{\partial x}$, $V_t = \frac{\partial V}{\partial t}$ are the partial derivatives of the value function.

Demonstration

Let u* be the optimal control and let us place ourselves at any point $(x_1;t_1)$ of the optimal trajectory. According to Bellman's optimality principle ("Any subpath of an optimal path is itself optimal"), the optimal control from $(x_1;t_1)$ is u*. Otherwise, the solution of problem (4.133) could be improved with another control from $(x_1;t_1)$.

The value function in $(x_1;t_1)$ is obtained with the control u*.

$$V(x_1,t_1) = \int_{t_1}^{t_f} L(x,u^*,t)dt + \varphi(x_f,t_f)$$

The trajectory from $(x_1;t_1)$ is broken down into two parts:
- the control u* from t_1 to $t_1+\delta t_1$ leads to $x_1+\delta x_1 = x_1 + f(x_1,u^*,t_1)\delta t_1$;
- the optimal path from $(x_1+\delta x_1;t_1+\delta t_1)$ has a cost $V(x_1+\delta x_1,t_1+\delta t_1)$.

The cost on the second part is obtained with the control u* (still according to Bellman's optimality principle).
The optimal cost from $(x_1;t_1)$ is the sum of these two contributions.

$$V(x_1,t_1) = \int_{t_1}^{t_1+\delta t_1} L(x,u^*,t)dt + V(x_1+\delta x_1,t_1+\delta t_1)$$

Let us expand each term to order 1 for δt_1 small.

$V(x_1,t_1) = L(x_1,u^*,t_1)\delta t_1 + V(x_1,t_1) + V_x^T\delta x_1 + V_t\delta t_1$ with $\delta x_1 = f(x_1,u^*,t_1)\delta t_1$

After simplification, we obtain: $L(x_1,u^*,t_1) + V_x^T f(x_1,u^*,t_1) + V_t = 0$

which can be written as: $H(x_1,u^*,V_x,t_1) + V_t = 0$

The point $(x_1;t_1)$ on the optimal trajectory is arbitrary. According to the PMP, the control u* minimizes the Hamiltonian at any point on the optimal path. This gives the equation HJB (4.134).

If we can solve the HJB equation, we can then define the adjoint by: $p = V_x$.
We thus retrieve the influence functions (4.25) and the adjoint equations.

Adjoint equations from HJB

Let us calculate the derivative of the adjoint.

$p = V_x \Rightarrow \dot{p} = \dot{V}_x = V_{xx}\dot{x} + V_{xt} = V_{xx}f(x,u^*,t) + V_{xt}$

The HJB equation gives

$V_t = -H(x,u^*,V_x,t) = -L(x,u^*,t) - V_x^T f(x,u^*,t)$

We derive with respect to x.

$V_{tx} = -L_x(x,u^*,t) - V_{xx}f(x,u^*,t) - V_x^T f_x(x,u^*,t)$

Replacing in $\dot{p}$, we obtain

$\dot{p} = -L_x(x,u^*,t) - V_x^T f_x(x,u^*,t) = -L_x(x,u^*,t) - p^T f_x(x,u^*,t)$

which is the adjoint equation: $\dot{p} = -H_x(x,u^*,p,t)$.

Non-differentiable solution
Equation HJB (4.134) assumes the value function to be of class C^1. This strong assumption is not necessarily satisfied as example 4-14 shows.

Example 4-14: Non-differentiable HJB solution

Consider the problem with fixed final time: $\min\limits_{-1\le u\le +1} J = x(t_f)$ with $\begin{cases}\dot{x} = xu \\ x(t_0) = x_0\end{cases}$

The HJB equation for this problem is: $\min\limits_{-1\le u\le +1} (V_x xu) + V_t = 0 \Rightarrow -|V_x x| + V_t = 0$

Let us look for the optimal control by examining the sign possibilities of x_0:

- if $x_0 < 0$, we should take $u = +1$ to have $\dot{x} < 0$ and thus minimize $x(t_f)$:

$$\Rightarrow \dot{x} = +x \Rightarrow x(t_f) = x_0 e^{+(t_f - t_0)}$$

- if $x_0 > 0$, we should take $u = -1$ to have $\dot{x} < 0$ and thus minimize $x(t_f)$:

$$\Rightarrow \dot{x} = -x \Rightarrow x(t_f) = x_0 e^{-(t_f - t_0)}$$

- if $x_0 = 0$, then $x(t_f) = 0$, $\forall u$.

The value function is thus defined by

$$V(x,t) = \begin{cases} xe^{+(t_f - t)} & \text{if } x < 0 \quad \rightarrow \quad V_x x = V < 0 \,,\; V_t = -V \\ 0 & \text{if } x = 0 \\ xe^{-(t_f - t)} & \text{if } x > 0 \quad \rightarrow \quad V_x x = V > 0 \,,\; V_t = +V \end{cases}$$

This function satisfies the HJB equation, but it is not differentiable in $x = 0$.
The HJB equation of this problem does not have a differentiable solution on $\mathbb{R}$.

Sub-gradient and over-gradient
To generalize the HJB equation to non-differentiable solutions, we define the sub- and over-gradient of a function.

A sub-gradient of $f : \mathbb{R}^n \to \mathbb{R}$ in x_0 is a vector $g \in \mathbb{R}^n$ satisfying locally

$$f(x) \le f(x_0) + g^T (x - x_0)$$

An over-gradient of $f : \mathbb{R}^n \to \mathbb{R}$ in x_0 is a vector $g \in \mathbb{R}^n$ satisfying locally

$$f(x) \ge f(x_0) + g^T (x - x_0)$$

The sub-differential $\partial f^-(x_0)$ of f in x_0 is the set of its sub-gradients.

The over-differential $\partial f^+(x_0)$ of f in x_0 is the set of its over-gradients.

These definitions are illustrated in figure 4-30.

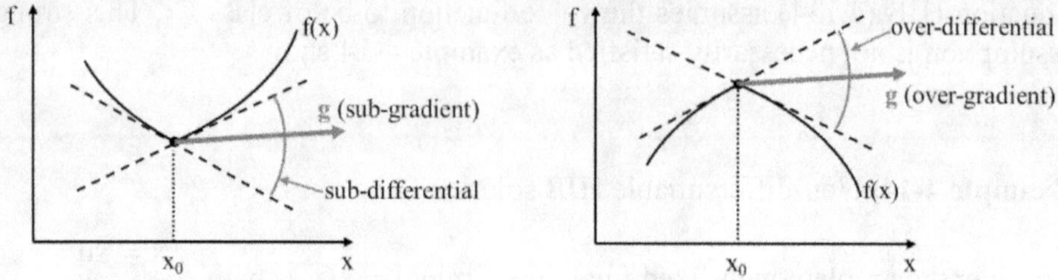

Figure 4-30: Sub- and over-gradient, sub- and over-differential.

If the function f is differentiable in x_0, then its gradient $\nabla f(x_0)$ is the unique sub- and over-gradient: $\partial f^-(x_0) = \partial f^+(x_0) = \{\nabla f(x_0)\}$.

Example 4-15: Sub-gradient and over-gradient

Consider the function

$$f(x) = \begin{cases} 0 & \text{if } x < 0 \\ \sqrt{x} & \text{if } 0 \le x \le 1 \\ 1 & \text{if } 1 < x \end{cases}$$

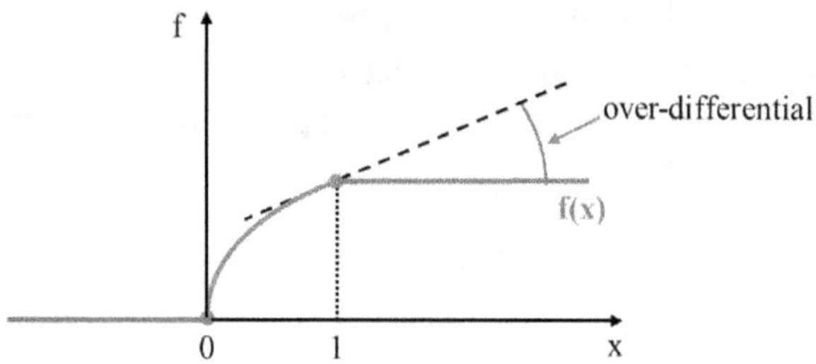

This function is derivable except in 0 and 1.

Let us calculate its sub-differential and over-differential in 0 and 1.

For $x_0 = 0$: $\begin{cases} \partial f^-(0) = [0;+\infty[\\ \partial f^+(0) = \varnothing \end{cases}$.

For $x_0 = 1$: $\begin{cases} \partial f^-(1) = \varnothing \\ \partial f^+(1) = [0;0,5] \end{cases}$.

Viscosity solution

Consider a partial differential equation (PDE): $F[x, f(x), \nabla f(x)] = 0$.

The unknown is the function $f : \mathbb{R}^n \to \mathbb{R}$.

The equation (PDE) is defined by the function $F : \mathbb{R}^n \times \mathbb{R} \times \mathbb{R}^n \to \mathbb{R}$.

A continuous function f is called:

- a sub-viscosity solution if: $F[x, f(x), g] \le 0$, $\forall x$, $\forall g \in \partial f^+(x)$;
- an over-viscosity solution if: $F[x, f(x), g] \ge 0$, $\forall x$, $\forall g \in \partial f^-(x)$;
- a viscosity solution if it is both a sub-viscosity and an over-viscosity solution.

Example 4-16: Viscosity solution

Consider the partial differential equation: $1 - |\nabla f(x)| = 0$ with $f : \mathbb{R} \to \mathbb{R}$.

For this PDE, the function F is defined as: $F(x, f, g) = 1 - |g|$.

A classic solution of class C^1 is: $f(x) = \pm x + c$.

Let us show that there is also the solution of viscosity: $f(x) = |x|$:

- this function is continuous;
- for $x \ne 0$, the function f is derivable and it satisfies the equation;
- for $x = 0$, we need to examine the sub- and over-gradients of f.

Sub-gradient in $x_0 = 0$: $\partial f^-(x_0) = [-1, +1]$

$\to$ $F(x_0, f(x_0), g) = 1 - |g| \ge 0$, $\forall g \in \partial f^-(x_0)$

$\to$ f is an over-viscosity solution in $x_0 = 0$.

Over-gradient in $x_0 = 0$: $\partial f^+(x_0) = \varnothing$

$\to$ f is a sub-viscosity solution in $x_0 = 0$.

f is therefore both a sub-viscosity and an over-viscosity solution.

The notion of a viscosity solution extends the set of solutions of the HJB equation to continuous functions. It can then be shown that the value function $V(x,t)$ is the unique viscosity solution of the HJB equation.

4.4 Optimality conditions of second order

This section presents the sufficient conditions of optimality derived from the auxiliary minimum problem and the conditions associated with singular arcs.

4.4.1 Auxiliary minimum problem

Consider a free final time problem with final constraints.

$$\min_{u,t_f} J = \int_{t_0}^{t_f} L(x,u,t)dt + \varphi(x_f,t_f) \quad \text{with} \begin{cases} \dot{x} = f(x,u,t) \\ x(t_0) = x_0 \\ \psi(x_f,t_f) = 0 \end{cases} \tag{4.135}$$

The augmented cost J_a is defined with the adjoint p and the multipliers λ.

$$J_a = \int_{t_0}^{t_f} \left[L + p^T(f - \dot{x})\right]dt + \varphi + \lambda^T\psi \tag{4.136}$$

We define the normal Hamiltonian H (with $p_0 = 1$) and functions Ω and Ψ.

$$\begin{cases} H = L + p^T f \\ \Omega = \varphi + \lambda^T\psi \\ \Psi = \dot{\Omega} + L \end{cases} \tag{4.137}$$

The optimality conditions are established from the first and second variations of the augmented cost considering variations $\delta x, \delta u, \delta p$, $\delta t_f, \delta x_f$. These variations must be feasible with respect to the final constraints $\psi(x_f,t_f) = 0$.
To generalize the results, we also consider variations δx_0 on the initial state and $\delta\psi_f$ on the final constraints which become: $\psi(x_f,t_f) = \delta\psi_f$. The conditions of optimality of an extremal will be established for $\delta x_0 = 0$ and $\delta\psi_f = 0$.

Property 4-8: First and second variation of the augmented cost

The first variation δJ_a is given by

$$\begin{aligned} \delta J_a = & \int_{t_0}^{t_f} \left[(H_x + \dot{p})^T\delta x + H_u^T\delta u + (f - \dot{x})^T\delta p\right]dt \\ & + (\Omega_x - p)_f^T\delta x_f + \Psi_f\delta t_f + \psi^T\delta\lambda + p_0^T\delta x_0 - \lambda^T\delta\psi_f \end{aligned} \tag{4.138}$$

The second variation $\delta^2 J_a$ in the neighborhood of an extremal is given by

$$\delta^2 J_a = \int_{t_0}^{t_f} \begin{pmatrix} \delta x \\ \delta u \end{pmatrix}^T \begin{pmatrix} H_{xx} & H_{ux} \\ H_{xu} & H_{uu} \end{pmatrix} \begin{pmatrix} \delta x \\ \delta u \end{pmatrix} dt + \begin{pmatrix} \delta x \\ \delta t \end{pmatrix}_f^T \begin{pmatrix} \Omega_{xx} & \Psi_x \\ \Psi_x^T & \dot{\Psi} \end{pmatrix} \begin{pmatrix} \delta x \\ \delta t \end{pmatrix}_f \quad (4.139)$$

Demonstration

The augmented cost (4.136) is put into the form: $J_a = \int_{t_0}^{t_f} \left(H - p^T \dot{x}\right) dt + \Omega$.

Let us show successively the formulas giving the first and second variation of the augmented cost.

First variation

We apply variations $\delta x, \delta u, \delta p, \delta x_f, \delta t_f, \delta\lambda$ and also $\delta x_0, \delta\psi_f$.

$$\delta J_a = \int_{t_0}^{t_f} (H_x^T \delta x + H_u^T \delta u + f^T \delta p - p^T \delta\dot{x} - \dot{x}^T \delta p) dt$$
$$+ (H - p^T \dot{x})_f + \Omega_x^T (\delta x_f + \dot{x}_f \delta t_f) + \Omega_t \delta t_f + \Omega_\lambda^T \delta\lambda - \lambda^T \delta\psi_f$$

including the variation of the integral bound and the contribution $\dot{x}_f \delta t_f$ on $x(t_f)$.

We integrate by parts the term: $\int_{t_0}^{t_f} p^T \delta\dot{x}\, dt = (p^T \delta x)_{t_f} - (p^T \delta x)_{t_0} - \int_{t_0}^{t_f} \dot{p}^T \delta x\, dt$.

By introducing the total derivative: $\dot{\Omega} = \Omega_x^T \dot{x} + \Omega_t$ and with $\Omega_\lambda = \psi$, we obtain

$$\delta J_a = \int_{t_0}^{t_f} \left[(H_x + \dot{p})^T \delta x + H_u^T \delta u + (f - \dot{x})^T \delta p \right] dt$$
$$+ (\Omega_x - p)_f^T \delta x_f + \Psi_f \delta t_f + \psi^T \delta\lambda + p_0^T \delta x_0 - \lambda^T \delta\psi_f$$

The stationarity conditions are (for δx_0 and $\delta\psi_f$ fixed)

$$\delta J_a = 0 \ , \ \forall \delta x, \delta u, \delta p, \delta x_f, \delta t_f, \delta\lambda \quad \Rightarrow \begin{cases} \dot{x} = f \\ \dot{p} = -H_x \\ H_u = 0 \end{cases} , \quad \begin{cases} \psi_f = 0 \\ \Psi_f = 0 \\ p_f = \Omega_x \end{cases} \quad \textbf{(equations E)}$$

<u>Second variation</u>

We apply again variations $\delta x, \delta u, \delta p, \delta x_f, \delta t_f, \delta\lambda$ and $\delta x_0, \delta\psi_f$ on δJ_a.

$$\delta^2 J_a = \int_{t_0}^{t_f}\Big[\delta x^T(H_{xx}\delta x + H_{ux}\delta u + H_{px}\delta p + \delta\dot{p}) + \delta u^T(H_{xu}\delta x + H_{uu}\delta u + H_{pu}\delta p)$$
$$+\delta p^T(f_x^T\delta x + f_u^T\delta u - \delta\dot{x})\Big]dt$$
$$+\Big[(H_x + \dot{p})^T\delta x + H_u^T\delta u + (f-\dot{x})^T\delta p\Big]_f \delta t_f \quad \text{(integral bound)}$$
$$+\delta x_f^T\left[\Omega_{xx}(\delta x + \dot{x}\delta t) + \Omega_{xt}\delta t + \Omega_{x\lambda}\delta\lambda - \delta p - \dot{p}\delta t\right]_f$$
$$+\delta t_f\left[\Psi_x^T(\delta x + \dot{x}\delta t) + \Psi_t\delta t + \Psi_u^T\delta u + \Psi_\lambda^T\delta\lambda\right]_f$$
$$+\delta\lambda^T(\psi_x^T\delta x + \dot{\psi}\delta t)_f + \delta p_0^T\delta x_0 - \delta\lambda^T\delta\psi_f$$

We integrate by parts the term: $\int_{t_0}^{t_f}\delta x^T\delta\dot{p}\,dt = (\delta x^T\delta p)_f - (\delta x^T\delta p)_0 - \int_{t_0}^{t_f}\delta\dot{x}^T\delta p\,dt$.

The variation $\delta^2 J_a$ is to be evaluated in the neighborhood of an extremal. Using equations (E) satisfied on the extremal and expressing the derivatives of Ω, Ψ

$$\begin{cases}\dot{\Psi} = \Psi_x^T\dot{x} + \Psi_t \\ \dot{\Omega} = \Omega_x^T\dot{x} + \Omega_t = \Omega_x^T f + \Omega_t \\ \Omega_\lambda = \psi \\ \Psi_\lambda = \dot{\psi}\end{cases}, \quad \begin{cases}(\dot{\Omega})_x = \dfrac{\partial\dot{\Omega}}{\partial x} = \Omega_{xx}^T f + \Omega_x^T f_x + \Omega_{tx} = \dot{\Omega}_x + \Omega_x^T f_x \\ \dot{\Omega}_x = \dfrac{d\Omega_x}{dt} = \Omega_{xx}^T f + \Omega_{tx}\end{cases}$$

we obtain after grouping the terms

$$\delta^2 J_a = \int_{t_0}^{t_f}\Big[\delta x^T H_{xx}\delta x + \delta x^T H_{ux}\delta u + \delta u^T H_{xu}\delta x + \delta u^T H_{uu}\delta u + 2\delta p^T(f_x^T\delta x + f_u^T\delta u - \delta\dot{x})\Big]dt$$
$$+\delta x_f^T(\Omega_{xx}\delta x + \Psi_x\delta t)_f + \delta t_f(\Psi_x^T\delta x + \dot{\Psi}\delta t)_f + 2\delta\lambda^T(\psi_x^T\delta x + \dot{\psi}\delta t - \delta\psi)_f$$

The first-order expansion of the state equation and of the constraint (considering the variation $\delta\psi_f$) gives: $\begin{cases}\dot{x} = f(x,u,t) \\ \psi(x_f, t_f) = \delta\psi_f\end{cases} \Rightarrow \begin{cases}\delta\dot{x} = f_x^T\delta x + f_u^T\delta u \\ \psi_x^T\delta x_f + \dot{\psi}\delta t_f = \delta\psi_f\end{cases}$

which eliminates the associated terms in $\delta^2 J_a$. There remains in matrix form

$$\delta^2 J_a = \int_{t_0}^{t_f}\begin{pmatrix}\delta x \\ \delta u\end{pmatrix}^T\begin{pmatrix}H_{xx} & H_{ux} \\ H_{xu} & H_{uu}\end{pmatrix}\begin{pmatrix}\delta x \\ \delta u\end{pmatrix}dt + \begin{pmatrix}\delta x \\ \delta t\end{pmatrix}_f^T\begin{pmatrix}\Omega_{xx} & \Psi_x \\ \Psi_x^T & \dot{\Psi}\end{pmatrix}\begin{pmatrix}\delta x \\ \delta t\end{pmatrix}_f$$

The optimality conditions of order 1 and 2 are: $\begin{cases}\delta J_a = 0 \\ \delta^2 J_a \geq 0\end{cases}$, $\forall \delta u, \forall \delta t_f$ feasible.

The variations δu and δt_f are feasible if they satisfy the final constraints.
The first-order condition leads to the extremal equations (4.121) recalled here.

$$\delta J_a = 0 \,,\, \forall \delta u, \forall \delta t_f \quad \Rightarrow \begin{cases}\dot{x} &= f \\ \dot{p} &= -H_x \\ H_u &= 0\end{cases} , \quad \begin{cases}\psi_f = 0 \\ \Psi_f = 0 \\ p_f = \Omega_x\end{cases} \tag{4.140}$$

To verify the second-order condition, we look for the minimum value that the second variation (4.121) can take for feasible variations δu and δt_0 in the neighborhood of an extremal. The deviations δx propagated to the final time must satisfy the final constraints (value $\delta \psi_f$).

This leads to the following **accessory minimum problem**.

$$\min_{\delta u, \delta t_f} \frac{\delta^2 J_a}{2} = \frac{1}{2}\int_{t_0}^{t_f} \begin{pmatrix}\delta x \\ \delta u\end{pmatrix}^T \begin{pmatrix}H_{xx} & H_{ux} \\ H_{xu} & H_{uu}\end{pmatrix}\begin{pmatrix}\delta x \\ \delta u\end{pmatrix} dt + \frac{1}{2}\begin{pmatrix}\delta x \\ \delta t\end{pmatrix}_f^T \begin{pmatrix}\Omega_{xx} & \Psi_x \\ \Psi_x^T & \dot{\Psi}\end{pmatrix}\begin{pmatrix}\delta x \\ \delta t\end{pmatrix}_f$$
$$\text{with } \begin{cases}\delta \dot{x} = f_x^T \delta x + f_u^T \delta u \\ \psi_x^T \delta x_f + \dot{\psi}\delta t_f = \delta \psi_f\end{cases} \tag{4.141}$$

The problem is formulated here in a general way for fixed variations δx_0 and $\delta \psi_f$ of the initial state and of the final constraints. The optimality conditions will be established by considering zero variations: $\delta x_0 = 0$ and $\delta \psi_f = 0$.
The derivatives of H, Ω and Ψ are evaluated on the reference extremal.
The variables are the deviations δu and δt_f from the reference extremal.

Property 4-9: Solution of the auxiliary minimum problem

The solution of the auxiliary minimum problem is the neighboring extremal defined by (4.124) and (4.125), and the cost variation at order 2 is given by

$$\delta^2 J_a = \int_{t_0}^{t_f} \delta v^T H_{uu} \delta v \, dt + \begin{pmatrix}\delta x_0 \\ \delta \psi_f\end{pmatrix}^T \begin{pmatrix}\bar{S} - \bar{M}\bar{N}^{-1}\bar{M}^T & \bar{M}\bar{N}^{-1} \\ \bar{N}^{-1}\bar{M}^T & -\bar{N}^{-1}\end{pmatrix}_{t_0} \begin{pmatrix}\delta x_0 \\ \delta \psi_f\end{pmatrix} \tag{4.142}$$
$$\text{with } \delta v \underset{def}{=} \delta u + H_{uu}^{-1}\left[(H_{xu} + f_u S)\delta x + f_u M \delta\lambda + f_u \alpha \delta t_f\right]$$

where the matrices $\bar{S}$, $\bar{M}$ and $\bar{N}$ are determined by equations (4.128) and (4.130).

Demonstration (from [R3])

Stage 1

Let us first show that the solution of (4.141) is the neighboring extremal associated with δx_0 and $\delta\psi_f$.

The auxiliary problem (4.141) is a problem with final constraints and with a parameter δt_f involved in the final cost. We write its optimality conditions by noting p_q the adjoint and H_q the Hamiltonian.

$$H_q = \frac{1}{2}\begin{pmatrix}\delta x\\ \delta u\end{pmatrix}^T \begin{pmatrix}H_{xx} & H_{ux}\\ H_{xu} & H_{uu}\end{pmatrix}\begin{pmatrix}\delta x\\ \delta u\end{pmatrix} + p_q^T(f_x^T\delta x + f_u^T\delta u)$$

The optimal control δu minimizes the Hamiltonian H_q.

$$\min_{\delta u} H_q \Rightarrow H_{uu}\delta u + H_{xu}\delta x + f_u p_q = 0 \Rightarrow \delta u = -H_{uu}^{-1}(H_{xu}\delta x + f_u p_q)$$

The adjoint equations are: $\dot{p}_q = -\dfrac{\partial H_q}{\partial \delta x} = -H_{xx}\delta x - H_{ux}\delta u - f_x p_q$.

We introduce the multipliers λ_q of the final constraints.

The transversality conditions (4.53) give: $p_q(t_f) = \Omega_{xx}\delta x_f + \Psi_x\delta t_f + \psi_x\lambda_q$.

The condition (4.35) on the parameter δt_f gives

$$\Psi_x^T\delta x_f + \dot{\Psi}\delta t_f + \dot{\psi}\lambda_q + \int_{t_0}^{t_f}\frac{\partial H_q}{\partial \delta t_f}dt = 0 \Rightarrow \Psi_x^T\delta x_f + \dot{\Psi}\delta t_f + \dot{\psi}\lambda_q = 0$$

Let us group these optimality conditions by noting $p_q \underset{\text{noted}}{=} \delta p$ and $\lambda_q \underset{\text{noted}}{=} \delta\lambda$.

$$\begin{cases}\delta\dot{x} = f_x^T\delta x + f_u^T\delta u\\ \delta\dot{p} = -H_{xx}\delta x - H_{ux}\delta u - f_x\delta p\\ \delta u = -H_{uu}^{-1}(H_{xu}\delta x + f_u\delta p)\end{cases} \text{ with } \begin{cases}\psi_x^T\delta x_f + \dot{\psi}\,\delta t_f = \delta\psi_f\\ \Omega_{xx}\delta x_f + \Psi_x\delta t_f + \psi_x\delta\lambda = \delta p(t_f)\\ \Psi_x^T\delta x_f + \dot{\Psi}\,\delta t_f + \dot{\psi}\,\delta\lambda = 0\end{cases}$$

We retrieve the equations of the neighboring extremals (4.124) and (4.125).

Stage 2

Next, let us calculate the value of $\delta^2 J_a$.

We need to calculate the value of $\delta^2 J_a$ given by (4.139) on the neighboring extremal associated with δx_0 and $\delta\psi_f$.

The following relation is used

$$\begin{pmatrix}\delta x\\ \delta u\end{pmatrix}^T\begin{pmatrix}H_{xx} & H_{ux}\\ H_{xu} & H_{uu}\end{pmatrix}\begin{pmatrix}\delta x\\ \delta u\end{pmatrix}=\delta v^T H_{uu}\delta v-\frac{d}{dt}\left[\begin{pmatrix}\delta x\\ \delta\lambda\\ \delta t_f\end{pmatrix}^T\begin{pmatrix}S & M & \alpha\\ M^T & N & \beta\\ \alpha^T & \beta^T & \gamma\end{pmatrix}\begin{pmatrix}\delta x\\ \delta\lambda\\ \delta t_f\end{pmatrix}\right]$$

$$\text{with}\quad \delta v \underset{def}{=} \delta u + H_{uu}^{-1}\left[(H_{xu}+f_u S)\delta x + f_u M\delta\lambda + f_u\alpha\delta t_f\right]$$

This relation is verified by deriving and using equations (4.126) and (4.128) giving the derivatives of δx and of the matrices $S, M, N, \alpha, \beta, \gamma$. The (very long) calculation is not detailed here. With this formula and by expressing the final term as a function of the matrices S, α, γ, the second variation (4.139) takes the form

$$\delta^2 J_a = \int_{t_0}^{t_f}\delta v^T H_{uu}\delta v\,dt-\left[\begin{pmatrix}\delta x\\ \delta\lambda\\ \delta t_f\end{pmatrix}^T\begin{pmatrix}S & M & \alpha\\ M^T & N & \beta\\ \alpha^T & \beta^T & \gamma\end{pmatrix}\begin{pmatrix}\delta x\\ \delta\lambda\\ \delta t_f\end{pmatrix}\right]_{t_0}^{t_f}+\begin{pmatrix}\delta x\\ \delta t_f\end{pmatrix}^T\begin{pmatrix}S & \alpha\\ \alpha^T & \gamma\end{pmatrix}\begin{pmatrix}\delta x\\ \delta t_f\end{pmatrix}(t_f)$$

which becomes by simplifying the terms in S, α, γ at t_f

$$\delta^2 J_a = \int_{t_0}^{t_f}\delta v^T H_{uu}\delta v\,dt+\begin{pmatrix}\delta x\\ \delta t_f\end{pmatrix}^T\begin{pmatrix}S & \alpha\\ \alpha^T & \gamma\end{pmatrix}\begin{pmatrix}\delta x\\ \delta t_f\end{pmatrix}(t_0)-\left[\begin{pmatrix}\delta x\\ \delta\lambda\\ \delta t_f\end{pmatrix}^T\begin{pmatrix}0 & M & 0\\ M^T & N & \beta\\ 0 & \beta^T & 0\end{pmatrix}\begin{pmatrix}\delta x\\ \delta\lambda\\ \delta t_f\end{pmatrix}\right]_{t_0}^{t_f}$$

The last term is expressed using (4.127): $\delta\psi_f = M^T\delta x + N\delta\lambda + \beta\delta t_f$.

$$\begin{pmatrix}\delta x\\ \delta\lambda\\ \delta t_f\end{pmatrix}^T\begin{pmatrix}0 & M & 0\\ M^T & N & \beta\\ 0 & \beta^T & 0\end{pmatrix}\begin{pmatrix}\delta x\\ \delta\lambda\\ \delta t_f\end{pmatrix}=2\delta\lambda^T(M^T\delta x+\beta\delta t_f)+\delta\lambda^T N\delta\lambda = 2\delta\lambda^T\delta\psi_f-\delta\lambda^T N\delta\lambda$$

The term $\delta\lambda^T\delta\psi_f$ is constant and $N(t_f)=0$ according to (4.128). There remains

$$\delta^2 J_a = \int_{t_0}^{t_f}\delta v^T H_{uu}\delta v\,dt+\begin{pmatrix}\delta x\\ \delta t_f\end{pmatrix}^T\begin{pmatrix}S & \alpha\\ \alpha^T & \gamma\end{pmatrix}\begin{pmatrix}\delta x\\ \delta t_f\end{pmatrix}(t_0)-\delta\lambda^T N(t_0)\delta\lambda$$

With the relations (4.130) giving $\delta t_f, \delta\lambda$ as a function of $\delta x_0, \delta\psi_f$, we have

$$\begin{pmatrix}\delta x\\ \delta t_f\end{pmatrix}^T\begin{pmatrix}S & \alpha\\ \alpha^T & \gamma\end{pmatrix}\begin{pmatrix}\delta x\\ \delta t_f\end{pmatrix}(t_0)-\delta\lambda^T N(t_0)\delta\lambda=\begin{pmatrix}\delta x_0\\ \delta\psi_f\end{pmatrix}^T\begin{pmatrix}\bar S-\bar M\bar N^{-1}\bar M^T & \bar M\bar N^{-1}\\ \bar N^{-1}\bar M^T & -\bar N^{-1}\end{pmatrix}_{t_0}\begin{pmatrix}\delta x_0\\ \delta\psi_f\end{pmatrix}$$

which leads to the formula (4.142).

4.4.2 Sufficient conditions of minimum

Consider an extremal of the problem (4.135), defined by equations (4.140).

The matrices $\bar{S}$, $\bar{M}$ and $\bar{N}$ associated with neighboring extremals are defined by equations (4.128) and (4.130). Theorem 4-10 gives sufficient conditions for the extremal to be a minimum.

Theorem 4-10: Sufficient conditions of local minimum

An extremal is a local minimum if it satisfies the following conditions:

- convexity condition (Legendre-Clebsch condition)

$$H_{uu} > 0 \ , \ \forall t \in [t_0 ; t_f] \tag{4.143}$$

- normality condition

$$\bar{N} < 0 \ , \ \forall t \in [t_0 ; t_f] \tag{4.144}$$

- no conjugate point condition (Jacobi condition)

$$\bar{S} - \bar{M}\bar{N}^{-1}\bar{M}^T \text{ finite} \ , \ \forall t \in [t_0 ; t_f] \tag{4.145}$$

Demonstration

It must be shown that, under conditions (4.143)-(4.145), the minimum of $\delta^2 J_a$ (4.142) is obtained for $\delta u = 0$ which corresponds to the nominal extremal.

If $H_{uu} > 0$ (convexity condition), $\delta^2 J_a$ (4.142) is minimum for $\delta v = 0$.
The matrix H_{uu} defined positive is invertible and the control variation δu is then given by: $\delta v = 0 \ \Rightarrow \ \delta u = -H_{uu}^{-1}\left[(H_{xu} + f_u S)\delta x + f_u M \delta\lambda + f_u \alpha \delta t_f\right]$.
If $\bar{N} < 0$ (normality condition), the matrix $\bar{N}$ is invertible. With the neighboring extremal equations (4.130), the variation δu is given in state feedback (4.131).
$\delta u = -H_{uu}^{-1}\left[\left(H_{xu} + f_u(\bar{S} - \bar{M}\bar{N}^{-1}\bar{M}^T)\right)\delta x + f_u \bar{M}\bar{N}^{-1}\delta\psi\right]$
If the matrix $\bar{S} - \bar{M}\bar{N}^{-1}\bar{M}^T$ remains finite, then the variations δu and δx remain zero on the reference extremal (for $\delta x_0 = 0$ and $\delta\psi = 0$). The only solution giving $\delta^2 J_a = 0$ is then the reference extremal ($\delta u = 0$) which is a local minimum.

Note on the normality condition

The matrix $\bar{N}$ is obtained by backward integration from equations (4.128).
If $H_{uu} > 0$, then $B = f_u^T H_{uu}^{-1} f_u \geq 0$ and $\dot{\bar{N}} = \bar{M}^T B \bar{M} \geq 0$ (reduced transport method).
If $\gamma(t_f) \geq 0$, then $\bar{N}(t_f) \leq 0$ because $\bar{N} = N - \dfrac{\beta^T\beta}{\gamma}$ and $N(t_f) = 0$.
We have then: $\bar{N}(t) \leq \bar{N}(t_f) \leq 0$. The normality condition strengthens this inequality $\bar{N} < 0$ to ensure that $\bar{N}$ is invertible.

Conjugate point

The extremal has a point conjugate point C at time t_c if: $\left(\bar{S} - \bar{M}\bar{N}^{-1}\bar{M}^T\right)(t_c) = \infty$.

The state feedback control (4.131) is then no longer necessarily zero at t_c. A pulse δu gives a non-zero derivative $\delta\dot{x}(t_c)$ and a feasible path different from the reference extremal OCF associated with $\delta x = 0$ (plotted in figure 4-31). This path reaches the final conditions $\psi = 0$ at a point F' different from the reference point F and with a cost variation $\delta^2 J_a$ zero on CF' (because $\delta v = 0$). There may then exist a direct path OF' not passing through C and of lower cost than the OCF path (because the intermediate point constraint C is removed).

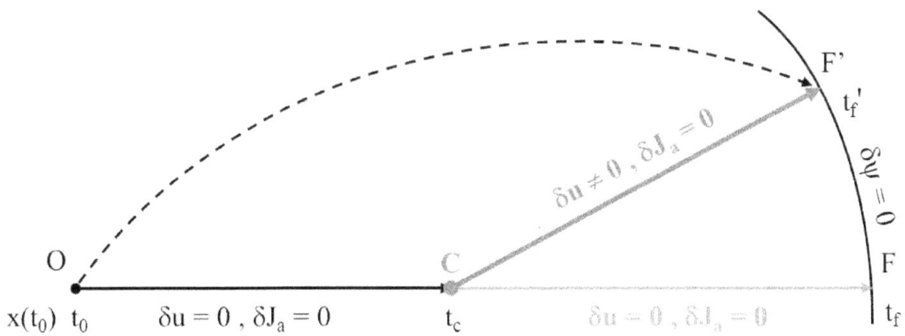

Figure 4-31: Conjugate point.

Example 4-17 (already solved by calculus of variations in example 3-5) illustrates the notion of conjugate points for the geodesics on a sphere.

Example 4-17: Geodesics on a sphere

We seek the minimum distance curves between points O and F on the unit sphere. The axes are chosen so that the two points are on the equator.

$$O\begin{pmatrix}\lambda_0 = 0\\ \varphi_0 = 0\end{pmatrix},\ F\begin{pmatrix}\lambda_f\\ \varphi_f = 0\end{pmatrix}$$

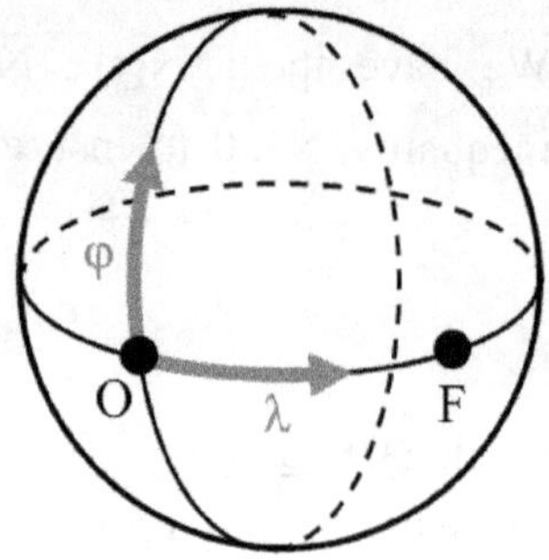

The problem is formulated with:

- the state vector: $x = \begin{pmatrix}x_1\\ x_2\end{pmatrix} = \begin{pmatrix}\lambda\\ \varphi\end{pmatrix}$;

- the control vector: $u = \begin{pmatrix}u_1\\ u_2\end{pmatrix} \rightarrow \begin{cases}\dot{\lambda} = u_1\\ \dot{\varphi} = u_2\end{cases}$.

The distance element is: $ds = \sqrt{d\varphi^2 + d\lambda^2\cos^2\varphi} = \sqrt{u_2^2 + u_1^2\cos^2 x_2}\,dt$.

Formulation of the control problem

$$\min_{u,t_f} J = \int_{t_0}^{t_f}\sqrt{u_2^2 + u_1^2\cos^2 x_2}\,dt \quad \text{with} \begin{cases}\dot{x}_1 = u_1\\ \dot{x}_2 = u_2\end{cases} \quad \text{and} \begin{cases}x(t_0) = x_O\\ x(t_f) = x_F\end{cases}$$

Search for extremals

The optimal control minimizes the Hamiltonian.

$$\min_{u_1,u_2} H = \sqrt{u_2^2 + u_1^2\cos^2 x_2} + p_1u_1 + p_2u_2$$

$$H_u = 0 \Rightarrow \begin{cases}u_1^2\cos^2 x_2(\cos^2 x_2 - p_1^2) - u_2^2p_1^2 = 0\\ u_1^2\cos^2 x_2(p_2^2) - u_2^2(1 - p_2^2) = 0\end{cases}$$

One possible solution (among others) is: $\begin{cases}x_2 = 0\ ,\ p_1 = 1\\ u_2 = 0\ ,\ p_2 = 0\end{cases}$.

This extremal is a great circle arc following the equator $(x_2 = \varphi = 0)$.
Let us check whether this extremal satisfies the sufficient minimum conditions.

Conditions of second order

Since the solution is along the equator, the problem can be reformulated with the variable λ instead of t. The state is reduced to a single component $\varphi(\lambda)$, which simplifies the study of the second-order conditions.

The reduced formulation is

$$\min_u J = \int_{\lambda_0}^{\lambda_f} \sqrt{u^2 + \cos^2\varphi}\, d\lambda \quad \text{with} \quad \varphi' = \frac{d\varphi}{d\lambda} = u \quad \text{and} \quad \begin{cases} \varphi(\lambda_0) = \varphi_0 = 0 \\ \varphi(\lambda_f) = \varphi_f = 0 \end{cases}$$

The extremal has the equation: $u = 0$, $\varphi = 0$.
The matrices A, B, C, S, M and N given by (4.126) and (4.128) are calculated.

Derivatives of f: $\quad f = u \ \Rightarrow\ f_\varphi = 0\,,\ f_u = 1$

Derivatives of H: $\quad H = \sqrt{u^2 + \cos^2\varphi} + pu = L + pu \quad \text{with} \quad L = \sqrt{u^2 + \cos^2\varphi}$

$$H_\varphi = L_\varphi \ ,\ H_u = L_u + p \qquad \text{with} \quad L_\varphi = \frac{-\sin 2\varphi}{2L} \ ,\ L_u = \frac{u}{L}$$

$$H_{\varphi\varphi} = -\frac{1}{2}\frac{2L\cos 2\varphi - L_\varphi \sin 2\varphi}{L^2} \ ,\ H_{u\varphi} = -u\frac{L_\varphi}{L^2} \ ,\ H_{uu} = \frac{L - uL_u}{L^2}$$

Values on the extremal: $\quad \begin{cases} u = 0 \\ \varphi = 0 \end{cases} \Rightarrow \begin{cases} L = 1 \\ L_\varphi = 0 \\ L_u = 0 \end{cases} \Rightarrow \begin{cases} H_{\varphi\varphi} = -1 \\ H_{\varphi u} = 0 \\ H_{uu} = 1 \end{cases}$

Matrices A, B, C: $\quad A = 0 \ ,\ B = 1 \ ,\ C = -1$

Matrices S, M, N

$$\begin{cases} \dot{S} = S^2 + 1 & \text{with } S(\lambda_f) = \Omega_{xx}(t_f) = 0 & \rightarrow S(\lambda) = -\tan(\lambda_f - \lambda) \\ \dot{M} = SM & \text{with } M(\lambda_f) = \psi_x(t_f) = 1 & \rightarrow M(\lambda) = \dfrac{1}{\cos(\lambda_f - \lambda)} \\ \dot{N} = M^T BM & \text{with } N(t_f) = 0 & \rightarrow N(\lambda) = -\tan(\lambda_f - \lambda) \end{cases}$$

State feedback control

$$\begin{cases} S - MN^{-1}M^T = \dfrac{1}{\tan(\lambda_f - \lambda)} \\ MN^{-1} = \dfrac{-1}{\sin(\lambda_f - \lambda)} \end{cases} \rightarrow \delta u = -\frac{\delta\lambda}{\tan(\lambda_f - \lambda)} + \frac{\delta\lambda_f}{\sin(\lambda_f - \lambda)}$$

A conjugate point appears if: $\lambda_f - \lambda = \pi \ \Rightarrow\ S - MN^{-1}M^T = \dfrac{1}{\tan(\lambda_f - \lambda)} = \infty$.

The extremal passes through a conjugate point if the angle OF reaches 180° (distance greater than a half circle). It is not optimal in this case, because the path in the opposite direction is shorter).

Derivatives of optimal the cost

Formulas (4.138) and (4.142) have been used to express the sufficient minimum conditions on the reference extremal $(\delta x_0 = 0 , \delta \psi_f = 0)$.

They also give the first and second derivatives of the optimal cost with respect to variations of the initial state δx_0 and the final constraints $\delta \psi_f$.

$$\begin{aligned} &\frac{\partial J}{\partial x}(t_0) = p(t_0) \ , \ \frac{\partial^2 J}{\partial x^2}(t_0) = \bar{S} - \bar{M}\bar{N}^{-1}\bar{M}^T \\ &\frac{\partial J}{\partial \psi} = -\lambda \ , \ \frac{\partial^2 J}{\partial \psi^2} = -\bar{N}^{-1} \ , \ \frac{\partial^2 J}{\partial x \partial \psi} = \bar{N}^{-1}\bar{M}^T \end{aligned} \tag{4.146}$$

4.4.3 Singular arcs

Consider a minimum time problem with bounded linear control.

$$\min_{-1 \le u \le +1} J = t_f \quad \text{with} \quad \dot{x} = f(x) + g(x)u \tag{4.147}$$

The state is of dimension n. We consider here a control u of dimension 1.

The Hamiltonian is expressed with the adjoint p and the multiplier p_0.

$$H = p_0 + p^T(f + ug) \tag{4.148}$$

The optimal control minimizes the Hamiltonian. It depends on the sign of the switching function Φ defined by

$$\Phi = H_u = p^T g(x) \ \rightarrow \begin{cases} u = +1 & \text{if } \Phi < 0 \\ u = -1 & \text{if } \Phi > 0 \end{cases} \ , \ u = \ ? \ \text{if} \ \Phi = 0 \tag{4.149}$$

Regular arcs with $u = \pm 1$ are called "bang" arcs.

A **singular arc** occurs if Φ is identically zero on an interval. The successive derivatives $\dot{\Phi}, \ddot{\Phi}, \cdots$ are then all zero. The order on the singular arc is determined by the first explicit derivative in u. The questions arise about the optimality of a singular arc, as well as its junction to regular arcs.

Kelley's condition

The following Kelley condition (or generalized Legendre-Clebsch condition) is a necessary condition for optimality of a singular arc of the system (4.147).

$$(-1)^k \frac{\partial}{\partial u}\left(\frac{d^{2k}}{dt^{2k}}\frac{\partial H}{\partial u}\right) \ge 0 \quad \Leftrightarrow \quad (-1)^k (\Phi^{(2k)})_u \ge 0 \ \text{ with } \ \Phi = H_u \tag{4.150}$$

The derivative $\Phi^{(2k)}$ is the first explicit derivative in u. It can be shown that this derivative is necessarily even. The integer k is called the **order** of the singular arc.

Lie derivatives and brackets

To study singular arcs, Lie derivatives and brackets are introduced.
A vector field X is a function of $\mathbb{R}^n$ in $\mathbb{R}^n$. The components of X are n functions of $\mathbb{R}^n$ in $\mathbb{R}$. The gradient of X noted ∇X is defined by

$$X = \begin{pmatrix} X_1 \\ \vdots \\ X_n \end{pmatrix} \quad \rightarrow \quad \nabla X = \left(\nabla X_1, \cdots, \nabla X_n\right) = \begin{pmatrix} \dfrac{\partial X_1}{\partial x_1} & \cdots & \dfrac{\partial X_n}{\partial x_1} \\ \vdots & \ddots & \vdots \\ \dfrac{\partial X_1}{\partial x_n} & \cdots & \dfrac{\partial X_n}{\partial x_n} \end{pmatrix} \tag{4.151}$$

The Lie derivative of a function $\varphi : \mathbb{R}^n \rightarrow \mathbb{R}$ along the vector field X is the function noted $L_X\varphi$ of $\mathbb{R}^n$ in $\mathbb{R}$ defined by

$$L_X\varphi \underset{def}{=} X^T\nabla\varphi \tag{4.152}$$

This function, also noted $L_X\varphi \underset{noted}{=} X\varphi$, is associated with the vector field X.

It represents the directional derivative of φ along X at any point $x \in \mathbb{R}^n$.
The Lie bracket of the vector fields X and Y is the operator defined by

$$[X,Y] \underset{def}{=} XY - YX \tag{4.153}$$

This operator consists in applying successively the Lie derivatives L_X and L_Y to functions $\varphi : \mathbb{R}^n \rightarrow \mathbb{R}$. It is equivalent to the Z operator defined by

$$Z = \nabla Y^T X - \nabla X^T Y \tag{4.154}$$

Demonstration (see [R17])

Let us apply the operator $[X,Y]$ to a function $\varphi : \mathbb{R}^n \rightarrow \mathbb{R}$.

$$[X,Y]\varphi = XY\varphi - YX\varphi = L_X(L_Y\varphi) - L_Y(L_X\varphi)$$

Developing the Lie derivatives defined by (4.152), we obtain

$$[X,Y]\varphi = L_X(Y^T\nabla\varphi) - L_Y(X^T\nabla\varphi) = X^T\nabla(Y^T\nabla\varphi) - Y^T\nabla(X^T\nabla\varphi)$$
$$= X^T(\nabla Y\,\nabla\varphi + \nabla^2\varphi\, Y) - Y^T(\nabla X\,\nabla\varphi + \nabla^2\varphi\, X)$$
$$= (X^T\nabla Y - Y^T\nabla X)\nabla\varphi \quad \text{because the Hessian } \nabla^2\varphi \text{ is symmetrical}$$
$$= (\nabla Y^T X - \nabla X^T Y)^T\nabla\varphi = L_Z\nabla\varphi = Z\varphi \quad \text{by posing } Z = \nabla Y^T X - \nabla X^T Y$$

The Lie bracket has the following properties, which are easily proven.

$$\begin{aligned} &[X,X] = 0 \\ &[X,Y] = -[Y,X] \\ &[X,[Y,Z]] + [Y,[Z,X]] + [Z,[X,Y]] = 0 \end{aligned} \tag{4.155}$$

The third equality is the Jacobi identity.

Derivative of a function along an extremal

Lie brackets are useful to express the successive derivatives $\dot\Phi, \ddot\Phi, \cdots$ of the switching function in a compact form. The system (4.147) has the state equation: $\dot x = f(x) + ug(x)$. By associating to the functions f and g the respective vector fields F and G on $\mathbb{R}^n$, the extremals of the problem (4.147) have the equations

$$\begin{cases} \dot x = H_p = f(x) + ug(x) \\ \dot p = -H_x = -[f_x(x) + ug_x(x)]p \end{cases} \Leftrightarrow \begin{cases} \dot x = F + uG \\ \dot p = -(\nabla F + u\nabla G)p \end{cases} \tag{4.156}$$

Consider a vector field Z and a function $\Psi : \mathbb{R} \to \mathbb{R}$ of the form

$$\Psi(t) = p(t)^T Z[x(t)] \tag{4.157}$$

The derivative of Ψ along an extremal is then expressed with the Lie bracket.

$$\dot\Psi = p^T[F + uG, Z] \underset{\text{noted}}{=} p^T[f + ug, Z] \tag{4.158}$$

Demonstration (see [R17])

The derivative of Ψ is: $\dot\Psi = \dot p^T Z + p^T \dot Z$ with $\begin{cases} \dot p = -(\nabla F + u\nabla G)p \\ \dot Z = \nabla Z^T \dot x = \nabla Z^T(F + uG) \end{cases}$

where the derivatives of x and p are taken along the extremal of equations (4.156).

The result is: $\dot{\Psi} = -p^T(\nabla F + u\nabla G)^T Z + p^T \nabla Z^T (F + uG)$
which is written using the Lie bracket: $\dot{\Psi} = p^T[F + uG, Z] \underset{\text{noted}}{=} p^T[f + ug, Z]$.

Derivatives of the switching function
Let us apply the formula (4.158) to calculate the successive derivatives of the switching function (4.149): $\Phi = p^T g = p^T G$.
The first derivative is given by

$$\Phi = p^T g \quad \rightarrow \dot{\Phi} = p^T[f + ug, g] = p^T[f, g] \tag{4.159}$$

The second derivative is given by

$$\begin{aligned} \dot{\Phi} = p^T[f,g] &\rightarrow \ddot{\Phi} = p^T[f + ug, [f,g]] \\ &\rightarrow \ddot{\Phi} = p^T[f,[f,g]] + up^T[g,[f,g]] \end{aligned} \tag{4.160}$$

The successive derivatives of Φ are all zero along a singular arc.
If $p^T[g,[f,g]] \neq 0$, the explicit second derivative in u determines the control.

$$u_{sing} = -\frac{p^T[f,[f,g]]}{p^T[g,[f,g]]} \tag{4.161}$$

The singular arc obtained with this control is of order 1.
If $p^T[g,[f,g]] = 0$, then we must continue to derive Φ until an explicit expression in u is obtained. The singular arc is of higher order.

Singular arc of order 1
The order on a singular arc of order 1 is determined by (4.161).
To be feasible, this control must satisfy the bounds: $-1 \le u_{sing} \le +1$.
To be optimal, it must satisfy the Kelley condition (4.150) for $k = 1$.

$$-\frac{\partial}{\partial u}\left(\frac{d^2}{dt^2}\frac{\partial H}{\partial u}\right) \ge 0 \iff -\frac{\partial}{\partial u}\ddot{\Phi} \ge 0 \iff p^T[g,[f,g]] \le 0 \tag{4.162}$$

Property 4-11: Junction between a singular arc of order 1 and a bang arc

A singular arc satisfying the Kelley condition in the strict sense can be connected to a bang arc $(u_{bang} = \pm 1)$ at a time τ such that: $-1 < u_{sing}(\tau) < +1$.

Demonstration (see [R17])

Suppose that the singular arc connects to a bang arc at time τ such that $-1 < u_{sing}(\tau) < +1$.

If the bang arc precedes the singular arc, it stops at time τ^-.
If the bang arc follows the singular arc, it starts at time τ^+.

Let us show that this junction satisfies the minimum principle at $\tau^\pm$.
The Hamiltonian is given by: $H = p_0 + p^T(f+ug) = p_0 + p^T f + u\Phi$.
The functions $\Phi = p^T g$ and $\dot{\Phi} = p^T[f,g]$ are continuous, since they depend only on x and p. They are zero in τ (on the singular arc), so also in $\tau^\pm$ by continuity.
The second derivative of Φ is given by (4.160): $\ddot{\Phi} = p^T[f,[f,g]] + up^T[g,[f,g]]$.
On the singular arc at τ: $\ddot{\Phi}(\tau) = p^T[f,[f,g]] + u_{sing} p^T[g,[f,g]] = 0$
On the arc bang at $\tau^\pm$: $\ddot{\Phi}(\tau^\pm) = p^T[f,[f,g]] + u_{bang} p^T[g,[f,g]]$

<u>1st case</u>: $u_{bang} = -1 < u_{sing}(\tau)$
We have: $\ddot{\Phi}(\tau^\pm) > \ddot{\Phi}(\tau)$ because $p^T[g,[f,g]] < 0$ (strict Kelley condition)
$\Rightarrow \ddot{\Phi}(\tau^\pm) > 0$ because $\ddot{\Phi}(\tau) = 0$ (on the singular arc)
Let us write the expansion of Φ to order 2 in τ using $\Phi(\tau)=0$ and $\dot{\Phi}(\tau)=0$.

$$\Phi(\tau^\pm) = \Phi(\tau) + (\tau^\pm - \tau)\dot{\Phi}(\tau) + \frac{1}{2}(\tau^\pm - \tau)^2\ddot{\Phi}(\tau^\pm) = \frac{1}{2}(\tau^\pm - \tau)^2\ddot{\Phi}(\tau^\pm) \Rightarrow \Phi(\tau^\pm) > 0$$

Let us apply the PMP at $\tau^\pm$.

$$\min_{-1\le u\le +1} H = p_0 + p^T f + u\Phi \quad \Rightarrow u(t^\pm) = -1 \text{ because } \Phi > 0$$

We obtain the control $u_{bang} = -1$, consistent with the assumption.
This junction is therefore possible.

<u>2nd case</u>: $u_{bang} = +1 > u_{sing}(\tau)$
The verification is carried out in a similar way.
The PMP at $\tau^\pm$ gives: $\Phi(\tau^\pm) < 0$ and $u(t^\pm) = +1$.

Example 4-18 illustrates a composite solution including a junction to a singular arc of order 1.

Example 4-18: Junction with a singular arc of order 1

Formulation

The problem is a double integrator in dimension 2 to be brought back to the origin by minimizing a quadratic cost. The initial state and the final time are fixed. The control is not bounded.

$$\min_u J = \frac{1}{2}\int_{t_0=0}^{t_f} x_1^2 dt \quad \text{with} \quad \begin{cases} \dot{x}_1 = x_2 + u \\ \dot{x}_2 = -u \end{cases} \quad \text{and} \quad \begin{cases} x_1(t_f) = 0 \\ x_2(t_f) = 0 \end{cases}$$

The Hamiltonian H is constant (autonomous system), with the adjoint (p_1, p_2).

$$H = \frac{1}{2}x_1^2 + p_1(x_2 + u) - p_2 u = \frac{1}{2}x_1^2 + p_1 x_2 + u\Phi \quad \rightarrow \quad \begin{cases} \dot{p}_1 = -x_1 \\ \dot{p}_2 = -p_1 \end{cases}$$

The switching function is defined by: $\Phi = p_1 - p_2$.

Search for singular arcs

A singular arc occurs if the function Φ remains identically zero.

$$\Phi = p_1 - p_2 = 0 \Rightarrow \begin{cases} \dot{\Phi} = -x_1 + p_1 = 0 \\ \ddot{\Phi} = -x_2 - u - x_1 = 0 \end{cases} \Rightarrow \begin{cases} p_1 = x_1 \\ u = -(x_1 + x_2) \end{cases}$$

The singular arc is of order 1, because the second derivative is explicit in u. This arc can be optimal, as it satisfies the Kelley necessary condition (4.162).

Indeed, for k = 1 : $(-1)^k \frac{\partial}{\partial u}\left(\frac{d^{2k}}{dt^{2k}}\frac{\partial H}{\partial u}\right) = -\frac{\partial}{\partial u}\ddot{\Phi} = 1 \geq 0$

According to property 4-11, this arc of order 1 can be connected to a regular arc.

The equations of the singular arc are obtained from the Hamiltonian.

$$H = \frac{1}{2}x_1^2 + p_1 x_2 + u\Phi = C^{te} \quad \text{with} \quad \begin{cases} \Phi = 0 \\ p_1 = x_1 \end{cases}$$

We obtain an arc of hyperbola of equation: $x_1(x_1 + 2x_2) = 2H$

having for asymptotes: $x_1 = 0$ and $x_1 + 2x_2 = 0$.

Let us determine the temporal evolution of x_1 and x_2:

- for x_1: $\begin{cases} \dot{p}_1 = -x_1 \\ p_1 = x_1 \end{cases} \Rightarrow \dot{x}_1 = -x_1 \quad \Rightarrow x_1(t) = ae^{-t}$

- for x_2: $x_2 = \dfrac{H}{x_1} - \dfrac{1}{2}x_1 \quad \Rightarrow x_2(t) = \dfrac{H}{a}e^{t} - \dfrac{1}{2}ae^{-t}$

<u>Searching for junctions</u>
Composite solutions connecting regular and singular arcs are considered.

The initial and final times and states are fixed: $\begin{cases} x_1(t_0) = x_{10} \\ x_2(t_0) = x_{20} \end{cases}$ and $\begin{cases} x_1(t_f) = 0 \\ x_2(t_f) = 0 \end{cases}$.

The control is not bounded and does not appear in the cost J. We can consequently apply "free" impulses δu moving the state from (x_1, x_2) to $(x_1 + \delta u, x_2 - \delta u)$. These instantaneous shifts follow the lines of equation: $x_1 + x_2 = C^{te}$.

We can thus define a solution composed of three arcs as follows:
- an impulse at t_0 changes (x_{10}, x_{20}) to $(x_{10} + \delta x_0, x_{20} - \delta x_0)$;
- the singular arc is travelled from t_0 to t_f with the control $u = -(x_1 + x_2)$;
- an impulse at t_f changes (x_{1f}, x_{2f}) to $(x_{1f} + \delta x_f, x_{2f} - \delta x_f) = (0, 0)$.

Let us note (y_{10}, y_{20}) and (y_{1f}, y_{2f}) the states at beginning and end of the singular arc. These endpoint conditions determine the equations of the singular arc.

$$\begin{cases} y_{10} = a \\ y_{20} = \dfrac{H}{a} - \dfrac{1}{2}a \end{cases} \Rightarrow \begin{cases} a = y_{10} \\ H = \dfrac{1}{2}y_{10}^2 + y_{10}y_{20} \end{cases} \Rightarrow \begin{cases} y_{1f} = y_{10}e^{-t_f} \\ y_{2f} = y_{20}e^{t_f} + y_{10}\,\text{sh}\,t_f \end{cases}$$

Let us write the junction conditions at t_0 and t_f.

$$\begin{cases} y_{10} + y_{20} = x_{10} + x_{20} \underset{\text{def}}{=} c & \text{at } t_0 \\ y_{1f} + y_{2f} = x_{1f} + x_{2f} = 0 & \text{at } t_f \end{cases} \Rightarrow \begin{cases} y_{1f} = y_{10}e^{-t_f} \\ y_{2f} = y_{20}e^{t_f} + y_{10}\,\text{sh}\,t_f \end{cases}$$

The state at the beginning and end of the singular arc is obtained.

$$y_{10} = 2c\frac{e^{2t_f}}{e^{2t_f} - 1} \;,\; y_{20} = c\frac{1 + e^{2t_f}}{1 - e^{2t_f}} \quad \text{with} \begin{cases} c = x_{10} + x_{20} \text{ fixed} \\ t_f \text{ fixed} \end{cases}$$

The cost of this solution is: $J = \dfrac{1}{2}\displaystyle\int_{t_0=0}^{t_f} x_1^2 dt = \dfrac{1}{2}\displaystyle\int_{t_0=0}^{t_f} y_{10}^2 e^{-2t} dt = \dfrac{c^2}{1 - e^{-2t_f}}$.

This solution is not necessarily the best. It must be compared with other solutions (combinations of regular and singular arcs) to establish its global optimality. However, we observe that a solution of the same type arriving earlier at the end point would be less good, as J is a decreasing function of t_f.

<u>Numerical application</u>

The initial state is set at: $\begin{cases} x_{10} = 1 \\ x_{20} = 3.28 \end{cases}$. The final time is set to: $t_f = 1.5$.

The initial and final states on the singular arc are: $\begin{cases} y_{10} \approx 9.01 \\ y_{20} \approx -4.72 \end{cases}$ and $\begin{cases} y_{1f} \approx 2 \\ y_{2f} \approx -2 \end{cases}$.

The cost is equal to: $J \approx 19.28$.
Figure 4-32 shows the trajectory in the phase plane with the initial impulse leading to the hyperbola (singular arc) and the final impulse leading to the origin.

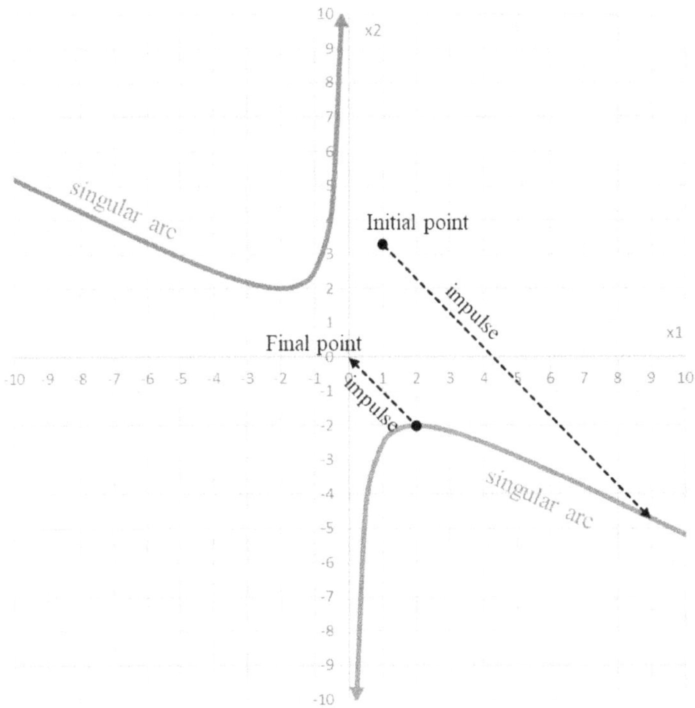

Figure 4-32: Composite solution pulse–singular arc–pulse.

Let us now turn our attention to singular arcs of order greater than 1.

Singular arc of higher order

A singular arc is of order greater than 1 if: $p^T[g,[f,g]] = 0$.

Since the second derivative of Φ is zero, we also have: $p^T[f,[f,g]] = 0$.

These very restrictive conditions are rarely met in practice.

However, there are systems for which we have: $[g,[f,g]] = 0$.

Let us calculate the derivatives of Φ in this case by applying formula (4.158).

To simplify the equations, we introduce the notation: $[X,Y] = ad_X(Y)$.

The operator ad_X (called adjoint operator) is associated with the vector field X.

Using the Jacobi identity (4.155), the derivatives of Φ are expressed as

$$\Phi^{(3)} = p^T ad_f^3(g) \quad , \quad \Phi^{(4)} = p^T ad_f^4(g) + up^T[g, ad_f^3(g)] \tag{4.163}$$

The singular arc is of order k if the first explicit derivative in u is $\Phi^{(2k)}$.

It can be shown that this derivative is necessarily even. The control on a singular arc of order k is determined by the cancellation of the derivative $\Phi^{(2k)}$.

To be feasible, this control must satisfy the bounds: $-1 \le u_{sing} \le +1$.

To be optimal, it must satisfy the Kelley condition (4.150).

$$(-1)^k \frac{\partial}{\partial u}\left(\frac{d^{2k}}{dt^{2k}}\frac{\partial H}{\partial u}\right) = (-1)^k p^T[g, ad_f^{2k-1}(g)] \ge 0 \tag{4.164}$$

Property 4-12: Junction between a singular arc of order 2 and a bang arc

A singular arc of order 2 satisfying the Kelley condition in the strict sense cannot connect to a bang arc $(u_{bang} = \pm 1)$ at a time τ such that $-1 < u_{sing}(\tau) < +1$.

Demonstration (see [R17])

Suppose that the singular arc connects to a bang arc at time τ such that $-1 < u_{sing}(\tau) < +1$

If the bang arc precedes the singular arc, it stops at time τ^-.
If the bang arc follows the singular arc, it starts at time τ^+.

Let us show that this junction does not satisfy the minimum principle at $\tau^\pm$.
The Hamiltonian is given by: $H = p_0 + p^T(f + ug) = p_0 + p^T f + u\Phi$
The functions $\Phi^{(i)} = p^T ad_f^i(g)$ $_{i=0,1,2,3}$ are continuous, because they depend only on x and p. They are zero in τ (on the singular arc), so also in $\tau^\pm$ by continuity.

The fourth derivative of Φ is given by (4.163): $\Phi^{(4)} = p^T ad_f^4(g) + up^T[g, ad_f^3(g)]$
On the singular arc at τ: $\Phi^{(4)}(\tau) = p^T ad_f^4(g) + u_{sing} p^T[g, ad_f^3(g)] = 0$
On the arc bang at $\tau^\pm$: $\Phi^{(4)}(\tau^\pm) = p^T ad_f^4(g) + u_{bang} p^T[g, ad_f^3(g)]$

1st case: $u_{bang} = -1 < u_{sing}(\tau)$
We have: $\Phi^{(4)}(\tau^\pm) < \Phi^{(4)}(\tau)$ because $p^T[g, ad_f^3(g)] > 0$ (strict Kelley condition)
$\Rightarrow \Phi^{(4)}(\tau^\pm) < 0$ because $\Phi^{(4)}(\tau) = 0$ (on the singular arc)
Let us write the expansion of Φ to order 4 in τ, using
$\Phi(\tau) = \dot{\Phi}(\tau) = \Phi^{(2)}(\tau) = \Phi^{(3)}(\tau) = 0$

$$\Phi(\tau^\pm) = \frac{1}{24}(\tau^\pm - \tau)^4 \Phi^{(4)}(\tau^\pm) \quad \Rightarrow \Phi(\tau^\pm) < 0.$$

Apply the PMP at $\tau^\pm$: $\min_{-1 \le u \le +1} H = p_0 + p^T f + u\Phi \Rightarrow u(t^\pm) = +1$ because $\Phi < 0$
This result is in contradiction with the assumption $u_{bang} = -1$.
This junction is therefore not possible.

2nd case: $u_{bang} = +1 > u_{sing}(\tau)$
The verification is done in a similar way. The PMP at $\tau^\pm$ gives
$\Phi(\tau^\pm) > 0$ and $u(t^\pm) = -1$.

A junction between a singular arc of order 2 and a bang arc is incompatible with the PMP. This type of situation leads to the phenomenon of **chattering** with an infinite number of commutations $(u_{bang} = \pm 1)$ on approaching the junction.

The Fuller problem illustrates this phenomenon, which can occur in concrete applications.

Example 4-19: Fuller's problem (from [R17])

Formulation

The Fuller problem (FP) is a double integrator in dimension 2, that has to be brought back to the origin by minimizing a quadratic cost. The final time is free and the control is bounded: $-1 \le u \le 1$. The (FP) problem is formulated as

$$\min_{u,t_f} J = \frac{1}{2}\int_{t_0}^{t_f} x_1^2 dt \quad \text{with} \quad \begin{cases} \dot{x}_1 = x_2 \\ \dot{x}_2 = u \end{cases}, \begin{cases} x_1(t_f) = 0 \\ x_2(t_f) = 0 \end{cases}, -1 \le u \le +1 \quad \textbf{(FP problem)}$$

The Hamiltonian H, with an adjoint (p_1, p_2) and an abnormal multiplier p_0, is

$$H = \frac{1}{2} p_0 x_1^2 + p_1 x_2 + p_2 u \quad \rightarrow \quad \begin{cases} \dot{p}_1 = -p_0 x_1 \\ \dot{p}_2 = -p_1 \end{cases}$$

The optimal control minimizes the Hamiltonian.

$$\min_u H \quad \rightarrow \quad \begin{cases} u = +1 \text{ if } p_2 < 0 \\ u = -1 \text{ if } p_2 > 0 \\ u = \ ? \text{ if } p_2 = 0 \end{cases}$$

To study the extremals, an equivalent minimum time problem is formulated.

Minimum time problem

Consider the following minimum time (MT) problem.

$$\min_{u,t_f} K = \int_{t_0}^{t_f} 1 dt \quad \text{with} \quad \begin{cases} \dot{x}_1 = x_2 \\ \dot{x}_2 = u \\ \dot{x}_3 = x_1^2/2 \end{cases}, \begin{cases} x_1(t_f) = 0 \\ x_2(t_f) = 0 \\ x_3(t_f) = 0 \end{cases}, -1 \le u \le +1 \quad \textbf{(MT problem)}$$

The Hamiltonian H_q, with an adjoint (q_1, q_2, q_3) and an abnormal multiplier q_0 is

$$H_q = q_0 + q_1 x_2 + q_2 u + \frac{1}{2} q_3 x_1^2 \quad \rightarrow \quad \begin{cases} \dot{q}_1 = -q_3 x_1 \\ \dot{q}_2 = -q_1 \\ \dot{q}_3 = 0 \end{cases}$$

We observe that the extremals of the (MT) and (FP) problems have the same equations in $(x_1; x_2)$, because the adjoint q_3 is constant and plays the same role as the abnormal multiplier p_0. The Hamiltonians H_q and H are therefore identical.

$$q_0 + q_1 x_2 + q_2 u + \frac{1}{2} q_3 x_1^2 = p_1 x_2 + p_2 u + \frac{1}{2} p_0 x_1^2 = 0 \,,\ \forall x_1, x_2 \Rightarrow \begin{cases} q_0 = 0 \\ q_3 = p_0 \end{cases}, \begin{cases} q_1 = p_1 \\ q_2 = p_2 \end{cases}$$

The adjoints (q_1, q_2) and (p_1, p_2) coincide. Moreover, the extremals of (FP) are abnormal extremals $(q_0 = 0)$ of (TM). This is because the solution of (FP) is unique (not shown here) and there exists only one final time to reach the origin. Let (u^*, t_f^*) be the solution to (FP), J^* its cost, and choose $x_3(t_0) = -J^* \le 0$. The solution of (FP) is then the solution of (TM) for this specific initial condition. Indeed, with the control u^* we obtain

$$x_3(t_f^*) = x_3(t_0) + \frac{1}{2}\int_{t_0}^{t_f^*} x_1^2 dt = -J^* + \frac{1}{2}\int_{t_0}^{t_f^*} x_1^2 dt = 0$$

<u>Singular extremals</u>

We look for the singular extremals of (MT) by using Lie brackets.

$$\begin{cases} \dot{x}_1 = x_2 \\ \dot{x}_2 = u \\ \dot{x}_3 = x_1^2/2 \end{cases} \rightarrow \dot{x} = f(x) + ug(x) \quad \text{with } f = \begin{pmatrix} x_2 \\ 0 \\ x_1^2/2 \end{pmatrix}, \quad g = \begin{pmatrix} 0 \\ 1 \\ 0 \end{pmatrix}$$

$$\rightarrow f_x = \begin{pmatrix} 0 & 0 & x_1 \\ 1 & 0 & 0 \\ 0 & 0 & 0 \end{pmatrix}, \quad g_x = \begin{pmatrix} 0 & 0 & 0 \\ 0 & 0 & 0 \\ 0 & 0 & 0 \end{pmatrix}$$

The first-order brackets are

$$[f,g] = g_x^T f - f_x^T g = \begin{pmatrix} -1 \\ 0 \\ 0 \end{pmatrix}, \quad [f,[f,g]] = \begin{pmatrix} 0 \\ 0 \\ x_1 \end{pmatrix}, \quad [g,[f,g]] = \begin{pmatrix} 0 \\ 0 \\ 0 \end{pmatrix}$$

Since the bracket $[g,[f,g]]$ is zero, there is no singular arc of order 1.

The second-order brackets are

$$\text{ad}_f^3(g) = [f,[f,[f,g]]] = \begin{pmatrix} 0 \\ 0 \\ x_2 \end{pmatrix}, \quad [g, \text{ad}_f^3(g)] = [g,[f,[f,[f,g]]]] = \begin{pmatrix} 0 \\ 0 \\ 1 \end{pmatrix}$$

The bracket $[g, \text{ad}_f^3(g)]$ does not cancel. Singular arcs are of order 2.

The Kelley condition $[g, \text{ad}_f^3(g)] > 0$ is satisfied. These singular arcs can therefore be optimal. To determine them, the switching function $\Phi = q_2 = p_2$ is derived using the adjoint equations and the state equations until the control u appears.

$$\Phi = p_2 \rightarrow \dot{\Phi} = -p_1 \rightarrow \ddot{\Phi} = p_0 x_1 \rightarrow \Phi^{(3)} = p_0 x_2 \rightarrow \Phi^{(4)} = p_0 u$$

The function Φ and all its derivatives are zero along a singular arc.

According to the PMP, the adjoint (p_1, p_2) and the abnormal multiplier p_0 cannot simultaneously be zero.

The only singular solution is therefore: $\begin{cases} p_1 = p_2 = 0 \\ x_1 = x_2 = 0 \quad \text{with} \quad p_0 \neq 0 \\ u_{sing} = 0 \end{cases}$

This singular arc of order 2 reduces to the origin which is the targeted end point.

According to property 4-12, there is no possible junction with a bang arc.
The origin can therefore only be reached by chattering, *i.e.* with an infinite number of switches $u = \pm 1$.

<u>Regular extremals</u>
All extremals of (FP) are regular (except the origin).

The solution of (FP) is a sequence of bang arcs defined by

$$\begin{cases} \dot{x}_1 = x_2 \\ \dot{x}_2 = u \end{cases}, \quad \begin{cases} \dot{p}_1 = -p_0 x_1 \\ \dot{p}_2 = -p_1 \end{cases}, \quad H = \frac{1}{2} p_0 x_1^2 + p_1 x_2 + p_2 u = 0 \ , \ u = \pm 1$$

If $p_0 = 0$ (abnormal case), then p_1 is constant and p_2 is linear. There is only one switch and the junction to the end point (singular arc of order 2) cannot be optimal. The bang arcs are therefore normal and we can pose $p_0 = 1$.

- First integrals

Bang arcs $(u = \pm 1)$ have the following two first integrals.

$I_1 = x_1 - \frac{1}{2} u x_2^2$ (Verification: $\dot{I}_1 = \dot{x}_1 - u x_2 \dot{x}_2 = x_2 - u^2 x_2 = 0$)

$I_2 = -p_1 - u x_1 x_2 + \frac{1}{3} x_2^3$

(Verification: $\dot{I}_2 = -\dot{p}_1 - u\dot{x}_1 x_2 - u x_1 \dot{x}_2 + x_2^2 \dot{x}_2 = x_1 - u x_2^2 - u^2 x_1 + u x_2^2 = 0$)

- Switching sense

Switching from $u = -1$ to $u = +1$ occurs for $x_2 < 0$.
Switching from $u = +1$ to $u = -1$ occurs for $x_2 > 0$.

Let us check this property relating to the sense of the commutations.
The switching sense depends on the sign of the first non-zero derivative of Φ.

At a switching time, we have: $\Phi = p_2 = 0 \ \Rightarrow \ H = \frac{1}{2} x_1^2 + p_1 x_2 = 0$.

If $p_1 = 0$, then $x_1 = 0$ (because $H = 0$) and $x_2 \neq 0$
(because the origin cannot be on a regular arc).
Then we have: $\Phi = \dot{\Phi} = \ddot{\Phi} = 0 \rightarrow \Phi^{(3)} = x_2$, determines the switching sense.

If $p_1 \neq 0$, then $x_1 \neq 0$ (otherwise $x_1 = x_2 = 0$) and $H = 0 \Rightarrow p_1 x_2 = -\dfrac{1}{2} x_1^2 < 0$.

The switching sense is determined by $\dot{\Phi} = -p_1$ which has the sign of x_2.

<u>Invariant extremals</u>

Consider the change of variable $(u, x_1, x_2, p_1, p_2) \rightarrow (v, y_1, y_2, q_1, q_2)$
defined with a parameter α by

$$v(t) = -u\left(\frac{t}{\alpha}\right) \ , \quad y_1(t) = -\alpha^2 x_1\left(\frac{t}{\alpha}\right) \ , \quad y_2(t) = -\alpha x_2\left(\frac{t}{\alpha}\right)$$

$$q_1(t) = -\alpha^3 p_1\left(\frac{t}{\alpha}\right) \ , \quad q_2(t) = -\alpha^4 p_2\left(\frac{t}{\alpha}\right)$$

<u>Remark</u>: we use here the notation (q_1, q_2) without any link with adjoints of (MT).

If (u, x_1, x_2, p_1, p_2) is an extremal

$$\begin{cases} \dot{x}_1 = x_2 \\ \dot{x}_2 = u \end{cases} , \quad \begin{cases} \dot{p}_1 = -x_1 \\ \dot{p}_2 = -p_1 \end{cases} , \quad H = \frac{1}{2} x_1^2 + p_1 x_2 + p_2 u = 0$$

then (v, y_1, y_2, q_1, q_2) is also an extremal

$$\begin{cases} \dot{y}_1 = y_2 \\ \dot{y}_2 = v \end{cases} , \quad \begin{cases} \dot{q}_1 = -y_1 \\ \dot{q}_2 = -q_1 \end{cases} , \quad H = \frac{1}{2} y_1^2 + q_1 y_2 + q_2 v = 0$$

This is verified directly by substitution.

The extremal is said to be point-invariant if
$(u, x_1, x_2, p_1, p_2)(t) = (v, y_1, y_2, q_1, q_2)(t) \ , \ \forall t$
We are looking for the smallest value of α giving an invariant extremal.
If t_k is a switching time, we have
$\Phi_2(t_k) = p_2(t_k) = q_2(t_k) = -\alpha^4 p_2(t_k / \alpha) = 0$
The switching times therefore form a geometric series.

$$t_0 < t_1 = \frac{t_0}{\alpha} < t_2 = \frac{t_1}{\alpha} < \cdots < t_f$$

Consider a bang arc $u = +1$ on $[t_0 ; t_1]$ followed by a bang arc $u = -1$ on $[t_1 ; t_2]$.

We use the first integral: $I_1 = x_1 - \frac{1}{2}ux_2^2$.

- On the arc from t_0 to t_1 : $x_1(t_1) - \frac{1}{2}x_2^2(t_1) = x_1(t_0) - \frac{1}{2}x_2^2(t_0)$

  By invariance of the extremal: $\begin{cases} x_1(t_0) = y_1(t_0) = -\alpha^2 x_1(t_0/\alpha) = -\alpha^2 x_1(t_1) \\ x_2(t_0) = y_2(t_0) = -\alpha\, x_2(t_0/\alpha) = -\alpha\, x_2(t_1) \end{cases}$

  Replacing $x_1(t_0)$ and $x_2(t_0)$: $x_1(t_1) = -\beta x_2^2(t_1)$ with $\beta = \frac{1}{2}\frac{\alpha^2 - 1}{\alpha^2 + 1}$

- On the arc from t_1 to t_2 : $x_1(t_2) + \frac{1}{2}x_2^2(t_2) = x_1(t_1) + \frac{1}{2}x_2^2(t_1)$

  By invariance of the extremal: $\begin{cases} x_1(t_1) = y_1(t_1) = -\alpha^2 x_1(t_1/\alpha) = -\alpha^2 x_1(t_2) \\ x_2(t_1) = y_2(t_1) = -\alpha\, x_2(t_1/\alpha) = -\alpha\, x_2(t_2) \end{cases}$

  Replacing $x_1(t_1)$ and $x_2(t_1)$: $x_1(t_2) = +\beta x_2^2(t_2)$ with $\beta = \frac{1}{2}\frac{\alpha^2 - 1}{\alpha^2 + 1}$

The following relations are obtained at the switching times t_1 and t_2.

$x_1(t_1) = -\beta x_2^2(t_1)$ with $x_2(t_1) > 0$ for switching from $u = +1$ to $u = -1$ at t_1

$x_1(t_2) = +\beta x_2^2(t_2)$ with $x_2(t_2) < 0$ for switching from $u = -1$ to $u = +1$ at t_2

The switching curves have the equations: $\begin{cases} C^+ : x_1 = +\beta x_2^2 \,,\ x_2 < 0 \\ C^- : x_1 = -\beta x_2^2 \,,\ x_2 > 0 \end{cases}$.

<u>Values of α and β</u>

Let us place ourselves on the arc from t_1 to t_2 with $u = -1$.

The Hamiltonian allows p_1 to be expressed in terms of x_1 and x_2.

$H = \frac{1}{2}x_1^2 + p_1 x_2 + p_2 u = 0$ with $p_2 = 0$ at each switching $\rightarrow$ $p_1 = -\frac{x_1^2}{2x_2}$

Then let us use the first integral: $I_2 = -p_1 - ux_1x_2 + \frac{1}{3}x_2^3$.

$$I_2 = \left(-p_1 + x_1x_2 + \frac{1}{3}x_2^3\right)(t_1) = \left(-p_1 + x_1x_2 + \frac{1}{3}x_2^3\right)(t_2)$$

with $\begin{cases} x_1(t_1) = -\beta x_2^2(t_1) \\ x_1(t_2) = +\beta x_2^2(t_2) \end{cases}$ and $p_1 = -\frac{x_1^2}{2x_2}$

By invariance of the extremal, we have
$x_2(t_1) = y_2(t_1) = -\alpha x_2(t_1/\alpha) = -\alpha x_2(t_2)$
Replacing $x_1(t_1)$ and $x_2(t_1)$ in the first integral I_2, we obtain

$$-\alpha^3\left(\frac{1}{2}\beta^2 - \beta + \frac{1}{3}\right) = \left(\frac{1}{2}\beta^2 + \beta + \frac{1}{3}\right) \quad \text{with } \beta = \frac{1}{2}\frac{\alpha^2 - 1}{\alpha^2 + 1} \Rightarrow \alpha^2 = \frac{1+2\beta}{1-2\beta}$$

which results in a second-degree equation in β^2.

$$\beta^4 + \frac{1}{12}\beta^2 - \frac{1}{18} = 0 \quad \to \beta = \sqrt{\frac{\sqrt{33}-1}{24}} \approx 0.44462 \text{ and } \alpha = \sqrt{\frac{1+2\beta}{1-2\beta}} \approx 4.13016$$

Summary
The invariant extremals allow the construction of a complete solution.

Switching times: $t_k = \alpha t_{k+1} \quad \to$ converging to $t_f = 0$

Switching curves: $\begin{cases} C^+ : x_1 = +\beta x_2^2 \,,\, x_2 < 0 \;\to\; \text{switching from } u = -1 \text{ to } +1 \\ C^- : x_1 = -\beta x_2^2 \,,\, x_2 > 0 \;\to\; \text{switching from } u = +1 \text{ to } -1 \end{cases}$

State feedback control: $\begin{cases} u = +1 \text{ if } x_1 < -\beta x_2^2 \,\text{sgn}(x_2) \\ u = -1 \text{ if } x_1 > -\beta x_2^2 \,\text{sgn}(x_2) \end{cases}$

with the constants: $\beta = \sqrt{\frac{\sqrt{33}-1}{24}} \approx 0.44462 \;,\; \alpha = \sqrt{\frac{1+2\beta}{1-2\beta}} \approx 4.13016$

This synthesis is obtained by connecting the bang arc associated with the invariant extremals. It can be shown [R17] that this synthesis is the optimal synthesis of the Fuller problem, which means that these invariant extremals are the only ones and that their junction is optimal.

Figure 4-33 shows the switching curves C^+ and C^- in the phase plane (x_1, x_2) and a dotted path with 4 successive bang arcs.

Unlike the double integrator problem (example 4-7), the switching curves are not optimal trajectories.

These curves are crossed an infinite number of times, thus causing chattering.

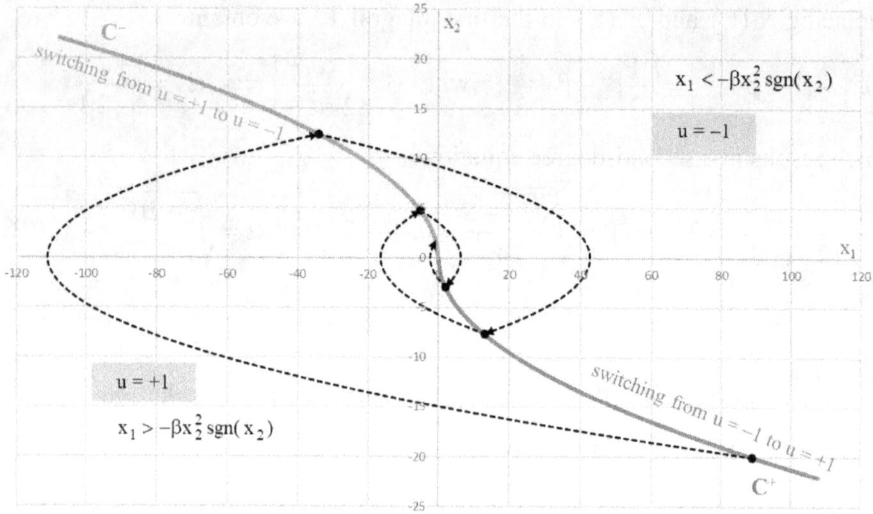

Figure 4-33: Optimal synthesis of the Fuller problem.

Figure 4-34 illustrates the chattering as the end point is approached. The initial conditions have been chosen on the switching curve C^+ : $x_1 = 88.92$, $x_2 = -20$. The first bang arc with $u = +1$ goes from t_0 to t_1.
From the equation of this initial arc: $x_2(t_1) - x_2(t_0) = u.(t_1 - t_0)$
and from the switching conditions in t_1 : $x_2(t_0) = -\alpha x_2(t_1)$, $t_0 = \alpha t_1$
the initial time (for $t_f = 0$) is determined: $t_0 = \dfrac{\alpha+1}{\alpha-1} x_2(t_0) \approx -32.779$.

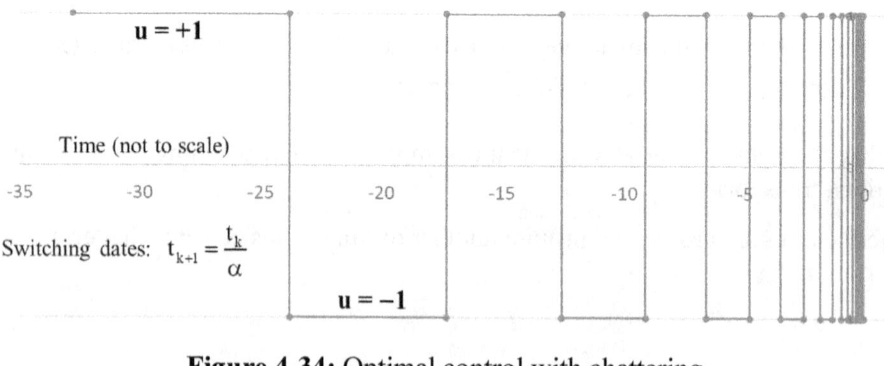

Figure 4-34: Optimal control with chattering.

<u>Value function</u>

The value function $V(x_1, x_2)$ of the problem (FP) can be obtained by solving the HJB equation.

$$\min_u H\left(x, u, \frac{\partial V}{\partial x}, t\right) + \frac{\partial V}{\partial t} = 0$$

with $H = \frac{1}{2}x_1^2 + p_1 x_2 + p_2 u$ and $\min_{-1 \le u \le +1} H \rightarrow u = -\text{sgn}\left(\frac{\partial V}{\partial x_2}\right)$

The HJB equation is in the form: $x_2 \frac{\partial V}{\partial x_1} - \left|\frac{\partial V}{\partial x_2}\right| + \frac{1}{2}x_1^2 = 0$.

It can be shown [R17] that the solution is

$$V_\pm(x_1, x_2) = \mp\frac{1}{15}x_2^5 + \frac{1}{3}x_1 x_2^3 \mp \frac{1}{2}x_1^2 x_2 + A\left(\frac{1}{2}x_2^2 \mp x_1\right)^{\frac{5}{2}}$$

$$\begin{cases} V = V_+ & \text{if } x_1 < -\beta x_2^2 \,\text{sgn}(x_2) \\ V = V_- & \text{if } x_1 > -\beta x_2^2 \,\text{sgn}(x_2) \end{cases} \quad \text{with } A = \frac{1}{5}\left(\frac{2}{3} + 2\beta + \beta^2\right)\left(\beta + \frac{1}{2}\right)^{-\frac{3}{2}} \approx 0.382$$

4.5 Conclusion

4.5.1 The key points

- The Pontryaguin minimum principle gives the necessary first-order conditions that the optimal control must satisfy;
- final, interior or path constraints induce additional conditions that can cause control discontinuities;
- determining the optimal control and trajectory requires solving a two-point boundary value problem;
- the transport method allows to determine the neighboring extremals of a reference solution and to construct a state feedback control;
- the optimal trajectory may contain singular arcs whose determination requires optimality conditions of order greater than 1.

4.5.2 To go further

- *Contrôle optimal – Théorie et applications (E. Trélat, Vuibert 2005)*

  This book presents the theory of optimal control in a very pedagogical way, while maintaining a very high level of rigor. The author is one of the world's leading specialists in the field. Numerous commented exercises help to understand how to apply the optimality conditions to concrete problems.

- *Les mathématiques du mieux-faire (J.B. Hiriart-Urruty, Ellipses 2008)*

  The second volume explains step by step the establishment of the principle of the minimum with all the demonstrations and many remarks in the margin which help the understanding.

- *Applied optimal control (A.E. Bryson, Y.C. Ho, Hemisphere Publishing Corporation 1975)*

  This book is aimed at the practical application of optimal control for engineers. The very progressive presentation in the chapters allows a good familiarization with the computational techniques, especially thanks to many very detailed examples.

- *Calculus of variations and optimal control theory (D. Liberzon, Princeton University Press, 2012)*

  This rather theory-oriented book explains the transition from calculus of variations to optimal control. The presentation is very didactic and clear and remains rigorous without excessive formalism. A book to be recommended to get familiar with the theoretical aspects of functional optimization.

- *Optimal control theory for applications (D.G. Hull, Springer, 2003)*

  This book is based on the variational approach to optimal control. It offers the interest of developing in a detailed way the optimality conditions associated with the different formulations (parameters, constraints...). Three application examples are used throughout the book to illustrate the results.

- *Geometric optimal control (H. Schättler, U. Ledzewicz, Springer, 2012)*

  This book deals with geometric methods of optimal control and investigate especially the properties of singular arcs. The focus is more on theory with proofs of results. Although the subject is difficult, the content remains accessible without excessive formalism.

5. Numerical methods in optimal control

The numerical solution of an optimal control problem requires the discretization of the state and the control. This chapter is an introduction to the usual approaches.

Section 1 introduces the transcription methods, which consist in formulating a finite-dimensional problem "equivalent" to the control problem. The two main choices concern the solution of the differential equations and the use or not of the optimality conditions of the control problem.

Section 2 is devoted to Runge-Kutta methods that solve a differential system by progressing through time steps. These single-step methods use only the current state at each step. The accuracy depends on the order of the method and can be controlled by setting the time step.

Section 3 is devoted to Adams methods. Like Runge-Kutta methods, the solution is assessed by time steps, but using several previous steps to construct the solution at the next step. These multi-step methods are more accurate, but more complex to implement.

Section 4 introduces collocation methods. The solution of the differential equation is approximated by polynomials satisfying conditions on their derivatives at certain points. The accuracy of these methods depends on the degree of the polynomials and the number of discretization points.

Section 5 presents the direct methods that reduce the problem to finite dimension. Depending on the method of solving the differential equations, either the control (Runge-Kutta) or the control and the state (collocation) are discretized. The discretized problem of large size is solved by a nonlinear programming algorithm or by a variational method based on the deviation equations.

Section 6 presents indirect methods that solve the optimality conditions of the minimum principle. The problem is reduced to a system of equations solved by the shooting method (Newton's method with state propagation and adjoint). This method is fast and accurate if one has a good initialization, which can be very difficult. The multiple shooting or the variational approaches alleviate this difficulty, but by increasing the size of the problem.

5.1 Transcription

We seek to solve a free final time control problem of the form

$$\min_{u,t_f} J(u) = \int_{t_0}^{t_f} L(x,u,t)dt + \varphi(x_f,t_f) \quad \text{with} \begin{cases} \dot{x}(t) = f(x,u,t) \, , \, x(t_0) = x_0 \\ \psi(x_f,t_f) = 0 \end{cases} \tag{5.1}$$

The state x is of dimension n, the control u is of dimension m, the number of final constraints is q. The problem is said in infinite dimension, because the unknown is a function u(t).

An analytical solution is rarely possible, even for simple problems. Numerical methods must be used which give an approximate solution in a finite number of dates. **Transcription** consists in formulating a finite-dimensional "equivalent" problem and solving it numerically. The finite-dimensional problem can be formulated in different ways.

The main choices are how to solve the differential equations (by propagation or collocation) and how to use the optimality conditions of the control problem (direct or indirect method).

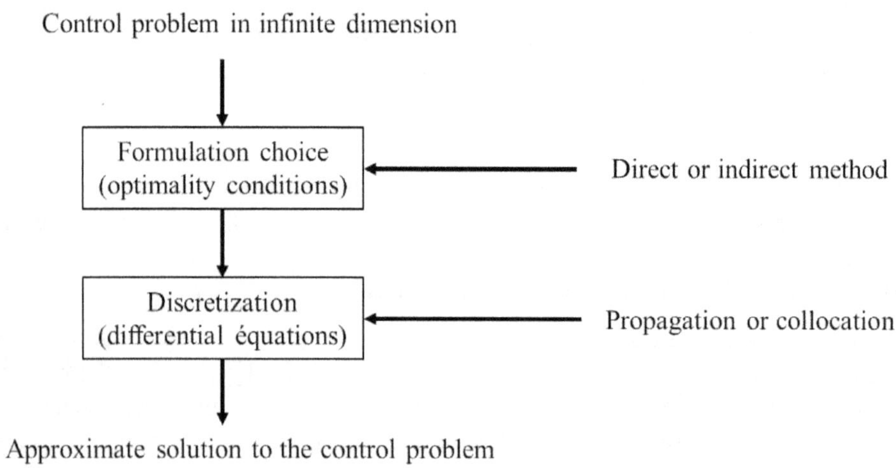

Figure 5-1: Transcription process.

5.1.1 Differential equations

A Cauchy problem or initial value problem (IVP) consists in solving a first order differential system with a given initial condition.

$$\begin{cases} \dot{x}(t) = f\left[x(t),t\right] \ , t_0 \le t \le t_f \\ x(t_0) = x_0 \end{cases} \tag{5.2}$$

The unknown is the function $x(t)_{t_0 \le t \le t_f}$ of dimension n.

The problem is reduced to a finite-dimensional problem by dividing the integration interval $[t_0 ; t_f]$ into N segments: $t_0 < t_1 < \cdots < t_i < \cdots < t_N = t_f$ and looking for the values taken by x at times $(t_i)_{i=0 \text{ to } N}$. The sequence of values $(x_i)_{i=0 \text{ to } N}$ is an approximate solution of the differential equation (5.2).

This solution can be constructed by propagation or by collocation.

<u>Propagation methods</u>

The propagation methods calculate the solution by progressing in time.

The value x_{i+1} is determined by quadrature formulas from the previous values:

- single-step methods (Runge-Kutta type) only use the previous value x_i;
- multi-step methods (Adams type) use the values of several previous steps $x_{i-k}, \cdots, x_i$ as shown in figure 5-2.

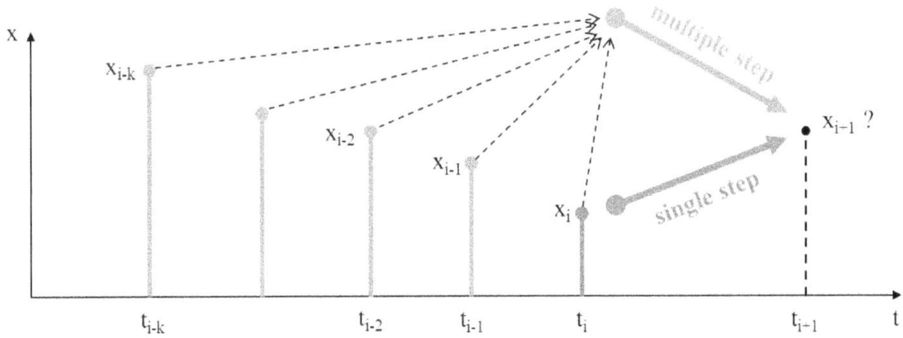

Figure 5-2: Single or multi-step propagation method.

Collocation methods

The collocation methods construct a piecewise polynomial solution. The unknown solution $x(t)$ is approximated by a polynomial $\xi(t)$ of degree K on each time interval $[t_i ; t_{i+1}]$. This polynomial satisfies connection conditions at the ends and its derivative satisfies the differential equation at intermediate times $(\tau_k)_{k=1 \text{ to } K}$. The solution accuracy depends on the polynomial degree and on the positioning of the intermediate points called collocation points (figure 5-3). The complete solution on $[t_0 ; t_f]$ is obtained globally by solving a large nonlinear system.

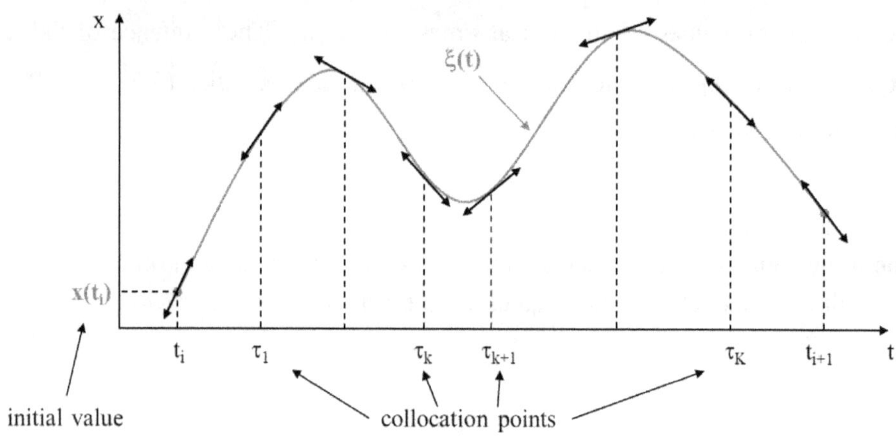

Figure 5-3: Collocation method.

Well-posed problem and conditioning

Solving problem (5.2) may be mathematically or numerically difficult.

The problem is said to be mathematically well-posed if the exact solution is unique and depends continuously on the initial condition x_0.

The problem is said to be numerically well-posed if the numerical solution is not sensitive to rounding errors.

The problem is said to be well-conditioned if an accurate numerical solution can be obtained without excessive reduction of the time step.

Example 5-1: Well-posed problem and conditioning

Example 1 Consider the differential equation: $\dot{x} = 2\sqrt{|x|}$ with $x(t_0 = 0) = \varepsilon$.

If $\varepsilon \neq 0$, a possible solution is: $x(t) = \begin{cases} -(t-\sqrt{\varepsilon})^2 & \text{if } t < \sqrt{\varepsilon} \\ +(t-\sqrt{\varepsilon})^2 & \text{if } t \geq \sqrt{\varepsilon} \end{cases}$

If $\varepsilon = 0$, there is an infinite number of solutions of the form

$$x(t) = \begin{cases} 0 & \text{for } t < \tau \\ (t-\tau)^2 & \text{for } t \geq \tau \end{cases}$$

The solution is not unique for $\varepsilon = 0$, and it is not continuous when $\varepsilon \to 0$.
The problem is mathematically ill-posed.

Example 2 Consider the differential equation: $\begin{cases} \dot{x} = ax + b \\ x(t_0 = 0) = x_0 \end{cases}$.

The general solution is: $x(t) = \frac{1}{a}\left[(ax_0 + b)e^{at} - b\right]$.

A rounding error ε on $x(t_0)$ produces an error εe^{at} on $x(t)$.

If $a > 0$, the error becomes significant (> 1) from $t > \frac{-\ln\varepsilon}{a}$.

The problem is numerically ill-posed for $a > 0$ if the time t_f is too large.

Example 3
Let us apply Euler's method with a time step h to the previous problem.

$$x_{i+1} = x_i + hf(x_i, t_i) = (1+ah)x_i + bh \quad \Rightarrow \quad \left(x_{i+1} + \frac{b}{a}\right) = (1+ah)\left(x_i + \frac{b}{a}\right)$$

The solution after n time steps is given by: $x_n = (1+ah)^n\left(x_0 + \frac{b}{a}\right) - \frac{b}{a}$.

If $a < 0$, the exact solution converges to

$$\lim_{t\to+\infty} x(t) = \lim_{t\to+\infty} \frac{1}{a}\left[(ax_0 + b)e^{at} - b\right] = -\frac{b}{a}$$

For the numerical solution to have the same limit, the step h must satisfy $|1+ah| < 1$.

If a is large, the time step must be reduced proportionally to obtain an accurate solution. The problem is ill-conditioned (steep problem).

5.1.2 Direct and indirect methods

The control problem (5.1) can be solved either "directly" by discretization or "indirectly" by using optimality conditions.

Direct methods

The direct methods transform the optimal control problem into an "equivalent" finite-dimensional problem. The interval $[t_0 ; t_f]$ is divided into N segments with times: $t_0 < t_1 < \cdots < t_i < \cdots < t_N = t_f$. The function $u(t)$ is approximated by its values $(u_i)_{i=0 \text{ to } N}$ at the times $(t_i)_{i=0 \text{ to } N}$ with an interpolation method to be chosen (constant, linear, splines...). We are thus reduced to a finite-dimensional optimization problem with $(N+1)\times m$ variables, where m is the dimension of the control vector u. The state differential equation is solved by propagation or by collocation, using or not the same discretization $(t_i)_{i=0 \text{ to } N}$. In the case of an integration by collocation, the states $(x_i)_{i=1 \text{ to } N}$ at times $(t_i)_{i=1 \text{ to } N}$ are added to the problem variables with associated collocation constraints. In this way, a large optimization problem is formulated and solved by classical nonlinear programming methods.

Direct methods have the advantage of decoupling optimization and simulation. It is thus possible to modify the dynamic model or the constraints without any specific coding effort. They do not presuppose a particular form of the solution (regular/singular arcs, active/inactive inequality constraints). The unknowns of the problem are physical variables (control law), which facilitates the initialization and interpretation of the results. Direct methods can be easily applied to many problems. Their main drawback is the size of the optimization problem (many variables and constraints) which can lead to slow and sometimes inaccurate convergence.

Indirect methods

The indirect methods try to solve the optimality conditions of problem (5.1). These conditions involve the adjoint $p(t)$ of same dimension n as the state $x(t)$ and the normal Hamiltonian defined by: $H(x,u,p,t) = L(x,u,t) + p^T f(x,u,t)$.
They lead to the following two-point boundary value problem.

$$\begin{cases} \dot{x} = H_p \;, & \psi(x_f, t_f) = 0 \\ \dot{p} = -H_x \;, & p_f = \varphi_{xf} + \psi_{xf}\lambda \\ H_u = 0 \;, & H_f = -\varphi_{tf} - \psi_{tf}\lambda \end{cases} \tag{5.3}$$

The differential system is integrated using the control obtained from $H_u = 0$.
The unknowns are the n components of the initial adjoint $p(t_0)$, the q multipliers λ of the final constraints and the final time t_f. The equations to be satisfied are the q final constraints, the n transversality conditions on the final adjoint and the transversality condition on the final Hamiltonian.
A system of equations of dimension $n+q+1$ is thus formulated which can be solved by a Newton method.

Indirect methods have the advantage of reducing the problem to a small system of equations. The solution of this system is generally very fast and precise, provided that a good initialization is available. The disadvantages are the difficulty of initialization (non-physical variables, high sensitivity), the need for a coding specific to the problem treated (adjoint equations, transversality conditions, structure of the solution). The difficulty of initialization can be prohibitive in certain cases. Indirect methods should be reserved for problems with a stable formulation (dynamics, constraints) and a known control structure. It is possible in this context to develop specific initialization strategies.

Table 5-1 compares the characteristics of direct and indirect methods.

	Direct method	Indirect method
Problem	Constrained Optimization	Problem at both ends
Unknown (dimensions)	Discrete control u_i (N×n), t_f (1)	Initial assistant, multipliers p_0 (n), λ (q), t_f (1)
Initialization	Easy (physical variables) Not very sensitive	Difficult (abstract variables) Very sensitive
Convergence	Slow (large size problem)	Fast (few unknowns)
Accuracy	According to discretization	Very precise
Coding	Non-intrusive if propagation Intrusive if collocation	Intrusive (adjoint) Problem specific

Table 5-1: Direct and indirect methods.

5.2 Runge-Kutta methods

The Runge-Kutta methods are single-step methods for numerically solving a first order differential system of the form

$$\dot{x}(t) = f[x(t),t] \ , \ t_0 \leq t \leq t_f \tag{5.4}$$

The state vector x is of dimension n and the function $f:\mathbb{R}^n \to \mathbb{R}^n$ is assumed to be of class C^1. The initial condition is given: $x(t_0)=x_0$. The interval $[t_0;t_f]$ is divided into N segments: $t_0 < t_1 < \cdots < t_i < t_{i+1} < \cdots < t_N = t_f$:

- the times t_i are the **nodes**;
- the segments $[t_i;t_{i+1}]$ are the integration **steps**;
- the unknowns are the values $x(t_i)$ noted x_i.

The differential equation is integrated on each segment $[t_i;t_{i+1}]$.

$$\dot{x} = f(x,t) \ \text{ on } \ [t_i;t_{i+1}] \ \Rightarrow \ x_{i+1} = x_i + \int_{t_i}^{t_{i+1}} f(x,t)dt \tag{5.5}$$

The integral is approximated by a quadrature formula. Starting from the given initial state x_0, we calculate successively x_1, then x_2 ... until x_N. This gives an approximate solution at N discretization times.

5.2.1 Quadrature formulas

A quadrature consists in approximating an integral by a sum. The interval $[t_i;t_{i+1}]$ is divided into K+1 segments: $t_i \leq \tau_1 \leq \tau_2 \leq \cdots \leq \tau_k \leq \tau_{k+1} \leq \cdots \leq \tau_K \leq t_{i+1}$.
The times $(\tau_k)_{k=1 \text{ to } K}$ are not necessarily distinct:

- the length of the interval $[t_i;t_{i+1}]$ is noted h_i: $h_i = t_{i+1} - t_i$;
- the times τ_k are of the form: $\tau_k = t_i + h_i\alpha_k$ with increasing coefficients $0 \leq \alpha_1 \leq \alpha_2 \leq \cdots \leq \alpha_k \leq \alpha_{k+1} \leq \cdots \leq \alpha_K \leq 1$;
- the segments $[\tau_k;\tau_{k+1}]$ are the integration **stages**.

The integral is approximated by a **quadrature formula** of the type

$$x_{i+1} - x_i = \int_{t_i}^{t_{i+1}} f(x,t)dt \approx h_i \sum_{k=1}^{K} \beta_k f(\xi_k,\tau_k) \tag{5.6}$$

This formula requires the states $(\xi_k)_{k=1\text{ to }K}$ at the intermediate times $(\tau_k)_{k=1\text{ to }K}$.
The intermediate states $(\xi_k)_{k=1\text{ to }K}$ are evaluated by a second quadrature.

$$\xi_k - x_i = \int_{t_i}^{\tau_k} f(x,t)dt \approx h_i \sum_{j=1}^{K} \gamma_{kj} f(\xi_j, \tau_j) \tag{5.7}$$

A Runge-Kutta method is thus defined as its combination of two quadrature methods with a particular choice of coefficients α_k, β_k and γ_{kj}.
The quadratures (5.6-7) are required to be exact if the function f is constant.

$$\begin{cases} x_{i+1} - x_i = (t_{i+1} - t_i)f = h_i \sum_{k=1}^{K} \beta_k f \;\Rightarrow\; \sum_{k=1}^{K} \beta_k = 1 \\ \xi_k - x_i = (\tau_k - t_i)f = h_i \sum_{j=1}^{K} \gamma_{kj} f \;\Rightarrow\; \sum_{j=1}^{K} \gamma_{kj} = \alpha_k \;,\; k = 1 \text{ to } K \end{cases} \tag{5.8}$$

<u>Explicit or implicit method</u>
The K intermediate values $(\xi_k)_{k=1\text{ to }K}$ satisfy a system of K equations.

$$\xi_k = x_i + h_i \sum_{j=1}^{K} \gamma_{kj} f(\xi_j, \tau_j) \;,\; k = 1 \text{ to } K \tag{5.9}$$

If the coefficients γ_{kj} are zero for $j \geq k$, the state ξ_k depends only on the previous states $(\xi_j)_{j<k}$. In this case, the ξ_k can be calculated successively and the method is said to be **explicit**.

Otherwise, one must solve the nonlinear system (5.9) of dimension K whose unknowns are $(\xi_k)_{k=1\text{ to }K}$. All values ξ_k are obtained simultaneously and the method is said to be **implicit**. This system of equations (5.9) is solved in two steps:
- the initialization (prediction) is done by an explicit method;
- iterations (correction) are performed by a fixed point or a Newton method.

An implicit method is more stable (as shown in example 5-2), but more difficult to implement as a nonlinear system must be solved at each time step.

When the states ξ_k have been determined, we obtain x_{i+1} by the quadrature (5.6).

Example 5-2: Stability of forward and backward Euler methods

Consider the differential equation: $\dot{x} = -x$ whose solution is: $x(t) = x_0 e^{-t}$.
The solution has a limit: $\lim\limits_{t \to +\infty} x(t) = 0$.

Let us compare the stability of the forward (explicit) and backward (implicit) Euler methods as a function of the step size h:

- the forward Euler method with a step h leads to
  $x_{i+1} = x_i - hx_i = (1-h)x_i$ (explicit formula)
  The method is stable if: $|1-h| < 1 \Rightarrow h < 2$ (then we retrieve $\lim\limits_{i \to +\infty} x_i = 0$);
- the backward Euler method with a step h leads to
  $x_{i+1} = x_i - hx_{i+1} \Rightarrow x_{i+1} = \dfrac{1}{1+h} x_i$ (by solving the implicit equation)
  The method is stable for any step h (and we obtain $\lim\limits_{i \to +\infty} x_i = 0$).

Implicit methods are generally more stable than explicit methods. They allow larger time steps at the cost of more calculations at each step to solve the equation giving x_{i+1}.

Butcher's array

A Runge-Kutta method is completely defined by the number of stages K (for the intermediate states) and the choice of coefficients $\alpha_k, \beta_k, \gamma_{kj}$. These coefficients are grouped in the Butcher array. Each row of the array corresponds to an integral approximated by a quadrature, as shown below.

$$\begin{array}{c|ccc} \alpha_1 & \gamma_{11} & \cdots & \gamma_{1K} \\ \vdots & \vdots & & \vdots \\ \alpha_K & \gamma_{K1} & \cdots & \gamma_{KK} \\ \hline & \beta_1 & \cdots & \beta_K \end{array}$$

$$\alpha \rightarrow \quad \tau_k = t_i + h_i \alpha_k$$

$$\gamma \rightarrow \quad \xi_k = x_i + h_i \sum_{j=1}^{K} \gamma_{kj} f(\xi_j, \tau_j) \quad , \quad k = 1 \text{ to } K$$

$$\beta \rightarrow \quad x_{i+1} = x_i + h_i \sum_{k=1}^{K} \beta_k f(\xi_k, \tau_k)$$

For an explicit method, the elements of the array on the diagonal and above are zero: $(\gamma_{kj})_{j \ge k} = 0$.

The simplest Runge-Kutta methods are:
- Euler explicit, implicit → K = 1 stage;
- modified Euler, Heun, Crank-Nicolson → K = 2 stages.

Explicit Euler method (or forward Euler method)
This one-stage method uses the following array and intermediate states.

$$\begin{array}{c|c} 0 & 0 \\ \hline & 1 \end{array} \qquad \begin{cases} \tau_1 = t_i \\ \xi_1 = x_i \end{cases} \rightarrow x_{i+1} = x_i + h_i f(\xi_1, \tau_1)$$

We obtain the explicit formula: $x_{i+1} = x_i + h_i f(x_i, t_i)$. This is equivalent to assuming that the solution is linear on $[t_i ; t_{i+1}]$ with the derivative taken at the known initial point $(x_i ; t_i)$, as illustrated in figure 5-4.

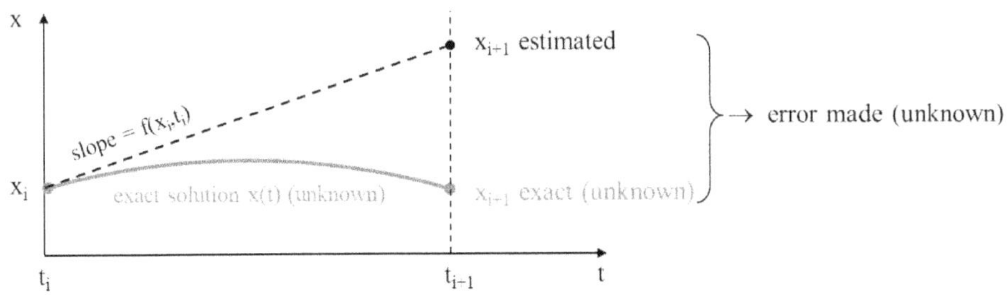

Figure 5-4: Explicit Euler method.

Implicit Euler method (or backwards Euler method)
This one-stage method uses the following array and intermediate states.

$$\begin{array}{c|c} 1 & 1 \\ \hline & 1 \end{array} \qquad \begin{cases} \tau_1 = t_i + h_i = t_{i+1} \\ \xi_1 = x_i + h_i f(\xi_1, \tau_1) \end{cases} \rightarrow x_{i+1} = x_i + h_i f(\xi_1, \tau_1)$$

We obtain the implicit formula: $x_{i+1} = x_i + h_i f(x_{i+1}, t_{i+1})$, which gives an equation in x_{i+1}. This is equivalent to assuming that the solution is linear on $[t_i ; t_{i+1}]$ with the derivative taken at the unknown end point $(x_{i+1} ; t_{i+1})$, as shown in figure 5-5.

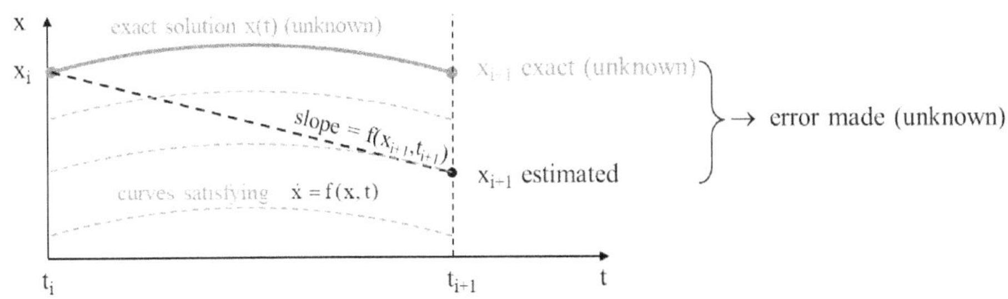

Figure 5-5: Implicit Euler method.

Modified Euler method (or midpoint method)
This one-stage method uses the following array and intermediate states.

0	0	0
1/2	1/2	0
	0	1

$$\begin{cases} \tau_1 = t_i \ , \ \tau_2 = t_i + \frac{1}{2} h_i \\ \xi_1 = x_i \ , \ \xi_2 = x_i + \frac{1}{2} h_i f(\xi_1, \tau_1) \end{cases} \rightarrow x_{i+1} = x_i + h_i f(\xi_2, \tau_2)$$

We obtain the explicit formula: $x_{i+1} = x_i + h_i f\left[x_i + \frac{1}{2} h_i f(x_i, t_i) \ , \ t_i + \frac{1}{2} h_i \right]$.

This is equivalent to assuming that the solution is linear on $[t_i ; t_{i+1}]$ with the derivative taken at the unknown midpoint $(\xi_2 ; \tau_2)$, as shown in figure 5-6.

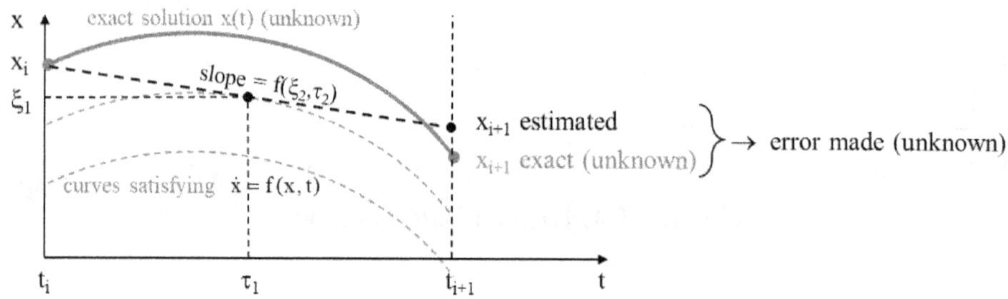

Figure 5-6: Modified Euler method.

Heun's method
This two-stage method uses the following array and intermediate states.

0	0	0
1	1	0
	1/2	1/2

$$\begin{cases} \tau_1 = t_i \ , \ \tau_2 = t_i + h_i = t_{i+1} \\ \xi_1 = x_i \ , \ \xi_2 = x_i + h_i f(\xi_1, \tau_1) \end{cases}$$

$$\rightarrow x_{i+1} = x_i + \frac{1}{2} h_i f(\xi_1, \tau_1) + \frac{1}{2} h_i f(\xi_2, \tau_2)$$

We obtain the explicit formula

$$x_{i+1} = x_i + \frac{1}{2} h_i f(x_i, t_i) + h_i f\left[x_i + \frac{1}{2} h_i f(x_i, t_i) \ , \ t_{i+1} \right]$$

This is equivalent to assuming a piecewise linear solution on $[t_i ; t_{i+1}]$ and calculating the integral by the trapezoidal method. The point ξ_2 is estimated at t_{i+1} by the explicit Euler method, as shown in figure 5-7.

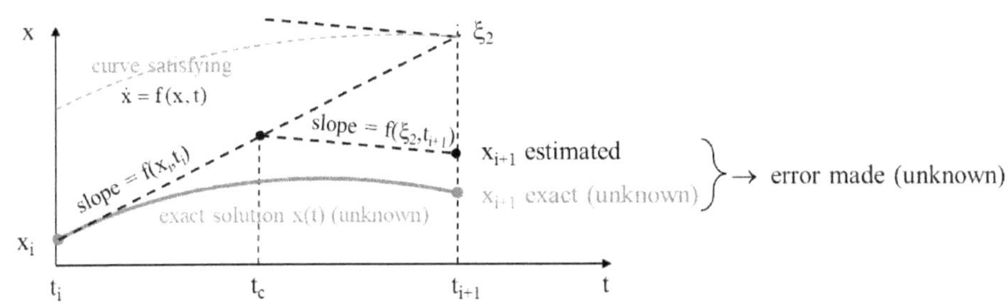

Figure 5-7: Heun's method.

Crank-Nicolson method (or trapezoidal method)
This two-stage method uses the following array and intermediate states.

0	0	0
1	1/2	1/2
	1/2	1/2

$$\begin{cases} \tau_1 = t_i \ , \ \tau_2 = t_i + h_i = t_{i+1} \\ \xi_1 = x_i \ , \ \xi_2 = x_i + \frac{1}{2}h_i f(\xi_1,\tau_1) + \frac{1}{2}h_i f(\xi_2,\tau_2) \end{cases}$$

$$\rightarrow \ x_{i+1} = x_i + \frac{1}{2}h_i f(\xi_1,\tau_1) + \frac{1}{2}h_i f(\xi_2,\tau_2)$$

We obtain the implicit formula: $x_{i+1} = x_i + \frac{1}{2}h_i f(x_i,t_i) + \frac{1}{2}h_i f(x_{i+1},t_{i+1})$

which gives an equation in x_{i+1}.

This is equivalent to assuming a piecewise linear solution on $[t_i;t_{i+1}]$ and calculating the integral by the trapezoidal method. The point $\xi_2 = x_{i+1}$ is such that the tangents from x_i and x_{i+1} meet at the midpoint, as shown in figure 5-8.

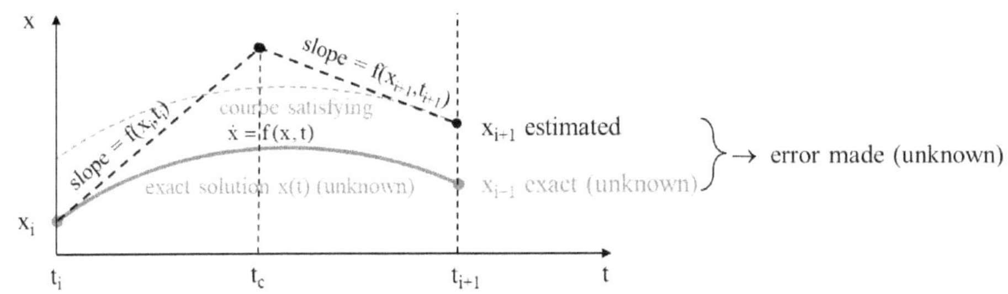

Figure 5-8: Crank-Nicolson method.

Example 5-3 compares the accuracy of these methods on a simple case.

Example 5-3: One and two-stage method

Consider the differential equation: $\dot{x} = x$ whose solution is: $x(t) = x_0 e^{t-t_0}$.
Let us apply the previous methods to solve this equation numerically on the interval from $t_0 = 0$ to $t_1 = 1$ with the initial condition: $x(t_0) = 1$.

The interval is divided into N equidistant steps of length: $h = \frac{1}{N}$.

Explicit Euler

$$x_{i+1} = x_i + hx_i \qquad \rightarrow x_{i+1} = (1+h)x_i \qquad \rightarrow x_N = (1+h)^N x_0$$

Implicit Euler

$$x_{i+1} = x_i + hx_{i+1} \qquad \rightarrow x_{i+1} = \frac{1}{1-h} x_i \qquad \rightarrow x_N = \left(\frac{1}{1-h}\right)^N x_0$$

Crank-Nicolson

$$x_{i+1} = x_i + h\frac{x_i + x_{i+1}}{2} \qquad \rightarrow x_{i+1} = \frac{2+h}{2-h} x_i \qquad \rightarrow x_N = \left(\frac{2+h}{2-h}\right)^N x_0$$

Table 5-2 gives the error made as a function of the number of steps.

Exact solution $x_1 = 2.71828$	1 step $h = 1$	2 steps $h = 0.5$	10 steps $h = 0.1$	100 steps $h = 0.01$	1000 steps $h = 0.001$	10 000 steps $h = 0.0001$
Euler explicit	2.00000	2.25000	2.59374	2.70481	2.71692	2.71815
Error	0.72	0.47	0.13	0.014	0.0014	0.00014
Euler implicit	Impossible	4.00000	2.86797	2.73100	2.71964	2.71842
Error		1.28	0.15	0.014	0.0014	0.00014
Crank-Nicolson	3.00000	2.78000	2.72055	2.71830	2.71828	2.71828
Error	0.28172	0.059	0.0023	$2.3\ 10^{-5}$	$2.3\ 10^{-7}$	$2.3\ 10^{-9}$

Table 5-2: Comparison of Euler and Crank-Nicolson methods.

We observe that the error is of the order of h for the Euler methods and of the order of h^2 for the Crank-Nicolson method. Consequently, a large number of steps is needed to obtain a very accurate solution. The error depends on the order of the method, as analyzed in section 5.2.2.

5.2.2 Error analysis

A Runge-Kutta method produces an approximate solution of the differential equation. This section introduces the concepts for quantifying the deviation between the approximate and the exact solution.

Consistency error

Consider the differential equation: $\dot{y} = f(y,t)$. By performing a time step from t_i to $t_{i+1} = t_i + h_i$ with a Runge-Kutta method from the initial condition $(t_i; x_i)$, we obtain the state x_{i+1} in the form

$$x_{i+1} = x_i + h_i \varphi(t_i, x_i, h_i) \tag{5.10}$$

We note $y_i(t)$ the exact (unknown) solution from the initial condition $(t_i; x_i)$. This exact solution takes the value $y_i(t_{i+1})$ at t_{i+1}.

The **consistency** error noted e_{i+1} is the deviation over one time step between the numerical and the exact solution from the same initial condition (figure 5-9).

$$e_{i+1} = y_i(t_{i+1}) - x_{i+1} \tag{5.11}$$

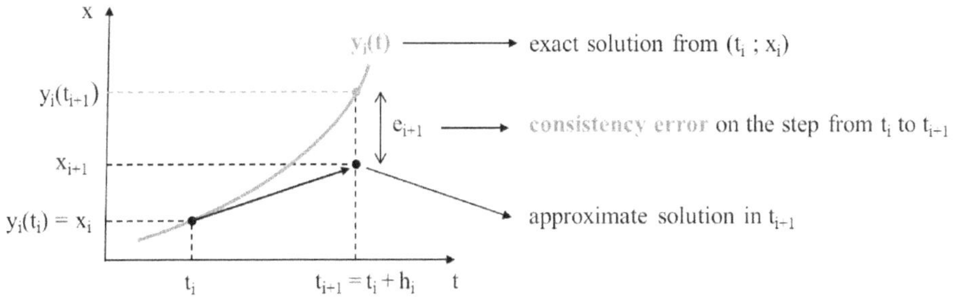

Figure 5-9: Consistency error.

Global error

The **global error** noted E_i is the deviation at t_i between the numerical solution x_i and the exact solution $y_0(t_i)$ obtained from the initial condition $x(t_0) = x_0$. It differs from the consistency error e_{i+1} at step h_i which assumes that the initial condition at t_i is reset on the exact solution $y_0(t_i)$ at that time.

This consistency error does not take into account the difference in initial value between $y_0(t_i)$ and x_i coming from previous steps. The global error comes from the accumulation of the consistency errors at each time step as well as the shift of the initial condition at each step as illustrated in figure 5-10.

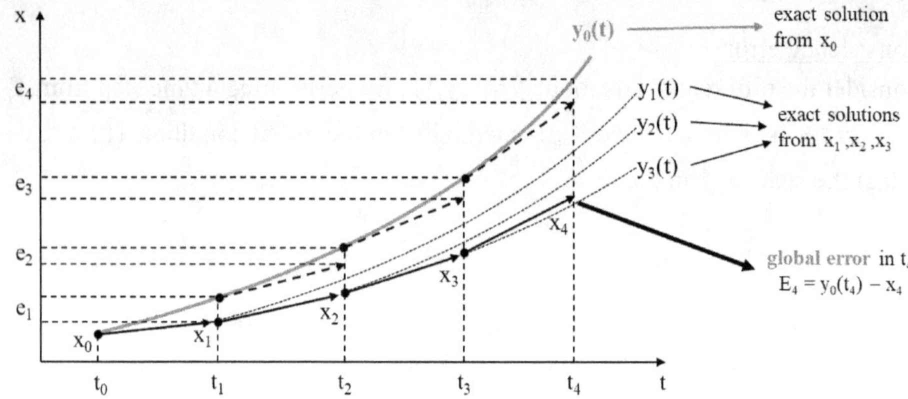

Figure 5-10: Global error.

<u>Consistency, stability, convergence and order</u>

A numerical method generates at each step consistency errors: $(e_i)_{i=1 \text{ to } N}$ and rounding errors: $(\varepsilon_i)_{i=1 \text{ to } N}$. We note:

- $(x_1, x_2, \cdots, x_N)$ the solution without rounding errors: $x_{i+1} = x_i + h_i\varphi(t_i, x_i, h_i)$;
- $(\tilde{x}_1, \tilde{x}_2, \cdots, \tilde{x}_N)$ the solution with rounding errors: $\tilde{x}_{i+1} = \tilde{x}_i + h_i\varphi(t_i, \tilde{x}_i, h_i) + \varepsilon_i$.

The initial point is also subject to rounding error: $\tilde{x}_0 = x_0 + \varepsilon_0$.

We define the notions of consistency, stability, convergence and order as follows:

- the method is **consistent** if: $$\sum_{i=1}^{N} |e_i| \xrightarrow[h_{max} \to 0]{} 0$$
  The sum of the consistency errors tends to 0 when the step size is reduced;
- the method is **stable** if: $$\max_{0 \le i \le N} |\tilde{x}_i - x_i| \le S \sum_{i=0}^{N} |\varepsilon_i|$$
  Rounding errors produce a bounded final error.
  The constant $S > 0$ is the stability constant of the method;

- the method is **convergent** if: $\max\limits_{0\le i\le N}|y_0(t_i)-x_i| \xrightarrow[h_{max}\to 0]{} 0$

  The exact solution can then be approached within the desired accuracy by reducing the step size;
- the method is **of order p** if the consistency error is upper bounded by Ch^{p+1} with a constant $C>0$.

These notions are linked by the following properties (see [R4], [R9]):

- the method defined by: $x_{i+1}=x_i+h_i\varphi(t_i,x_i,h_i)$ is of order p if

$$\frac{\partial^k\varphi}{\partial h^k}(t,x,0)=\frac{1}{k+1}\frac{d^kf}{dt^k}(x,t) \quad , \quad k=0 \text{ to } p-1 \tag{5.12}$$

- a method is consistent if it is at least of order 1: $\varphi(t,x,0)=f(x,t)$;
- a consistent and stable method is convergent;
- if the function $\varphi(t,x,h)$ is Lipschitzian in x, then the method is stable.

  The stability constant is $S=e^{L(t_f-t_0)}$ where L is the Lipschitz constant of φ;
- the global error of a stable method of order p is upper bounded by $\max\limits_{0\le i\le N}|y_0(t_i)-x_i|\le SCTh_{max}^p$ where $T=t_f-t_0$, S is the stability constant of φ and C is the constant associated with the consistency error in h^{p+1}.

Proof of property (5.12) (see [R4])

The consistency error is the difference over a time step between:

- the exact solution satisfying: $\dot{y}=f(y,t)$ with $y(t_i)=x_i$;
- the approximate solution obtained by: $x_{i+1}=x_i+h_i\varphi(t_i,x_i,h_i)$.

We write the Taylor expansion of y and x with respect to the step in $h_i=0$:

- expansion of y to order p+1 in $h_i=0$

$$y(t_{i+1})=y(t_i)+\sum_{k=1}^{p+1}\frac{h_i^k}{k!}\frac{d^ky}{dt^k}(t_i)+o(h_i^{p+1})=x_i+\sum_{k=1}^{p+1}\frac{h_i^k}{k!}\frac{d^{k-1}f}{dt^{k-1}}(x_i,t_i)+o(h_i^{p+1})$$

using $y(t_i)=x_i$ and $\frac{dy}{dt}(t_i)=f\left[y(t_i),t_i\right]=f(x_i,t_i)$

- expansion of x to order p in $h_i=0$

$$x_{i+1}=x_i+h_i\left[\sum_{k=0}^{p}\frac{h_i^k}{k!}\frac{\partial^k\varphi}{\partial h^k}(t_i,x_i,0)+o(h_i^p)\right]$$

The consistency error is the difference: $e_{i+1}=y(t_{i+1})-x_{i+1}$.

The term in h^{k+1} in the expansion of e_{i+1} is

$$\frac{h^{k+1}}{k!}\left[\frac{1}{k+1}\frac{d^k f}{dt^k}(x_i,t_i)-\frac{\partial^k \varphi}{\partial h^k}(t_i,x_i,0)\right]$$

The method is of order p if: $|e_{i+1}|=|y(t_{i+1})-x_{i+1}|\le Ch_i^{p+1}$.

The terms of order less than p must be zero, which leads to the relations (5.12).

Example 5-4 gives the order of the simplest methods.

Example 5-4: Methods of order 1 and 2

Let us look for the order of the explicit Euler, midpoint and Heun methods.
For this purpose, we have to express the function $\varphi(x,t,h)$, derive it with respect to the step h and check up to which order p the relation (5.12) is satisfied. The formulas $x_{i+1}=x_i+h_i\varphi(t_i,x_i,h_i)$ for these methods have been determined above.

Explicit Euler: $x_{i+1}=x_i+h_i f(x_i,t_i)$

$$\varphi(t,x,h)=f(x,t) \qquad \rightarrow \varphi_h(t,x,0)=0$$

The relation (5.12) does not hold for $k=1$. The method is of order $p=1$.

Midpoint: $x_{i+1}=x_i+h_i f\left[x_i+\frac{1}{2}h_i f(x_i,t_i)\ ,\ t_i+\frac{1}{2}h_i\right]$

$$\varphi(t,x,h)=f\left[x+\frac{1}{2}hf(x,t)\ ,\ t+\frac{1}{2}h\right]$$

$$\rightarrow \varphi_h(t,x,0)=\frac{1}{2}f_x f+\frac{1}{2}f_t=\frac{1}{2}\dot{f}$$

The relation (5.12) holds for $k=1$. The method is of order $p=2$.

Heun: $x_{i+1}=x_i+\frac{1}{2}h_i f(x_i,t_i)+\frac{1}{2}h_i f\left[x_i+h_i f(x_i,t_i)\ ,\ t_{i+1}\right]$

$$\varphi(t,x,h)=\frac{1}{2}f(x,t)+\frac{1}{2}f\left[x+hf(x,t)\ ,\ t+h\right]$$

$$\rightarrow \varphi_h(t,x,0)=\frac{1}{2}f_x f+\frac{1}{2}f_t=\frac{1}{2}\dot{f}$$

The relation (5.12) holds for $k=1$. The method is of order $p=2$.

Step adjustment

Consider a stable method of order p over N time steps from t_0 to t_f.

The theoretical error is bounded, depending on the step: $h_{max} = \max(h_1, \cdots, h_N)$.

$$x_{i+1} = x_i + h_i \varphi(t_i, x_i, h_i) \quad \rightarrow \quad \max_{0 \le i \le N} |y_0(t_i) - x_i| \le SCTh_{max}^p$$

The rounding errors $(\varepsilon_1)_{i=1 \text{ to } N}$ generate an additional bounded error.

$$\tilde{x}_{i+1} = \tilde{x}_i + h_i \varphi(t_i, \tilde{x}_i, h_i) + \varepsilon_i \quad \rightarrow \quad \max_{0 \le i \le N} |\tilde{x}_i - x_i| \le S \sum_{i=0}^{N} |\varepsilon_i| \le SN\varepsilon_{max}$$

The total error (theoretical + rounding) is bounded by

$$\max_{0 \le i \le N} |y_0(t_i) - \tilde{x}_i| \le S\left[CTh_{max}^p + N\varepsilon_{max}\right] \underset{\text{noted}}{=} E_{max}$$

With equidistant steps: $h = \frac{T}{N}$, the total error is: $E_{max} = ST\left[Ch^p + \frac{\varepsilon}{h}\right]$

The optimal step that minimizes the total error is then

$$h_{opt} = \left[\frac{\varepsilon}{pC}\right]^{\frac{1}{p+1}} \tag{5.13}$$

Example 5-5: Optimal step for the midpoint method

Consider the differential equation: $\dot{x} = x$ whose solution is: $y(t) = y_0 e^{t-t_0}$.

The midpoint method is applied from $t_0 = 0$ to $t_f = 1$ starting from: $x(t_0) = 1$.

The consistency error on the step $[t_i; t_{i+1}]$ with the initial condition x_i is the difference between the exact and the approximate solution.

Exact: $$y(t_{i+1}) = y(t_i)e^{t_{i+1}-t_i} = x_i e^{h_i} = x_i\left[1 + h_i + \frac{h_i^2}{2} + \frac{h_i^3}{6} + o(h_i^3)\right]$$

Approximate: $$x_{i+1} = x_i + h_i f\left[x_i + \frac{h_i}{2} f(x_i, t_i),\ t_i + \frac{h_i}{2}\right] = x_i\left[1 + h_i + \frac{h_i^2}{2}\right]$$

The error is in h^3: $e_{i+1} = y(t_{i+1}) - x_{i+1} = \frac{x_i}{6} h_i^3 + o(h_i^3)$. The method is of order 2.

The interval is divided into N equidistant steps: $h = \frac{t_f - t_0}{N} = \frac{1}{N}$.

The consistency error is bounded by: $|e_{i+1}| \le Ch^3$ with $C=e/6$ (because $x_i \le e$).

The optimal step size is (5.13): $h_{opt} = \left[\dfrac{\varepsilon}{pC}\right]^{\frac{1}{p+1}}$ where ε is the machine accuracy.

Numerical application

Assume that the calculation accuracy is: $\varepsilon = 10^{-7}$. The tables below give the total error obtained as a function of the number of steps N varying from 10 to 10^6 (first table), then refined from N = 100 to N = 1 000 (second table).

Number of steps	Total error
10	$4.20\ 10^{-3}$
100	$5.37\ 10^{-5}$
1000	$8.35\ 10^{-5}$
10 000	$8.60\ 10^{-4}$
100 000	$8.58\ 10^{-3}$
1000 000	$8.45\ 10^{-2}$

Number of steps	Total error
100	$5.37\ 10^{-5}$
150	$3.17\ 10^{-5}$
200	$2.84\ 10^{-5}$
250	$2.84\ 10^{-5}$
300	$2.87\ 10^{-5}$
350	$3.40\ 10^{-5}$
400	$3.63\ 10^{-5}$
450	$4.15\ 10^{-5}$
500	$4.40\ 10^{-5}$

The optimal step is determined by

$$\begin{cases} p=2 \\ C=e/6 \end{cases} \rightarrow h_{opt} = \left[\frac{\varepsilon}{pC}\right]^{\frac{1}{p+1}} = 4.8 \cdot 10^{-3} \text{ for } \varepsilon = 10^{-7}$$

which gives an optimal number of steps: N = 208.

Figure 5-11 shows the error as a function of the number of steps.

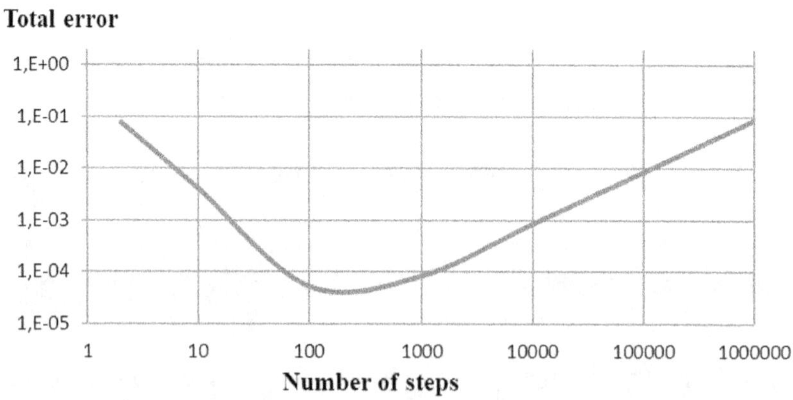

Figure 5-11: Error as a function of the number of steps.

5.2.3 Order conditions

The order of a Runge-Kutta method depends on the coefficients α_k, β_k and γ_{kj}. Here we analyze the conditions on these coefficients to obtain a given order in the case of explicit Runge-Kutta methods: $(\gamma_{kj})_{j\geq k} = 0$. Each intermediate point ξ_k of the quadrature depends only on the previous points $(\xi_j)_{j=1 \text{ to } k-1}$.

$$x_{i+1} = x_i + h_i \sum_{k=1}^{K} \beta_k f_k \quad \text{with} \quad f(\xi_j, \tau_j) \underset{\text{noted}}{=} f_j \quad \text{and} \quad \begin{cases} \tau_k = t_i + h_i \alpha_k \\ \xi_k = x_i + h_i \sum_{j=1}^{k-1} \gamma_{kj} f_j \end{cases} \tag{5.14}$$

The method is of order 1 at least if

$$\sum_{j=1}^{k-1} \gamma_{kj} = \alpha_k \quad , \quad \sum_{k=1}^{K} \beta_k = 1 \tag{5.15}$$

The method is of order 2 at least if

$$\sum_{k=1}^{K} \alpha_k \beta_k = \frac{1}{2} \tag{5.16}$$

The method is of order 3 at least if

$$\sum_{k=1}^{K} \alpha_k^2 \beta_k = \frac{1}{3} \quad , \quad \sum_{k=1, j<k}^{K} \alpha_j \beta_k \gamma_{kj} = \frac{1}{6} \tag{5.17}$$

The method is of order 4 at least if

$$\begin{aligned} &\sum_{k=1}^{K} \alpha_k^3 \beta_k = \frac{1}{4} \quad , \quad \sum_{k=1, j<k}^{K} \alpha_j \alpha_k \beta_k \gamma_{kj} = \frac{1}{8} \\ &\sum_{k=1, j<k}^{K} \alpha_j^2 \beta_k \gamma_{kj} = \frac{1}{12} \quad , \quad \sum_{k=1, j<k}^{K} \alpha_i \beta_k \gamma_{kj} \gamma_{ji} = \frac{1}{12} \end{aligned} \tag{5.18}$$

Demonstration (see [R4], [R9])

These relations are established from (5.12): $\dfrac{\partial^k \varphi}{\partial h^k}(t, x, 0) = \dfrac{1}{k+1} \dfrac{d^k f}{dt^k}(x, t)$.

We calculate the derivatives with respect to h of the function φ defined by

$$\varphi(t_i, x_i, h) = \sum_{k=1}^{K} \beta_k f(\xi_k, \tau_k) \quad \text{with} \quad \xi_k = x_i + h \sum_{j=1}^{k-1} \gamma_{kj} f(\xi_j, \tau_j) \quad \text{and} \quad \tau_k = t_i + h\alpha_k$$

Value of φ $(k=0)$

In $h=0$, we have: $\xi_k = x_i$ and $\tau_k = t_i$.

The value of φ in $h=0$ is: $\varphi(t,x,0)=\sum_{k=1}^{K}\beta_k f(x,t)$.

The condition: $\varphi(t,x,0) = f(x,t)$ gives: $\sum_{k=1}^{K}\beta_k = 1$.

This condition has already been imposed on the quadratures (5.8). The conditions (5.15) ensure that the quadratures are exact if the function f is constant.

First derivative of φ $(k=1)$

The formula for φ is derived with respect to h.

$$\frac{\partial\varphi}{\partial h} = \sum_{k=1}^{K}\beta_k \frac{\partial f(\xi_k,\tau_k)}{\partial h} = \sum_{k=1}^{K}\beta_k\left[f_x \frac{\partial\xi_k}{\partial h} + f_t \frac{\partial\tau_k}{\partial h}\right]$$

$$\text{with} \quad \frac{\partial\xi_k}{\partial h} = \sum_{j=1}^{k-1}\gamma_{kj}f(\xi_j,\tau_j) + h\frac{\partial}{\partial h}\sum_{j=1}^{k-1}\gamma_{kj}f(\xi_j,\tau_j) \quad \text{and} \quad \frac{\partial\tau_k}{\partial h} = \alpha_k$$

$$\text{In } h=0: \quad \frac{\partial\xi_k}{\partial h} = \sum_{j=1}^{k-1}\gamma_{kj}f(\xi_j,\tau_j) = \left(\sum_{j=1}^{k-1}\gamma_{kj}\right)f(x,t) = \alpha_k f(x,t)$$

$$\text{because} \sum_{j=1}^{k-1}\gamma_{kj} = \alpha_k \quad \text{(order 1)}$$

The result is: $\frac{\partial\varphi}{\partial h}(t,x,0) = \sum_{k=1}^{K}\beta_k(f_x f\alpha_k + f_t\alpha_k) = \left(\sum_{k=1}^{K}\alpha_k\beta_k\right)\frac{df}{dt}$.

The condition: $\frac{\partial\varphi}{\partial h}(t,x,0) = \frac{1}{2}\frac{df}{dt}(x,t)$ gives the condition (5.16): $\sum_{k=1}^{K}\alpha_k\beta_k = \frac{1}{2}$.

Second derivative of φ $(k=2)$

Again, the formula for φ is derived with respect to h.

$$\frac{\partial^2\varphi}{\partial h^2} = \sum_{k=1}^{K}\beta_k \frac{\partial^2 f(\xi_k,\tau_k)}{\partial h^2} = \sum_{k=1}^{K}\beta_k\left[f_{xx}\left(\frac{\partial\xi_k}{\partial h}\right)^2 + 2f_{xt}\frac{\partial\xi_k}{\partial h}\frac{\partial\tau_k}{\partial h} + f_{tt}\left(\frac{\partial\tau_k}{\partial h}\right)^2 + f_x\frac{\partial^2\xi_k}{\partial h^2}\right]$$

$$\text{with} \quad \frac{\partial^2\xi_k}{\partial h^2} \underset{\text{in } h=0}{=} 2\frac{\partial}{\partial h}\sum_{j=1}^{k-1}\gamma_{kj}f(\xi_j,\tau_j) \quad \text{and} \quad \frac{\partial f(\xi_j,\tau_j)}{\partial h} = f_x\frac{\partial\xi_j}{\partial h} + f_t\frac{\partial\tau_j}{\partial h} \underset{\text{in } h=0}{=} \alpha_j\frac{df}{dt}$$

$$\text{In } h=0: \quad \frac{\partial^2\xi_k}{\partial h^2} = 2\left(\sum_{j=1}^{k-1}\alpha_j\gamma_{kj}\right)\frac{df}{dt} \quad \text{and} \quad \frac{\partial\xi_k}{\partial h} = \alpha_k f(x,t)$$

The result is

$$\frac{\partial^2 \varphi}{\partial h^2}(t,x,0) = \left(\sum_{k=1}^{K} \alpha_k^2 \beta_k\right)(f_{xx}f^2 + 2f_{xt}f + f_{tt}) + 2\left(\sum_{k=1,j<k}^{K} \alpha_j \beta_k \gamma_{kj}\right)(f_x^2 f + f_x f_t)$$

The condition: $\frac{\partial^2 \varphi}{\partial h^2}(t,x,0) = \frac{1}{3}\frac{d^2 f}{dt^2}$ with $\frac{d^2 f}{dt^2} = f_{xx}f^2 + 2f_{xt}f + f_{tt} + f_x(f_x f + f_t)$

gives the conditions (5.17): $\sum_{k=1}^{K} \alpha_k^2 \beta_k = \frac{1}{3}$ and $\sum_{k=1,j<k}^{K} \alpha_j \beta_k \gamma_{kj} = \frac{1}{6}$.

The calculation of the third derivative is similar and leads to the fourth order conditions (5.18). This very long calculation is not detailed here.

A Runge-Kutta method is defined by the **number of stages** K (number of intermediate evaluations of f) and the choice of α_k, β_k and γ_{kj}. The conditions (5.15) to (5.18) allow to choose these coefficients to obtain a given order.

Let us look at the most common possibilities for the 1 to 4 stages methods.

<u>Explicit methods with 1 stage</u> $(K=1)$

The method is explicit: $\gamma_{11} = 0$

Order 1 conditions: $\alpha_1 = 0$, $\beta_1 = 1$

We obtain the explicit Euler method whose array is: $\begin{array}{c|c} 0 & 0 \\ \hline & 1 \end{array}$

<u>Explicit methods with 2 stages</u> $(K=2)$

The method is explicit: $\gamma_{11} = \gamma_{12} = \gamma_{22} = 0$

Order 1 conditions: $\begin{cases} \gamma_{11} + \gamma_{12} = \alpha_1 \\ \gamma_{21} + \gamma_{22} = \alpha_2 \\ \beta_1 + \beta_2 = 1 \end{cases} \Rightarrow \begin{cases} \alpha_1 = 0 \\ \gamma_{21} = \alpha_2 \underset{\text{noted}}{=} \alpha \end{cases}$

Order 2 conditions: $\alpha_1 \beta_1 + \alpha_2 \beta_2 = \frac{1}{2} \Rightarrow \beta_2 = \frac{1}{2\alpha}$, $\beta_1 = 1 - \frac{1}{2\alpha}$

This results in methods of order 2 with a parameter α associated with the array opposite.
The midpoint method corresponds to $\alpha = 0.5$.
The Heun method of order 2 corresponds to $\alpha = 1$.

$$\begin{array}{c|cc} 0 & 0 & 0 \\ \alpha & \alpha & 0 \\ \hline & 1-\dfrac{1}{2\alpha} & \dfrac{1}{2\alpha} \end{array}$$

<u>Methods with 3 or 4 stages</u> (K = 3 or 4)
The most commonly used 3- or 4-stage methods are the following:

- explicit 3-stage Heun of order 3

$$\begin{array}{c|ccc} 0 & 0 & 0 & 0 \\ 1/3 & 1/3 & 0 & 0 \\ 2/3 & 0 & 2/3 & 0 \\ \hline & 1/4 & 0 & 3/4 \end{array}$$

- implicit 3-stage Hermite-Simpson of order 3

$$\begin{array}{c|ccc} 0 & 0 & 0 & 0 \\ 1/2 & 5/24 & 1/3 & -5/24 \\ 1 & 1/6 & 2/3 & 1/6 \\ \hline & 1/6 & 2/3 & 1/6 \end{array}$$

- explicit 4-stage Runge of order 3

$$\begin{array}{c|cccc} 0 & 0 & 0 & 0 & 0 \\ 1/2 & 1/2 & 0 & 0 & 0 \\ 1 & 0 & 1 & 0 & 0 \\ 1 & 0 & 0 & 1 & 0 \\ \hline & 1/6 & 2/3 & 0 & 1/6 \end{array}$$

- explicit 4-stage 3/8 method of order 4

$$\begin{array}{c|cccc} 0 & 0 & 0 & 0 & 0 \\ 1/3 & 1/3 & 0 & 0 & 0 \\ 2/3 & -1/3 & 1 & 0 & 0 \\ 1 & 1 & -1 & 1 & 0 \\ \hline & 1/8 & 3/8 & 3/8 & 1/8 \end{array}$$

- explicit 4-stage Runge-Kutta of order 4

$$\begin{array}{c|cccc} 0 & 0 & 0 & 0 & 0 \\ 1/2 & 1/2 & 0 & 0 & 0 \\ 1/2 & 0 & 1/2 & 0 & 0 \\ 1 & 0 & 0 & 1 & 0 \\ \hline & 1/6 & 2/6 & 2/6 & 1/6 \end{array}$$

This last method, called the "classical Runge-Kutta method of order 4", is widely used in practice, as it is easy to program and extremely robust. This is due to its stability constant $S = e^{L(t_f - t_0)}$, where L is the Lipschitz constant of the function f.

The detailed formulas of the Runge-Kutta method of order 4 are as follows.

$$\begin{cases} k_1 = h_i f(x_i, t_i) \\ k_2 = h_i f\left(x_i + \frac{1}{2}k_1, t_i + \frac{1}{2}h_i\right) \\ k_3 = h_i f\left(x_i + \frac{1}{2}k_2, t_i + \frac{1}{2}h_i\right) \\ k_4 = h_i f(x_i + k_3, t_{i+1}) \end{cases} \rightarrow \ x_{i+1} = x_i + \frac{1}{6}(k_1 + 2k_2 + 2k_3 + k_4) \quad (5.19)$$

The quadratures associated with the Butcher array can be interpreted as follows.

0	0	0	0	0		
1/2	1/2	0	0	0	→ left rectangles	$\int_0^{1/2} f(t)dt \approx \frac{1}{2}f(0)$
1/2	0	1/2	0	0	→ right rectangles	$\int_0^{1/2} f(t)dt \approx \frac{1}{2}f\left(\frac{1}{2}\right)$
1	0	0	1	0	→ mid point	$\int_0^{1} f(t)dt \approx f\left(\frac{1}{2}\right)$
	1/6	2/6	2/6	1/6	→ Simpson	$\int_0^{1} f(t)dt \approx \frac{1}{6}f(0) + \frac{2}{6}f\left(\frac{1}{2}\right) + \frac{2}{6}f\left(\frac{1}{2}\right) + \frac{1}{6}f(1)$

5.2.4 Embedded methods

Error estimation

To estimate the error on a step from t_0 to t_1, the calculation is made in two ways:

- the integration in a single step of length $2h = t_1 - t_0$ gives a solution x_1;
- the integration in two successive steps of length h gives a solution x_2.

If the method is of order p, the consistency error is Ch^{p+1}. The consistency errors e_1, e_2 between the solutions x_1, x_2 and the exact solution $y(t_1)$ are then

$$\begin{cases} e_1 = y(t_1) - x_1 = \ C(2h)^{p+1} + o(2h^{p+1}) \\ e_2 = y(t_1) - x_2 = 2C(h)^{p+1} + o(h^{p+1}) \end{cases} \quad (5.20)$$

By observing the difference δ between x_2 (2 steps of h) and x_1 (1 step of 2h), we can assess the constant C and choose a step h_{max} to obtain a given error δ_{max}.

$$\delta = x_2 - x_1 = 2Ch^{p+1}(2^p - 1) + o(h^{p+1}) \quad \rightarrow \quad h_{max} \approx \left(\frac{\delta_{max}}{\delta}\right)^{\frac{1}{p+1}} h \quad (5.21)$$

The deviation δ also allows the elimination of the term in h^{p+1} in x_2 (5.20) giving a solution $\tilde{x}_2$ of order p + 1 from x_2 (local extrapolation method).

$$\delta = 2Ch^{p+1}(2^p - 1) \quad \rightarrow \quad \tilde{x}_2 = x_2 + \frac{\delta}{2^p - 1} \quad \text{with} \quad y(t_1) - \tilde{x}_2 = o(h^{p+1}) \quad (5.22)$$

<u>Embedded method</u>

The embedded methods Runge-Kutta use this double evaluation process at each step to estimate the error and adapt the integration step.

A K-stage method performs K evaluations of f to construct the solution. If the coefficients of the method are well chosen, it is possible to reuse these evaluations to construct a second solution of different order (embedded method). The difference between the two solutions gives an error estimate without further evaluation. The step size is then adjusted so that the error remains within a given tolerance. The evaluations are also used to construct a dense output. The dense output is an interpolation to a given order of the solution over the time step. It allows a discretized solution to be generated without necessarily going through all the intermediate times.

There are numerous ways of constructing embedded methods. These methods are denoted as p(p') where p and p' are the respective orders of the computed solution and the embedded solution to control the error. It can be shown that at least 6 steps are needed to construct a embedded method of order greater than 4.

The most commonly used methods of order 4 and 5 are listed below. For each method, the column β_k corresponds to the calculated solution and the column β_k^* to the error estimate.

- Cash-Karp method of order 5(4)

α_k	γ_{kj}					β_k	β_k^*
0	0	0	0	0	0	$\frac{37}{378}$	$\frac{2825}{27648}$
$\frac{1}{5}$	$\frac{1}{5}$	0	0	0	0	0	0
$\frac{3}{10}$	$\frac{3}{40}$	$\frac{9}{40}$	0	0	0	$\frac{250}{621}$	$\frac{18575}{48384}$
$\frac{3}{5}$	$\frac{3}{10}$	$-\frac{9}{10}$	$\frac{6}{5}$	0	0	$\frac{125}{694}$	$\frac{13525}{55296}$
1	$-\frac{11}{54}$	$\frac{5}{2}$	$-\frac{70}{27}$	$\frac{35}{27}$	0	0	$\frac{277}{14336}$
$\frac{7}{8}$	$\frac{1631}{55296}$	$\frac{175}{512}$	$\frac{575}{13824}$	$\frac{44275}{110592}$	$\frac{253}{4096}$	$\frac{512}{1771}$	$\frac{1}{4}$

- Fehlberg method of order 4(5)

α_k	γ_{kj}					β_k	β_k^*
0	0	0	0	0	0	$\frac{25}{216}$	$\frac{16}{135}$
$\frac{1}{4}$	$\frac{1}{4}$	0	0	0	0	0	0
$\frac{3}{8}$	$\frac{3}{32}$	$\frac{9}{32}$	0	0	0	$\frac{1408}{2565}$	$\frac{6656}{12825}$
$\frac{12}{13}$	$\frac{1932}{2197}$	$-\frac{7200}{2197}$	$\frac{7296}{2197}$	0	0	$\frac{2197}{4104}$	$\frac{28561}{56430}$
1	$\frac{439}{216}$	-8	$\frac{3680}{513}$	$-\frac{845}{4104}$	0	$-\frac{1}{5}$	$-\frac{9}{50}$
$\frac{1}{2}$	$-\frac{8}{27}$	2	$-\frac{3544}{2565}$	$\frac{1859}{4104}$	$-\frac{11}{40}$	0	$\frac{2}{55}$

The most successful higher order methods include:
- the 5(4) method with 7 stages of Dormand-Prince (DOPRI5 software);
- the 8(6) method with 16 stages of Dormand-Prince (DOPR86 software);
- the 8(5,3) method with 17 stages of Dormand-Prince (DOP853 software).

The DOP853 software includes a fifth-order error estimate and another third-order error estimate (which improves the estimate) as well as a dense seventh-order output. The reference work for the development of these methods is [R9].

Extrapolation method

An alternative approach to controlling the integration error is to apply a method of order p on the interval $[t_i;t_{i+1}]$ with n_k steps equal to h_i/n_k. By varying the number of steps $(n_1 < n_2 < \cdots < n_K)$, K evaluations of x_{i+1} are constructed.

A Richardson extrapolation (Volume I section 1.2) allows then to calculate the limit solution when the step h tends to 0. We thus obtain an extrapolated solution x_{i+1} of order $p+K-1$. This method makes it possible to obtain a very accurate numerical solution at the cost of a larger number of function evaluations.

The Gragg-Bulirsch-Stoer (GBS) methods Gragg-Bulirsch-Stoer (GBS) methods use this principle with the step defined by a Romberg sequence $(n_k = 2k)$ or by a harmonic sequence $(n_k = k)$. The ODEX software based on this choice allows to obtain solutions up to order 18.

5.3 Adams methods

Adams methods are multi-step methods for numerically solving a first order differential system of the form

$$\dot{x}(t) = f\left[x(t),t\right] \quad , t_0 \le t \le t_f \tag{5.23}$$

It is assumed that the solution has been calculated at K + 1 previous times $t_{i-K} < t_{i-K+1} < \cdots < t_{i-1} < t_i$.

The values of x at these times are: $x_{i-K}, x_{i-K+1}, \cdots, x_{i-1}, x_i$

The values of f at these times are: $f_{i-K}, f_{i-K+1}, \cdots, f_{i-1}, f_i$

The value of x at time t_{i+1} is defined by: $x_{i+1} = x_i + \int_{t_i}^{t_{i+1}} f(x,t)dt$.

To calculate the integral, we replace f by its interpolation polynomial P or P*:

- the polynomial P is based on the K + 1 known values $(f_k)_{k=i-K \text{ to } i}$ at $(t_k)_{k=i-K \text{ to } i}$;
- the polynomial P* additionally uses the unknown value f_{i+1} at time t_{i+1} .

These two polynomials are shown in figure 5-12.

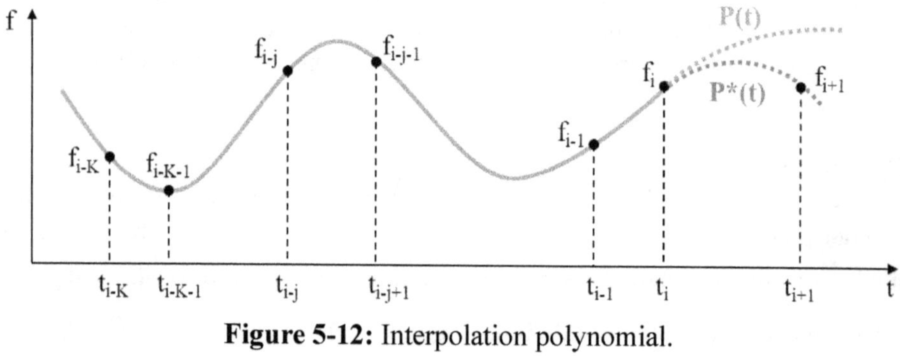

Figure 5-12: Interpolation polynomial.

The Adams-Bashford methods (section 5.3.1) use the polynomial P, which depends only on the known previous values of the function. These methods are explicit. The Adams-Moulton methods (section 5.3.2) use the polynomial P*, which depends on the unknown value of the function at the new time. These methods are implicit.

Multi-step methods require an initialization of the first K + 1 points by a single-step method. An explicit Runge-Kutta method of the same order as the chosen Adams method is usually used.

5.3.1 Adams-Bashford methods

The polynomial P of degree K takes the values $(f_k)_{k=i-K \text{ to } i}$ at times $(t_k)_{k=i-K \text{ to } i}$.
It is formed from Lagrange polynomials $(L_k)_{k=0 \text{ to } K}$ associated with $(t_k)_{k=i-K \text{ to } i}$.

$$L_k(t) = \prod_{\substack{j=0 \\ j\neq k}}^{K} \frac{t-t_{i-j}}{t_{i-k}-t_{i-j}} \quad \rightarrow \quad L_k(t_{i-j}) = \begin{cases} 0 & \text{if } j \neq k \\ 1 & \text{if } j = k \end{cases}$$

$$P(t) = \sum_{k=0}^{K} f_{i-k} L_k(t) \quad \rightarrow \quad P(t_{i-j}) = f_{i-j} \tag{5.24}$$

The Adams-Bashford method with K + 1 steps approximates x_{i+1} by

$$x_{i+1} \approx x_i + \int_{t_i}^{t_{i+1}} P(t)dt = x_i + h_i \sum_{k=0}^{K} \beta_k f_{i-k} \quad \text{with} \quad \beta_k = \frac{1}{h_i} \int_{t_i}^{t_{i+1}} L_k(t)dt \tag{5.25}$$

This explicit formula gives x_{i+1} as a function of the previous K + 1 values.
This method noted AB_{K+1} is of order K + 1. Its stability constant for a Lipschitzian function of constant L is: $S = e^{bL(t_f - t_0)}$ with $b = \sum_{k=0}^{K} |\beta_k|$.
The stability is low as the number of steps K increases (because b and S become large). Let us detail the Adams-Bashford methods with 1, 2, 3 and 4 steps:

- method AB_1

The interpolation polynomial is constant: $P(t) = f_i$.
Formula (5.25) gives: $x_{i+1} = x_i + h_i f_i$ (explicit Euler method);

- method AB_2

The interpolation polynomial is affine: $P(t) = f_i + \dfrac{f_i - f_{i-1}}{t_i - t_{i-1}}(t - t_i)$.

Its integral on $[t_i ; t_{i+1}]$ gives: $\displaystyle\int_{t_i}^{t_{i+1}} P(t)dt = h_i \left[f_i + \frac{h_i}{2h_{i-1}}(f_i - f_{i-1}) \right]$.

The steps $h_{i-1} = t_i - t_{i-1}$ and $h_i = t_{i+1} - t_i$ are not necessarily identical.
With a constant step size h, formula (5.25) gives: $x_{i+1} = x_i + h\dfrac{3f_i - f_{i-1}}{2}$;

- method AB_3 with constant step

Formula (5.25) gives: $x_{i+1} = x_i + h\frac{23f_i - 16f_{i-1} + 5f_{i-2}}{12}$;

- method AB_4 with constant step

Formula (5.25) gives: $x_{i+1} = x_i + h\frac{55f_i - 59f_{i-1} + 37f_{i-2} - 9f_{i-3}}{24}$.

The Nyström method is a variant of the Adams-Bashford method in which the point x_{i+1} is calculated from the point x_{i-1} (instead of x_i).
With a constant step size h, formula (5.25) becomes

$$x_{i+1} \approx x_{i-1} + \int_{t_{i-1}}^{t_{i+1}} P(t)dt = x_{i-1} + h\sum_{k=0}^{K}\beta_k f_{i-k} \quad \text{with } \beta_k = \frac{1}{h}\int_{t_{i-1}}^{t_{i+1}} L_k(t)dt \qquad (5.26)$$

- Nyström method with 1 or 2 steps

Formula (5.26) gives: $x_{i+1} = x_{i-1} + 2hf_i$ (midpoint method);

- Nyström method with 3 steps

Formula (5.26) gives: $x_{i+1} = x_{i-1} + h\frac{7f_i - 2f_{i-1} + f_{i-2}}{3}$.

5.3.2 Adams-Moulton methods

The polynomial P* of degree K + 1 takes values $(f_k)_{k=i-K \text{ to } i+1}$ at times $(t_k)_{k=i-K \text{ to } i+1}$
The point $(t_{i+1};f_{i+1})$ is not known (we are looking for x_{i+1}). P* is expressed with the K + 2 Lagrange polynomials $(L_k^*)_{k=-1 \text{ to } K}$ associated with $(t_k)_{k=i-K \text{ to } i+1}$.

$$\begin{aligned} L_k^*(t) &= \prod_{\substack{j=-1 \\ j\neq k}}^{K} \frac{t - t_{i-j}}{t_{i-k} - t_{i-j}} &&\rightarrow L_k^*(t_{i-j}) = \begin{cases} 0 & \text{if } j \neq k \\ 1 & \text{if } j = k \end{cases} \\ P^*(t) &= \sum_{k=-1}^{K} f_{i-k} L_k^*(t) &&\rightarrow P^*(t_{i-j}) = f_{i-j} \end{aligned} \qquad (5.27)$$

The Adams-Moulton method with K + 1 steps approximates x_{i+1} by

$$x_{i+1} \approx x_i + \int_{t_i}^{t_{i+1}} P^*(t)dt = x_i + h_i\sum_{k=-1}^{K}\beta_k^* f_{i-k} \quad \text{with } \beta_k^* = \frac{1}{h_i}\int_{t_i}^{t_{i+1}} L_k^*(t)dt \qquad (5.28)$$

The method is implicit, as the second member depends on x_{i+1} (index $k=-1$).
To obtain x_{i+1}, we have to solve the following nonlinear equation.

$$x_{i+1} = x_i + h_i \sum_{k=0}^{K} \beta_k^* f_{i-k} + h_i \beta_{-1}^* f(x_{i+1}, t_{i+1}) \tag{5.29}$$

The resolution is done by prediction-correction:

- a prediction ξ_{i+1} is made by an explicit method (*e.g.* AB_{K+1});
- the function is evaluated in (ξ_{i+1}, t_{i+1}): $\varphi_{i+1} = f(\xi_{i+1}, t_{i+1})$;
- the correction is performed by (5.28): $x_{i+1} = x_i + h_i \sum_{k=0}^{K} \beta_k^* f_{i-k} + h_i \beta_{-1}^* \varphi_{i+1}$;
- the function is evaluated in (x_{i+1}, t_{i+1}): $f_{i+1} = f(x_{i+1}, t_{i+1})$.

If necessary, the formula (5.28) can be iterated to increase the accuracy (fixed point method). A single correction is generally sufficient if the prediction is of order K + 1.

This method is called PECE (Prediction-Evaluation-Correction-Evaluation).
The PEC variant omits the evaluation of f_{i+1} and keeps the value φ_{i+1}.
The calculation is faster, but slightly less stable.

This method noted AM_{K+1} is of order $K+2$. Its stability constant for a Lipschitzian function of constant L is: $S^* = e^{b^* L (t_f - t_0)}$ with $b^* = \sum_{k=0}^{K} |\beta_k^*|$.

The stability of AM_{K+1} is better than that of AB_{K+1}, because $b^* < b$.

Let us detail the formulas of the Adams-Moulton methods with 1, 2, 3 and 4 steps:

- method AM_1

The interpolation polynomial is affine: $P^*(t) = f_i + \dfrac{f_{i+1} - f_i}{t_{i+1} - t_i}(t - t_i)$.

Its integral on $[t_i, t_{i+1}]$ gives: $\displaystyle\int_{t_i}^{t_{i+1}} P^*(t)dt = h_i \frac{f_i + f_{i+1}}{2}$.

Formula (5.28) gives: $x_{i+1} = x_i + h_i \dfrac{f_i + f_{i+1}}{2}$.

We retrieve the Crank-Nicolson method (or trapezoidal method).

The equation to solve to obtain x_{i+1} is: $x_{i+1} - \dfrac{1}{2} h_i f(x_{i+1}, t_{i+1}) = x_i + \dfrac{1}{2} h_i f_i$.

A PECE method can be applied using an AB_1 predictor (= Euler) of order 1. The algorithm is then identical to the Heun method:

- prediction (Euler): $\xi_{i+1} = x_i + h_i f_i$
- correction: $x_{i+1} = x_i + \frac{1}{2}h_i f_i + \frac{1}{2}h_i \varphi_{i+1}$

- method AM_2

The interpolation polynomial is of degree 2.

$$P^*(t) = f_{i+1}\frac{(t-t_{i-1})(t-t_i)}{h_i(h_{i-1}+h_i)} - f_i\frac{(t-t_{i-1})(t-t_{i+1})}{h_{i-1}h_i} + f_{i-1}\frac{(t-t_i)(t-t_{i+1})}{h_{i-1}(h_{i-1}+h_i)}$$

Integrating it on $[t_i ; t_{i+1}]$ gives

$$x_{i+1} = x_i + h_i\frac{3h_{i-1}+2h_i}{6(h_{i-1}+h_i)}f_{i+1} + h_i\frac{3h_{i-1}+h_i}{6h_{i-1}}f_i - h_i\frac{h_i^2}{6h_{i-1}(h_{i-1}+h_i)}f_{i-1}$$

A constant step size h is usually chosen.

Formula (5.28) gives: $x_{i+1} = x_i + h_i\left(\frac{5}{12}f_{i+1} + \frac{8}{12}f_i - \frac{1}{12}f_{i-1}\right)$;

- method AM_3 with constant step

Formula (5.28) gives: $x_{i+1} = x_i + h\frac{9f_{i+1}+19f_i-5f_{i-1}+f_{i-2}}{24}$;

- method AM_4 with constant step

Formula (5.28) gives: $x_{i+1} = x_i + h\frac{251f_{i+1}+646f_i-264f_{i-1}+106f_{i-2}-19f_{i-3}}{720}$.

Like the Nyström method (section 5.3.1), the Milne-Simpson method is a variant of Adams-Moulton in which the point x_{i+1} is calculated from the point x_{i-1} (instead of x_i). With a constant step size, the formula (5.28) becomes

$$x_{i+1} \approx x_{i-1} + \int_{t_i}^{t_{i+1}} P^*(t)dt = x_{i-1} + h\sum_{k=-1}^{K}\beta_k^* f_{i-k} \quad \text{with } \beta_k^* = \frac{1}{h}\int_{t_{i-1}}^{t_{i+1}} L_k^*(t)dt \quad (5.30)$$

- Milne-Simpson 1-stage method

Formula (5.30) gives: $x_{i+1} = x_{i-1} + 2hf_i$ (midpoint method);

- Milne-Simpson 2-stage method

Formula (5.30) gives: $x_{i+1} = x_{i-1} + h\frac{f_i+4f_{i-1}+f_{i-2}}{3}$.

5.4 Collocation methods

Collocation methods search for a piecewise polynomial solution of a first order differential system of the form

$$\begin{cases} \dot{x}(t) = f\left[x(t),t\right] \ , t_0 \le t \le t_f \\ x(t_0) = x_0 \end{cases} \tag{5.31}$$

The interval $[t_0 ; t_f]$ is divided into N segments with times

$t_0 < t_1 < \cdots < t_i < t_{i+1} < \cdots < t_N = t_f$

The solution on each segment is approximated by a polynomial of degree K.

$$\xi(t) = a_0 + a_1 t + a_2 t^2 + \cdots + a_K t^K \quad , \quad t_i \le t \le t_{i+1} \tag{5.32}$$

The K + 1 coefficients $a_0, a_1, \cdots, a_K$ must be determined so that the polynomial is close to the exact solution of the differential equation on the segment $[t_i ; t_{i+1}]$.

5.4.1 Collocation conditions

To determine the K+1 coefficients of the polynomial (5.32), we choose K times $(\tau_k)_{k=1 \text{ to } K}$ in $[t_i ; t_{i+1}]$ and impose the K + 1 conditions

$$\begin{cases} \xi(t_i) \ = x(t_i) \\ \dot{\xi}(\tau_k) = f(\xi_k, \tau_k) \underset{\text{noted}}{=} f_k \quad , \quad k = 1 \text{ to } K \end{cases} \tag{5.33}$$

The times $(\tau_k)_{k=1 \text{ to } K}$ called collocation points or nodes are freely chosen between t_i and t_{i+1} as in figure 5-3: $t_i \le \tau_1 < \tau_2 < \cdots < \tau_{K-1} < \tau_K \le t_{i+1}$.

The conditions (5.33) require that the derivatives of the polynomial satisfy the differential equation $\dot{x} = f(x,t)$ at the collocation points and that the polynomial takes the known initial value $x(t_i)$ in t_i.

<u>Link to Runge-Kutta methods</u>

The derivative of the polynomial (5.32) is a polynomial of degree K – 1 taking values $(f_k)_{k=1 \text{ to } K}$ at times $(\tau_k)_{k=1 \text{ to } K}$.

This derivative polynomial can be formed from the K Lagrange interpolation polynomials $(L_j)_{j=1 \text{ to } K}$ associated with $(\tau_k)_{k=1 \text{ to } K}$.

$$L_j(t)=\prod_{\substack{k=1\\k\neq j}}^{K}\frac{t-\tau_k}{\tau_j-\tau_k} \quad \rightarrow L_j(\tau_k)=\begin{cases}0 \text{ if } k\neq j\\ 1 \text{ if } k=j\end{cases}$$
$$\dot{\xi}(t)=\sum_{j=1}^{K} f_j L_j(t) \quad \rightarrow \dot{\xi}(\tau_k)=f_k \tag{5.34}$$

By integrating the derived polynomial $\dot{\xi}(t)$ on $[t_i;\tau_k]$, we obtain

$$\xi(\tau_k)=\xi(t_i)+\sum_{j=1}^{K} f_j \int_{t_i}^{\tau_k} L_j(t)dt \quad \text{with } \xi(t_i)=x(t_i) \tag{5.35}$$

Now suppose that we apply a K-stage Runge-Kutta method with the coefficients α_k, β_k and γ_{kj}. Formulas (5.6) and (5.7) give

$$x_{i+1}=x_i+h_i\sum_{k=1}^{K}\beta_k f(\xi_k,\tau_k) \text{ with } \xi_k=x_i+h_i\sum_{j=1}^{K}\gamma_{kj} f(\xi_j,\tau_j) \ , \ k=1 \text{ to } K \tag{5.36}$$

By choosing the coefficients α_k, β_k and γ_{kj} such that

$$\tau_k=t_i+\alpha_k h_i \ , \ \beta_k=\frac{1}{h_i}\int_{t_i}^{t_{i+1}} L_k(t)dt \ , \ \gamma_{kj}=\frac{1}{h_i}\int_{t_i}^{\tau_k} L_j(t)dt \tag{5.37}$$

we retrieve the values of the polynomial $\xi(t)$ at the times $(\tau_k)_{k=1 \text{ to } K}$ and t_{i+1}.
A collocation method of degree K can thus be considered as a particular implicit Runge-Kutta method with K stages. It will have the same properties of order and stability.

5.4.2 Collocation points

The collocation polynomial $\xi(t)$ can be expressed with the K + 1 interpolation polynomials of Lagrange $(L_j)_{j=0 \text{ to } K}$ associated with $(\tau_k)_{k=0 \text{ to } K}$, and noting by convention $\tau_0=t_i$ and $\xi_0=x(t_i)$.

$$L_j(t)=\prod_{\substack{k=0\\k\neq j}}^{K}\frac{t-\tau_k}{\tau_j-\tau_k} \quad \rightarrow L_j(\tau_k)=\begin{cases}0 \text{ if } k\neq j\\ 1 \text{ if } k=j\end{cases}$$
$$\xi(t)=\sum_{j=0}^{K} \xi_j L_j(t) \quad \rightarrow \xi(\tau_k)=\xi_k \tag{5.38}$$

The unknowns are the values $\xi_k = \xi(\tau_k)$ taken by the polynomial at the collocation times. The initial condition is imposed: $\xi_0 = x(t_i)$. The collocation conditions at times τ_k form a nonlinear system of dimension K.

$$\sum_{j=0}^{K} \xi_j \dot{L}_j(\tau_k) = f(\xi_k, \tau_k) \ , \ k = 1 \text{ to } K \tag{5.39}$$

The solution of equation (5.31) is approximated by a sequence of N polynomials of degree K of the form (5.32) defined on the successive segments $[t_i; t_{i+1}]$ and satisfying the conditions (5.39).

The quality of the approximation depends on the number N of segments, the degree K of the polynomials and the positioning of the collocation points. Many variants are possible and are detailed in reference [R2].

Local or global collocation

Local collocation methods consist in fixing the degree K of the polynomials and varying the number N of segments. Conversely, the global collocation methods consist in fixing the number N of segments (possibly reduced to a single segment) and varying the degree K of the polynomials. The discretization (value of N or K) is progressively refined until a stable solution with a given error tolerance is obtained.

Standard or orthogonal collocation

The standard collocation uses equidistant times $(\tau_k)_{k=1 \text{ to } K}$ in the interval $[t_i; t_{i+1}]$ usually with polynomials of degree 3. This approach can be easily applied to many problems (section 5.4.3).

The orthogonal collocation takes as collocation times τ_k the roots of a polynomial of a family of orthogonal polynomials (Legendre, Chebyshev). This positioning increases the accuracy of the quadrature formulas.

A frequent choice is to use the Legendre polynomials P_K (of degree K) defined on $[-1;+1]$, coming back to the integration interval $[t_i; t_{i+1}]$ by: $\dfrac{t - t_i}{t_{i+1} - t_i} = \dfrac{\tau + 1}{2}$.

For an infinite final time, a global collocation can be applied passing from the interval $[-1;+1]$ to the integration interval $[t_0; +\infty]$ by: $t - t_0 = \dfrac{1 + \tau}{1 - \tau}$.

The main methods for positioning the collocation points are as follows:

- Legendre-Gauss or Legendre-Gauss-Radau method

The Legendre-Gauss method defines the times $(\tau_k)_{k=1\text{ to }K}$ as the K roots of P_K, plus the initial time $\tau_0=-1$. The Legendre-Gauss-Radau method is similar, taking the K roots of $(P_{K-1}+P_K)$.

The collocation polynomial $\xi(t)$ is defined on $[-1;+1]$ by: $\xi(\tau)=\sum_{j=0}^{K}\xi_j L_j(\tau)$.

The collocation conditions form a system of dimension K + 1.

$$\dot{\xi}(\tau_k)=f\left[\xi(\tau_k),\tau_k\right] \ , \ k=0 \text{ to } K \tag{5.40}$$

The final state at t_{i+1} is then calculated for $\tau=+1$;

- Legendre-Gauss-Lobatto method

The times $(\tau_k)_{k=1\text{ to }K-1}$ are the K – 1 roots of the derived polynomial $\dot{P}_K$, plus the initial time $\tau_0=-1$ and the final time $\tau_K=+1$.

The collocation polynomial $\xi(t)$ is defined on $[-1;+1]$ by: $\xi(\tau)=\sum_{j=0}^{K}\xi_j L_j(\tau)$.

The collocation conditions form a system of dimension K + 1.

$$\dot{\xi}(\tau_k)=f\left[\xi(\tau_k),\tau_k\right] \ , \ k=0 \text{ to } K \tag{5.41}$$

The final state at t_{i+1} is obtained directly by solving the system, as $\tau_K=+1$. This is the main difference with the Legendre-Gauss method. Pseudo-spectral methods use a global (on $[t_0;t_f]$) orthogonal (generally with Legendre-Gauss-Lobatto points) collocation.

5.4.3 Collocation of degree 3

The simplest form of collocation uses polynomials of degree 3.

$$\xi(t)=a_0+a_1t+a_2t^2+a_3t^3 \ \text{ on } \ [t_0;t_1] \tag{5.42}$$

We return to the interval $[0;1]$ by the change of variable: $s=\dfrac{t-t_0}{t_1-t_0}$.

$$\xi(s)=\alpha_0+\alpha_1 s+\alpha_2 s^2+\alpha_3 s^3 \ \text{ on } \ [0\,;1] \tag{5.43}$$

The derivative of ξ with respect to s is denoted ξ'.

$$\xi' = \frac{d\xi}{ds} = \frac{d\xi}{dt}\frac{dt}{ds} = Tf(\xi,t) \quad \text{with} \quad t_1 - t_0 \underset{\text{noted}}{=} T \tag{5.44}$$

The collocation conditions in variable s on the interval $[0; 1]$ are of the form

$$\xi'(s) = \alpha_1 + 2\alpha_2 s + 3\alpha_3 s^2 = Tf(\xi,t) \tag{5.45}$$

The unknowns are the coefficients $(\alpha_0,\alpha_1,\alpha_2,\alpha_3)$. To determine them, three collocation points (s_1,s_2,s_3) are chosen, usually equidistant:

- the point $s_1 = 0$ corresponds to the initial time t_0;
- the point $s_2 = 0.5$ corresponds to the middle time t_c;
- the point $s_3 = 1$ corresponds to the final time t_1.

The collocation conditions (5.33) give a system of dimension 4.

$$\begin{cases} \xi_0 = x_0 \\ \xi'_0 = Tf(\xi_0,t_0) \\ \xi'_c = Tf(\xi_c,t_c) \\ \xi'_1 = Tf(\xi_1,t_1) \end{cases} \Rightarrow \begin{cases} \alpha_0 = x_0 \\ \alpha_1 = Tf(\alpha_0,t_0) \\ \alpha_1 + \alpha_2 + \frac{3}{4}\alpha_3 = Tf\left(\alpha_0 + \frac{1}{2}\alpha_1 + \frac{1}{4}\alpha_2 + \frac{1}{8}\alpha_3, t_c\right) \\ \alpha_1 + 2\alpha_2 + 3\alpha_3 = Tf\left(\alpha_0 + \alpha_1 + \alpha_2 + \alpha_3, t_1\right) \end{cases} \tag{5.46}$$

This system is reduced by expressing the unknowns $(\alpha_0,\alpha_1,\alpha_2,\alpha_3)$ as a function of $(\xi_0,\xi_1,\xi'_0,\xi'_1)$.

$$\begin{cases} \xi_0 = \alpha_0 \\ \xi'_0 = \alpha_1 \\ \xi_1 = \alpha_0 + \alpha_1 + \alpha_2 + \alpha_3 \\ \xi'_1 = \alpha_1 + 2\alpha_2 + 3\alpha_3 \end{cases} \Rightarrow \begin{cases} \alpha_0 = \xi_0 \\ \alpha_1 = \xi'_0 \\ \alpha_2 = -3\xi_0 - 2\xi'_0 + 3\xi_1 - \xi'_1 \\ \alpha_3 = 2\xi_0 + \xi'_0 - 2\xi_1 + \xi'_1 \end{cases} \tag{5.47}$$

The conditions of collocation in t_0, t_1 give ξ'_0, ξ'_1 as a function of ξ_0, ξ_1.

$$\begin{cases} \xi'_0 = T\dot{\xi}_0 = Tf(\xi_0,t_0) \underset{\text{noted}}{=} Tf_0 \\ \xi'_1 = T\dot{\xi}_1 = Tf(\xi_1,t_1) \underset{\text{noted}}{=} Tf_1 \end{cases} \rightarrow \begin{cases} \alpha_0 = \xi_0 \ , \ \alpha_1 = f_0 \\ \alpha_2 = -3\xi_0 + 3\xi_1 - 2f_0 - f_1 \\ \alpha_3 = 2\xi_0 - 2\xi_1 + f_0 + f_1 \end{cases} \tag{5.48}$$

The coefficients $(\alpha_0,\alpha_1,\alpha_2,\alpha_3)$ thus depend only on the values of (ξ_0,ξ_1):

- the initial value is known: $\xi_0 = x_0$;
- the unknown final value ξ_1 must satisfy the collocation condition in t_c.

$$\xi'_c = T\dot{\xi}_c = Tf(\xi_c, t_c) \quad \text{with} \begin{cases} \xi_c = \alpha_0 + \frac{1}{2}\alpha_1 + \frac{1}{4}\alpha_2 + \frac{1}{8}\alpha_3 \\ \xi'_c = \alpha_1 + \alpha_2 + \frac{3}{4}\alpha_3 \end{cases} \tag{5.49}$$

By replacing the coefficients $(\alpha_0, \alpha_1, \alpha_2, \alpha_3)$ given by (5.48), we are reduced to an equation with unknown ξ_1. This equation is called the defect equation.

$$Tf\left[\frac{1}{2}(\xi_0 + \xi_1) + \frac{1}{8}(f_0 - f_1)\,,\, t_c\right] + \frac{3}{2}(\xi_0 - \xi_1) + \frac{1}{4}(f_0 + f_1) = 0 \tag{5.50}$$

Figure 5-13 summarizes the process of determining the polynomial of degree 3.

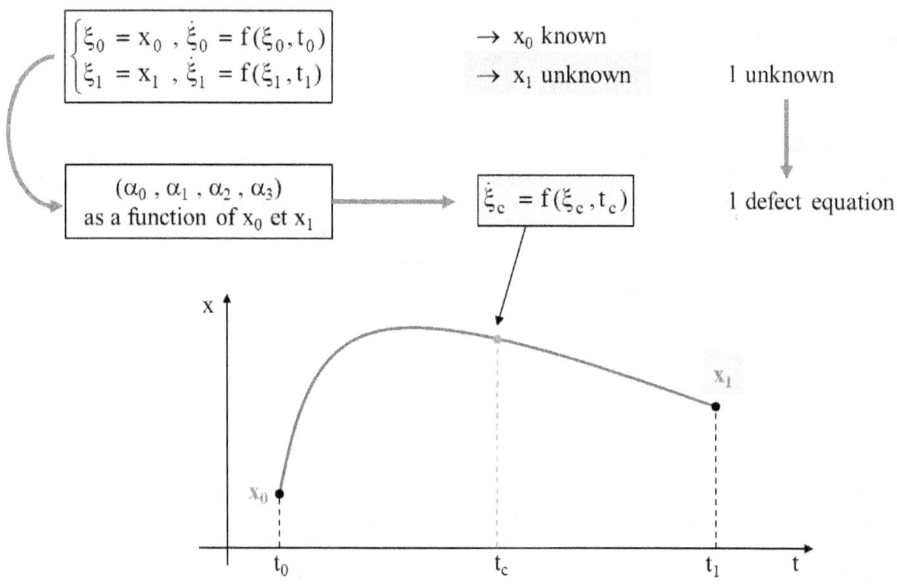

Figure 5-13: Collocation polynomial of degree 3.

Example 5-6: Collocation of degree 3

Consider the differential equation: $\dot{x} = x$ on the interval from $t_0 = 0$ to $t_1 = 1$.
We seek the collocation polynomial of degree 3 for the initial condition: $x(t_0) = 1$.

The defect equation associated with the midpoint is

$$\mathrm{T}f\left[\frac{1}{2}(\xi_0+\xi_1)+\frac{1}{8}(f_0-f_1)\,,\,t_c\right]+\frac{3}{2}(\xi_0-\xi_1)+\frac{1}{4}(f_0+f_1)=0 \quad \text{with} \begin{cases} T=t_1-t_0=1 \\ f(x,t)=x \end{cases}$$

Replacing: $\begin{cases} f_0 = Tf(\xi_0,t_0)=T\xi_0=\xi_0 \\ f_1 = Tf(\xi_1,t_1)=T\xi_1=\xi_1 \end{cases}$, the equation reduces to: $19\xi_0-7\xi_1=0$.

This gives the solution: $\xi_1=\frac{19}{7}\xi_0\approx 2.71429$ for the initial condition: $\xi_0=1$.

This solution is close to the exact solution: $x(t_1)=e^{t_1-t_0}=e\approx 2.71828$.

The collocation polynomial on $[0;1]$ is: $\xi(t)=1+t+0.42857t^2+0.28571t^3$.

5.4.4 Collocation of degree 5

A more accurate solution can be obtained with polynomials of degree 5.

$$\xi(t)=a_0+a_1t+a_2t^2+a_3t^3+a_4t^4+a_5t^5 \quad \text{on} \quad [t_0;t_1] \tag{5.51}$$

We return to the interval $[0;1]$ by the change of variable: $s=\frac{t-t_0}{t_1-t_0}$.

$$\xi(s)=\alpha_0+\alpha_1s+\alpha_2s^2+\alpha_3s^3+\alpha_4s^4+\alpha_5s^5 \quad \text{on} \quad [0\,;1] \tag{5.52}$$

The collocation conditions in variable s on the interval $[0;1]$ are of the form

$$\xi'(s)=\alpha_1+2\alpha_2s+3\alpha_3s^2+4\alpha_4s^3+5\alpha_5s^4=Tf(\xi,t) \tag{5.53}$$

The unknowns are the coefficients $(\alpha_0,\alpha_1,\alpha_2,\alpha_3,\alpha_4,\alpha_5)$. To determine them, five collocation points are chosen (s_1,s_2,s_3,s_4,s_5). The interpolation error is minimized by choosing orthogonal Gauss-Lobatto points.

$$\begin{cases} s_1=0\ ,\ s_2=\frac{1}{2}\left(1-\sqrt{\frac{3}{7}}\right)\approx 0.173 \\ s_3=\frac{1}{2}\ ,\ s_4=\frac{1}{2}\left(1+\sqrt{\frac{3}{7}}\right)\approx 0.827\ ,\ s_5=1 \end{cases} \tag{5.54}$$

These points correspond to the times $t_0,t_{g1},t_c,t_{g2},t_1$.

The collocation conditions (5.33) give a system of dimension 6.

$$\begin{cases} \xi_0 = \alpha_0 \\ \xi'_0 = \alpha_1 & \text{with } \xi'_0 = T\dot{\xi}_0 = Tf(\xi_0, t_0) \\ \xi_1 = \alpha_0 + \alpha_1 + \alpha_2 + \alpha_3 + \alpha_4 + \alpha_5 \\ \xi'_1 = \alpha_1 + 2\alpha_2 + 3\alpha_3 + 4\alpha_4 + 5\alpha_5 & \text{with } \xi'_1 = T\dot{\xi}_1 = Tf(\xi_1, t_1) \\ \xi_c = \alpha_0 + \frac{1}{2}\alpha_1 + \frac{1}{4}\alpha_2 + \frac{1}{8}\alpha_3 + \frac{1}{16}\alpha_4 + \frac{1}{32}\alpha_5 \\ \xi'_c = \alpha_1 + \alpha_2 + \frac{3}{4}\alpha_3 + \frac{1}{2}\alpha_4 + \frac{5}{16}\alpha_5 & \text{with } \xi'_c = T\dot{\xi}_c = Tf(\xi_c, t_c) \end{cases} \tag{5.55}$$

This linear system gives $(\alpha_0, \alpha_1, \alpha_2, \alpha_3, \alpha_4, \alpha_5)$ as a function of (ξ_0, ξ_c, ξ_1):

- the initial value is known: $\xi_0 = x_0$;
- the values ξ_c, ξ_1 must satisfy the collocation conditions in t_{g1}, t_{g2}.

This reduces to a system of two equations with two unknowns.

Figure 5-14 summarizes the process of determining the polynomial of degree 5.

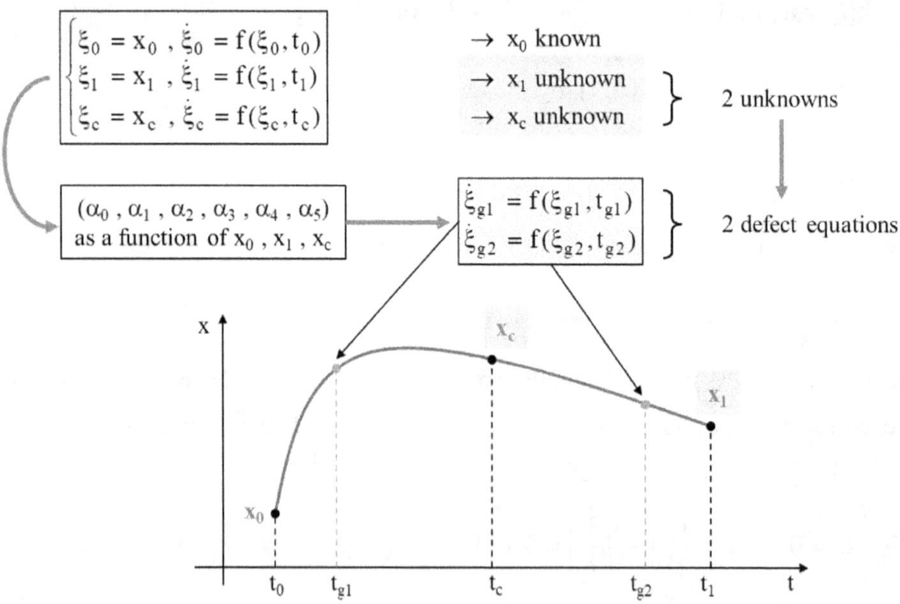

Figure 5-14: Collocation polynomial of degree 5.

5.5 Direct methods

Direct methods consist in discretizing the control problem in order to formulate an "equivalent" finite-dimensional constrained optimization problem. This problem can then by solved either by a nonlinear programming algorithm or by a variational method.

5.5.1 Discretization

We consider the problem with a fixed final time in Mayer form (final cost). Any control problem can be put into this form by a change of variables.

$$\min_u J = \varphi(x_f) \quad \text{with} \quad \begin{cases} \dot{x} = f(x,u,t) \ , \ x(t_0) = x_0 \\ \psi(x_f) = 0 \end{cases} \tag{5.56}$$

The PMP necessary conditions of optimality are expressed with the adjoint p, the multipliers $\nu \in \mathbb{R}^q$ of the q final constraints and the Hamiltonian defined by: $H(x,u,p,t) = p^T f(x,u,t)$.

$$\begin{cases} \dot{x} = H_p \\ \dot{p} = -H_x \\ H_u = 0 \\ \psi(x_f) = 0 \\ p_f = \varphi_{xf} + \psi_{xf}\nu \end{cases} \Rightarrow \begin{cases} \dot{x} = f \\ \dot{p} = -f_x^T p \\ f_u^T p = 0 \\ \psi(x_f) = 0 \\ p_f = \varphi_{xf} + \psi_{xf}\nu \end{cases} \tag{5.57}$$

Let us discretize the state and control on N intervals: $t_0 < t_1 < \cdots < t_i < \cdots < t_N = t_f$. The values of state $(x_i)_{i=0 \text{ to } N}$ and control $(u_i)_{i=0 \text{ to } N}$ define an approximation to the exact solution of the problem (5.56). Considering a piecewise constant control on each interval $[t_i ; t_{i+1}]$, the state equation $\dot{x} = f(x,t)$ can be approximated by

$$\dot{x}_i \approx \frac{x_{i+1} - x_i}{h_i} = f(x_i, u_i, t_i) \quad \text{with} \quad h_i = t_{i+1} - t_i \tag{5.58}$$

Equations (5.58) are constraints linking the values $(x_i)_{i=0 \text{ to } N}$ and $(u_i)_{i=0 \text{ to } N}$. These N constraints are of the form

$$C_i(u_i, x_i, x_{i+1}) = x_{i+1} - x_i - h_i f(x_i, u_i, t_i) = 0 \tag{5.59}$$

We thus formulate a finite-dimensional constrained optimization problem.

$$\min_{u_i,x_i} F = \varphi(x_N) \quad \text{s.t.} \begin{cases} C_i(u_i,x_i,x_{i+1}) = 0 \ , \ i = 0 \text{ to } N-1 \\ \psi(x_N) = 0 \end{cases} \tag{5.60}$$

This problem has $2N+1$ variables (x_0 is fixed) and $N+q$ constraints.

<u>Optimality conditions</u>

The KKT optimality conditions of problem (5.60) are expressed introducing the N multipliers $(\lambda_i)_{i=0 \text{ to } N-1}$ of the constraints $(C_i)_{i=0 \text{ to } N-1}$, the q multipliers $\nu \in \mathbb{R}^q$ of the q final constraints and the Lagrangian defined by

$$L(u_i,x_i,\lambda_i,\nu) = \varphi(x_N) + \sum_{i=0}^{N-1} \lambda_i^T C_i(u_i,x_i,x_{i+1}) + \nu^T \psi(x_N).$$

$$\begin{cases} \nabla_{u_i} L = 0 \\ \nabla_{x_i} L = 0 \\ \nabla_{x_N} L = 0 \\ \nabla_{\lambda_i} L = 0 \\ \nabla_{\nu} L = 0 \end{cases} \Rightarrow \begin{cases} f_u^T(x_i,u_i,t_i)\lambda_i = 0 \\ \lambda_{i-1} - \lambda_i - h_i f_x^T(x_i,u_i,t_i)\lambda_i = 0 \\ \varphi_x(x_N) + \psi_x(x_N)\nu + \lambda_{N-1} = 0 \\ C_i(u_i,x_i,x_{i+1}) = 0 \\ \psi(x_N) = 0 \end{cases} \quad \text{for } i = 0 \text{ to } N-1 \tag{5.61}$$

These optimality conditions form a two-point boundary value discrete system.

The state propagates forwards from x_0: $\quad x_{i+1} = x_i + h_i f(x_i,u_i,t_i)$.

The adjoint propagates backwards from λ_{N-1}: $\begin{cases} \lambda_{N-1} = -\varphi_x(x_N) - \psi_x(x_N)\nu \\ \lambda_{i-1} = \lambda_i + h_i f_x^T(x_i,u_i,t_i)\lambda_i \end{cases}$.

Let us look for the limit of the KKT conditions when the step h_i tends to 0.

The multipliers $(\lambda_i)_{i=0 \text{ to } N-1}$ become a function $\lambda(t)$ and we get

$$\begin{cases} f_u^T(x_i,u_i,t_i)\lambda_i = 0 \\ \lambda_{i-1} - \lambda_i - h_i f_x^T(x_i,u_i,t_i)\lambda_i = 0 \\ \varphi_x(x_N) + \psi_x(x_N)\nu + \lambda_{N-1} = 0 \\ C_i(u_i,x_i,x_{i+1}) = 0 \\ \psi(x_N) = 0 \end{cases} \xrightarrow[h_i \to 0]{} \begin{cases} f_u^T(x,u,t)\lambda = 0 \\ \dot{\lambda} = -f_x^T(x,u,t)\lambda \\ \lambda(t_f) = -\varphi_x(x_f) - \psi_x(x_f)\nu \\ \dot{x} = f(x,u,t) \\ \psi(x_f) = 0 \end{cases} \tag{5.62}$$

The optimality conditions (5.57) of the control problem can be retrieved by setting $\lambda = -p$. The multipliers (λ_i) represent thus approximations of the adjoint p at the times (t_i). The opposite sign comes from the choice of the abnormal multiplier $(p_0 = 1)$ used to define the Hamiltonian.

Discretization choice

The discretization can be done in different ways. The control represented by its values $(u_i)_{i=0 \text{ to } N}$ at the times $(t_i)_{i=0 \text{ to } N}$ can be interpolated as piecewise constant, piecewise linear, splines... The differential state equation can be solved by propagation or collocation.

With a propagation approach, the state is calculated from x_0 by applying the control defined by the values $(u_i)_{i=0 \text{ to } N}$. A Runge-Kutta type method of appropriate order (depending on the desired accuracy) is chosen using or not the same discretization times as the control. When it is known from experience that the solution control does not have large variations, a rough discretization $(t_i)_{i=0 \text{ to } N}$ can be applied to limit the number of variables, while applying an accurate integration method with a large number of steps.

With a collocation approach, the state is also discretized and the variables $(x_i)_{i=1 \text{ to } N}$ become additional unknowns of the optimization problem. The constraints (5.59) are replaced by the collocation conditions. Thus, a large optimization problem is formulated that can be solved by standard nonlinear programming methods or by the variational method presented in the next section. We observe that each collocation constraint is on a segment $[t_i ; t_{i+1}]$ and involves only the "local" variables u_i, x_i and x_{i+1}. The Jacobian matrix of these constraints is sparse and has non-zero elements only near the diagonal. Optimization algorithms specific to large-scale problems exploit this characteristic (especially when calculating gradients) in order to reduce the calculation time.

5.5.2 Variational approach

We consider a control problem with q final constraints.

$$\min_{u,t_f} J = \int_{t_0}^{t_f} L(x,u,t)dt + \varphi(x_f,t_f) \quad \text{with} \quad \begin{cases} \dot{x} = f(x,u,t) \; , \; x(t_0) = x_0 \\ \psi_j(x_f,t_f) = 0 \; , \; j = 1 \text{ to } q \end{cases} \tag{5.63}$$

Suppose that we choose arbitrary values for the final time t_f and for the control $u(t)_{t_0 \le t \le t_f}$. This solution will not optimal nor it will satisfy the constraints.
The variational method uses the first variations δJ and $\delta\psi$ to find corrections δt_f and $\delta u(t)_{t_0 \le t \le t_f}$ improving the solution.

Variation in cost and constraints

The variation of the cost J is expressed with the adjoint p_0 and the Hamiltonian H_0 (4.17). The notation p_0 is used here to denote the cost adjoint (not to confuse with the abnormal multiplier).

$$\delta J = [L+\dot{\varphi}]_{t_f}\delta t_f + \int_{t_0}^{t_f} H_{0u}^T \delta u dt \quad \text{with} \begin{cases} \dot{p}_0 & = -H_{0x} \\ p_0(t_f) = & \varphi_x(x_f,t_f) \\ H_0(x,u,p_0,t) = L(x,u,t) + p_0^T f(x,u,t) \end{cases} \tag{5.64}$$

The same technique is applied to the variation of the constraint ψ_j. For this, we define an adjoint p_j and a Hamiltonian H_j associated with the constraint.

The formula (5.64) is used in a similar way by replacing the final cost φ by ψ_j and the integral cost L by 0.

$$\delta\psi_j = [\dot{\psi}_j]_{t_f}\delta t_f + \int_{t_0}^{t_f} H_{ju}^T \delta u dt \quad \text{with} \begin{cases} \dot{p}_j & = -H_{jx} \\ p_j(t_f) = & \psi_{jx}(x_f,t_f) \\ H_j(x,u,p_j,t) = p_j^T f(x,u,t) \end{cases} \tag{5.65}$$

We note $P=(p_1,\cdots,p_q)\in\mathbb{R}^{n\times q}$ the matrix of adjoints and $G=(H_1,\cdots,H_q)^T\in\mathbb{R}^q$ the vector of Hamiltonians. The equations (5.65) can be put in matrix form

$$\begin{cases} \dot{p}_j & = -H_{jx} \\ p_j(t_f) = & \psi_{jx}(x_f,t_f) \\ H_j(x,u,p_j,t) = p_j^T f(x,u,t) \end{cases} \rightarrow \begin{cases} \dot{P} = -G_x = -f_x P \\ P(t_f) = \psi_x(x_f,t_f) \\ G = P^t f \end{cases} \tag{5.66}$$

$$\delta\psi_j = [\dot{\psi}_j]_{t_f}\delta t_f + \int_{t_0}^{t_f} H_{ju}^T \delta u dt \rightarrow \delta\psi = [\dot{\psi}]_{t_f}\delta t_f + \int_{t_0}^{t_f} G_u^T \delta u dt$$

Assume that the initial solution $(t_f;u)$ gives a value of the constraints $\psi_f \neq 0$. The objective is to find corrections $(\delta t_f;\delta u)$ in order to cancel the constraints ψ, while minimizing the functional J.

Local minimization problem

The local minimization problem is formulated as

$$\min_{\delta u, \delta t_f} \delta J \quad \text{s.t.} \quad \delta\psi = -\psi_f \tag{5.67}$$

This problem in this form has no solution, because the variations δJ and $\delta\psi$ are linear in terms of δt_f and δu. To obtain a bounded correction, we convexify the problem by adding quadratic terms in δt_f and δu.

$$\min_{\delta u, \delta t_f} \delta J_c \underset{def}{=} \delta J + \frac{1}{2} b(\delta t_f)^2 + \frac{1}{2} \int_{t_0}^{t_f} \delta u^T W \delta u \, dt \quad \text{s.t.} \quad \delta\psi = -\psi_f \tag{5.68}$$

The positive real b and the positive symmetric matrix $W \in \mathbb{R}^{m \times m}$ are weights to control the magnitude of the corrections δt_f and δu.

The quadratic-linear problem (5.68) is solved by introducing the multipliers $\lambda = (\lambda_1, \cdots, \lambda_q)^T \in \mathbb{R}^q$ of the constraints and forming the augmented cost J_a.

$$\begin{aligned} J_a = \delta J_c + \lambda^T(\delta\psi + \psi_f) = & \frac{1}{2} b(\delta t_f)^2 + [L + \dot{\varphi} + \lambda^T \dot{\psi}]_{t_f} \delta t_f \\ & + \frac{1}{2} \int_{t_0}^{t_f} \delta u^T W \delta u dt + \int_{t_0}^{t_f} \left(H_{0u}^T + \lambda^T G_u^T \right) \delta u dt + \lambda^T \psi_f \end{aligned} \tag{5.69}$$

The first and second variations of J_a are given by

$$\begin{aligned} \delta J_a &= \left(b\delta t_f + [L + \dot{\varphi} + \lambda^T \dot{\psi}]_{t_f} \right) \delta(\delta t_f) + \int_{t_0}^{t_f} \left(\delta u^T W + H_{0u}^T + \lambda^T G_u^T \right) \delta(\delta u) dt \\ \delta^2 J_a &= b\delta(\delta t_f)^2 + \int_{t_0}^{t_f} \delta(\delta u)^T W \, \delta(\delta u) dt \end{aligned} \tag{5.70}$$

Calculation of the corrections

The conditions for a local minimum are: $\delta J_a = 0$ and $\delta^2 J_a \geq 0$.

The second variation is always positive, because $b > 0$ and W is positive.

The cancellation of the first variation gives the corrections $(\delta t_f ; \delta u)$.

$$\begin{cases} \delta u = -W^{-1}\left(H_{0u} + G_u \lambda \right) \\ \delta t_f = -\dfrac{1}{b} [L + \dot{\varphi} + \lambda^T \dot{\psi}]_{t_f} \end{cases} \tag{5.71}$$

The corrections $(\delta t_f;\delta u)$ produce variations $\delta\psi$ of the constraints.

$$\delta\psi = -\frac{1}{b}[L+\dot\varphi+\lambda^T\dot\psi]_{t_f}[\dot\psi]_{t_f} - \int_{t_0}^{t_f} G_u^T W^{-1}(H_{0u}+G_u\lambda)dt \tag{5.72}$$

These variations are equal to $-\psi_f$ (5.68), which determines the multipliers λ.

$$\lambda = \left(\frac{1}{b}[\dot\psi]_{t_f}[\dot\psi]_{t_f}^T + \int_{t_0}^{t_f} G_u^T W^{-1} G_u\, dt\right)^{-1}\left(\psi_f - \frac{1}{b}[L+\dot\varphi]_{t_f}[\dot\psi]_{t_f} - \int_{t_0}^{t_f} G_u^T W^{-1} H_{0u}\, dt\right) \tag{5.73}$$

Adjustment of the corrections

If the linearized model is reliable, the corrections (5.71) will produce variations in cost and constraints $\delta\hat J$ and $\delta\hat\psi$ (expected variations).

$$\begin{aligned}\delta\hat J &= [L+\dot\varphi]_{t_f}\delta t_f + \int_{t_0}^{t_f} H_{0u}^T\delta u dt = -\frac{1}{b}[L+\dot\varphi+\lambda^T\dot\psi]_{t_f}[L+\dot\varphi]_{t_f} - (I_{JJ}+I_{J\psi}\lambda)\\ \delta\hat\psi &= [\dot\psi]_{t_f}\delta t_f + \int_{t_0}^{t_f} G_u^T\delta u dt = -\frac{1}{b}[L+\dot\varphi+\lambda^T\dot\psi]_{t_f}[\dot\psi]_{t_f} - (I_{\psi J}+I_{\psi\psi}\lambda)\end{aligned} \tag{5.74}$$

where we have introduced the **matrices of influence** $I_{\psi\psi}$, $I_{\psi J}$, I_{JJ} defined by

$$I_{\psi\psi} = \int_{t_0}^{t_f} G_u^T W^{-1} G_u\, dt\ ,\ I_{\psi J} = \int_{t_0}^{t_f} G_u^T W^{-1} H_{0u}\, dt = I_{J\psi}^T\ ,\ I_{JJ} = \int_{t_0}^{t_f} H_{0u}^T W^{-1} H_{0u}\, dt \tag{5.75}$$

If the observed variations $(\delta J_r;\delta\psi_r)$ differ from the expected variations $(\delta\hat J;\delta\hat\psi)$, the corrections have to be reduced to remain in the linear domain. This is done:
- either by changing the weights b and W;
- or by aiming at a partial correction of the constraints: $\delta\psi = -\varepsilon\psi_f$, $0<\varepsilon<1$;
- or by partially applying the control correction: $\alpha\delta u$, $0<\alpha<1$.

The optimal solution is obtained when the corrections $(\delta t_f;\delta u)$ and the variations $(\delta J;\delta\psi)$ are zero (or small enough). We have then the following relations.

$$\left.\begin{aligned}\delta u = 0 &\rightarrow H_{0u}+G_u\lambda = 0\\ \delta t_f = 0 &\rightarrow [L+\dot\varphi+\lambda^T\dot\psi]_{t_f} = 0\end{aligned}\right\}\rightarrow \text{PMP with} \left[\begin{aligned} H &\underset{def}{=} H_0+G\lambda = L+p^Tf\\ p &\underset{def}{=} p_0+P\lambda\end{aligned}\right. \tag{5.76}$$

$$\delta\psi = 0 \rightarrow \lambda = -I_{\psi\psi}^{-1}I_{\psi J}\quad ,\quad \delta J = 0 \rightarrow I_{JJ} - I_{J\psi}I_{\psi\psi}^{-1}I_{\psi J} = 0$$

The first two conditions correspond to the PMP optimality conditions.

Implementation

The calculation of the corrections $(\delta t_f ; \delta u)$ requires a forward and a backward integration:

- forward integration of the state x with the control u gives the final state $x(t_f)$;
- the adjoints p_0 and $(p_j)_{j=1 \text{ to } q}$ are assessed at t_f by the derivatives φ_x and ψ_x;
- backward integration of the adjoints yields the matrices $I_{\psi\psi}$, $I_{\psi J}$ and I_{JJ};
- the corrections $(\delta t_f ; \delta u)$ are calculated by

$$\begin{cases} \delta u = -W^{-1}\left(H_{0u} + G_u \lambda\right) \\ \delta t_f = -\dfrac{1}{b}[L + \dot{\varphi} + \lambda^T \dot{\psi}]_{t_f} \end{cases} \tag{5.77}$$

$$\text{with } \lambda = \left(\frac{1}{b}[\dot{\psi}]_{t_f}[\dot{\psi}]_{t_f}^T + I_{\psi\psi}\right)^{-1}\left(\psi_f - \frac{1}{b}[L + \dot{\phi}]_{t_f}[\dot{\psi}]_{t_f} - I_{\psi J}\right)$$

Figure 5-15 summarizes the calculation stages at each iteration. This method of "directly" minimizing the variation of J is called the direct variational method.

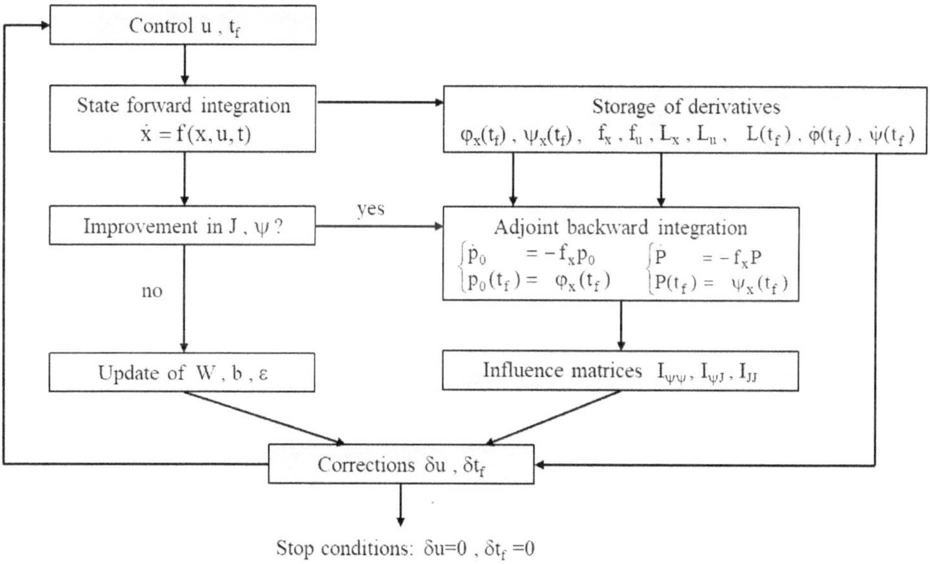

Figure 5-15: Direct variational method.

Example 5-7 illustrates an iteration of the direct variational method.

Example 5-7: Direct variational method

Let us retrieve the flat earth orbiting problem (examples 4-5 and 4.13).
We seek the trajectory maximizing the final horizontal velocity at a fixed altitude and final time. The acceleration of constant modulus a is oriented by the angle θ.

Noting $v = v_z$ the vertical velocity, the problem is formulated as

$$\min_\theta J = -v_x(t_f) = -a\int_{t_0}^{t_f}\cos\theta\, dt \quad \text{with} \quad \begin{cases} \dot{z} = v \\ \dot{v} = a\sin\theta - g \end{cases} \quad \text{and} \quad \begin{cases} z(t_f) = z_c \\ v(t_f) = 0 \end{cases}$$

This problem has been solved in example 4-5. Recall that the optimal control follows a linear tangent law of the form: $\tan\theta = \tan\theta_0 - ct$.

To apply the direct variational method, we define the adjoints and Hamiltonians associated with the cost and constraints respectively.
The functions defining cost, dynamics and constraints are

$$\begin{cases} \varphi(z,v) = 0 \\ L(z,v,\theta) = -\cos\theta \end{cases}, \quad f(z,v,\theta) = \begin{pmatrix} v \\ a\sin\theta - g \end{pmatrix}, \quad \psi(z,v) = \begin{pmatrix} z - z_c \\ v \end{pmatrix}$$

Cost adjoint

The adjoint p_0 and the Hamiltonian H_0 are defined by (5.64).

$$\left.\begin{aligned} H_0 = L + p_0^T f = -\cos\theta + p_{0z}v + p_{0v}(a\sin\theta - g) &\rightarrow \begin{cases} \dot{p}_{0z} = 0 \\ \dot{p}_{0v} = -p_{0z} \end{cases} \\ \varphi(z,v) = 0 &\rightarrow \begin{cases} p_{0z}(t_f) = 0 \\ p_{0v}(t_f) = 0 \end{cases} \end{aligned}\right\} \rightarrow p_0(t) = \begin{pmatrix} 0 \\ 0 \end{pmatrix}$$

The Hamiltonian H_0 reduces to: $H_0 = -\cos\theta$.

Altitude constraint adjoint

The adjoint p_1 and the Hamiltonian H_1 are defined by (5.65).

$$\left.\begin{aligned} H_1 = p_1^T f = p_{1z}v + p_{1v}(a\sin\theta - g) &\rightarrow \begin{cases} \dot{p}_{1z} = 0 \\ \dot{p}_{1v} = -p_{1z} \end{cases} \\ \psi_1(z,v) = z - z_c &\rightarrow \begin{cases} p_{1z}(t_f) = 1 \\ p_{1v}(t_f) = 0 \end{cases} \end{aligned}\right\} \rightarrow p_1(t) = \begin{pmatrix} 1 \\ t_f - t \end{pmatrix}$$

Velocity constraint adjoint

The adjoint p_2 and the Hamiltonian H_2 are defined by (5.65).

$$\left.\begin{array}{ll} H_2 = p_2^T f = p_{2z} v + p_{2v}(a\sin\theta - g) \rightarrow & \begin{cases} \dot{p}_{2z} = 0 \\ \dot{p}_{2v} = -p_{2z} \end{cases} \\ \psi_2(z,v) = v \rightarrow & \begin{cases} p_{2z}(t_f) = 0 \\ p_{2v}(t_f) = 1 \end{cases} \end{array}\right\} \rightarrow \quad p_2(t) = \begin{pmatrix} 0 \\ 1 \end{pmatrix}$$

Influence matrices

The adjoints are grouped in matrix form (5.66) in order to calculate the influence matrices (5.75). Let us assume for this example that the weights b and W are unitary $(b = 1, W = I)$ and that the variational method is initialized with a constant control $\theta(t) = \theta_c$.

With this constant control, the influence matrices can be calculated analytically.

Matrices P and G: $\quad P = \begin{pmatrix} 1 & 0 \\ t_f - t & 1 \end{pmatrix} \quad , \quad G = P^t f = \begin{pmatrix} 1 & t_f - t \\ 0 & 1 \end{pmatrix} \begin{pmatrix} v \\ a\sin\theta - g \end{pmatrix}$

Matrices H_{0u} and G_u: $H_0 = -\cos\theta \rightarrow H_{0u} = \sin\theta \quad , \quad G_u^T = a\cos\theta \begin{pmatrix} t_f - t \\ 1 \end{pmatrix}$

Matrix $I_{\psi\psi}$: $\quad I_{\psi\psi} = \int_{t_0}^{t_f} G_u^T W^{-1} G_u \, dt = \frac{1}{6} a^2 t_f \cos^2\theta \begin{pmatrix} 2t_f^2 & 3t_f \\ 3t_f & 6 \end{pmatrix}$

Matrix $I_{\psi J}$: $\quad I_{\psi J} = \int_{t_0}^{t_f} G_u^T W^{-1} H_{0u} \, dt = \frac{1}{2} a t_f \sin\theta\cos\theta \begin{pmatrix} t_f \\ 2 \end{pmatrix}$

Matrix I_{JJ}: $\quad I_{JJ} = \int_{t_0}^{t_f} H_{0u}^T W^{-1} H_{0u} \, dt = t_f \sin^2\theta$

Control correction

We calculate the value of the constraints for the initial constant control $\theta(t) = \theta_c$.

$$\begin{cases} v(t_f) = (a\sin\theta_c - g)t_f \\ z(t_f) = \frac{1}{2}(a\sin\theta_c - g)t_f^2 \end{cases} \rightarrow \psi_f = \begin{pmatrix} z(t_f) - z_c \\ v(t_f) \end{pmatrix} = \begin{pmatrix} \frac{1}{2}(a\sin\theta_c - g)t_f^2 - z_c \\ (a\sin\theta_c - g)t_f \end{pmatrix}$$

The formulas (5.77) give the multipliers λ and the control correction $\delta\theta$ to achieve a full constraint correction: $\delta\psi = -\psi_f$.

$$\lambda = I_{\psi\psi}^{-1}\left(\psi_f - I_{\psi J}\right) = \frac{1}{a^2 t_f^3 \cos^2\theta_c}\begin{pmatrix} -12z_c \\ 6z_c t_f + [a\sin\theta_c(1-\cos\theta_c) - g]t_f^3 \end{pmatrix}$$

$$\delta\theta = -W^{-1}\left(H_{0u} + G_u\lambda\right) = \frac{g - a\sin\theta_c}{a\cos\theta_c} + \frac{6z_c}{at_f^3\cos\theta_c}(t_f - 2t)$$

Let us assume that we choose θ_c to reach the final velocity: $a\sin\theta_c - g = 0$.

The corrected control at the first iteration is: $\theta(t) = \theta_c + \dfrac{6z_c}{at_f^3\cos\theta_c}(t_f - 2t)$.

If the correction is small compared to θ_c : $\tan\theta(t) \approx \tan\theta_c + \dfrac{6z_c}{at_f^3\cos\theta_c}(t_f - 2t)$.

<u>Comparison to the optimal control</u>

The optimal control is of the linear tangent form: $\tan\theta(t) = \tan\theta_0 - ct$.

Note θ_m the angle at half flight: $t_m = \frac{1}{2}t_f$. We have: $\tan\theta_m = \tan\theta_0 - \frac{1}{2}ct_f$.

The optimal control is expressed as: $\tan\theta(t) = \tan\theta_m + \frac{1}{2}c(t_f - 2t)$.

The similarity with the variational corrected control can be seen.
The initial constant control θ_c satisfying the final velocity constraint is close to the "average" value θ_m and the corrective term is linear in $(t_f - 2t)$. The corrected control approaches thus the optimal control.

5.6 Indirect methods

Indirect methods consist in solving the two-point boundary value problem associated with the PMP optimality conditions. This system of nonlinear equations can be solved either by a shooting method or by a variational method.

5.6.1 Shooting method

We consider a free final time control problem.

$$\min_{u,t_f} J = \int_{t_0}^{t_f} L(x,u,t)dt + \varphi(x_f,t_f) \quad \text{with} \quad \begin{cases} \dot{x} = f(x,u,t) \ , \ x(t_0) = x_0 \\ \psi_j(x_f,t_f) = 0 \ , \ j = 1 \text{ to } q \end{cases} \tag{5.78}$$

The optimality conditions of the PMP are expressed with the adjoint p, the multipliers λ of the final constraints and the normal Hamiltonian defined by $H(x,u,p,t) = L(x,u,p,t) + p^T f(x,u,t)$.

The optimal control minimizes the Hamiltonian: $\min\limits_u H \;\rightarrow\; u^*(x,p,t)$.

With this control $u^*(x,p,t)$, the Hamiltonian becomes a function of x and p. For simplicity, we keep the notation H to denote the function $H\left[x, u^*(x,p,t), p, t\right]$.

The state and the adjoint satisfy the following two-point boundary value problem

$$\begin{cases} \dot{x} = \;\; H_p(x,p) \\ \dot{p} = -H_x(x,p) \end{cases} , \quad \begin{cases} x(t_0) \quad = x_0 \\ \psi(x_f,t_f) = 0 \\ p(t_f) \;=\; \left[\varphi_x + \psi_x\lambda\right](x_f,t_f) \\ H(t_f) = -\left[\varphi_t + \psi_t\lambda\right](x_f,t_f) \end{cases} \tag{5.79}$$

The unknowns of this system are the initial adjoint $p(t_0)$, the final time t_f and the multipliers λ. The corresponding equations are the transversality conditions on $p(t_f)$, on $H(t_f)$ and the constraints $\psi(x_f,t_f)$.

The two-point boundary value problem (5.79) shown in figure 5-16 forms a system of nonlinear equations of dimension $n+q+1$.

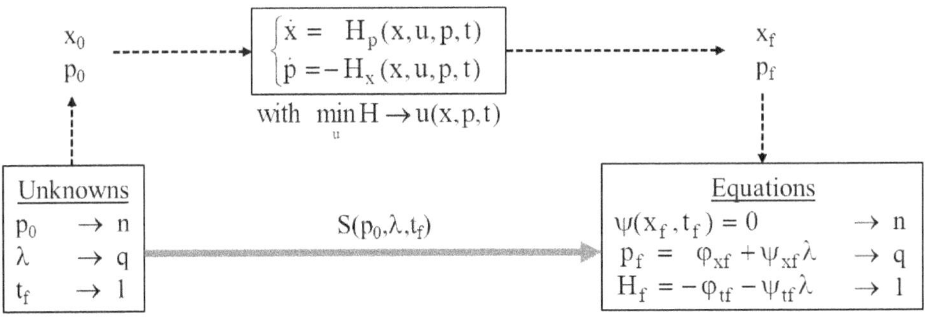

Figure 5-16: Two-point boundary value problem.

The system to be solved is noted in a compact way: $S(p_0,\lambda,t_f) = 0$.

The function $S : \mathbb{R}^{n+q+1} \rightarrow \mathbb{R}^{n+q+1}$ giving the values of the second members is called the **shooting function**. Its value measures the deviation from the solution, the target being zero.

The shooting problem consists in finding the unknowns p_0, λ, t_f to reach the target, as shown in figure 5-17.

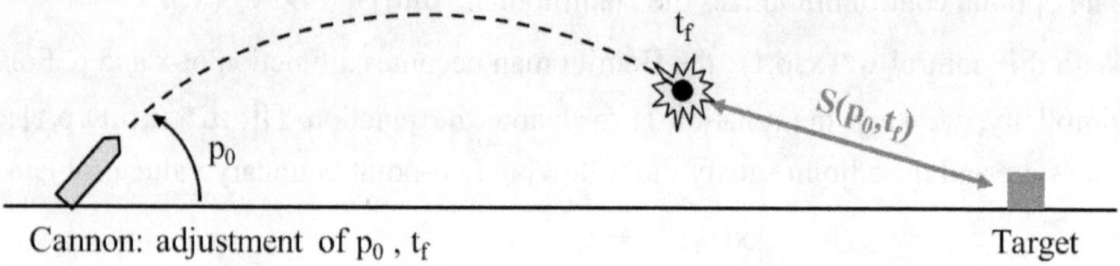

Figure 5-17: Shooting problem.

The shooting method consists in solving the system of nonlinear equations $S(p_0, \lambda, t_f) = 0$ by a Newton-type method. The critical points are the numerical calculation of the derivatives and the initialization of the adjoint p_0.

<u>Derivatives of the differential system</u>

The evaluation of the shooting function $S(p_0, \lambda, t_f)$ requires the numerical integration of the differential equations of the state and the adjoint. To apply a Newton-type method, the gradient of the shooting function with respect to the initial adjoint p_0 must be evaluated.

Noting $z = (x, p)$, the differential system is of the form: $\dot{z} = F(z, t)$.

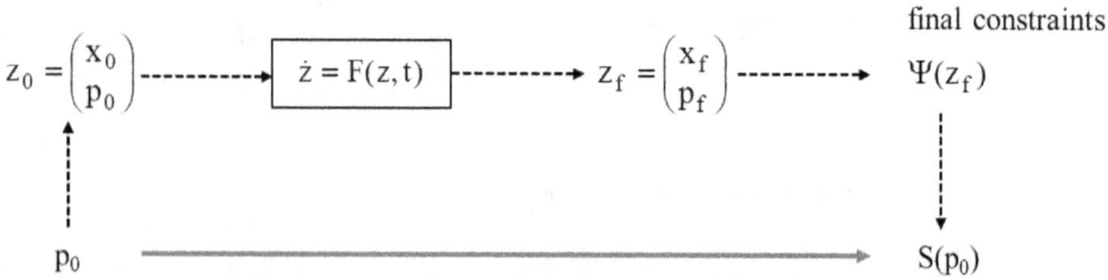

Figure 5-18: Calculation of the shooting function.

Three methods can be considered for calculating the transition matrix $\dfrac{\partial z_f}{\partial z_0}$.

- External derivation

The external derivation consists in applying an increment δ_k on the kth component of the initial state z_0, integrating until t_f and observing the resulting variation of the final state z_f. By successively varying each component as in figure 5-19, we obtain the set of partial derivatives forming the transition matrix $\frac{\partial z_f}{\partial z_0}$.

This calculation requires $n+1$ successive integrations of the differential system.

$$z_0 \dashrightarrow \boxed{\dot{z} = F(z,t)} \dashrightarrow z_f(z_0)$$

$$z_0+\delta_1 \dashrightarrow \boxed{\dot{z} = F(z,t)} \dashrightarrow z_f(z_0+\delta_1) \quad \rightarrow \quad \frac{\partial z_f}{\partial (z_0)_1} \approx \frac{z_f(z_0+\delta_1)-z_f(z_0)}{\delta_1}$$

$$z_0+\delta_k \dashrightarrow \boxed{\dot{z} = F(z,t)} \dashrightarrow z_f(z_0+\delta_k) \quad \rightarrow \quad \frac{\partial z_f}{\partial (z_0)_k} \approx \frac{z_f(z_0+\delta_k)-z_f(z_0)}{\delta_k}$$

$$z_0+\delta_n \dashrightarrow \boxed{\dot{z} = F(z,t)} \dashrightarrow z_f(z_0+\delta_n) \quad \rightarrow \quad \frac{\partial z_f}{\partial (z_0)_n} \approx \frac{z_f(z_0+\delta_n)-z_f(z_0)}{\delta_n}$$

Figure 5-19: External derivation.

The external derivation is simple to implement, as the integrator of the differential system is used as a "black box". However, this method requires an appropriate choice of increments δ_k (to be adapted according to the problem) and it may give inaccurate derivatives if a variable-step integrator is used.

Example 5-8: Effect of a variable integration step on the derivatives

Consider the differential equation: $\dot{x} = f(x)$ with initial condition: $x(t_0) = x_0$.

Let us apply Euler's method to two consecutive steps of lengths h_1 and h_2.

$$\begin{cases} x_1 = x_0 + f(x_0)h_1 \\ x_2 = x_1 + f(x_1)h_2 \end{cases} \Rightarrow x_2 = x_0 + f(x_0)h_1 + f\left[x_0 + f(x_0)h_1\right]h_2$$

By expanding f to order 1 (for h_1 small), and noting $f_0 = f(x_0)$ and $f_1 = f(x_1)$, we obtain: $x_2 = x_0 + f_0(h_1 + h_2 + f'_0 h_1 h_2)$.

Let us use this formula to calculate the derivative $\frac{\partial x_2}{\partial x_0}$ by finite difference and the effect of a step change:

- derivative by finite difference

An increment δx_0 on x_0 produces a variation noted $\delta_x x_2$ on x_2.
By expanding to order 1 the formula giving x_2 with

$$\begin{cases} f(x_0+\delta x_0) \approx f_0 + f'_0\,\delta x_0 \\ f'(x_0+\delta x_0) \approx f'_0 + f''_0\,\delta x_0 \end{cases} \text{ we get: } \delta_x x_2 = \delta x_0\left[1+f_0(h_1+h_2)+(f'^2_0+f_0 f''_0)h_1 h_2\right]$$

- step variation

Let us vary the step h_1 by $+\delta h$ and the step h_2 by $-\delta h$.
The solution x_2 varies by: $\delta_h x_2 = f_0 f'_0\,\delta h(h_2 - h_1 - \delta h)$.

Let us compare these variations $\delta_x x_2$ and $\delta_h x_2$ assuming equal steps: $h_1 = h_2$.

$$\begin{cases} \delta_x x_2 = \delta x_0\left[1+2hf_0+h^2(f'^2_0+f_0 f''_0)\right] \\ \delta_h x_2 = \delta h^2\left[-f_0 f'_0\right] \end{cases}$$

Assuming that the values and derivatives of f are of the magnitude of 1, it is observed that a variable time step can distort the evaluation of the derivative if $\delta h \sim \sqrt{\delta x_0}$. This integrator-dependent step variation can occur between the nominal evaluation (x_0) and the evaluation with increment $(x_0+\delta x_0)$.
In practice, if a small increment δx_0 is chosen, it is preferable to use a fixed-step integrator to calculate the derivative by finite difference.

- Internal derivation

The internal derivation consists in making simultaneously the $n+1$ integrations. A vector Z of size $(n+1)\times n$ is propagated, which contains $n+1$ copies of the vector z:
- the copy number 0 is the nominal solution initialized with z_0;
- the copy number k is the perturbed solution initialized with $z_0+\delta_k$.

This internal derivation process is illustrated in figure 5-20.

$$Z_0 = \begin{pmatrix} z_0 \\ z_0+\delta_1 \\ \vdots \\ z_0+\delta_k \\ \vdots \\ z_0+\delta_n \end{pmatrix} \dashrightarrow \dot{Z} = \begin{pmatrix} F(z,t) \\ F(z,t) \\ \vdots \\ F(z,t) \\ \vdots \\ F(z,t) \end{pmatrix} \dashrightarrow Z_f = \begin{pmatrix} z_f(z_0) \\ z_f(z_0+\delta_1) \\ \vdots \\ z_f(z_0+\delta_k) \\ \vdots \\ z_f(z_0+\delta_n) \end{pmatrix} \rightarrow \frac{\partial z_f}{\partial (z_0)_k}$$

Figure 5-20: Internal derivation.

The transition matrix $\frac{\partial z_f}{\partial z_0}$ is obtained by a single integration of the Z vector.

Since the $n+1$ integrations are simultaneous, they use the same time step and do not exhibit any consistency error, even if the integrator uses a variable step. However, this method requires modifying the implementation of the differential system in order to propagate the state vector Z of dimension $(n+1)\times n$.

- Variational derivative

The variational equations of the system $\dot{z} = F(z,t)$ are the differential equations of the deviations. They are obtained by an expansion to order 1.

$$\dot{z}_{nom} + \delta\dot{z} = F(z_{nom} + \delta z, t) \Rightarrow \delta\dot{z} = \frac{\partial F}{\partial z}\delta z \quad \text{with } F_z = \frac{\partial F}{\partial z}(z_{nom}, t) \quad (5.80)$$

The matrix of this linear system is evaluated on the nominal solution z_{nom}.
Integrating (5.80) from an initial deviation δz_0 gives the final deviation δz_f.
For $\delta z_0 = \left(\overset{1}{0}, \overset{2}{0}, \cdots, \overset{k}{1}, \cdots, \overset{n}{0} \right)$, the final deviation δz_f is the derivative $\frac{\partial z_f}{\partial (z_0)_k}$.
The n partial derivatives are obtained by integrating the matrix Z of size $n\times n$.

$$\dot{Z} = \frac{\partial F}{\partial z} Z \quad \text{with } Z_0 = I \quad (5.81)$$

This method is more accurate than the external or internal derivation, but it requires on the one hand to precompute and store the matrix F_z, and on the other hand to implement and propagate the variational equations that are specific to the given problem.

Example 5-9: Derivatives of a differential system

Consider the differential equation: $\dot{x} = -\alpha x^2$ with: $x(t=0) = x_0$.
We seek the derivative of $x(t)$ with respect to the initial condition.

Exact derivative

The exact solution is: $x = \dfrac{x_0}{\alpha x_0 t + 1}$ and its derivative is: $\dfrac{\partial x}{\partial x_0} = \dfrac{1}{(\alpha x_0 t + 1)^2}$.

External derivative

An increment δx_0 perturbs the solution by δx : $x + \delta x = \dfrac{x_0 + \delta x_0}{\alpha(x_0 + \delta x_0)t + 1}$.

The result is: $\delta x = \dfrac{\delta x_0}{\left[\alpha(x_0 + \delta x_0)t + 1\right](\alpha x_0 t + 1)}$.

The external derivative tends to the exact value as the increment δx_0 decreases. The error with respect to the exact derivative depends on the increment δx_0 which in practice cannot be too small (to be adapted according to the problem and the numerical integration method).

Variational derivative

The variational equation is: $\delta\dot{x} = -2\alpha x_{nom} \delta x$ with $x_{nom} = \dfrac{x_0}{\alpha x_0 t + 1}$.

This equation is integrated: $\delta\dot{x} = -\dfrac{2\alpha x_0}{\alpha x_0 t + 1} \delta x \rightarrow \dfrac{\delta x}{\delta x_0} = \dfrac{1}{(\alpha x_0 t + 1)^2}$.

The variational derivative gives the exact derivative with respect to the initial condition. In practice, the analytical solution is not available and the variational equation must be numerically integrated from a unitary initial condition.

If the control problem depends on parameters α (hybrid problem), the derivatives of the differential system with respect to α are calculated in a similar way by one of the three previous methods.

The variational equations are then of the form: $\delta\dot{z} = \dfrac{\partial F}{\partial z}\delta z + \dfrac{\partial F}{\partial \alpha}\delta\alpha$.

The final time can also be treated as a parameter by changing the variable.

The normalized time τ varying between 0 and 1 is defined as

$$\tau = \frac{t - t_0}{t_f - t_0} = \alpha(t - t_0) \;\rightarrow\; \frac{dz}{d\tau} = \frac{1}{\alpha}\frac{dz}{dt} = \frac{1}{\alpha}F(z,\alpha,t) \tag{5.82}$$

The control problem is then at fixed final time $(\tau_f = 1)$ with the parameter α to be optimized.

<u>Sensitivity to initial conditions</u>

Evaluating the shooting function implies to integrate the Hamiltonian system

$$\begin{cases} \dot{x} = H_p(x,u,p,t) \\ \dot{p} = -H_x(x,u,p,t) \end{cases} \quad \text{with} \quad H(x,u,p,t) = L(x,u,t) + p^T f(x,u,t) \tag{5.83}$$

Let us write the equations of this system with the deviations $(\delta x;\delta p)$ in the vicinity of a reference trajectory.

$$\begin{cases} \delta\dot{x} = H_{px}^T\delta x + H_{pp}^T\delta p = f_x^T\delta x \\ \delta\dot{p} = -H_{xx}^T\delta x - H_{xp}^T\delta p = -f_x^T\delta p - H_{xx}^T\delta x \end{cases} \tag{5.84}$$

The derived matrices of f and H are evaluated on the reference path. We observe that the linear equations in δx and δp share the same matrix f_x^T, but with opposite sign. The negative eigenvalues of f_x^T have a convergent effect on the equation in δx and a divergent effect on the equation in δp. The effect of positive eigenvalues is the opposite. It is therefore not possible for the evolutions of δx and δp to be simultaneously stable. The Hamiltonian system is by nature <u>very sensitive</u> to changes in initial conditions.

Due to this intrinsic characteristic, the radius of convergence of a shooting method is quite small around the solution and the initial adjoint p_0 requires a very accurate initialization. In some cases (especially when the propagation time is large), numerical errors related to the integration method may be sufficient to cause divergence, even when the initial adjoint corresponds to the optimal solution.

The following quadratic-linear example illustrates this high sensitivity.

Example 5-10: Sensitivity of the Hamiltonian system

Consider the quadratic-linear control problem with fixed final time and state.

$$\min_u J = \frac{1}{2}\int_{t_0}^{t_f} u^2 dt \quad \text{s.t.} \quad \dot{x} = ax + bu \quad \text{with} \quad \begin{cases} x(t_0) = x_0 \ , \ t_0 = 0 \\ x(t_f) = x_f \ , \ t_f \text{ fixé} \end{cases}$$

This problem is solved analytically:

- Hamiltonian: $H = \frac{1}{2}u^2 + p(ax + bu)$;
- control: $H_u = u + bp = 0 \quad \rightarrow \quad u = -bp$;
- adjoint: $\dot{p} = -H_x = -ap \quad \rightarrow \quad p = p_0 e^{-at}$;
- state: $\dot{x} = ax + bu = ax - b^2 p_0 e^{-at}$

$$\rightarrow \quad x = x_0 e^{at} + p_0 \frac{b^2}{2a}(e^{-at} - e^{at}).$$

The unknown of the shooting problem is the initial adjoint p_0.
The final constraint is on the state $x(t_f)$. The shooting function is defined by

$$S(p_0) = x(t_f) - x_f = p_0 \frac{b^2}{2a}(e^{-at_f} - e^{at_f}) + x_0 e^{at_f} - x_f$$

The derivative of S is given by: $\dfrac{dS}{dp_0} = \dfrac{b^2}{2a}(e^{-at_f} - e^{at_f})$.

This exponential function of t_f takes a strong value if the final time is large, independently of the sign of a. Even a small deviation from p_0 in the vicinity of the solution can then produce a very high value of the shooting function, or even its numerical explosion.

Multiple shooting

The intrinsic sensitivity of the Hamiltonian system makes the shooting method numerically inapplicable in some cases, especially if the final time is large.

The **multiple shooting** method aims to reduce this sensitivity by splitting the trajectory into N segments: $t_0 < t_1 < \cdots < t_i < \cdots < t_N = t_f$.

Each segment $[t_i ; t_{i+1}]$ is independently integrated starting from a freely chosen initial state x_i and initial adjoint p_i. The integration results in conditions at the end of segment i, noted $x(t_{i+1}^-)$ and $p(t_{i+1}^-)$. These final conditions differ from the conditions (x_{i+1}, p_{i+1}) considered at the beginning of the next segment.

$$\begin{cases} \dot{x} = H_p \ , \ x(t_i^+) = x_i \\ \dot{p} = -H_x \ , \ p(t_i^+) = p_i \end{cases} \xrightarrow{\text{integration from } t_i \text{ to } t_{i+1}} \begin{cases} x(t_{i+1}^-) \\ p(t_{i+1}^-) \end{cases} \tag{5.85}$$

The variables $(x_i, p_i)_{i=1 \text{ to } N}$ become additional unknowns of the shooting problem.

We seek to satisfy the continuity conditions on the state and the adjoint between successive segments.

$$\begin{cases} x(t_i^-) = x_i \\ p(t_i^-) = p_i \end{cases} , \quad i = 1 \text{ to } N-1 \tag{5.86}$$

Let $z = (x, p)$ be the 2n-dimensional vector formed by the state and the adjoint. The additional $(N-1) \times 2n$ unknowns are $(z_1, z_2, \cdots, z_{N-1})$ to which the continuity conditions are imposed: $\Delta z_i = z(t_i^-) - z_i = 0$ as shown below.

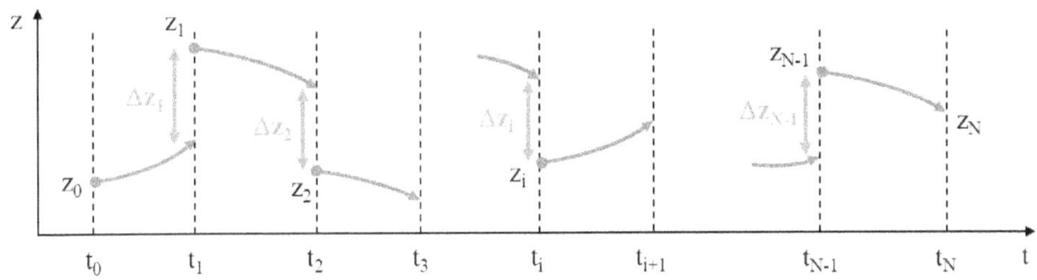

Figure 5-21: Multiple shooting method.

The simple shooting method is associated with the two-point boundary value problem (5.79).

$$S(p_0, \lambda, t_f) = \begin{pmatrix} \psi(x_f) \\ p_f - (\varphi_{xf} + \psi_{xf}\lambda) \\ H_f + (\varphi_{tf} + \psi_{tf}\lambda) \end{pmatrix} \begin{matrix} \to q \\ \to n \\ \to 1 \end{matrix} \tag{5.87}$$

The multiple shooting function on N segments is defined by

$$S_N(p_0,\lambda,t_f,z_1,\cdots,z_{N-1}) = \begin{pmatrix} S(p_0,\lambda,t_f) \\ \Delta z_1 \\ \vdots \\ \Delta z_{N-1} \end{pmatrix} \begin{matrix} \to n+q+1 \\ \to 2n \\ \vdots \\ \to 2n \end{matrix} \tag{5.88}$$

The multiple shooting problem is the system of dimension $n+q+1+(N-1)\times 2n$.

$$S_N(p_0,\lambda,t_f,z_1,\cdots,z_{N-1}) = 0 \tag{5.89}$$

The splitting of the trajectory into several independently integrated segments has the effect of reducing the sensitivity of the shooting function. This is achieved at the expense of increasing the dimension of the shooting problem. The choice of the number of segments results from a sensitivity/dimension trade-off.

5.6.2 Variational approach

Let us retrieve the control problem (5.78) with q final constraints.

$$\min_{u,t_f} J = \int_{t_0}^{t_f} L(x,u,t)dt + \varphi(x_f,t_f) \quad \text{with} \quad \begin{cases} \dot{x} = f(x,u,t) \ , \ x(t_0) = x_0 \\ \psi_j(x_f,t_f) = 0 \ , \ j = 1 \text{ to } q \end{cases} \tag{5.90}$$

The optimality conditions given by the PMP yield the Hamiltonian system

$$\begin{cases} \dot{x} = \ \ H_p \\ \dot{p} = -H_x \quad \text{with} \quad H = L + p^T f \\ H_u = 0 \end{cases} \tag{5.91}$$

and the final conditions (transversality and constraints):

$$\begin{cases} \Omega_x = p \quad \text{with} \quad \Omega = \varphi + \lambda^T \psi \\ \Psi = 0 \quad \text{with} \quad \Psi = \Omega_t + H = \dot{\Omega} + L \quad \text{at } t_f \\ \psi = 0 \end{cases} \tag{5.92}$$

Let us take an arbitrary solution defined by a control $u(t)_{t_0 \le t \le t_f}$, a final time t_f and multipliers λ and perform the following calculations:

- the forward integration (5.91) of the state x with the control u gives $x(t_f)$;
- the final adjoint $p(t_f)$ is evaluated by (5.92) with the multipliers λ;
- the backward integration (5.91) of the adjoint allows to evaluate $H_u(t)_{t_0 \le t \le t_f}$.

If the solution (u,t_f,λ) is not optimal, the conditions $(H_u=0,\ \Psi=0,\ \psi=0)$ are not met.

The variational method consists in looking for corrections $(\delta u,\delta t_f,\delta\lambda)$ using the equations (5.91) and (5.92) expanded to order 1.
These variational equations (or neighboring equations) deal with the deviations $(\delta x,\delta p,\delta u,\delta t_f,\delta\lambda)$.

$$\begin{cases} \dot{x} = H_p \to \delta\dot{x} = H_{px}^T\delta x + H_{pu}^T\delta u + H_{pp}^T\delta p = f_x^T\delta x + f_u^T\delta u \\ \dot{p} = -H_x \to \delta\dot{p} = -H_{xx}^T\delta x - H_{xu}^T\delta u - H_{xp}^T\delta p = -H_{xx}\delta x - H_{ux}\delta u - f_x\delta p \\ H_u = 0 \to \delta H_u = H_{ux}^T\delta x + H_{uu}^T\delta u + H_{up}^T\delta p = H_{xu}\delta x + H_{uu}\delta u + f_u\delta p \end{cases}$$
$$\begin{cases} \Omega_x = p \to \delta p_f = \Omega_{xx}\delta x_f + \Omega_{xt}\delta t_f + \Omega_{x\lambda}\delta\lambda = \Omega_{xx}\delta x_f + \Omega_{xt}\delta t_f + \psi_x\delta\lambda \\ \Psi = 0 \to \delta\Psi = \Psi_x^T\delta x_f + \Psi_t^T\delta t_f + \Psi_\lambda^T\delta\lambda = \Psi_x^T\delta x_f + \Psi_t^T\delta t_f + \dot{\psi}^T\delta\lambda \\ \psi = 0 \to \delta\psi = \psi_x\delta x_f + \psi_t\delta t_f \end{cases} \quad (5.93)$$

The first stage is to reduce the variational system (5.93).

<u>Reduction of the variational system</u>

We note (H_{uc},Ψ_c,ψ_c) the values obtained with the arbitrary initial solution (u,t_f,λ). Since the objective is to cancel (H_u,Ψ,ψ), we look for correction $(\delta u,\delta t_f,\delta\lambda)$ such that

$$\begin{cases} \delta H_u = -H_{uc} \\ \delta\Psi = -\Psi_c \\ \delta\psi = -\psi_c \end{cases} \Leftrightarrow \begin{cases} H_{xu}\delta x + H_{uu}\delta u + f_u\delta p = -H_{uc} \\ \Psi_x^T\delta x_f + \Psi_t^T\delta t_f + \dot{\psi}^T\delta\lambda = -\Psi_c \\ \psi_x\delta x_f + \psi_t\delta t_f = -\psi_c \end{cases} \quad (5.94)$$

The equation in δH_u gives the control correction δu as a function of $(\delta x,\delta p)$.

$$\delta u = -H_{uu}^{-1}(H_{uc} + H_{xu}\delta x + f_u\delta p) \quad (5.95)$$

Replacing δu in the differential equations (5.93) yields

$$\begin{cases} \delta\dot{x} = f_x^T\delta x + f_u^T\delta u = A\delta x - B\,\delta p + DH_{uc} \\ \delta\dot{p} = -H_{xx}\delta x - H_{ux}\delta u - f_x\delta p = -C\delta x - A^T\delta p + EH_{uc} \end{cases} \quad (5.96)$$

with the matrices A, B, C, D, E given by

$$\begin{cases} A = f_x^T - f_u^T H_{uu}^{-1} H_{xu} \\ B = f_u^T H_{uu}^{-1} f_u \\ C = H_{xx} - H_{ux} H_{uu}^{-1} H_{xu} \end{cases} , \quad \begin{cases} D = -f_u^T H_{uu}^{-1} \\ E = -H_{ux} H_{uu}^{-1} \end{cases} \tag{5.97}$$

The endpoint conditions of the differential system (5.96) are as follows:

- the initial state change δx_0 is zero, as $x(t_0) = x_0$ is fixed;
- the final adjoint variation δp_f is given by (5.93).

$$\begin{cases} \delta x_0 = 0 \\ \delta p_f = \Omega_{xx} \delta x_f + \Omega_{xt} \delta t_f + \psi_x \delta \lambda \end{cases} \tag{5.98}$$

Equations (5.96) and (5.98) define a two-point boundary value system. For given corrections $(\delta x_f, \delta t_f, \delta\lambda)$, the equations (5.96) in $(\delta x, \delta p)$ can be integrated backwards to obtain the value of δx_0. The objective is then to find the unknowns $(\delta x_f, \delta t_f, \delta\lambda)$ to satisfy the last two equations of (5.94).

$$\begin{cases} \delta x_0 = 0 \\ \Psi_x^T \delta x_f + \Psi_t^T \delta t_f + \dot{\psi}^T \delta\lambda = -\Psi_c \\ \psi_x \delta x_f + \psi_t \delta t_f = -\psi_c \end{cases} \tag{5.99}$$

Solving the variational system

To solve the system (5.99), we try to express δx_0 in terms of $(\delta x_f, \delta t_f, \delta\lambda)$. The linear differential system (5.96) is non-homogeneous because of the term in δH_{uc}. Its general solution is of the form

$$\delta x(t) = X(t)\delta x(t_f) + P(t)\delta p(t_f) + \xi(t) \tag{5.100}$$

The matrices $X(t)$ and $P(t)$ are the unit solution matrices obtained by integrating the homogeneous system (with $H_{uc} = 0$) backwards from unit final conditions. For each integration, the vectors $x(t_f)$ and $p(t_f)$ are zero, except for one component of x (to generate the matrix X) or p (to generate the matrix P). The particular solution $\xi(t)$ is obtained by integrating (5.96) backwards from $x(t_f) = 0$, $p(t_f) = 0$ and with $\delta H_u = -H_{uc}$ (targeted correction on H_u).

With (5.100), the change in initial state is expressed as a function of $(\delta x_f, \delta p_f)$.

$$\delta x_0 = X(t_0)\delta x_f + P(t_0)\delta p_f + \xi(t_0) \underset{\text{noted}}{=} X_0 \delta x_f + P_0 \delta p_f + \xi_0 \qquad (5.101)$$

Replacing δp_f with its expression (5.98), equations (5.99) yield a linear system in terms of the unknowns $(\delta x_f, \delta t_f, \delta\lambda)$.

$$\begin{pmatrix} X_0 + P_0\Omega_{xx} & P_0\Omega_{xt} & P_0\psi_x \\ \Psi_x^T & \Psi_t^T & \dot{\psi}^T \\ \psi_x^T & \psi_t & 0 \end{pmatrix} \begin{pmatrix} \delta x_f \\ \delta t_f \\ \delta\lambda \end{pmatrix} + \begin{pmatrix} \xi_0 \\ 0 \\ 0 \end{pmatrix} = \begin{pmatrix} 0 \\ -\Psi_c \\ -\psi_c \end{pmatrix} \qquad (5.102)$$

Control correction

Solving the linear system (5.102) gives $(\delta x_f, \delta t_f, \delta\lambda)$. The evolution of the deviations $(\delta x, \delta p)$ is obtained with the unit solutions (X, P, ξ) already calculated starting from δx_f and δp_f given by (5.98). The control correction δu can be deduced from (5.95). The application of these corrections produces actual variations $(\delta H_{ur}, \delta\Psi_r, \delta\psi_r)$ which may differ from the expected ones (5.94).

If the difference is too large, the corrections must be reduced in order to remain in the linear domain. This can be done:

- by aiming at a partial correction $(\delta H_u, \delta\Psi, \delta\psi) = -\varepsilon(H_{uc}, \Psi_c, \psi_c)$, $0 < \varepsilon < 1$;
- or by partially applying the control correction: $\alpha\delta u$, $0 < \alpha < 1$.

A constant or time-increasing α coefficient can be chosen to limit the sensitivity to corrections applied at the beginning of the trajectory.
The optimal solution is obtained when the corrections $(\delta u, \delta t_f, \delta\lambda)$ and the values of (H_u, Ψ, ψ) are zero or small enough.

Implementation

The corrections $(\delta u, \delta t_f, \delta\lambda)$ are calculated in the following stages:

- the forward integration of the state x by (5.91) with the control u gives $x(t_f)$;
- the final adjoint $p(t_f)$ is evaluated by (5.92) with the multipliers λ;
- backward integration of the adjoint by (5.91) gives A, B, C, D, E (5.97);
- the unit matrices X, P are calculated by backward integration of (5.96);
- the linear system (5.102) is solved to find $(\delta x_f, \delta t_f, \delta\lambda)$;
- the deviations $(\delta x, \delta p)$ are calculated with the unit solutions (X, P, ξ);
- the control correction δu is given by (5.95).

These corrections are accepted or not depending on the differences between the observed and expected variations. Figure 5-22 summarizes the calculation stages at each iteration. This method for reaching the PMP optimality conditions is called the indirect variational method.

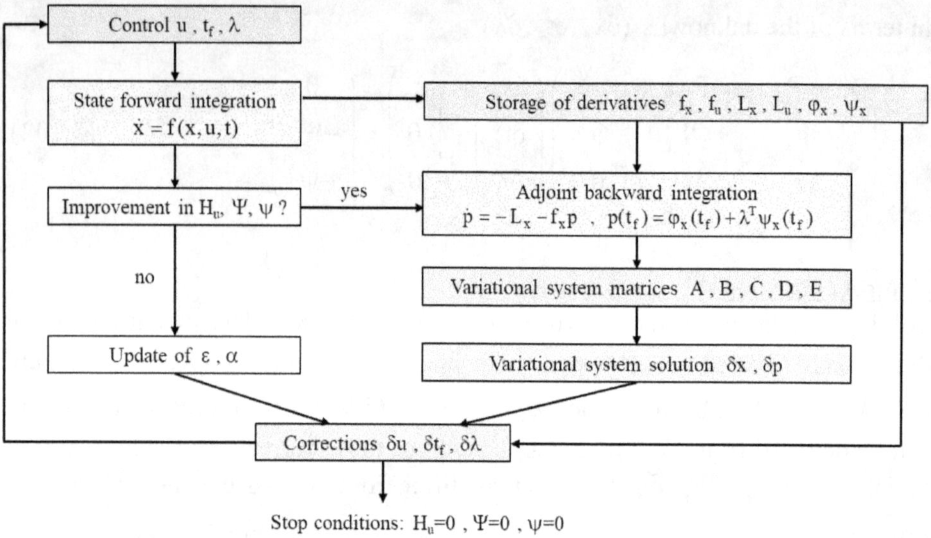

Figure 5-22: Indirect variational method.

An iteration of the indirect variational method can be interpreted as solving the auxiliary minimum problem (section 4.4.1) formulated by expanding the cost to order 2, the dynamics and the constraints to order 1.

$$\min_{\delta u, \delta t_f} \frac{\delta^2 J_a}{2} = \frac{1}{2} \int_{t_0}^{t_f} \begin{pmatrix} \delta x \\ \delta u \end{pmatrix}^T \begin{pmatrix} H_{xx} & H_{ux} \\ H_{xu} & H_{uu} \end{pmatrix} \begin{pmatrix} \delta x \\ \delta u \end{pmatrix} dt + \frac{1}{2} \begin{pmatrix} \delta x \\ \delta t \end{pmatrix}_f^T \begin{pmatrix} \Omega_{xx} & \Psi_x \\ \Psi_x^T & \dot{\Psi} \end{pmatrix} \begin{pmatrix} \delta x \\ \delta t \end{pmatrix}_f$$

$$\text{s.t.} \begin{cases} \delta \dot{x} = f_x^T \delta x + f_u^T \delta u & , \ \delta x_0 \text{ fixed} \\ \delta \psi = \psi_x^T \delta x_f + \dot{\psi} \delta t_f & , \ \delta \psi \text{ fixed} \\ \delta \Psi = \Psi_x^T \delta x_f + \dot{\Psi} \delta t_f & , \ \delta \Psi \text{ fixed} \end{cases} \quad (5.103)$$

As with the shooting methods, the numerical difficulties arise from the high sensitivity of the adjoint equations. Similarly to the multiple shooting approach, differential dynamic programming (DDP) methods alleviate these difficulties by splitting the trajectory into several segments with junction conditions.

5.7 Conclusion

5.7.1 The key points

- Differential equations can be solved by propagation methods (Runge-Kutta type) or collocation methods approximating the solution by polynomials;
- direct methods transform the control problem into a finite-dimensional constrained optimization problem. They do not use the optimality conditions of the minimum principle;
- indirect methods solve the two-point boundary value problem arising from the optimality conditions of the minimum principle;
- direct methods can be easily applied to any type of problem (dynamics, control and constraints). Their disadvantage is the compromise to be found between the resolution time (size of the discretized problem) and the desired accuracy of the solution;
- indirect methods give a very fast and accurate solution. Their disadvantage is the difficulty of implementation (adjoint equations) and of use (initialization).

5.7.2 To go further

- *Analyse numérique et équations différentielles (J.-P. Demailly, EDP Sciences 2006)*
  This book presents in detail the numerical methods for solving differential equations. The properties of the methods are demonstrated in a rigorous manner, while remaining at a very accessible mathematical level. The presentation is very clear and the reading is very pleasant.
- *Solving ordinary differential equations (E. Hairer, S.P. Norsett, G. Wanner, Springer 1993)*
  This two-volume book is an essential reference written by the best specialists in the field. It provides an exhaustive presentation of the numerical methods for solving differential equations with their history, demonstrations of their properties, detailed examples of applications and numerical comparisons. The book is divided into short, almost independent chapters, allowing direct access to the information required. The listings of the different softwares developed by the authors are given in the appendix and can be downloaded from their website.

- *Practical methods for optimal control and estimation using nonlinear programming (J.T. Betts, Siam 2010)*
  This book is devoted to numerical methods in optimal control. After a review of nonlinear programming algorithms, it compares direct and indirect methods. The advantages and disadvantages of both approaches are discussed in detail, with examples of applications and many practical tips to overcome difficulties. The author is a well-known specialist in direct methods (and does not hide his preference for them). This book is recommended for anyone wishing to go deeper into the subject and apply these direct methods.
- *A survey of numerical methods for optimal control (A.V. Rao, AAS 09-334 Preprint 2010)*
  This article gives a comprehensive overview of numerical methods in optimal control. It discusses successively the propagation and collocation methods for differential equations, then the direct and indirect approaches. The presentation is very clear and easy to read. This article gives a very good overview of the methods.
- *Applied optimal control (A.E. Bryson, Y.C. Ho, Hemisphere Publishing Corporation 1975)*
  This book is aimed at the practical application of optimal control for engineers. The very progressive presentation in the chapters makes it possible to become familiar with the computational techniques, especially thanks to many very detailed examples. One chapter is devoted to numerical methods, especially those based on the variational approach.

Index

Short bibliography

[R1] Optimisation discrète
(A. Billionnet, Dunod 2007)

[R2] Practical methods for optimal control and estimation using nonlinear programming
(J.T. Betts, Siam 2010)

[R3] Applied optimal control
(A.E. Bryson, Y.C. Ho, Hemisphere Publishing Corporation 1975)

[R4] Analyse numérique et équations différentielles
(J.-P. Demailly, EDP Sciences 2006)

[R5] Calculus of variations with applications
(G.M. Ewing, Dover 1985)

[R6] Précis de recherche opérationnelle
(R. Faure, B. Lemaire, C. Picouleau, Dunod 2009)

[R7] Calculus of variations
(I.M. Gelfand, S.V. Fomin, Dover 1963)

[R8] Programmation linéaire
(C. Guéret, C. Prins, M. Sevaux, Eyrolles 2003)

[R9] Solving ordinary differential equations
(E. Hairer, S.P. Norsett, G. Wanner, Springer 1993)

[R10] Les mathématiques du mieux-faire
(J.B. Hiriart-Urruty, Ellipses 2008)

[R11] Optimal control theory for applications
(D.G. Hull, Springer, 2003)

[R12] Algorithmes de graphes
(P. Lacomme, C. Prins, M. Sevaux, Eyrolles 2007)

[R13] The variational principles of mechanics
(C. Lanczos-Urruty, Dover 1949)

[R14] Calculus of variations and optimal control theory
(D. Liberzon, Princeton University Press 2012)

[R15] Programmation mathématique
(M. Minoux, Lavoisier 2008, 2e édition)

[R16] A survey of numerical methods for optimal control
(A.V. Rao, AAS 09-334 Preprint 2010)

[R17] Geometric optimal control
(H. Schättler, U. Ledzewicz, Springer 2012)

[R18] Stochastic optimization
(J.J. Schneider, S. Kirkpatrick, Springer 2006)

[R19] Contrôle optimal – Théorie et applications
(E. Trélat, Vuibert 2005)

www.ingramcontent.com/pod-product-compliance
Lightning Source LLC
Chambersburg PA
CBHW071409230926
51846CB00001B/39
9782759831630